The Evolution of Insect Mating Systems

WITHDRAWN
University of Southampton

The Evolution of Insect Mating Systems

Edited by

DAVID M. SHUKER AND
LEIGH W. SIMMONS

The Evolution of Insect Mating Systems. Edited by David M. Shuker and Leigh W. Simmons.
© The Royal Entomological Society 2014. Published 2014 by Oxford University Press.

OXFORD
UNIVERSITY PRESS

Great Clarendon Street, Oxford, OX2 6DP,
United Kingdom

Oxford University Press is a department of the University of Oxford. It furthers the University's objective of excellence in research, scholarship, and education by publishing worldwide. Oxford is a registered trade mark of Oxford University Press in the UK and in certain other countries

© The Royal Entomological Society 2014

The moral rights of the authors have been asserted

First Edition published in 2014
Impression: 1

All rights reserved. No part of this publication may be reproduced, stored in a retrieval system, or transmitted, in any form or by any means, without the prior permission in writing of Oxford University Press, or as expressly permitted by law, by licence or under terms agreed with the appropriate reprographics rights organization. Enquiries concerning reproduction outside the scope of the above should be sent to the Rights Department, Oxford University Press, at the address above

You must not circulate this work in any other form
and you must impose this same condition on any acquirer

Published in the United States of America by Oxford University Press
198 Madison Avenue, New York, NY 10016, United States of America

British Library Cataloguing in Publication Data
Data available

Library of Congress Control Number: 2014933927

ISBN 978–0–19–967802–0 (hbk.)
ISBN 978–0–19–967803–7 (pbk.)

Printed and bound in Great Britain by
Bell & Bain Ltd., Glasgow

Links to third party websites are provided by Oxford in good faith and for information only. Oxford disclaims any responsibility for the materials contained in any third party website referenced in this work.

00642974

This book is published on the occasion of the Royal Entomological Society's International Symposium on "The Evolution of Insect Mating Systems" in St Andrews, September 4-6 2013. All symposium speakers contributed to this volume.

Preface

Perhaps rather few books are truly canonical, helping to create and define a field and thus leave a lasting legacy. *The Evolution of Insect Mating Systems*, written by Randy Thornhill and John Alcock, and published in 1983, is certainly one of them. A crucial book in the emergence of behavioural ecology, Thornhill and Alcock brought mating systems and sexual selection alive for many, including ourselves as youthful PhD students starting out on our research careers. In their preface, Thornhill and Alcock promoted the virtues of insects as model systems for studying behavioural and evolutionary biology. Their hope was that by explaining the complex and often strange sex lives of insects, when viewed within an evolutionary framework, they would persuade their readers to both share their passion for entomology and the power of the 'adaptationist' approach in making sense of the world generally. This approach certainly worked on us, shaping our thinking and careers to this day. Many of the authors in this edited volume share a similar experience and a similar sense of gratitude.

The book was visionary. Thornhill and Alcock unashamedly argued that all of insect behaviour must have adaptive value. Often their observations led to the generation of hypotheses, rather than concrete proof of adaptive value. Their adaptationist approach was not met with unanimous approval, as is evident from Wade's (1984) review of Thornhill and Alcock (1983). Nonetheless, the book provided a gold mine of ideas that spawned a generation of insect behavioural ecologists, a mine that is far from exhausted. The researchers that were so profoundly influenced by *The Evolution of Insect Mating Systems* have, over the last thirty years, demonstrated beyond doubt that Thornhill and Alcock's enthusiasm was well justified.

While much in the original book remains relevant today, our aim in this edited volume is to bring together some of the World's leading insect researchers in order to update the empirical and conceptual scope of the original book. The developments made over the last thirty years do necessitate a shift in focus, in particular by moving away from the male–female dichotomy in terms of descriptions of mating systems (a dichotomy that many of us see as misleading), and by including some of the rich literature on behavioural mechanisms that have helped shape our view of behavioural function in new and exciting ways. Moreover, a number of fields have emerged since 1983. For example, we now more fully appreciate the role of females in shaping mating systems, and evolutionary conflicts of interest between males and females are now recognized as a strong source of selection. The role of parasites in sexual selection has also become evident, and in consequence the rise of the new field of ecological immunology has been driven forward by studies on insects and their mating systems.

Thornhill and Alcock's *The Evolution of Insect Mating Systems* is one of the great achievements of the functional, behavioural–ecological approach, but thirty years is a long time. We hope this update will both renew that achievement, and reveal many new questions to inspire a new generation of insect behaviour researchers, as we were inspired all those years ago.

D.M.S.
L.W.S.

Notes and Acknowledgements

Chapter 1—Benjamin B. Normark

I wish to thank Laura Ross, Jan Engelstädter, Francisco Úbeda, Jack Werren, and Norman Johnson for enlightening conversations on this topic, and Dave Shuker and Leigh Simmons for editing that improved the manuscript.

Chapter 2—David M. Shuker

My thinking about sexual selection has been shaped by many friends and colleagues over the years, including Mike Siva-Jothy, Tom Day, Andre Gilburn, Roger Butlin, Mike Ritchie, Allen Moore, Nina Wedell, Tom Tregenza and Nathan Bailey. Joan Roughgarden has also made me think harder about some aspects of sexual selection than I probably wanted to, but I appreciate our discussions even if we do not always agree. Mark Kirkpatrick and Leigh Simmons made many useful comments on earlier versions of the manuscript, and I am also grateful for my postgraduate students Liam Dougherty, Becky Boulton and Ginny Greenway for their perspectives on what I was trying to say. Ginny also helped organise the references, for which I am very grateful. Finally, I owe a great debt of thanks to Sue Healy, Leigh Simmons, Lucy Nash and Ian Sherman for their support, encouragement and patience.

Chapter 3—Hanna Kokko, Hope Klug, and Michael D. Jennions

We thank the editors of this book, Leigh Simmons and David Shuker, for the invitation to write this chapter and for their constructive feedback, and the Australian Research Council and the Academy of Finland for funding.

Chapter 4—Michael G. Ritchie and Rodger K. Butlin

We wish to thank our collaborators for extensive help and discussion, and would like to make a special mention of our friend and mentor, Godfrey M. Hewitt.

Chapter 5—Patricia J. Moore

I need to thank my most important mating system collaborator, Allen Moore. Allen is one of the first generation of young evolutionary biologists whose research was stimulated by *The Evolution of Insect Mating Systems*. And through our own 30 years of conversations, arguments, and research collaborations, he has inspired me to make my own small contribution to evolutionary ecology despite (or perhaps because of) my beginnings as a cell and developmental biologist.

Chapter 6—Douglas J. Emlen

I wish to thank Kerry Bright, Bruno Buzatto, Alison Perkins, David Shuker, and Leigh W. Simmons for comments on this chapter, and the National Science Foundation (OEI-0,919,781) for funding.

Chapter 7—Bruno A. Buzatto, Joeseph L. Tomkins and Leigh W. Simmons

J.L.T. and L.W.S. were supported by Fellowships from the Australian Research Council. B.A.B. was funded by an International Postgraduate Research Scholarship, and by the University of Western Australia. We thank Wade Hazel, Jane Brockmann, and Mark Elgar for valuable comments on a draft of this chapter, and Caio Carneiro de Mendonça and William Roscito for help in processing figures.

Chapter 8—John Hunt and Scott K. Sakaluk

We thank Clarissa House, James Rapkin and Sarah Lane for providing valuable feedback on an earlier draft of this chapter. J.H. was funded by the Natural Environment Research Council and a University Royal Society Fellowship. S.K.S. was funded by the National Science Foundation and a Visiting Professorship from The Leverhulme Trust.

Chapter 9—Rhonda R. Snook

I wish to thank Tommaso Pizzari for early discussion about the chapter and NERC for funding my work on the evolutionary and genetic consequences of polyandry.

Chapter 10—Leigh W. Simmons

After finishing this chapter I was reminded of R. L. Smith's 'Story of Three Flies' in his Forward to T. R. Birkhead and A. P. Møller's 1998 edited volume *Sperm Competition and Sexual Selection*. Smith too saw work on yellow dung flies and *Drosophila* as pivotal in the discovery and subsequent development of sperm competition research. His third fly, the screwworm fly *Cochlionmyia hominovorax* takes its place in the history of sperm competition research because it was the first pest species to be controlled using the sterile male release technique, a technique adopted by Parker, and many since, to study patterns of sperm utilization following double matings. Thornhill and Alcock (1983) had a tremendous influence on my early thinking about insect mating systems, and kickstarted my career. I have since had the pleasure of meeting and working with John Alcock and our times spent together are deeply treasured. Much of my work on sperm competition over the years has been inspired by Geoff Parker, to whom I owe an immeasurable debt of gratitude. My work is supported by the Australian Research Council.

Chapter 11—Göran Arnqvist

The writing of this chapter was supported by the European Research Council (AdG-294333) and the Swedish Research Council (621-2010-5266). I am grateful to Mark Kirkpatrick, David Shuker and Leigh Simmons for constructive comments on an earlier version of this contribution.

Chapter 12—Per T. Smiseth

I thank James Gilbert, Mathias Kölliker, Dave Shuker and Leigh Simmons for valuable comments on the chapter, and Jon Carruthers, Chris Goforth, Hannes Günther, Joël Meunier and Ray Wilson for permission to use their photographs.

Chapter 13—Marlene Zuk and Nina Wedell

Marlene Zuk is grateful to the editors for their invitation to contribute to the volume, and particularly to Leigh Simmons for longstanding collaboration, ideas and shared adventures. Many other colleagues and students have provided valuable feedback and discussion, particularly Nathan Bailey, Robin Tinghitella, Gita Kolluru, and John Rotenberry. The National Science Foundation (US) has supported her research. Nina Wedell thanks David Shuker and Leigh Simmons for the opportunity to participate in this project and for their insightful comments and feedback, and to Tom Price for many years of discussing the importance of selfish genes. The Royal Society (Wolfson Award) and NERC supported her research.

Chapter 14—Boris Baer

I wish to thank Leigh Simmons and David Shuker for continuous stimulating discussions over the last years on sexual selection and Koos Boomsma for continuous stimulating discussions about social insect reproduction. I am also grateful to Elisabeth Tibbetts for providing me insights into her work on paper wasps and providing me with pictures.

Contents

Contributors xiii

1 **Modes of reproduction** 1
Benjamin B. Normark

2 **Sexual selection theory** 20
David M. Shuker

3 **Mating systems** 42
Hanna Kokko, Hope Klug, and Michael D. Jennions

4 **The genetics of insect mating systems** 59
Michael G. Ritchie and Roger K. Butlin

5 **Reproductive physiology and behaviour** 78
Patricia J. Moore

6 **Reproductive contests and the evolution of extreme weaponry** 92
Douglas J. Emlen

7 **Alternative phenotypes within mating systems** 106
Bruno A. Buzatto, Joseph L. Tomkins, and Leigh W. Simmons

8 **Mate choice** 129
John Hunt and Scott K. Sakaluk

9 **The evolution of polyandry** 159
Rhonda R. Snook

10 **Sperm competition** 181
Leigh W. Simmons

11 **Cryptic female choice** 204
Göran Arnqvist

12 **Parental care** 221
Per T. Smiseth

13 Parasites and pathogens in sexual selection 242
Marlene Zuk and Nina Wedell

14 Sexual selection in social insects 261
Boris Baer

15 *The Evolution of Insect Mating Systems* 275
John Alcock and Randy Thornhill

References 279
Index 335

Contributors

John Alcock School of Life Sciences, Arizona State University, Tempe, AZ 85287-4701, USA. j.alcock@asu.edu

Göran Arnqvist Animal Ecology, Department of Ecology and Genetics, Uppsala University, Norbyvägen 18 D, SE-752 36 Uppsala, Sweden. goran.arnqvist@ebc.uu.se

Boris Baer Centre for Integrative Bee Research (CIBER), ARC Centre of Excellence in Plant Energy Biology, Bayliss building M 316, and the Centre for Evolutionary Biology, School of Animal Biology (M092), University of Western Australia, Crawley, 6009, Australia. boris.baer@uwa.edu.au

Roger K. Butlin Department of Animal and Plant Sciences, University of Sheffield, Sheffield S10 2TN UK, and Sven Lovén Centre—Tjärnö, University of Gothenburg, S-452 96 Strömstad, Sweden. r.k.butlin@sheffield.ac.uk

Bruno A. Buzatto Centre for Evolutionary Biology, School of Animal Biology (M092), University of Western Australia, Crawley, 6009, Australia. bruno.buzatto@uwa.edu.au

Douglas J. Emlen Division of Biological Sciences, 104 Health Science Building, University of Montana, Missoula, MT 59812, USA. doug.emlen@mso.umt.edu

John Hunt Centre for Ecology and Conservation, College of Life and Environmental Sciences, University of Exeter, Cornwall Campus, Penryn TR10 9EZ, UK. J.Hunt@exeter.ac.uk

Michael D. Jennions Division of Evolution, Ecology and Genetics, Research School of Biology, Australian National University, Canberra ACT 0200, Australia. michael.jennions@anu.edu.au

Hope Klug Department of Biological and Environmental Sciences, University of Tennessee Chattanooga, 215 Holt Hall, Dept 2653, 615 McCallie Avenue, Chattanooga, TN 37403 USA. hope-klug@utc.edu

Hanna Kokko Centre of Excellence in Biological Interactions, Division of Evolution, Ecology and Genetics, Research School of Biology, Australian National University, Canberra ACT 0200, Australia. hanna.kokko@anu.edu.au

Patricia J. Moore Department of Entomology, University of Georgia, Athens, GA 30602-2603, USA. pjmoore@uga.edu

Benjamin B. Normark Department of Biology and Graduate Program in Organismic and Evolutionary Biology, University of Massachusetts Amherst, Amherst, MA 01003, USA. bnormark@ent.umass.edu

Michael G. Ritchie School of Biology, University of St Andrews, St Andrews, Fife KY16 9TH, UK. mgr@st-andrews.ac.uk

Scott K. Sakaluk Behavior, Ecology, Evolution and Systematics Section, School of Biological Sciences, Illinois State University, Normal, IL 61790-4120, USA. sksakal@ilstu.edu

David M. Shuker School of Biology, University of St Andrews, Harold Mitchell Building, St Andrews, Fife KY16 9TH, UK. david.shuker@st-andrews.ac.uk

Leigh W. Simmons Centre for Evolutionary Biology, School of Animal Biology (M092), University of Western Australia, Crawley 6009, Australia. leigh.simmons@uwa.edu.au

Per T. Smiseth Institute of Evolutionary Biology, School of Biological Sciences, University of Edinburgh, West Mains Road, Edinburgh EH9 3JT, UK. per.t.smiseth@ed.ac.uk

Rhonda R. Snook Department of Animal and Plant Sciences, University of Sheffield, Alfred Denny Building, Western Bank, Sheffield S10 2TN, UK r.snook@sheffield.ac.uk

Randy Thornhill Department of Biology, University of New Mexico, Albuquerque, NM 87131-0001, USA. rthorn@unm.edu

Joseph L. Tomkins Centre for Evolutionary Biology, School of Animal Biology (M092), University of Western Australia, Crawley 6009, Australia. joseph.tomkins@uwa.edu.au

Nina Wedell Centre for Ecology and Conservation, College of Life and Environmental Sciences, University of Exeter, Cornwall Campus, Penryn TR10 9FE, UK. N.Wedell@exeter.ac.uk

Marlene Zuk Department of Ecology, Evolution and Behavior, 1987 Upper Buford Circle, University of Minnesota, St Paul, MN 55108, USA. mzuk@umn.edu

CHAPTER 1

Modes of reproduction

Benjamin B. Normark

1.1 Introduction

Before we explore the diversity of insect mating systems, let us acknowledge that some insects reproduce without mating at all. And even in species that mate, some offspring may receive only their mother's genes. Most of this book is devoted to considering the strategies by which insects can get more of their genes into the next generation. But the strategy you employ depends upon the rules of the game you are playing, and in this first chapter we consider how the rules of the game—in this case the laws of genetics—vary. Thornhill and Alcock (1983) started their 'Modes of Reproduction' chapter by reciting the names of some of the diverse genetic systems of insects, 'pedogenesis, polyembryony, thelytoky, gynogenesis, heterogony, facultative deuterotoky . . . arrhenotokous parthenogenesis' and joked that the list may tempt readers to take up the study of mammals (Box 1.1). But the great allure of insects for researchers is their staggering diversity, and in the area of modes of reproduction, they do not disappoint. Insects arguably possess a greater variety of genetic systems than all other animals combined.

Box 1.1 GLOSSARY OF TERMS ENCOUNTERED IN ANY DISCUSSION OF MODES OF INSECT REPRODUCTION

Apomixis, apomictic parthenogenesis
Production of eggs by mitosis instead of meiosis, resulting in clonal female offspring.

Arrhenotoky, arrhenotokous parthenogenesis
Development of males from unfertilized eggs.

Automixis, automictic parthenogenesis
Parthenogenetic egg production that involves meiosis.

Diplodiploidy
Sexual system in which both males and females are diploid.

Gynogenesis
Mode of reproduction in which sperm is required to initiate development but the sperm genome is excluded from the offspring. Also called sperm-dependent parthenogenesis or pseudogamy.

continued

The Evolution of Insect Mating Systems. Edited by David M. Shuker and Leigh W. Simmons.
© The Royal Entomological Society 2014. Published 2014 by Oxford University Press.

Box 1.1 *Continued*

Haplodiploidy
Any system in which males transmit only maternal genes. Arrhenotoky and paternal genome elimination are modes of haplodiploidy. (Alternatively, some authors apply the term haplodiploidy only to arrhenotoky.)

Hybridogenesis
Mode of reproduction in which parthenogenetic females mate with males from a sexual lineage to produce diploid female offspring, which in turn eliminate the paternal genome when producing their own eggs, so that only maternal genomes are transmitted across generations.

Parthenogenesis
Development of an unfertilized egg. Usually refers to development of female offspring from unfertilized eggs (thelytoky), but may also be applied to development of male offspring (arrhenotoky).

Paternal genome elimination
Genetic system in which males develop from diploid zygotes but the paternal genome is excluded from their sperm.

Thelytoky, thelytokous parthenogenesis
Development of female offspring from unfertilized eggs.

Parallel clonality
Genetic system known only in the little fire ant, in which both males and females are clonally propagated, and sexually produced individuals are sterile workers.

Premeiotic doubling
Form of automixis in which the genome replicates prior to meiosis.

Gamete duplication
Form of automixis in which diploidy is restored by duplication of the gamete's genome.

Pseudogamy
Gynogenesis.

Cyclic parthenogenesis
Life cycle that includes an alternation between sexual and parthenogenetic generations.

Tychoparthenogenesis
Parthenogenetic development of a small percentage of unfertilized eggs.

Social hybridogenesis
Mode of reproduction in which queen ants produce daughter queens parthenogenetically and daughter workers sexually. (May also be applied to systems in which queen ants mate with both conspecific and heterospecific males, and hybrid offspring develop into workers.)

1.2 Parthenogenesis

1.2.1 *Types of parthenogenesis*

1.2.1.1 *Apomixis vs automixis*

In more than a thousand species of insects, reproduction occurs without any mating at all. In these *obligately parthenogenetic* populations, unmated females lay eggs that all develop into females. The evolutionary transition from sexuality to obligate parthenogenesis requires a transition from making haploid eggs to making diploid ones, and we can classify different kinds of parthenogenesis according to how diploidy is established in the egg.

The simplest way for a female to make a diploid egg is to use mitosis instead of meiosis. That is, to use simple cell division (the same kind of cell division that happens during growth) instead of the special and complicated chromosome-recombining-and-ploidy-reducing cell division that sexual animals use to make gametes. The resulting offspring has a near-perfect replica of the mother's genome. This simple system is called *apomixis*, and it seems to be the most widespread form of obligate parthenogenesis in insects (Suomalainen et al. 1987).

The other ways of making a diploid egg are all a little more complicated than this. All involve meiosis and go under the collective name of *automixis*. Some forms of automixis are genetically equivalent to apomixis, in that the mother's genome is transmitted essentially intact to the daughter. For instance, in the Australian grasshopper *Warramaba virgo*, the genome replicates prior to meiosis, so that when meiosis occurs, it results in eggs with a full diploid set of chromosomes (*premeiotic doubling*). At the other extreme, one form of automixis (*gamete duplication*) involves producing a normal haploid egg and then doubling its genome, so that the offspring's entire genome is homozygous. Between these extremes (retaining all vs none of the heterozygosity found in the mother), there are several other types of automixis involving fusion of the egg with one of the polar bodies, resulting in the loss of some but not all of the mother's heterozygosity. A common misconception about these intermediate types of automixis is that they are genetically equivalent to self-fertilization (which is common in plants): 50% probability of loss of heterozygosity at all loci in each generation, leading to loss of all heterozygosity within a few generations. But self-fertilization brings together the products of different meioses. Stranger things happen in automictic fusion, which brings together the products of a single meiosis. Depending on exactly which nuclei fuse, the result can be total loss of heterozygosity around centromeres and its retention near telomeres, or the opposite. Empirical work on the long-term consequences of this phenomenon in automictic insects is lacking, but work on analogous mating systems in fungi shows that heterozygosity can skyrocket in these systems, due to positive feedback between accumulation of recessive lethals in low-recombination regions and evolution of ever-lower rates of crossing over in such regions (Hood and Antonovics 2004). A detailed classification of the different mechanisms of parthenogenesis and their distribution across insects is given by Suomalainen et al. (1987).

1.2.1.2 *Gynogenesis, hybridogenesis, and related systems*

Parthenogenesis literally means 'virgin birth'. Many researchers have therefore avoided applying the term to modes of reproduction that involve mating, even when the sperm makes no genetic contribution to the offspring. Some insects use apomixis or automixis to furnish their eggs with complete diploid genomes, but still require sperm to trigger the onset of development. This system goes by the name of *gynogenesis* or *pseudogamy*. Only a few gynogenetic insect species are known, scattered across several orders (Coleoptera, Lepidoptera, Hemiptera, and Collembola; Figure 1.1), but gynogenesis often goes undetected without careful genetic study, so the actual number of gynogenetic insect species may be much larger. The study of gynogenesis is interesting from the point of view of the evolution of mating systems, because mating does still occur, but its costs and benefits are very different from mating in ordinary sexual populations—especially for males. A gynogenetic population is in effect a parasite upon a sexual 'host' population (Lehtonen et al. 2013).

Another kind of sperm-dependent clonal reproduction is found in the *Bacillus rossius* species complex of Mediterranean stick insects (Figure 1.1). It is known as *hybridogenesis*,

Figure 1.1 Gynogenetic and hybridogenetic insects. Two species for which no good images are available are here represented by closely similar sexual species (c, e). (a) Gynogenetic *Alsophila pometaria* (Lepidoptera; photo by Jim Troubridge). (b) Gynogenetic *Ips acuminatus* (Coleoptera; photo by Claude Schott). (c) *Oncychiurus* sp., representing gynogenetic *Oncychiurus procampatus* (Collembola; photo by Sean McVey). (d) *Ribautodelphax pungens* (Hemiptera; photo by Gernot Kunz). (e) *Bacillus rossius*, sexual parent of gynogenetic *Bacillus rossius-grandii* and hybridogenetic *Bacillus rossius-grandii benazzii* (Phasmatodea; photo by Vladimír Motyčka).

or hemiclonal inheritance. The hybridogenetic lineages consist entirely of females. To reproduce, they must mate with males of a related sexual population. The hybridogenetic females produce haploid eggs, which are fertilized by the males and develop into diploid progeny. But those progeny are all females, and when they produce their own eggs, they do so by eliminating their father's genome and using their mother's haploid genome intact. Thus, in terms of transmission genetics and ecology, hybridogenesis is much like gynogenesis, in that it requires a sexual host population to mate with, and the males of that population acquire no genuine paternity (at least no grandpaternity, as it were). It has been more than 20 years since the discovery of hybridogenesis in *Bacillus*, and no other cases in insects have yet come to light. Only a few other cases are known in nature (in frogs and fish). But the intriguing complications of the system have attracted a good deal of attention from evolutionary biologists (Beukeboom and Vrijenhoek 1998; Archetti 2005; Som and Reyer 2007; Lehtonen et al. 2013).

1.2.1.3 *Facultative and cyclic parthenogenesis*

Many insect species have a life cycle that includes both sexual reproduction and parthenogenesis. In aphids, for instance, there is typically one sexual generation per year, usually in the autumn, followed by a number of parthenogenetic generations in the spring and summer. This is an example of cyclic parthenogenesis. Four other insect clades are characterized by cyclic parthenogenesis: oak and maple gall wasps, *Micromalthus* beetles, and two clades of cecidomyiid midges (Normark 2003).

In many sexual insect species, if a female is experimentally prevented from mating, she will lay unfertilized eggs, and a small percentage of these will hatch and produce viable offspring, a phenomenon known as *tychoparthenogenesis*. A large percentage of mayfly species apparently have this ability (Funk et al. 2010), along with many Orthoptera and Phasmatodea, and at least a few Diptera and Lepidoptera (Suomalainen et al. 1987; Schwander et al. 2010; Matsuura 2011). Occasional parthenogenetic reproduction by unmated females also occurs in both major groups of social insects. In a number of species of termites (Isoptera), winged females that do not mate may still initiate a colony by laying unfertilized eggs that develop into an all-female brood. Often, a pair of unmated females will found a colony together (Matsuura 2011). In a number of ant species, when the queen within a colony dies, some workers (which as a rule in ants are all unmated females) lay diploid eggs that develop into replacement queens (Cheron et al. 2011; Matsuura 2011).

In some insect species, mated females may produce some combination of sexual and parthenogenetic progeny. We do not have many examples of these, either because they are truly rare or because they are difficult to detect. But a fascinating set of examples has recently come to light in the social insects. In termites, the eggs within a colony are typically the sexual progeny of a single king and queen. But in some termites of the genus *Reticulitermes*, queens produce a few eggs parthenogenetically. These queens' sexual eggs develop as usual into workers and alate reproductives. The parthenogenetically produced eggs develop into secondary queens (Matsuura et al. 2009; Vargo et al. 2012). A colony may have up to several hundred secondary queens, all of whom mate with the primary king (Matsuura 2011). Somewhat analogous cases have also been found in ants, in which queens use sperm to fertilize eggs that develop into workers, but also produce some diploid eggs parthenogenetically, which develop into reproductive females. In termites, mated queens only use parthenogenesis to produce secondary queens within the same colony, while the alate reproductives that found new colonies are all produced sexually; in some ant lineages, queens produce all reproductive females parthenogenetically and use sex only to produce sterile workers (Fournier et al. 2005; Foucaud et al. 2010; Matsuura 2011; Leniaud et al. 2012). Thus the ant system is probably best regarded as a form of obligate parthenogenesis, since new colonies are produced parthenogenetically. Indeed if we think of a colony as analogous to an individual organism, this system is analogous to hybridogenesis (parthenogenetic germline, sexually produced soma), and is sometimes called *social hybridogenesis* (Leniaud et al. 2012), though somewhat confusingly that term has also been applied to entirely sexual systems in which queens use sperm of conspecific males to produce reproductives and sperm of heterospecific males to produce sterile workers (Cahan 2003).

1.2.2 *A thousand clones: incidence and phylogenetic distribution of obligate parthenogenesis*

The approximately 1,100 species thought to reproduce by obligate parthenogenesis are scattered across the phylogeny of insects and other hexapods, but are more concentrated in some orders than others (Figure 1.2). This total includes all the species for which there is evidence of at least one parthenogenetic lineage; it includes a few hundred species that are also known to include one or more sexual populations under the same species name.

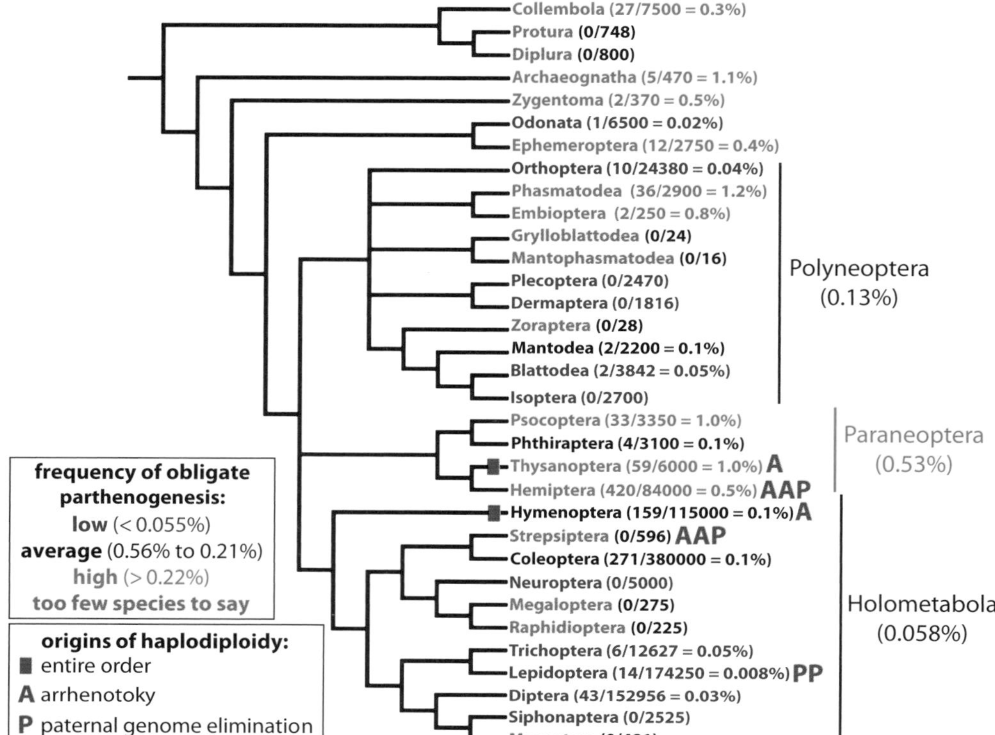

Figure 1.2 Distribution of obligate parthenogenesis and haplodiploidy across the phylogeny of insects and other hexapods. I have found published references to obligate parthenogenesis in 1,098 hexapod species, out of a total of about one million described, yielding an average incidence of about 0.11% across the entire class. Orders with more than twice the expected number of species with obligate parthenogenesis (more than 0.22% of described species) are listed in red. Those with fewer than half the expected number of obligate parthenogens (less than 0.055%) are listed in blue. But if the order has so few species that the expected number of obligate parthenogens (assuming a frequency of 0.11%) is zero or one, that order is listed in grey. Orders that have approximately the expected proportion of parthenogens (between 0.056% and 0.21%) are listed in black. Total numbers of species per order are from Chapman (2009), supplemented by Resh and Cardé (2009). References and further information on origins of haplodiploidy are given in Normark (2003, 2004b). Numbers of species with obligate parthenogenesis are based on too many references to list here. Many of the references are included in the supplemental online material of Normark (2003). Insect phylogeny is from Trautwein et al. (2012).

Because there are about a million hexapod species in total, the evidence at hand suggests that obligate parthenogenesis is rare: about 0.1% are obligately parthenogenetic, whereas 99.9% are sexual.

1.2.3 *Understanding the distribution of obligate parthenogenesis*

1.2.3.1 *Theories of sex*

Why is parthenogenesis so much rarer than sex? This question has motivated the modern study of the evolutionary biology of sex. It is a difficult problem in part because the success of sex is a paradox. Parthenogenesis is a more efficient form of reproduction than sex. In species with separate sexes, outcrossing, and no paternal care (a category that includes many insect species, but see Chapter 12), only one sex gathers resources from the environment and invests them in offspring: females. A population that consists of 100% females should in principle produce twice as many offspring as a 50% male population of the same size (Lively and Lloyd 1990). This twofold cost of sex (or, more precisely, cost of males) poses a daunting hurdle for hypotheses postulating an advantage for sexual reproduction. Not only must sexual offspring have higher fitness than parthenogenetic offspring: their fitness has to be, on average, twice as high.

Sex may also have additional costs (Lehtonen et al. 2012; Meirmans et al. 2012). Sex destroys existing combinations of alleles—the very combinations that enabled the previous generation to survive and reproduce—and creates new, untested combinations of alleles (Otto 2009). Here is another paradox of sex: why destroy successful genotypes? A half-century of theoretical work on this problem has limned the outlines of an answer. Evolutionary biologists had classically expected natural selection to result in a close fit between genotype and environment, with high-fitness genotypes at high frequency. But it turns out that, in a world like that, sex is useless. Sex only pays in worlds that have a poor fit of genotype to environment. As we shall see in Chapter 8, another famous mating-systems paradox—the lek paradox—hinges on the same conundrum. Extremely choosy females are behaving as if good genotypes were rare. Haven't they read R. A. Fisher? So what disrupts the fit between genotype and environment? Something must be changing too fast for natural selection to keep up: either the environment, or the genome (by mutation), or some combination of the two (Kondrashov 1993; Howard and Lively 1994).

We can almost take for granted that environments change, and yet the twofold cost of sex puts stringent constraints on the kind of environmental challenge that is required to make sex pay: it must be lethal, rapidly changing, and—given the prevalence of sex in the world—ubiquitous. If not literally lethal, it must prevent reproduction or have some other huge fitness effect. Genotypes with high fitness in one generation must have low fitness in the next generation. The most plausible candidate for the job of making the world sufficiently dystopian is infectious disease: pathogens and parasites (Jaenike 1978; Hamilton 1980; Hartfield and Keightley 2012) (see Chapter 13).

Whenever different loci are under selection at once, whether positive selection favouring new resistance alleles or negative selection against new deleterious mutant alleles, parthenogenesis causes interference between selection at different loci, reducing the effectiveness of selection, whereas sex reduces selective interference between loci (Felsenstein 1974). The greater the number of loci under selection at the same time, the greater is the advantage to sex (Hartfield and Keightley 2012). Thus it is plausible that both environmental and genomic challenges—parasites and harmful mutations—contribute to making sex useful (Howard and Lively 1994).

1.2.3.2 *The standard model of the evolution of obligate parthenogenesis: high extinction rate*

The ecological correlates of obligate parthenogenesis have long been scrutinized for clues to the adaptive significance of sexuality, and the many obligately parthenogenetic insects have played a major role in the literature on this topic (Vandel 1928; Glesener and Tilman 1978; Bell 1982). The prevailing paradigm in most of this literature is roughly as follows: parthenogenetic lineages originate at low or moderate frequency from within sexual populations; they can be strikingly successful in the short term but are at high risk of extinction and rarely persist for long, which is why they have such a scattered ('tippy') phylogenetic distribution. The ecological distribution of parthenogenetic lineages provides clues to

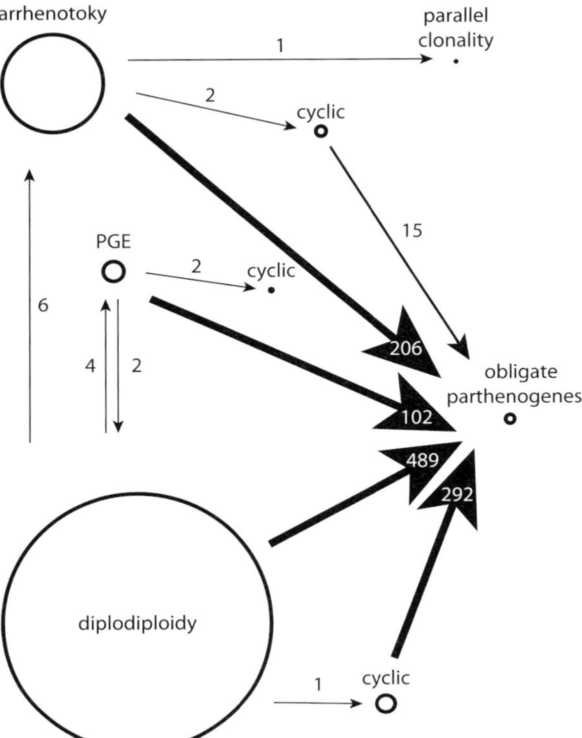

Figure 1.3 Relative abundance of different genetic systems across insect species, and the numbers of independent transitions that have occurred between them during insect evolution. The area of each circle is proportional to the number of species with the given genetic system. Systems marked 'cyclic' consist of cyclic alternation between their ancestral sexual system (diplodiploidy, arrhenotoky, or paternal genome elimination) and parthenogenesis. Facultative parthenogenesis is here grouped together with its sexual system, which in almost all cases is diplodiploidy. More taxonomic detail is given in Figure 1.2.

the adaptive significance of sex, reflecting environments in which the benefits of sex are reduced (Williams 1975; Glesener and Tilman 1978; Bell 1982) or in which the costs of sex are elevated (Cuellar 1994; Normark 2006; Meirmans et al. 2012). The preponderance of the evidence seems to support this paradigm (Figure 1.3), but it has never been rigorously tested, and it is a tempting target for iconoclasts (Schwander and Crespi 2009b, Johnson et al. 2011).

An alternative explanation for the scattered distribution of parthenogenesis is frequent reversion of obligate parthenogenesis back to sexuality (Williams 1975; FitzJohn et al. 2009). In a few groups of insects, especially aphids and *Wolbachia*-infected parasitic wasps, researchers have succeeded in experimentally inducing a reversion from parthenogenesis back to sexuality, a 'proof of concept' that it can happen (Takada 1988; Stouthamer et al. 1990). But for many other obligately parthenogenetic insects there are significant cytogenetic barriers, such as odd ploidy and loss of the Y chromosome (Bull and Charnov 1985). And all obligately parthenogenetic lineages should be subject to the genetic decay of sex-specific pathways, structures, and behaviours (Schurko et al. 2009; Lehmann et al. 2011). Indeed, even in aphids and *Wolbachia*-infected parasitic wasps, some obligate parthenogens are recalcitrant to experimental attempts to revert them, and appear to be permanently parthenogenetic (Takada 1988; Russell and Stouthamer 2011).

Analytical tools are at hand that should enable a more rigorous testing and parameterization of this standard model than has previously been possible (FitzJohn et al. 2009; Magnuson-Ford and Otto 2012). But two issues will prevent any quick resolution of the perennial controversies. First, there is the problem of singling out a group in which parthenogenesis is particularly frequent, analysing the evolutionary dynamics of parthenogenesis in that group, and concluding that parthenogenetic lineages are not nearly as extinction-prone as the conventional wisdom would have it. In order to avoid the bias intrinsic to such an approach, it will be necessary to analyse a fine-grained (species-level) phylogeny for a truly large number of species, which remains a daunting task. Second, there is the problem of assigning a prior probability to reversion from parthenogenesis to sex. This conundrum is analogous to the problem of inferring reversions from specialist to generalist or from wingless to winged (Stone and French 2003; Stireman 2005). Derived states that arise very frequently (parthenogenesis, specialization, winglessness) are likely to be mistaken for the ancestral state unless an asymmetry in the transition rates is posited a priori.

1.2.3.3 *Ecological correlates of obligate parthenogenesis and the role of population size*

There are several ecological correlates of parthenogenesis. In pairwise comparisons with close sexual relatives, the parthenogen is more likely to occur at higher latitude, higher altitude, in a more arid habitat, or on islands. In each case, the parthenogen is found in the biotically more depauperate part of the range, which has often been interpreted as providing support for the hypothesis that sex is an adaptation for dealing with pathogens and parasites (Bell 1982), though of course several other interpretations are possible (Cuellar 1994; Lively 2011). Before discussing some recent interpretations of these patterns, we focus primarily on another apparent correlate of parthenogenesis that has received less attention: population size.

For most of the five decades during which biologists have been actively investigating the evolutionary biology of sex, theorists have tended to disregard population size as an important consideration. Muller (1964) argued that small asexual populations would go extinct due to the accumulation of deleterious mutations, but the mechanism that he invoked seemed to apply only to the smallest populations and was not widely regarded as a general explanation (Maynard Smith 1978; Bell 1982). The next influential hypothesis about deleterious mutations to emerge was by Kondrashov (1982), applying to all populations regardless of size. More recently Kondrashov's hypothesis has been largely refuted by a steady drumbeat of genetic studies that have failed to find any evidence of the synergistic epistasis between deleterious mutations that the hypothesis required (Otto and Gerstein 2006; Hartfield and Keightley 2012). Meanwhile, the environmental hypotheses (such as the parasite, or Red Queen hypothesis) could in principle operate in populations of any size. Thus, until recently, population size did not seem to be important.

More recently, the potential importance of population size is coming to light. The selective interference between different loci that reduces the effectiveness of selection in parthenogens only operates in finite populations, and the size of the population matters, with sex becoming less advantageous at large population size (Poon and Chao 2004; Hartfield et al. 2010). This effect is expected to occur irrespective of the source of selection on multiple loci, whether due to mutations, or parasites, or some other source, but the more loci and the stronger the selection, the larger the population size at which sex remains useful (Hartfield and Keightley 2012).

Thus, the best available current theory yields the expectation that sex should be more important for smaller populations than for larger ones, and thus that very large populations may be asexual. Is this expectation supported by observation? There is little direct information about insect population size in nature, but we do have information on whether insect species are pests. Insects that reach high abundance in habitats used by humans (and this covers a lot of ground, including forests used for timber) are likely to have been recorded as a pest. Insects that do not reach high abundance in such habitats are unlikely to have been recorded as a pest. The literature also contains information about other correlates of population size, such as geographic range and host range.

Hoffman et al. (2008) showed that pests are more likely to be parthenogenetic than related non-pests. They interpreted this as a consequence of the properties of agricultural habitats. Essentially they added one more item to the list of environmental correlates of parthenogenesis (high latitude, high altitude, xeric, island, *agricultural*), one that is similar to the other correlates in reflecting reduced biotic diversity. But close examination of Hoffman et al.'s list of parthenogenetic pests reveals that only a few are pests of agriculture per se. They are mostly pests of horticulture, forestry, orchard crops, and households (Normark and Johnson 2011). Most of the plant-feeding pests listed by Hoffman et al. (2008) are highly polyphagous, able to feed on very long lists of host species: they seem, if anything, unusually well equipped to deal with biotic diversity (Figure 1.4). What's going on here?

Recently, a radically different interpretation of the connection between pest status and parthenogenesis has been presented (Normark and Johnson 2011; Ross et al. 2013). Here, pest status is interpreted as an index of high abundance. Current theory predicts a decreasing utility of sex at very large population size (Hartfield et al. 2010; Hartfield and Keightley 2012). As expected, abundant species (pests) are more likely than less abundant species (non-pests) to be parthenogenetic. The probable mechanism is that abundant parthenogens have a longer expected mean time to extinction, and thus are more likely to be observed before they go extinct. Another process likely to contribute to the pattern is that

Figure 1.4 Polyphagous parthenogenetic pests. (a) *Saissetia coffeae* (Hemiptera; 105 host families; photo by Chen Kuntsan). (b) *Orgyia antiqua* (Lepidoptera; 24 host families; photo by Walter Schön).

more abundant species have a higher number of new parthenogenetic lineages arising within them, whether by spontaneous mutation or by hybridization. The implication is that pests are not specialized denizens of human-altered habitats, but are globally and anciently abundant. Human-altered habitats have 'sampled' them in proportion to their pre-existing abundance, rather than having caused that abundance. This is a reversal of the usual interpretation of the causal connection between abundance and parthenogenesis (that parthenogenesis is the cause and abundance is the effect).

The extremely broad niches of many parthenogens (Lynch 1984) are also causally entangled with large population size: ability to use a broader range of resources permits larger population size, whereas larger population size (which increases the efficiency of natural selection) permits adaptation to a broader range of resources (Normark and Johnson 2011). Most of these highly abundant, broad-niche species are sexual, or mixed sexual and parthenogenetic, supporting the hypothesis that parthenogenesis is an effect rather than a cause of abundance. Another ecological correlate of this syndrome is female flightlessness, with usually wind-borne larval dispersal to new host plants. This combination of high abundance, broad environmental tolerance, and essentially random dispersal means that these species often have very large geographic ranges and are constantly turning up in new and marginal habitats, which may help explain some of the other ecological correlates.

An almost precisely opposite claim about the relationship of abundance to parthenogenesis is sometimes made: that it characterizes low-density populations and is an adaptation that obviates the problem of finding mates in such populations (Cuellar 1994; Schwander et al. 2010). In a recent empirical survey of the documented causes of female mating failure, Rhainds (2010) found that female mating success *declines* with increasing population density. He suggested that this 'counter-intuitive trend' results in part from interference between females in mate attraction. Yet another perspective on the relationship of parthenogenesis to population density is provided by Lively (2011). He summarizes the ecological correlates of parthenogenesis as characteristic of 'r-selected' populations, and presents modelling results showing that the cost of males is maximal in growing populations and is much lower in populations at carrying capacity.

Perhaps the contrast between high-abundance populations and low-density populations is more apparent than real. There are several reasons why abundant generalist species may be adapted to rapid population growth, even if their average abundance has been high for a geologically long period of time. First, hyperabundant polyphagous species have a long history of having expanded their population size by colonizing new hosts and new areas. Second, many polyphages are flightless herbivores of long-lived woody plants, each individual plant is host to a deme, and metapopulations persist by continually 'infecting' new individual hosts. Thus the host life cycle imposes boom-and-bust demography on the insect. Third, highly abundant species often experience 'niche destruction', as pathogens and predators pile onto the abundant resource (Holt 2009); in insects this is best seen in the boom-and-bust ('outbreak') dynamics of temperate forest pests (Nothnagle and Schultz 1987; Barbosa et al. 1989). Finally, the high abundance (large global population size) of parthenogenetic insects is mediated more by their ubiquity (large geographic and host range) than by their local density—though species considered to be pests are generally those that undergo at least occasional outbreaks to high density. In most cases the density of these insects is regulated by natural enemies, rather than by competition for resources (Van Driesche et al. 2008). Their high birth rates and high death rates may keep them in a region of Lively's (2011) parameter space in which the cost of males is high.

1.2.3.4 *Persistence and evolution of ancient parthenogenetic insects*

The fact that parthenogenetic lineages are scattered across the tree of life implies that they do not persist long enough to radiate into large clades. But how long do they actually persist, and in what respects do they evolve? We still do not have systematic answers to these questions, but the accumulation of clues continues. A recent review lists just a few yet-unrefuted potential ancient parthenogenetic insect species, in the stick insect genus *Timema* and the beetle genus *Calligrapha* (Schurko et al. 2009). Recent work has provided further evidence that the *Timema* species are genuinely parthenogenetic, though it casts some doubt on how ancient they are—their ages were estimated by assuming a 'standard' rate of molecular evolution but it turns out that the rate is accelerated in these parthenogenetic lineages (Schwander et al. 2011; Henry et al. 2012). The accelerated rate of molecular evolution in parthenogenetic *Timema* is consistent with the hypothesis of a genome-wide accumulation of mildly deleterious mutations, and tends to support a mutational or mixed mutational–environmental hypothesis of sex as against a purely parasite-driven model of the advantage of sex. Evidence for a similar speed-up has been seen in a few other parthenogenetic taxa (Paland and Lynch 2006; Neiman et al. 2010), including some insects (Normark and Moran 2000; Kraaijeveld et al. 2012).

1.2.3.5 *Origins of parthenogenesis*

1.2.3.5.1 A neglected problem

Whereas the standard model assumes that the rarity of parthenogenesis is mediated by the tendency of parthenogenetic lineages to go extinct, an alternative hypothesis is that its rarity is due to severe constraints upon its rate of origin (Nunney 1989; Engelstadter 2008). Other features of the phylogenetic and ecological distribution of parthenogenesis could also be interpreted as reflecting peaks in its rate of origin rather than troughs in its rate of extinction. This approach has not been anywhere near as popular with evolutionary biologists as the extinction-driven standard model. Interest in parthenogenetic lineages in recent decades has been stoked by the idea that they provide a window into the Mystery of Sex. If parthenogenetic lineages go extinct at a high rate, it implies that sex confers protection against extinction. The idea has an almost literary allure—Love or Death—which helps to explain its enduring fascination. In contrast, the idea that the distribution of sex is limited by constraints on its origins is harder to frame in epic-sounding terms. If parthenogenesis does not elevate the extinction rate, then sex is more like an arbitrary convention of no particular utility. The idea that (almost) all parthenogenetic lineages are rapidly killed off by *some* process implies a universal challenge faced by (almost) all organisms. The quest to identify this challenge is a bid for the further unification of the biological sciences. But constraints on the origins of parthenogenesis (and the loopholes in those constraints that allow them to originate) are arcane details of cytogenetics and development—they will almost certainly differ in different taxa. This argument does not mean that they are not important, or even in some cases decisive. Instead it implies that their importance is likely to have been underestimated.

1.2.3.5.2 Modes of egg activation

The entry of sperm into the egg triggers the onset of development in most vertebrates and marine invertebrates, and in some insects (see Section 1.2.1.2). But some groups of insects have other modes of egg activation. In many wasps, and in *Drosophila*, eggs are activated by mechanical stimulation as they are squeezed through the oviduct or ovipositor (Horner and Wolfner 2008). In some stick insects, eggs are activated by exposure to air (Went 1982). It seems probable that parthenogenesis would arise more frequently in groups in which egg activation is not dependent upon sperm, but egg activation has been studied in few species, and its connection to parthenogenesis has never been the subject of a comparative study.

1.2.3.5.3 Hybrid vs non-hybrid origins

Origins of obligately parthenogenetic lineages have historically been best studied in vertebrates, where essentially all are of hybrid origin (Neaves and Baumann 2011). For insects, there is evidence that hybrid origins are typical of obligately parthenogenetic Phasmatodea (Morgan-Richards and Trewick 2005; Ghiselli et al. 2007; Milani et al. 2010), Coleoptera (Normark and Lanteri 1998; Stenberg et al. 2003; Gomez-Zurita et al. 2006; Kajtoch and Lachowska-Cierlik 2009), and at least a few Orthoptera (Kearney and Moussalli 2003) and Hemiptera (Normark 2003). There are also a few well-studied cases (also in Phasmatodea, Orthoptera, and Hemiptera) in which evidence of hybrid origin is lacking (Schwander and Crespi 2009a; Andersen et al. 2010; Kolics et al. 2012), and there are clearly some important non-hybrid modes of origins of parthenogenesis in insects, including loss of the sexual phase in cyclically parthenogenetic aphids, and symbiont-induced parthenogenesis. But for the great majority of obligately parthenogenetic insects we still have no evidence one way or another on whether hybridization is involved in the origin.

The connection between hybridization and parthenogenesis is not fully understood, but it seems likely that hybridization is a frequent disruptive factor in reproduction and development, and that as such it increases the probability of origin of reproductive and developmental novelty. In particular, interspecific chromosome complements may be less likely to pair as homologues, which may tend to prevent meiosis and promote the production of diploid eggs. It is also possible that the pattern of hybrid origin is driven in part by a moderating effect on extinction rate: a hybrid parthenogenetic lineage may differ ecologically from both parental species and thus may be buffered against competitive exclusion by either.

1.2.3.5.4 Symbiont-induced parthenogenesis

A dramatic advance in our understanding of one mode of origin of parthenogenesis occurred in the 1990s with the demonstration of symbiont-induced parthenogenesis in Hymenoptera. Stouthamer et al. (1990) reported that parthenogenetic *Trichogramma* wasps were 'cured' of parthenogenesis and reverted to sexuality when fed antibiotics. It turned out that the parthenogenetic *Trichogramma* harboured intracellular *Wolbachia* bacteria, which are transmitted by females to their offspring, through their eggs. By turning male offspring into females, the *Wolbachia* harboured by *Trichogramma* were transmitted by 100% of a female's offspring rather than just 50%. Theorists had previously hypothesized that such a phenotype in a cytoplasmically transmitted element would be favoured by natural selection (Hamilton 1979), but the direct experimental demonstration of the reality of this phenomenon led to an explosion of research into parthenogenesis induction and other sex-ratio-distorting phenotypes of intracellular bacterial endosymbionts of insects (Zchori-Fein et al. 1998; Huigens et al. 2000; Hunter et al. 2003; Giorgini et al. 2010; Adachi-Hagimori et al. 2011). Chapter 13 discusses how cytoplasmically transmitted endosymbionts and other selfish genetic elements have had a major impact on the evolution of insect mating systems.

1.2.3.5.5 Molecular detail

Pondering the origins of parthenogenesis may not have the Theory-of-Everything allure of pondering their extinction, but, as case studies, origins are more susceptible to definitive solution than extinctions. The origin of parthenogenesis in the Cape honeybee has been plausibly linked to a deletion in a flanking region of a CP2 transcription factor homologous to the *gemini* (genitalia missing) locus in *Drosophila*, which results in alternative splicing of the protein and a queen-like phenotype (Jarosch et al. 2011). This transcription factor also interacts with a protein involved in meiotic spindle formation, and thus the same modification may also have been sufficient to induce diploid egg production. The impressive discovery of this mechanism nonetheless underlines how multifarious and taxon-specific the modes of origin of parthenogenesis are likely to be: the modified protein interferes with the usual self-sterilization of honeybee workers, a process absent in non-eusocial insects. The Cape honeybee is an infectious social parasite, taking over colonies of ordinary sexual honeybees (Neumann et al. 2011). This seems to be a unique situation, as none of the several parthenogenetic ants have precisely this habit—though there is an interesting case of apparently stable cohabitation of parthenogenetic socially parasitic ants and their parthenogenetic hosts (Dobata et al. 2009). We do not have quite as complete an understanding of the origin and mechanism of parthenogenesis in any other insect, though parthenogenesis-inducing bacteria and a few other well-characterized systems (Sandrock and Vorburger 2011) offer plausible routes to progress in the near future.

1.3 Haplodiploidy

1.3.1 *Arrhenotoky vs paternal genome elimination*

Parthenogenesis has arisen thousands of times in most animal phyla and many families of plants, and it is genetically equivalent to forms of asexual reproduction frequent throughout bacteria, protists, and fungi. But there is another mode of reproduction that is more of an insect speciality: haplodiploidy. Of the twenty known origins, ten are in insects (Figures 1.2, 1.3, and 1.5). In the broad sense in which the term is used here, haplodiploidy refers to any system in which males transmit only the genes that they received from their mothers. Often the term is used more restrictively to apply only to the simplest and most widespread form of haplodiploidy: *arrhenotoky*, in which males develop from haploid unfertilized eggs. In terms of transmission genetics, this is identical to *paternal genome elimination* (PGE), in which males develop from diploid zygotes, but the paternal genome is excluded from their sperm. The precise fate of the paternal genome varies a good deal in different systems of PGE. The paternal chromosomes may be lost early in embryogenesis, so that males are haploid, as in some armoured scale insects; or they may be silenced, appearing as a blob of heterochromatic chromosomes in male tissues, as in most scale insects and in *Hypothenemus* bark beetles; or the paternal chromosomes may function normally in males and be eliminated only at spermatogenesis, as in sciarid and cecidomyiid flies (Brun et al. 1995; Burt and Trivers 2006; Ross et al. 2010).

1.3.2 *Evolution of haplodiploidy*

Two things are striking about the repeated evolutionary origins of haplodiploidy. First, there is the asymmetry of the system itself: females completely monopolize the parentage of one of the sexes. Second, there is the fact that in twenty origins, the symmetry always

Figure 1.5 Haplodiploid insects. (a) *Ferrisia virgata*, a paternal-genome-eliminating species; photo by Scott Justis. (b) *Polistes dominula*, an arrhenotokous species; photo by David Shuker.

breaks in the same direction: it is always the females that monopolize the parentage of sons. There are no observations of males monopolizing the parentage of daughters. In recent years the other logical possibilities—females monopolizing the parentage of daughters and males monopolizing the parentage of sons—have been discovered in an ant species, the little fire ant, *Wasmannia auropunctata* (Foucaud et al. 2010).

S. W. Brown (1964) modelled the fate of an allele that causes females to have sons lacking paternal genes. He showed that such an allele would increase rapidly in frequency, an effect he called 'automatic frequency response', and compared to meiotic drive. Brown was a scale insect cytogeneticist who had long studied the varied fates of paternal chromosomes in male scale insects. The shutdown and elimination of paternal chromosomes in these insects had never been satisfactorily explained, and Brown was the first to interpret this phenomenon in terms of what we would now call genetic conflict. The idea of 'selfish' genes was somewhat older than this, but had previously been applied only to the odd B chromosome or driving allele making mischief (Östergren 1945; Crow 1988; Burt and Trivers 2006) (see also Chapter 13). Here Brown was explaining the fundamental laws of heredity that have governed thousands of species for millions of years as a legacy of genetic conflict. And his hypothesis has been highly influential (Hartl and Brown 1970; Bull 1979; Burt and Trivers 2006). Other adaptive hypotheses for the evolution of haplodiploidy have invoked inbreeding or deleterious mutation clearance (Normark 2004b). A challenge for all of the hypotheses has been accounting for the ecological correlates of the origins of haplodiploidy.

1.3.3 *Ecological correlates of haplodiploidy*

Several origins of haplodiploid systems apparently occurred in insects whose larvae live under the bark of dead trees. This arguably applies to all three origins in Coleoptera (one in micromalthid beetles and two in bark beetles), both origins in Diptera (sciarid fungus gnats and cecidomyiid wood midges), the origin in Hymenoptera (sawflies, later diversifying into wasps and relatives), and with less confidence also the origin in Thysanoptera (thrips). Ecologically rather different are the three origins in Hemiptera, all in lineages that suck sap from live trees (whiteflies, iceryine scale insects, and neococcoid scale insects). Hamilton (1978) argued that the under-bark habit led to gregarious broods, which in turn led to inbreeding, which is in many ways conducive to the origin of haplodiploidy. Genetically, haplodiploidy and inbreeding are both very sensitive to (and hence exert strong selection against) recessive deleterious mutations. More to the point for Hamilton, inbreeding exerts strong selection for a female-biased sex ratio, and haplodiploidy provides a mechanism for maternal control of the sex ratio that can respond to such selection. Two of the ten origins of haplodiploidy in insects (the two origins in bark beetles) are clearly associated with inbreeding, but for the other eight origins there is no evidence of inbreeding and some evidence of outbreeding (flying males). Unlike the situation with mutational-or-environmental theories of sex, Hamilton's inbreeding hypothesis and Brown's maternal transmission hypothesis do not synergize with each other: maternal transmission advantage occurs only under outcrossing (Smith 2000). The ecological evidence seems to weigh against the mutation-clearance hypothesis, since the origins mostly occur in dark, protected, low-radiation, low-mutation-risk environments.

Brown's maternal transmission hypothesis confers an unconditional advantage on any female that can produce fatherless sons. So why isn't every species haplodiploid? And why has haplodiploidy arisen in groups associated with the woody parts of plants? An obstacle for Brown's hypothesis is that the fitness of male haploids needs to be greater than 50% that of male diploids for the haplodiploidy allele to invade (Smith 2000). Hamilton's observation that haplodiploidy arises in groups with gregarious broods potentially reduces the seriousness of this problem: if there is competition between siblings for resources, then the death of one sibling may benefit others, reducing the fitness cost of the new trait (Lively and Johnson 1994).

One new trait that repeatedly invades insect populations with gregarious broods is male-killing—this is a trait encoded in the genomes of maternally transmitted endosymbiotic bacteria (Randerson et al. 2000). Male-killing bacteria suicidally kill their male hosts, a 'somatic' trait that frees up resources for the female siblings that constitute the endosymbionts' 'germline' (Chapter 13). Male haploidy may have originally been a lethal trait encoded by male-killing bacteria (i.e. the bacteria kill males by making them haploid) (Normark 2004b); tolerance for male haploidy could have evolved later, and have spread rapidly due to Brown's maternal transmission advantage (Úbeda and Normark 2006; Kuijper and Pen 2010).

Many authors have hypothesized that arrhenotoky (in which unfertilized eggs develop into haploid males) evolved from a system of PGE (Schrader and Hughes-Schrader 1931; Cruickshank and Thomas 1999), and the haploidizing male-killer hypothesis (Normark 2004b) follows this tradition. John Werren (personal communication) has pointed out that a high frequency of male killers would itself exert strong selection for each of the two characteristics of arrhenotoky: parthenogenetic production of offspring (since the rarity of males may leave many females unmated) and the maleness of those parthenogenetic offspring (since the rarer sex has higher average reproductive success). Thus the mere presence of male killers in a population could have created the conditions for a direct origin of arrhenotoky in uninfected females. Complicating the situation further is the frequent association of male-killing bacteria with inbreeding: in a population with a high frequency of male killers it may be advantageous to mate with your brother, if you have one, because you may not find any other males (Majerus 2003; Normark 2004a). In summary, the ecological evidence suggests that male-killing bacteria played a role in the origins of haplodiploidy, but it remains unclear precisely what that role may have been.

1.3.4 *Consequences of haplodiploidy for the evolution of mating systems: conflict over exclusion of sperm or sperm genomes*

'For me at least merely pondering the absurdity of the reproductive system of male-haploidy has made me at times break into loud guffaws and also cry, so infinitely teasing are its problems', wrote W. D. Hamilton (2001). Haplodiploidy seems to have endless surprising consequences ranging from genetics to social evolution, but we now focus on one consequence for the evolution of mating systems. Haplodiploidy adds an arena of conflict between the members of a mated pair: from what proportion of a female's eggs will sperm be excluded (or will the sperm genome be eliminated)? A male benefits by fertilizing more of his mate's eggs. His optimal strategy is to fertilize all of them, even though in an arrhenotokous species the result is an all-female brood. Under PGE, a male would benefit if his sperm's genomes could resist being eliminated in all of his mate's offspring—again resulting in an all-female brood. Thus males in haplodiploid species have interests parallel to those of maternally transmitted endosymbionts. Indeed, male-killing would be a winning strategy for sperm in a haplodiploid population (Ross et al. 2011).

In most haplodiploid species, females seem to have remained in control of sex determination, but there are some interesting apparent exceptions. All three of the reported cases of hermaphroditism in insects occur in the arrhenotokous scale insect tribe Iceryini (Ross et al. 2010). *Icerya purchasi,* the cottony cushion scale, is usually described as a self-fertilizing hermaphrodite. Anatomically female individuals harbour a haploid tissue that develops into mature sperm that fertilize their eggs. But it turns out that this tissue develops from supernumerary haploid sperm pronuclei that were present in the egg from which the anatomically female individual developed (Royer 1973; Gardner and Ross 2011). Thus rather than describing this individual as a hermaphrodite, it may be more appropriate to describe her as a female who is infected by a maternally transmitted male tissue. What would happen if a male in an arrhenotokous population produced an ejaculate that included spermatogenic stem cells, capable of dividing and differentiating into mature sperm in the female's reproductive tract? The female's ability to protect half her eggs from fertilization might be compromised, and the male might gain access to the entire egg trove. Such an ability could spread rapidly in a population and might result in something like the *Icerya purchasi* mating system that we see today (Normark 2009; Gardner and Ross 2011).

Haplodiploidy in general, and the PGE system of scale insects in particular, seems to be evolutionarily labile, with frequent origins of novel systems (Ross et al. 2010). This evolutionary instability has been interpreted as resulting from alternate resolutions of the genetic conflicts intrinsic to the system (Herrick and Seger 1999). For instance, male diploidy has re-originated twice from PGE in scale insects, with sperm or paternally imprinted chromosomes in those lineages having apparently evolved resistance to elimination. Interestingly, one of the two neo-diplodiploid lineages—the scale insect family Stictococcidae—is one in which at least some species have a placenta: the embryo splits into two, and part of it migrates to digest the ovariole's vitellogenic tissue before rejoining the rest of the embryo

(Buchner 1965). It is tempting to see parallels with mammals, in which early expression of paternally imprinted genes in the placenta affects maternal physiology in ways that apparently benefit the embryo at the expense of the mother (Burt and Trivers 2006). These stictococcid scale insects also appear to have endosymbiont-based sex determination, in which embryos with symbionts develop into females, and embryos that lack symbionts develop into males (Buchner 1965). Lacking symbionts, males are unable to feed, and are entirely reliant on their mother for nutrition via the placenta.

Another strange case of a genetic system innovation in which males 'broke out' (Herrick and Seger 1999) of some of the constraints of haplodiploidy is provided by the little fire ant, *Wasmannia auropunctata*. Unlike other male hymenopterans, little fire ants contribute their genomes to sons. Indeed they turn the tables on females, excluding the *maternal* genome and producing a son whose nuclear genome is a clone of the father's genome (Foucaud et al. 2010). However, the females are still able to control fertilization and still protect some eggs against fertilization—these are diploid eggs that develop into reproductive female clones of the mother. So both males and females reproduce clonally (in Figure 1.3 this system is termed parallel clonality). But the great majority of eggs are fertilized to give rise to diploid workers with one maternal genome and one paternal genome (Foucaud et al. 2010). This is perhaps the strangest of the several systems of 'social hybridogenesis' in which ants mate with heterospecifics to make mule-like sterile hybrid workers (Leniaud et al. 2012).

It is fairly clear on theoretical grounds that PGE is contrary to the interest of the paternal genome, and that anything the paternal genome can do to resist said elimination will be

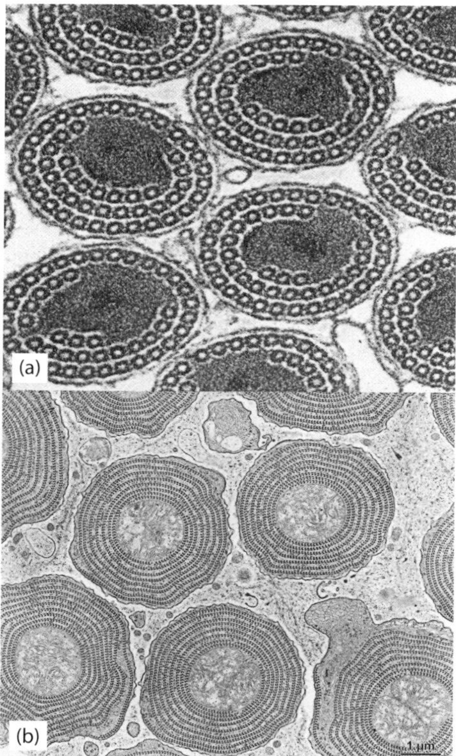

Figure 1.6 Proliferation of microtubules in the sperm axonemes of paternal-genome-eliminating lineages. (a) Diaspidid scale insect (*Unaspis euonymi*; reprinted with permission from Robison 1972, Fig. 1.6). (b) Cecidomyiid fly (*Asphondylia ruebsaameni*; courtesy Romano Dallai).

favoured by natural selection. What form such resistance might take is less clear. One clue is to be had from a comparative study of the sperm ultrastructure of PGE insects. Most insects, like most other animals, have a '9 + 2' arrangement of microtubules in the sperm axoneme, with a ring of nine paired microtubules surrounding two central microtubules (Jamieson et al. 1999). But in three of the five clades of PGE insects we see a radical deviation from this pattern: sperm with hundreds of microtubules (Jamieson et al. 1999; Normark 2009). In scale insect sperm, the hundreds of microtubules are neatly arranged in spirals or concentric circles. In the sperm of some sciarid and cecidomyiid flies there are hundreds of pairs of microtubules packed more haphazardly into the axoneme (Figure 1.6). Why this explosion of microtubules? We don't know, but the number of microtubules in the sperm axoneme is identical to the number of microtubules in the centriole, which is involved in organizing the mitotic and meiotic spindle. In many diplodiploid species, the sperm cell contributes a centriole to the zygote, but in arrhenotokous and parthenogenetic species, the egg is the only source of centrioles for the zygote. It is tempting to speculate that properties of centrioles may affect whether a genome is eliminated; that there has sometimes been conflict over maternal vs paternal inheritance of centrioles; and that large centrioles have had an advantage in being transmitted in preference to that of the other parent, in protecting paternal chromosomes from elimination, or both (Normark 2009).

1.4 Conclusions

The strategy you adopt depends upon the rules of the game you are playing. Sometimes the best strategy is to cheat. Cheater alleles, such as meiotic drivers, are usually rare deviations from the amazingly fair and symmetrical lottery of Mendelian genetics. But from time to time, the cheaters take over. Parthenogenesis and haplodiploidy are systems arising when a female (or even a bacterium inside that female) breaks free of the rules and, in so doing, invents an entirely new game. Parthenogenesis is a simplification of the game, and, despite perennial attempts to find a future in parthenogenesis, it probably lacks one. Haplodiploidy is a complication of the game—one that has many, many different futures.

CHAPTER 2

Sexual selection theory

David M. Shuker

2.1 Introduction

Much of what we observe of insects are their sexual lives, with many of the most obvious behaviours (such as swarming or calling on a summer's evening) associated with trying to obtain mates. The study of insect mating systems has therefore revolved around the study of insect reproductive behaviour, in particular mating behaviour (Thornhill and Alcock 1983). While some aspects of mating behaviour, including primary sexual function, are shaped by natural selection, so many aspects of mating behaviour are shaped by Darwin's other evolutionary process: sexual selection (Andersson 1994; Figure 2.1). In this chapter we focus on sexual selection—what it is and what we understand the action of sexual selection to be in the light of several decades of theoretical exploration of the different processes of sexual selection. Many of the chapters that follow consider the empirical study of sexual selection in insects, and some extend the theory base outlined here. However, in this chapter we outline the basic concepts at the heart of sexual selection. In so doing, we also try not to lose sight of the interaction between sexual and natural selection; indeed one of the main themes of this chapter is that understanding how natural and sexual selection overlap and interact is at the heart of many of the remaining puzzles over the nature and extent of sexual selection. For instance, all of the various physiological, morphological, and behavioural traits involved in reproduction must trade off with each other for an organism's ultimately finite resource pool. Many of these trade-offs will be under both natural and sexual selection. Many of the concepts in this chapter will be familiar, and were already current in Thornhill and Alcock's book. However, in the years since 1983 our knowledge of sexual selection has perhaps stretched the applicability of some simple dichotomies (for example, male–male competition versus female choice, or the 'Fisher process' versus 'good-genes' sexual selection). Yet accepting an increasingly nuanced view of sexual selection is not the same as rejecting the overall conceptual logic, as some have argued for, nor indeed has it negated the usefulness of attempting to break selective processes down into their constituent parts.

The theory base of sexual selection is now represented by a rich and voluminous literature. As with many of the chapters in this book, a book-length treatment in its own right would be possible. This chapter therefore focuses on the conceptual foundations of sexual selection and some of the insights that the various models and the resulting discussions about the processes of sexual selection have revealed. The mathematics underlying these models take second place to trying to understand some of those insights, not least the

The Evolution of Insect Mating Systems. Edited by David M. Shuker and Leigh W. Simmons.
© The Royal Entomological Society 2014. Published 2014 by Oxford University Press.

Figure 2.1 Sexual selection in insects. This may involve the evolution of elaborate secondary sexual characters, selection for the control of resources needed by the opposite sex, as well as episodes of pre- and post-copulatory male–male competition and female choice: (a) elaborate display traits in longhorn moths; (b) resource defence and sperm competition in yellow dung flies; (c) prolonged copulation and mate-guarding in seed bugs; (d) scramble competition for mates in grasshoppers. Photographs: D. M. Shuker.

conceptual scope of sexual selection. Inevitably, much of the nuance of modern sexual selection theory will be glossed over. Key sources for students wishing to track the modern development of the field include the two important books edited by Bateson (1983) and Bradbury and Andersson (1987), followed by the influential review papers of Maynard Smith (1991) and Kirkpatrick and Ryan (1991). The state of the field in the early 1990s in terms of the theoretical models of mate choice was thoroughly reviewed by Andersson (1994), a book that remains the key text for any student of sexual selection. More recent reviews include Andersson and Iwasa (1996), Kokko et al. (2002), Mead and Arnold (2004), Andersson and Simmons (2006), Kokko et al. (2006), and Kuijper et al. (2012), much of which we will touch on here. However, reviews can become somewhat self-perpetuating, and much is to be gained by returning to the original literature. Furthermore, post-copulatory sexual selection has long been realized as a key arena for sexual selection in insects (Parker 1970a). Although pre- and post-copulatory sexual selection may differ in mechanistic details, and such details may influence the outcomes of theoretical models (as reviewed by Kuijper et al. 2012, for example), here we consider pre- and post-copulatory episodes of sexual selection together for the most part. The key works for post-copulatory sexual selection in insects are Smith (1984), Eberhard (1996), and Simmons (2001), and Chapters 10 and 11 provide more detail on the relevant theory for post-copulatory sexual selection. Finally, those interested in the historical development of the field can refer to

Cronin (1991), Andersson (1994), Birkhead (2010) and Milam (2010), although much remains to be done in understanding how sexual selection theory developed following Darwin, in both social and scientific contexts (e.g. Dawson 2007).

2.2 What is sexual selection?

Sexual selection arises from competition for access to mates and their gametes (Andersson 1994; see also the discussion in Shuker 2010). From the outset, three key concepts need to be kept in mind. First, gametes need to be a limiting resource. If gametes are not limiting then there can be no competition for those gametes. In addition, gametes may be limiting in terms of *quantity* (not enough gametes to go around) or *quality* (not enough high-quality gametes to go around). That gametes can vary in quality means that sexual selection can occur in strictly monogamous species, where individuals of one sex compete to pair up with members of the other sex that will provide high-quality gametes. Second, by gaining access to gametes of a particular individual, other benefits may accrue (for example, parental care may be provided for offspring or there may be a nuptial gift—Chapter 12). Failure to secure access to gametes leads to failure to secure other benefits. Thus sexual selection occurs over access to mates and gametes, even if those mates and gametes then provide benefits that increase productivity (and so are naturally selected). This means that natural and sexual selection are often intimately linked when competition for mates directly influences fecundity. Third, sexual selection occurs *within* a sex. While this might seem obvious given the above definition, the focus on mate choice (where sexual selection in one sex arises as a result of interacting with the other sex) may lead to confusion regarding where and on which sex sexual selection has actually occurred.

Darwin (1859, 1871) developed the idea of sexual selection as it was clear that numerous traits expressed by plants and animals could not be explained in terms of natural selection, that is, in terms of success in surviving and in being fecund. Large weapons or extravagant ornaments (often expressed in males) made little sense in terms of selection to limit predation or maximize fecundity, but Darwin saw that such traits could make evolutionary sense in terms of gaining access to mates, without which a long life and a sizeable supply of gametes are of little use. This was his justification for creating the concept of sexual selection. It is worth emphasizing again the close links between natural and sexual selection though. For instance, even traits that are avowedly 'sexually selected' (the 'antlers' of antler flies for instance; Chapter 6) are predicted to evolve in size or extravagance only up to the point where the natural selection cost of the trait is balanced by the sexual selection benefits, and no further. However, the concept of sexual selection as something separate from other forms of natural selection has had enormous utility and explanatory power (Andersson 1994) and has so far been worth keeping.

The way in which competition for mates arises and the form it takes are strongly influenced by an organism's mating system (e.g. Trivers 1972; Emlen and Oring 1977; Shuster and Wade 2003; Kokko et al. 2006; Chapter 3). We address the causes and consequences of mating system evolution in the Chapter 3, but a few important concepts need to be outlined here. First, competition for mates can be in terms of the number of mates (including whether or not a mate is obtained at all), or the quality of those mates. Second, obtaining mates (i.e. copulating) is only the first step: we know that post-copulatory processes may mean that copulations need not lead to fertilization events (Parker 1970a; Chapters 10 and 11). The crucial resource therefore is the gametes of the opposite sex, although for brevity in this chapter and elsewhere in the book we often short-hand this with 'mates'.

Third, competition for mates does not require separate sexes (e.g. anisogamy) as competition for fertilizations, and thus sexual selection, can occur in isogamous species (such as yeast: Rogers and Greig 2009). This point often appears to be lost on recent critiques of sexual selection and mating systems theory. Fourth, the form of competition for mates often varies between the sexes. As has been well known from Darwin onwards, males are typically more overt in their competition for mates, with intra-sexual competition often a crucial component of male fitness (Andersson 1994). Male fitness tends to be limited by access to mates and their gametes, whereas female fitness tends to be limited to a greater extent by access to resources. The basis for this difference is in part the nature of their gametes: males generally produce copious, individually cheap sperm, whereas females produce fewer, more individually expensive, eggs. This means that the access to gametes is more likely to be limiting for males than females. However, asymmetries may also arise and be reinforced by other aspects of the organism's biology or mating system. It is an empirical fact that males are more likely to compete actively for mates, express elaborate secondary sexual characters, and to be the object of inter-sexual mate choice, whereas females are more likely to compete for ecological resources and to be the choosy sex in terms of mate choice (Andersson 1994). However, it is also an empirical fact that such 'sex roles' are neither ubiquitous across all species, nor fixed within species. Ecology can reverse reproductive asymmetries, providing important tests of mating system theory. Moreover, even within a species with 'traditional' sex roles, males may often be choosy too (e.g. Bonduriansky 2001) and there is now a large body of evidence to suggest that males tailor the number of sperm they transfer due to a variety of factors (Simmons 2001; Kelly and Jennions 2011; Chapter 10). While individual sperm may be relatively cheap compared to an egg, males may invest as much, or more, in total reproductive effort as females, which will impact the sex roles adopted by males and females.

Recent critiques of sexual selection theory have questioned this emphasis on competition, preferring to view interactions between males and females in a more co-operative light (e.g. Roughgarden et al. 2006). That not all interactions within a sex or between the sexes in terms of mating and other aspects of reproduction involve conflict is clear: males and females have to co-operate to some extent during copulation for sperm transfer to occur, and there are many examples of co-operative breeding. However, evolution by natural selection is ultimately competitive: organisms compete for genetic representation in the next generation. The strategies (or mechanisms) organisms use will be a mix of co-operation and conflict, but the selective process itself is competitive.

This inherent nature of selection should not mislead us when thinking about how sexual selection may arise. For instance, in truly monogamous species, males may compete for access to fertilizations by increasing the number of gametes that become available to them, perhaps by providing nuptial gifts to the female or by taking on parenting duties that mean the female has energy available for further reproduction. Such mechanisms of competition for gametes are well known in polygamous species such as scorpionflies, where nuptial gift provision influences access to gametes (see Chapter 11 for more details). Similarly, male seminal proteins that increase rates of ovulation and oviposition in *Drosophila melanogaster* also increase the access to a female's gametes for a male (Chapters 4 and 9). In these latter two examples, these male traits are typically thought be under sexual selection, but in the monogamous case it is perhaps less clear, as selection seems to be more associated with fecundity. However, if gametes are a limiting resource, competition to obtain access to as many of those gametes as possible, either by force or through co-operation, is occurring among males and the traits associated with that access are best considered as being under sexual selection.

2.2.1 *Sexual selection and natural selection*

Sexual selection encompasses a particular set of processes that can influence an organism's fitness, processes involved in the competition to obtain mates and subsequent fertilizations. As such, sexual selection is part of a broader set of forms of selection, which together make up what might be termed *broad-sense* natural selection (Endler 1986). However, since the Modern Synthesis, partitioning out processes of selection and giving them names may not seem that necessary or useful: at the population genetic level, DNA sequences change in frequency in the population as a result of competition over genetic representation in the next generation. Fitness variation will arise in countless different ways, encompassing things we call sexual selection, kin selection, and so forth, but at the genetic level it might not matter what causes fitness variation, merely that it is present. However, to understand how evolutionary processes shape phenotypes, it has proved useful to classify particular kinds of selection (as exemplified by the success of the behavioural ecological approach of the original Thornhill and Alcock 1983; see also Davies et al. 2012). The downside is that it may lead to arguments over the definitions of things which are all fundamentally just correlations between phenotypes and inclusive fitness.

In this chapter we consider that all processes of selection can be collected under the overall umbrella of natural selection. Within that conceptual space, we can identify *narrow-sense* natural selection as processes such as viability and fecundity selection, and sexual selection as competition-for-mates selection. Thus sexual selection is part of a *broad-sense* conception of Darwinian natural selection. However, this partition is as much a convenience to help us understand evolutionary processes as it is some sort of monumental edifice. As just mentioned, at the genetic level such distinctions are somewhat moot. Perhaps more importantly, we should also expect that the overall association between a trait and fitness will be hard to separate out into different natural and sexual selection components. In practice, natural and sexual selection may often be indistinguishable, and rightfully so. For example, if access to mates only comes with longevity (e.g. surviving as a territory-holding or harem-holding male for as long as possible) then natural and sexual selection will closely coincide, and it becomes more of a subjective decision where we consider competition for mates to start and other aspects of an organism's life-history (such as survival to adulthood) to finish. The usefulness of the different concepts comes from making us recognize that insects and other organisms need to survive to reproduce but also may need to compete for gametes to reproduce, and that these processes may come into conflict, as originally noted by Darwin. However, these processes may also be closely related or even effectively one and the same.

The definition of sexual selection outlined here is that which has developed over the last four or so decades of intensive research (Andersson 1994, Shuker 2010). However, recently the idea that sexual selection should only be considered a result of variation in access to mates has been questioned, most notably in terms of animals (primarily vertebrates) where females compete among themselves for resources other than gametes which influence reproductive success (Clutton-Brock 2007, 2009; Rosvall 2011). Insects provide similar examples, of course, such as female *Goniozus legneri* wasps competing for hosts to parasitize (Goubault et al. 2008). One argument made to support integrating female–female resource competition into sexual selection is that sexual selection in females has been underappreciated. While much of the revival of interest in sexual selection since the 1970s has been driven by an increasing realization of the active agency of females in sexual selection (first in terms of pre-copulatory mate choice, and then in terms of post-copulatory processes; e.g. Bateson 1983 and Eberhard 1996, respectively), it may still be the case that inherent gender biases of researchers have underplayed the role of sexual selection in the evolutionary biology of females across many species. However, by shifting the definition of sexual selection away from being merely the result of competition for fertilizations to include other aspects of reproduction, the already intimate association between natural and sexual selection becomes complete, and the two become one and the same (a point made by Clutton-Brock 2007). This is because, barring kin selection, all traits influence fitness through their effects on individual reproductive success and so all traits eventually become involved in competition (directly or indirectly) for resources that will contribute to reproductive success. The difference between natural and sexual selection then collapses. Future researchers may decide that the distinction between natural and sexual selection is no longer helpful; until then, in this book we continue to value the different perspectives on the functions of traits provided by natural and sexual selection.

2.3 What is sexual conflict?

The importance of the interaction between different components of natural and sexual selection has most clearly come to the fore over the last two decades in one particular context: sexual conflict. Sexual conflict arises when the two sexes have different evolutionary optima for a given trait (Parker 1979; Holland and Rice 1998; Chapman et al. 2003a; Arnqvist and Rowe 2005). In other words, natural and/or sexual selection favour a different trait value in males or females. The concept of sexual conflict has become closely associated with sexual selection both in terms of the rather general sense that sexual selection may contribute to patterns of sexually antagonistic selection, but also in terms of the fact that sexual conflict may be associated with important traits such as female multiple mating (polyandry: Chapter 9) and that sexual conflicts may influence patterns of non-random mating (Section 2.6).

We often talk about sexual conflict as a process in its own right. However, sexual conflict is in fact a consequence of various interacting selection pressures. As such, when we use the short-hand of saying that sexual conflict 'causes' this or that, what we really mean is that the end-result of multiple components of natural and/or sexual selection favours certain trait values in males and other trait values in females. Importantly, although here we are considering sexual conflict in a chapter on sexual selection, no doubt some (perhaps even most) occasions of sexual conflict will arise due to conflicting patterns of natural selection (e.g. Fairbairn et al. 2007). Moreover, sexual conflict is most definitely *not* synonymous with sexual selection, despite suggestions to the contrary (Carranza 2010). However, sexual conflict certainly has received a lot of attention in terms of mating behaviour and related traits (e.g. Arnqvist and Rowe 2005; Tregenza et al. 2006 and papers therein).

Whereas the definition of sexual conflict is quite simple, organizing our thoughts about sexual conflict can be quite complicated. For instance, the trait over which there is conflict (the 'conflict trait') may be the same trait in each sex (e.g. body size), or it may be a trait expressed in one sex only (e.g. male coercion behaviour). In the latter case, selection may then favour an evolutionary response in the other sex, which may involve a completely different trait (e.g. female resistance behaviour), such that the conflict now involves two separate traits, one expressed in each sex. In such cases, we often then talk about sexual conflict over the key behavioural or fitness 'outcome' (in our example, this would be mating rate, so mating rate becomes the conflict trait), even though the traits underlying this outcome may be very different in the two sexes. Ultimately the underlying conflict trait is fitness, such that male fitness is maximized by a certain set of trait values expressed across males and females, with female fitness maximized by an alternative set of trait values. Useful short-hand (for instance when developing theoretical models) should not mean we become bogged down over what is or isn't a conflict trait.

The other key part of the sexual conflict conceptual space is the idea of intra-locus versus inter-locus conflict. Intra-locus conflict describes the situation in which the same genetic locus influences the conflict trait in males and females. Most straightforwardly, one allele may have high fitness in one sex and low fitness in the other, whereas another allele may have the opposite pattern: together such alleles are called sexually antagonistic alleles (Rice 1987). More generally, all that is needed is that alleles have different fitness effects in the two sexes (including having no fitness effect in one sex). Inter-locus conflict refers to when the conflict trait is influenced by different loci in each sex. Although this distinction is useful for theoretical models and for organizing our thoughts, it is based inevitably on a cartoon view of the genetics of many or indeed most traits. Most likely, at any one time there will be segregating variation in multiple genes influencing a conflict trait, and sometimes there may be sexually antagonistic alleles at one locus within a polygenic context, sometimes not. As such, the distinction between intra- and inter-locus conflict may be at best transient. That said, sexually antagonistic variation has been mapped (e.g. sexually antagonistic variation is more likely to be associated with genes on sex chromosomes: Gibson et al. 2002) but the simple distinction is probably best kept for heuristic value and should not be taken too literally.

We now consider the different forms of sexual selection before reviewing the theoretical foundations of those different forms.

2.4 Forms of sexual competition

By definition, all forms of sexual selection derive from competition for mates *within* a sex. However, following Darwin, traditionally sexual selection has been divided into intra-sexual competition and inter-sexual choice. The former appears straightforward: individuals of one sex compete to gain access to, or monopolize, mates and their gametes. There are many forms this could take in insects, including contests involving weapons (Chapter 6) and sperm competition (Chapter 10). Importantly, intra-sexual competition for mates may occur in the first instance independently of the phenotypes of the opposite sex.

Inter-sexual choice also involves competition within a sex, but in this case the competition is mediated by a phenotype of the other sex. More specifically, mate choice arises when any trait in one sex leads to non-random mating success in the other sex (Halliday 1983; Chapter 8). The 'choice' trait need not be a cognitive, human-like 'decision', rather it can be behavioural, physiological, morphological, or even a life-history decision (e.g. a developmental trajectory leading to early adult emergence, as seen in protandrous insects: Thornhill and Alcock 1983). This definition has been hugely important for the field as it removes the need for active or cognitive decision-making to underlie the 'choice' (and so avoids anthropomorphic concerns about the tastes, aesthetic or otherwise, of our study organisms). Sexual selection via inter-sexual choice therefore involves members of one sex (the *chosen* sex) competing against each other via the action of one or more choice phenotypes expressed by the other sex (the *choosy* sex).

Of course, intra- and inter-sexual selection will likely interact. First, the choosy sex may prefer to mate with winners of intra-sexual competition—over display sites, for example—such that the two forms of selection are effectively indistinguishable (this has been appreciated for a long time, although is sometimes overlooked thanks to the contemporary emphasis on mate choice: Halliday 1978,1983). Second, contests for mates may become more intense in the presence of certain choosers. This may have no bearing on the outcome, but if it does, so that winners of intra-sexual competition vary non-randomly with the phenotype of the choosy sex, then both intra- and inter-sexual selection will act together, although not necessarily in the same direction or favouring the same trait. The distinction between the two forms of sexual selection may therefore be difficult to resolve in empirical studies, and this is perhaps most apparent in the insect post-copulatory sexual selection literature, where males compete for fertilizations within the reproductive tracts of females, which may vary in ways that favour different sperm phenotypes (e.g. swimming speed over number; Chapters 10 and 11).

Importantly, inter-sexual mate choice can arise even if there is no selection on the choice trait itself. Mate choice by definition must lead to sexual selection in the chosen sex, but must it also lead to sexual selection in the choosy sex? For there to be *sexual selection* on the choosy sex, choice has to influence the non-random mating success of the choosy sex. This could occur if, by choosing a particular mate, that mate is now excluded from the mating pool for such a time that other members of the choosy sex cannot copulate with what would otherwise be a preferred mate (i.e. the gametes of that preferred male become a limiting resource for the other females). In other words, choice can also become a weapon in the arsenal of within-sex competition for access to mates. Choice as part of intra-sexual competition among the choosy sex also reinforces the importance of the *identity* as well as the *quantity* of mates and gametes. Sexual selection on preferences is perhaps more likely when mate *quality* is more important than *how many* mates. For instance, if all females choose and manage to mate with the same preferred male, then there is no sexual selection on the choosy sex. The extent to which there is *natural selection* on mate choice will be discussed in more detail below.

2.5 Intra-sexual contest theory

In his seminal book, Andersson outlined five mechanisms of competition for mates (Andersson 1994). Excluding mate choice for now (see Section 2.6), these are scrambles, endurance rivalry, contests, and sperm competition. Scramble competition for mates should select for mate-searching phenotypes and also efficient handling of mates (rapid courtship, insemination, and so forth). The evolution of many of these traits is conceptually straightforward, with analogous examples in non-sexual traits associated with ecological competition for resources. For example, mate search effort on a patch can be modelled using the classic approach from optimal foraging theory, the marginal value theorem (e.g. Parker 1978; see also Figure 10.1).

Endurance rivalry describes the fact that mating and fertilization success may simply be a function of surviving and being in the mating pool for as long as possible. Although data on endurance rivalry are very hard to collect (especially for insects in the wild; territory-holding damselflies and dragonflies perhaps offer some of the best examples), it may be the case that endurance rivalry is the most widespread mechanism of sexual selection. Endurance rivalry selects for traits that increase survival, and so will be closely aligned with (or act in parallel to) other forms of natural selection. Again, endurance rivalry offers little in the way of theoretical problems, above and beyond those associated with all life-history allocation decisions that influence trade-offs between growth, survival and reproduction (e.g. Stearns 1992).

The third mechanism is physical contest. Contests for access to mates may vary from the largely ritualistic through to the out-and-out fatal. Contests will select for attributes such as large size, weaponry, musculature, or grappling morphologies (Chapter 6). Contests may well co-occur with other mechanisms, such as endurance rivalry to determine mating success (e.g. fighting to maintain a place in the breeding pool) or mate choice (e.g. fighting to maintain control of a harem or of resources that the other sex requires or uses). Fighting to maintain territories of one form or another is a common component of sexual selection in many male insects (Thornhill and Alcock 1983). In addition, contests may well select for alternative mating tactics (AMTs), whereby tactics evolve to circumvent costly physical fights: AMTs are considered in theoretical and empirical detail in Chapter 7. Contest competition has engendered the greatest theoretical interest and will be considered below.

The final mechanism of competition highlighted by Andersson (1994) was sperm competition, perhaps best considered today more inclusively as post-copulatory sexual competition, influenced by both sexes. Post-copulatory sexual selection has given rise to its own rich theory base, some of which has been tested in exquisite detail (e.g. Simmons et al. 1999; Parker and Simmons 2000). For a full treatment see Chapters 10 and 11.

2.5.1 *Physical contests*

When animals come together to fight, the outcome (and so the evolutionary consequence) may depend upon investment made either before the fight (for instance investment in traits such as weapons or large body size) or investment made during the interaction (for instance the energetic expenditure of a prolonged ritualistic display or physical fight). In terms of the former, the evolution of weaponry and other traits involved in direct mating contests has been successfully modelled, and is relatively uncontroversial (Parker 1983a; Maynard Smith and Brown 1986; for more see Chapter 6). One important insight from these models is that investment in weapons may be evolutionarily unstable, for instance leading to cycles of higher and lower investment in weapons (as envisaged by the concept of 'arms races': Dawkins and Krebs 1979; Parker 1983a), or in the evolution of polymorphisms. The latter are often considered under the umbrella of AMTs and are considered in further detail in Chapter 7.

When the outcome depends on what happens during the contest, game theory has again proved an invaluable tool in understanding the logic of such contests, including those over mates (classic early studies include Maynard Smith and Price 1973; Maynard Smith 1974; Maynard Smith and Parker 1976; Maynard Smith 1982; Enquist and Leimar 1983; Grafen 1987). The two central concepts for understanding animal contests are Resource Holding Power (RHP: the ability of an animal to defend a resource or to fight, which might be influenced by investment in structures such as weapons) and Resource Value (RV: the value of winning a fight, which may well differ between contestants and also vary for a given contestant with ecological and life-history context). The decision to engage in a contest may vary with the contestant's (perceived) RHP, the RHP of the opponent (both of which may change during the contest), plus the RV to the contestant (which again may change during a fight). A rich theoretical literature has built up around this framework, with one of the key factors being how contestants obtain information about both RHP and RV, and how this information influences contest behaviour and outcomes. Although the theoretical models are not themselves controversial, how empiricists have made use of them and the conclusions they have drawn from them are more contentious (Elwood and Arnott 2012). In particular, it is often assumed that individuals assess and compare the fighting ability of their opponents (Enquist and Leimar 1983; Bradbury and Vehrencamp 2011).

The simplest theoretical models of animal contests assume that the contestants obtain no information about their opponent (e.g. the canonical hawk–dove game of Maynard Smith and Price 1973). In this game, the decision to escalate and play 'hawk' is determined by the costs of displaying, the cost of being injured in a fight, and the resource value. If RV is high, then playing hawk may always be favoured and so all interactions involve fights, even if fights are extremely costly (Grafen 1987). In some games, such as the 'war of attrition without assessment' and the 'energetic war of assessment', animals are assumed to have some idea of their own resource-holding power, but do not assess the RHP of their opponents. Individuals are predicted to carry on the fight until the accrual of costs reaches a threshold set by the RV, with individual RHP determining how costs accumulate such that individuals with lower RHP (e.g. smaller body size) reach their costs threshold first. Arnott and Elwood (2009) termed these games 'pure self-assessment', and these models predict that fight duration will be positively correlated with the RHP of the loser (Figure 2.2).

In a related model, the 'cumulative assessment model' (Payne 1998), individuals have information about their own RHP, but in this scenario they also inflict costs on their opponent, such that the RHP of both contestants influences contest duration, but again there is no assessment of the opponent. Both pure self-assessment and cumulative assessment therefore require no cognitive processes for assessing or comparing opponents.

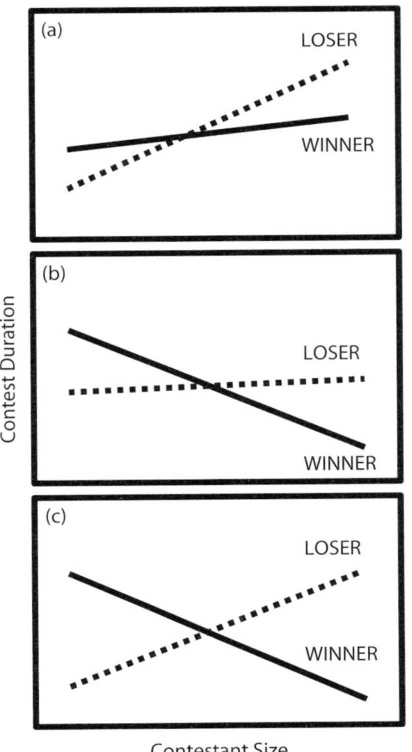

Figure 2.2 Predicted relationships between opponent size and contest duration for different models of animal contests. Winners' relationships are represented by solid lines, and losers' relationships are represented by dotted lines. Contestant body size is considered here to correlate positively with resource-holding potential (RHP: i.e. 'competitive ability'). (a) Under pure self-assessment, a positive correlation between a loser's body size and contest duration is predicted, with a weak positive correlation or no relationship predicted between a winner's body size and contest duration. (b) Under opponent-only assessment, a negative correlation is predicted between a winner's body size and contest duration, with no correlation for loser's body size. (c) Under cumulative assessment and sequential assessment models, a negative correlation is predicted between a winner's body size and contest duration, with a positive correlation predicted for a loser's body size and contest duration. After Elwood and Arnott (2012).

Models where contestants do assess each other come in two forms. The first are 'opponent-only assessment' models (Arnott and Elwood 2009; Elwood and Arnott 2012). In this scenario, individuals base their behaviour on their opponent's RHP only, leading to a negative correlation between the winner's RHP and contest duration, and no relationship between the loser's RHP and contest duration (Figure 2.2). The second are 'mutual assessment' models, such as the 'sequential assessment model' of Enquist and Leimar (1983, 1987). In these models, contestants know something about their own RHP, but also gain information about the RHP of their opponents, potentially rather cheaply at first, gaining more expensive but more reliable information about differences in RHP as the contest escalates in intensity. Resource value can also play a role, for instance with contests over resources or mates with high RV requiring more accurate estimates of the difference in RHP before one contestant knows it has the lower RHP and so will lose. Similarly, if the differences in RHP are large, they should become reliably apparent earlier in the contest. As such, a negative relationship between the difference in RHP and contest duration has often been used as a way of showing that mutual assessment was occurring (Figure 2.2). Unfortunately, however, similar patterns can be generated in contests underpinned by less cognitively complex assessment (Elwood and Arnott 2012).

Assuming various forms of self- and opponent-assessment may be problematic in two ways. First, it is unlikely that the individuals of many organisms can (literally) look at the same trait in themselves that they might use in an opponent to estimate RHP; how many stalk-eyed flies have seen their own eye-stalks? As such, organisms may not literally be comparing two similar things, rather individuals may be integrating different sources of information (e.g. I feel strong today, but he looks big). Such integration of information need require very little in terms of 'cognitive complexity' (Elwood and Arnott 2012). Second, when it comes to assessing resource value, again complex comparisons need not always be invoked. For instance, when two resources are involved (as in hermit crab contests for shells), although the perceived value of both resources may influence investment in the contest, it does not require that the two resources be compared, rather that the information about the two resources is combined in some way (for instance with two thresholds).

The difficulties that can arise are apparent in an example of male–male contests in stalk-eyed flies. Panhuis and Wilkinson (1999) showed that in contests between males where they square up to each other, eye-stalk to eye-stalk, contest duration was negatively correlated with size difference. They interpreted this as evidence for a sequential assessment model, with male flies assessing their own and their opponent's eye-stalk lengths. However, Brandt and Swallow (2009) later compared the duration of contests in terms of the winner's and loser's eye-stalks and showed that contest duration was correlated with the loser's eye-stalk length only. This is indicative of a model of pure self-assessment, as the winner does not influence the outcome. As such, eye-stalk length should correlate with some internally assessed measure of RHP in these flies. A similar example comes from the sea anemone *Actinia equina*, which lacks a central nervous system and presumably has limited 'complex cognition'. While contests did show a negative relationship between the difference in RHP and contest duration, as predicted by mutual assessment, again only the loser's RHP was correlated (positively) with contest duration, suggesting pure self-assessment (Rudin and Briffa 2011).

In summary, intra-sexual contests are probably the most widespread form of sexual selection, and in that sense the most important. Nevertheless, that may not be apparent from the literature, including recent reviews of sexual selection (for example Kuijper et al. 2012 focus almost entirely on mate choice); put simply, mate choice is more sexy. Part of the reason for the popularity of mate choice might be the impression that contests for mates are more intuitive to understand and so are well characterized in theoretical terms. As outlined above, however, such an impression may not be wholly accurate and much remains to be done linking plausible theoretical models with realistic mechanisms. It is important that sexual selection does not become fully synonymous with mate choice and that the varied ways in which competition for mates within a sex play out (independently of mate choice) remains fully appreciated and a continued focus of theoretical and empirical research. However, it is true that mate choice remains more of a puzzle.

2.6 Inter-sexual mate choice

Ever since Darwin, mate choice has been controversial (Cronin 1991; Chapter 8). There are two components to the evolutionary impact of mate choice. The first is the evolution of the *chosen* sex, in particular the evolution of traits associated with the non-random mating preferences of the choosy sex. By definition, the response to selection in the chosen sex is the result of sexual selection: access to mates is non-random thanks to some attribute of the choosy sex favouring some individuals of the chosen sex as mates over others. As discussed above, the effects of mate choice may interact with natural selection or other processes of mate competition (which might lead to sexual conflict). Despite these complications, the picture for the chosen sex is relatively straightforward, at least at face value.

Consider a mate preference in females that is completely non-heritable (that is, either all females express the same preference, or that variation in preference is environmental or non-additive genetic in origin). In this case, sexual selection will favour males with the preferred trait (the 'ornament'), and those males will leave more progeny in the next generation than unornamented males. If that preference is 'open-ended', extreme male trait values may come to be favoured, without any form of indirect sexual selection (such as the 'Fisher process', Section 2.7.1) acting. There is nothing complicated or controversial about these dynamics, and nothing in females evolves. Traits such as female adult emergence time may be largely influenced by local environmental conditions, and so any traits that allow males to better access those emerging females, such as their own rapid development to emerge before females (protandry), may be selected thanks to a female trait that may have little or no genetic variation.

The second component has been far more contentious: the evolution of the traits that enact the choosing in the choosy sex. This is usually phrased in terms of the benefits of being choosy. There are several explanations for why choice traits evolve, some of them involving natural selection, some involving sexual selection, and some involving otherwise non-adaptive processes. In addition, it is important to recognize that the evolutionary origin of traits may involve different processes than those which favour their maintenance or further development. The origin of mate preferences was of some concern at the beginning of the modern burst of interest in the evolution of mate choice in the late 1970s and early 1980s (e.g. Heisler 1984a, 1985), but has received less attention more recently, perhaps as a result of ideas about sensory bias (see below), although it is relevant for the notion of a null hypothesis for mate choice (Prum 2010; see below).

2.6.1 *The origin of mate preferences: adaptive and non-adaptive side-effects*

A mate preference arises from some phenotypic attribute in one sex that leads to non-random mating in the other. There are several ways such a phenotypic attribute can arise without the preference itself being directly selected. For ease of exposition, we shall consider that males have an ornament that becomes the target of a female mate preference.

First, a male ornament may arise by mutation that by chance happens to attract females by taking advantage of some underlying *sensory bias* in females (e.g. Ryan 1990; Ryan and Rand 1990; Ryan and Keddy-Hector 1992; Endler and Basolo 1998). The idea of a sensory bias has been explored by a number of authors, and phrases such as *sensory traps* or *sensory exploitation* are also in the literature (West-Eberhard 1983; Endler and Basolo 1998; Fuller et al. 2005), but sensory bias may be the least loaded of those terms. The typical example is that a red-coloured ornament appears in males, and females have an underlying psychophysical bias towards being attracting to red objects, and so red males gain increased access to females. Importantly, any trait may represent a sensory bias, including visual, auditory or chemical traits. Indeed, these may not even be obvious 'sensory' traits, for instance if females are attracted to fast-moving objects for some reason.

Second, the above scenario may be slightly altered to allow the female bias to have arisen as a result of selection in another context. Again, the usual example is to imagine that the preference for red evolved (potentially quite a long time ago) due to natural selection for efficient foraging on red-coloured food (such as berries, or red prey). The mate preference is therefore a pleiotropic side-effect of another component of natural selection. The possibilities here for natural selection to favour phenotypes that then bias the mating success of members of the opposite sex as a side-effect are nearly endless, from the design of reproductive tracts through to foraging behaviour. However, as Dawkins and Guilford (1996) point out, these sensory biases may influence the evolution of mate choice in other important ways, for instance by reducing search costs for mates (which is relevant for the outcome of models of indirect selection; see below). This means that arguing for or against 'adaptive' choice may be difficult.

The third way in which mate preferences may arise is as a result of sexual conflict over mating. As outlined above, we may expect sexual conflict to be widespread, including sexual conflict over mating (Parker 1979; Arnqvist and Rowe 2005). The typical scenario is that males, as the limited sex, may be sexually selected to attempt to coerce or force females into mating, even if it is not in the evolutionary interest of those females to mate. So far, this is just within-male competition for access to female gametes. However, natural selection on females may favour the evolution of a trait that allows them to resist male mating attempts. If there is variation among males in the ability to overcome this resistance trait, then this female trait leads to non-random mating success among males, and is thus the source of a mate preference. The evolutionary consequences of female resistance traits and the resulting (co)evolution between male coercion and female resistance has been encapsulated in a number of models, including the so-called 'evolutionary chases' of Parker (1979), the 'chase-away' models of Holland and Rice (1998), and further models such as Gavrilets et al. (2001), Cameron et al. (2003), and Rowe et al. (2005). We shall return to these theoretical models when we explore the co-evolution of preferences and ornaments.

2.6.2 *The origin of mate preferences: happenstance*

Alternatively, a mate preference may just arise by chance, such that a mutation arises in females that generates a phenotype that leads to non-random mating success among males. If this is not the result of some inherent bias in female sensory architecture, or is not favoured in another context, it is unclear how likely this preference would be to spread in females in and of itself, rather than being lost by genetic drift. However, if the preference rose to an appreciable frequency, indirect selection may favour its spread and maintenance in the population (see below).

2.6.3 *The origin of mate preferences: adaptive preferences*

Finally there may arise a preference, which, from the outset, gains a direct fitness advantage. For instance, a preference may evolve in females resulting in males that are good parents or that provide large nuptial gifts being favoured as mates, such that females gain direct benefits from imposing non-random mating on males. These direct benefits lead to the production of more offspring for the preference-bearing females. Any direct fitness advantage accruing to females from expressing a mate preference is considered to be *naturally selected*. However, the female preference can also mean that the preference-bearing females exclude, i.e. outcompete, other females from obtaining these direct benefits. The mate preference therefore mediates sexual selection among females. As such, it may be extremely difficult to partition out natural and sexual selection in such cases, but such complexity should not surprise us.

2.7 The evolutionary maintenance of a preference

Mate preferences that arose as side-effects of other components of selection can be maintained by those self-same phenomena. Likewise, preferences favoured by direct benefits can similarly be maintained by natural selection (e.g. Price et al. 1993; Iwasa and Pomiankowski 1999). However, once we have a preference and a preferred ornament, an important evolutionary dynamic arises, a dynamic that has come to dominate sexual selection research: preference-ornament co-evolution.

2.7.1 *The Fisher process*

Regardless of how a preference arises, once there is a mate preference for an ornament, if there is additive genetic variation underlying both preference and ornament, a genetic covariance begins to build up between the preference and the ornament (Fisher 1930; Lande 1981; Kirkpatrick 1982; Prum 2010; Kuijper et al. 2012). This genetic covariance means that so-called *indirect selection* acts on both the preference and the ornament. If we consider the preference first, a female mating with an attractive male will produce sons that are also attractive. These sons will have a mating advantage (as they are attractive) and so pass on more copies of their genes than unattractive males. Importantly, though, some of those genes that are transmitted by sons are also the genes that influence the female mate preference. Thus the genes for the preference will also gain that transmission advantage. Indirect selection therefore describes the transmission advantage obtained through attractive offspring. Equally, the male ornament benefits from the same indirect benefit by being associated with the female preference. This means that a positive co-evolutionary feedback loop forms, with linkage disequilibria forming between genes influencing preferences and genes influencing ornaments, generating the genetic covariance.

The canonical models of the process were presented by Lande (1981) and Kirkpatrick (1982) for quantitative genetic and two-locus population models respectively (Figure 2.3). These models mathematically formalized the insight of Fisher, and the positive feedback between a preference and an ornament has been termed the 'Fisher process'. These models had a huge impact and revolutionized the field of sexual selection. One notable feature

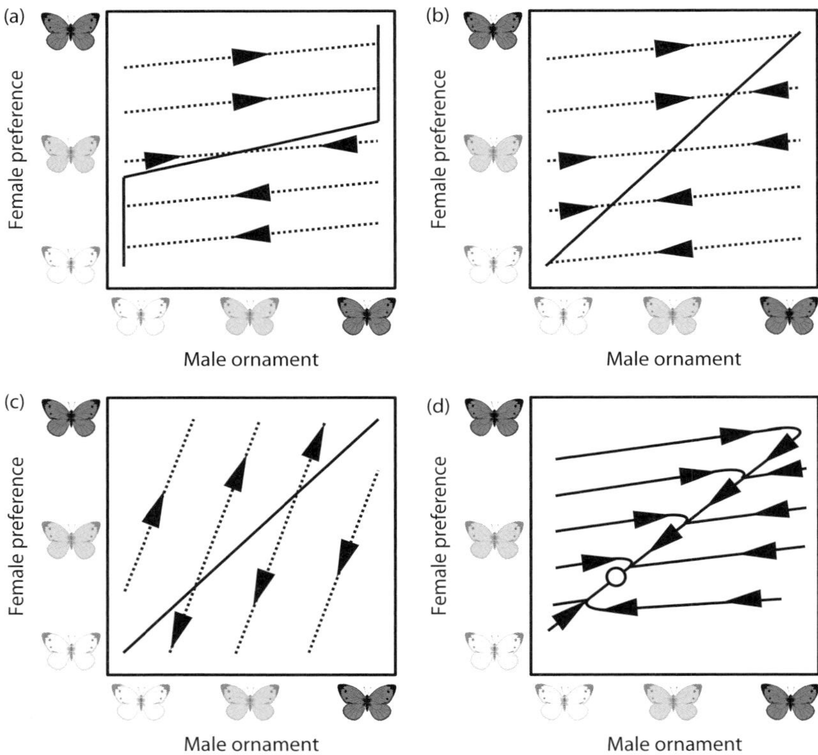

Figure 2.3 The co-evolution of mate preferences and ornaments under the Fisher process. Consider a species of butterfly in which females vary in their mate preference for male wing coloration. Natural selection favours pale males. The Fisher process describes how the build-up of a genetic covariance between a mate preference and an ornament that influences mating success can drive the evolution of both the ornament and preference. (a) Kirkpatrick's (1982) population genetic model showed that linkage disequilibrium between a mate preference and an ornamental trait could drive the exaggerated evolution of the ornament and co-evolution of the mate preference. The model predicted a line of equilibria where natural and sexual selection on the ornament reached a balance. The dotted lines show evolutionary trajectories. (b) Lande (1981) provided a quantitative genetic model of the Fisher process, again revealing a line of equilibria, where natural and sexual selection on the ornament balance. (c) The stability of the line of equilibria depended on the relationship between the genetic covariance between the female preference and the male ornament (Cov_{Apo}) and the additive genetic variance in the male ornament (V_{Ao}, i.e. Cov_{Apo}/V_{Ao}). If the slope of the line of equilibria > Cov_{Apo}/V_{Ao}, then the line of equilibria is stable (as shown in panel b), whereas if the slope < Cov_{Apo}/V_{Ao}, then the evolutionary trajectories drive the co-evolution of preference and ornament away from the line of equilibria (panel c). This was interpreted as encapsulating the 'runaway' aspect of Fisher's verbal description of the process. (d) Further modelling revealed that when expressing a mate preference is costly, the line of equilibria collapsed to a single point (shown here as a white dot). Whether this point is at the natural selection optimum for male wing coloration (i.e. no development of the trait under sexual selection), or at a point where natural selection and sexual selection are balanced, depended on whether there was a 'mutation bias' acting on the trait, with mutation tending to erode the ornament (in our example, mutations would be more likely to make male wings paler). If there is such a mutation bias (which might not be unreasonable) then this rescues the Fisher process and allows sexual selection to exaggerate the male ornament (Pomiankowski et al. 1991).

of both of these models was the presence under certain circumstances of a neutral line of equilibria between preference and ornament values, along which sexual selection on the trait was balanced by natural selection acting against that trait. At the time, there was much interest in whether or not a 'runaway' process would result from the genetic covariance between ornament and preference, with a runaway being the evolution of an ever-increasing ornament and an ever more extreme preference. Given the extravagance of some male ornaments, such runaway evolution seemed to encapsulate the mechanism for developing outrageously elaborated traits. For example, in the Lande model, the line of equilibria was unstable if the slope of the line was smaller than the additive genetic covariance between ornament and preference (Cov_{Apo}) divided by additive genetic variance (V_{Ao}) in the ornament (Figure 2.3), leading to a continual co-evolution between trait and preference that fitted the concept of runaway. On the other hand, if the line of equilibria is stable (slope $> Cov_{Apo}/V_{Ao}$), then further preference and trait evolution can occur by drift. Today, researchers are perhaps less focused on whether or not the conditions for runaway sexual selection occur, focusing instead on the initial presence or absence of the genetic covariance between preference and ornament (Kuijper et al. 2012).

The next key step in the analysis of the Fisher process was to include a cost to expressing a preference (a reasonable addition). By adding a cost of choice, the line of equilibria collapses to a point (limiting runaway evolution or the role of genetic drift in preference and ornament evolution; Pomiankowski 1987; see also Figure 2.3). However, the important paper by Pomiankowski et al. (1991) showed that the Fisher process may continue if there is a mutation bias acting on the ornament—that is, mutations are more likely to reduce or 'harm' ornament elaboration, which seems reasonable (Figure 2.3d). Similarly, the Fisher process may continue if there is a migration bias, with incoming migrants being less elaborate (Day 2000). It remains an empirical question whether such biases in nature are sufficient in magnitude to maintain the Fisher process (for a discussion of male-biased mutation see Ellegren 2007).

In addition to the importance of a cost of choice, there is also another important caveat, namely that the females should not pay too great a fecundity cost for mating with an attractive male, as the indirect benefits of producing attractive offspring may then not outweigh the direct fitness benefits of producing more, but less attractive sons. More technically: is female fecundity maximized at evolutionary equilibrium? This caveat is the reason why the so-called 'sexy sons' hypothesis is problematic. As originally formulated (from work on red-winged blackbirds: Weatherhead and Robertson 1979), females mating with attractive males may suffer reduced fecundity, as those males provide less parental care. However, these costs may be balanced by the indirect benefits of producing attractive male offspring. The extent to which the proposed mechanism could work was a source of some controversy and is still empirically unresolved (Kirkpatrick 1985, 1986; Curtsinger and Heisler 1988, 1989; Pomiankowski et al. 1991; Andersson 1994). In recent years though, the term 'sexy sons' has begun to encompass the Fisher process more generally (e.g. Cameron et al. 2003).

The logic of the Fisher process has been shown using other modelling frameworks and so seems relatively robust to alternative genetic formulations, although different assumptions and sets of benefits and costs influence the range of evolutionary outcomes (most recently reviewed by Kuijper et al. 2012). Detailed coverage is not possible here. However, it seems plausible that the Fisherian co-evolution will be a near-universal feature of preference and ornament evolution whenever additive genetic variance in both traits exists, regardless of whether the ornament is also correlated with other aspects of fitness (see Section 2.7.2). As such, the Fisher process is relatively uncontroversial.

2.7.2 *'Good-genes' sexual selection*

What happens if the ornament is genetically correlated with other components of fitness? If such a genetic correlation exists, it is easy to see that females mating with ornamented males produce male offspring that are both attractive and of high genetic quality (they have 'good genes'). Moreover, if those fitness effects are also expressed in daughters, then they produce high-quality daughters too. Thus indirect selection progresses via transmission advantages gained by both the production of attractive sons and high-quality sons and daughters. Simple? Well, not really.

The first thing to note about good-genes indirect selection is that there have to be other fitness components. In other words, natural selection has to be occurring in the population; there have to be 'good genes'. Tests of good-genes sexual selection that do not show natural selection in the population first are therefore testing the impossible. Second, how likely is it that an ornament will be genetically correlated with fitness variation? This is one of the key stumbling blocks for good-genes effects. Early theoretical treatments modelled a preference, an ornament, and a viability trait. Some models showed a good-genes effect (e.g. Pomiankowski 1987; Iwasa et al. 1991) and some did not (e.g. Kirkpatrick 1986).

One solution to the problem of how a trait maintains its association with fitness is that of condition dependence. First suggested in the context of the sometimes acrimonious debate over 'handicaps' (see below), our current view of condition dependence comes from a hugely influential paper by Rowe and Houle (1996). Ornaments are condition dependent if expression of the ornament reflects the overall condition of the bearer; 'condition' is taken to mean some genetic component of an organism's fitness. Although Rowe and Houle introduced their idea of genic capture as a solution to another problem for indirect selection (the lek paradox: Borgia 1979), the idea is perhaps more general (e.g. Lorch et al. 2003). Rowe and Houle argued that as traits became more elaborated (e.g. larger), more loci would become associated with their expression, and so any segregating variation at those loci would contribute to the overall genetic variance underlying the trait. However, we can also flip this around and suggest that ornaments that are by their original nature associated with as large a fraction as possible of genome-wide fitness variation will be the ones most likely to be favoured by indirect good-genes effects. This is most easily seen in terms of traits that are not so much ornaments as they are other indicators of fitness, such as athletic courtship displays, body size, or age, all of which might co-vary with total fitness (ideas put forward by Maynard Smith 1991, for example).

2.7.3 *A different historical perspective*

The link between an ornament or indicator and overall fitness has been at the heart of the debate over good genes from the very beginning, including in ways that are less explicitly genetic. One influential debate has concerned what information a trait in one sex signals to the other sex. Central to this debate were the ideas of Amotz Zahavi, who argued that male traits evolved to signal genetic quality by being 'handicaps' to those males, i.e. only a genetically fit male could carry around a large ornament. Thus the so-called 'handicap principle' became a short-hand for good-genes sexual selection for many years. This chapter does not dwell on the development and controversies of the idea of handicaps (see Andersson 1994), although concepts such as epistatic, revealing, and conditional handicaps were widespread in the literature in the 1980s and 1990s and drove much of the theoretical modelling. Instead we merely note that the key issue is whether or not genetic quality—some or all components of total fitness—co-varies with expression of an ornament or indicator trait. What keeps a trait 'honest' (the key question back in the 1980s) is really a question of what maintains that genetic covariance.

2.7.4 *Good genes and compatible genes*

The rationale for good-genes selection outlined in Section 2.7.3 assumes that fitness effects are additive, such that what is a fit allele in one individual is also a fit allele in another (see also Chapter 9). However, there are numerous cases where we might expect this not to be true, for instance if there are epistatic effects on fitness, or if fitness varies across environments (genotype by environment interactions: e.g. Ingleby et al. 2010). Here we focus on the idea of genetic compatibility, whereby mate choice arises in terms of the degree of genetic compatibility between mating partners, such that there are no inherently genetically superior individuals in the population (Neff and Pitcher 2005; Puurtinen et al. 2009). The current interest in the potential for genetic compatibility to shape mating behaviour and mate preferences is partly thanks to two influential contributions by the Zehs, particularly in the context of the evolution of polyandry (Zeh and Zeh 1996, 1997; see also Tregenza and Wedell 2000; Colegrave et al. 2002; Zeh and Zeh 2003; Chapter 9). If mate choice arises in order to produce genetically fit offspring via the compatibility of parental genotypes, in many cases we might expect different males to be favoured by different females (and vice versa). This should not deplete genetic variation in the same way as other forms of mate choice where the genetic benefits are additive.

One of the best-studied systems for choice in terms of genetic compatibility comes from vertebrates rather than insects, namely the major histocompatibility complex (MHC), and there is good evidence for non-random mating with respect to MHC genotype (reviewed by Tregenza and Wedell 2000; Neff and Pitcher 2005). More generally, genetic compatibility is only plausible under certain circumstances. Within populations, inbreeding avoidance is the most likely, whereas between diverging populations, mate choice to limit outbreeding (or hybridization) is also plausible. One clear example of the former in insects can be found in Hymenoptera with single-locus Complementary Sex Determination (sl-CSD; Cook 1993; Heimpel and de Boer 2008). In a number of Hymenoptera, sex is determined by the zygosity of a single locus: heterozygotes become females, while hemizygotes and homozygotes develop as males. Hymenoptera are haplodiploid (Chapter 1), so normally this system produces the standard pattern of diploid females and haploid males. However, if there is inbreeding, diploid individuals may be generated that are homozygous at the CSD locus, producing diploid males which are generally non-viable or infertile. Unsurprisingly, inbreeding avoidance behaviours have evolved in these species (Godfray and Cook 1997). Otherwise, particular genetic architectures are needed, such as inversion systems, which reduce or prevent recombination, allowing the build-up of linked co-adapted genes (Tregenza and Wedell 2000). It is not yet clear how often these systems arise.

2.7.5 *Sexually antagonistic chase-away models*

As introduced earlier, there may be sexual conflict between males and females over whether mating takes place. There may also be aspects of mating (such as genital morphology or seminal products) that increase male fitness but are harmful or deleterious to females (reviewed by Arnqvist and Rowe 2005; Chapter 11). Setting to one side the question as to whether 'harm' is actually selected for (e.g. Johnstone and Keller 2000), selection may favour females that evolve 'resistance' traits (for instance that limit male coercion or limit the extent of male harm). These resistance traits will be favoured by the direct, natural selection benefit of reducing harm. However, such traits may also generate non-random mating success among males, generating selection for males able to overcome female resistance. Hence there is mate choice for male coerciveness, and, as we have seen, indirect selection may then come to act as females that mate with good coercers produce sons that are also good coercers. This form of male–female co-evolution has been termed an 'evolutionary chase' by Parker (1979) and 'chase-away' sexual selection by Holland and Rice (1998), both paying homage to Fisher and also emphasizing that selection 'for' preferences in these kinds of scenarios might be somewhat non-intuitive (the preference arises from selection not to mate, for instance). Important aspects of these models include whether female resistance is a threshold trait (Gavrilets et al. 2001) or whether female resistance evolves as a general form of insensitivity to males (Rowe et al. 2005). In the latter case, sexual co-evolution is not an inevitable outcome, and indeed the short-hand notion of sexual conflict being characterized by endless cycles of adaptation and counter-adaptation by males and females may not prove to be the rule (Parker 1979; Lessells 2006). Males and females may in fact be a general hotch-potch of mostly resolved sexual conflicts. Finally, as with other forms of indirect selection, the extent to which the costs imposed on females by males are balanced by indirect benefits is unclear, but indirect effects certainly influence how evolution might proceed in the theoretical models (e.g. Härdling and Karlsson 2009).

2.7.6 *Summary of indirect models of sexual selection*

In summary, indirect sexual selection based either solely on the Fisher process, on good genes, or sexually antagonistic chase-away—though not genetic compatibility—share the feature that a Fisherian mating advantage (or more formally, a genetic covariance between mating preference and ornament) will arise. Whether this is fair grounds to call the Fisher process 'the null hypothesis' for indirect selection (Prum 2010) is discussed below. Either way, in many cases of indirect sexual selection, the Fisher process will be an integral part of the process—therefore testing 'different' models of indirect sexual selection is problematic. Indeed, it has been argued that Fisher and good-genes processes form a 'sexual selection continuum' (Kokko et al. 2002). Finally, mathematically it is possible to divorce Fisherian and good-genes sexual selection from one another, as Grafen did to prove that a good-genes process could occur without a Fisherian mating advantage (Grafen 1990). However, in those models, the benefit of mate choice to females was an increase in fecundity (i.e. a direct benefit), and so the models were perhaps more akin to a 'good parent' model (e.g. Price et al. 1993; Iwasa and Pomiankowski 1999) where a male advertises his ability to enhance female direct fitness, rather than being a model of the indirect sexual selection (where selection on preference arises from the fitness of offspring) more usually associated with Fisherian and good-genes sexual selection.

2.8 Mutual mate choice

Both sexes may have traits that lead to non-random mating success in the opposite sex. Put another way, not all individuals in each sex have free access to the gametes of the other sex. When this occurs, we can say that there is mutual mate choice. There are several possible benefits associated with mutual mate choice, including benefits associated with gamete quality, genetic compatibility, or the timing of pair formation (for a review see Kraaijeveld et al. 2007). Mutual mate choice may be assortative (e.g. both sexes having similar preferences for large body size—recently reviewed by Jiang et al. 2013) or the two sexes may have preferences for very different traits (e.g. males may prefer larger females, and females may prefer males with an elaborate courtship display). With the emphasis on sexual selection being stronger in males than in females (and females being choosy and males being chosen), consideration of mutual mate choice often focuses on monogamous species (e.g. Kirkpatrick et al. 1990) but it can occur in polygamous species as well, although the conditions for its evolution are more restricted (Kokko and Johnstone 2002; Kokko and Jennions 2008).

2.9 The status of indirect selection: the lek paradox and the strength of indirect selection

The *lek paradox* refers to the fact that, for some species, the choosy sex appears to get nothing from mating except genes carried in sperm, and so any benefits of choice would seem to have to be genetic, indirect benefits that accrue through offspring fitness. However, we might expect both natural selection and then the action of mate choice to deplete genetic variation in fitness, leaving nothing to drive forward a good genes process (Borgia 1979; Taylor and Williams 1982). Put simply, with depleted genetic variation in fitness, why does mate choice still occur? The simplest solution is that there are direct benefits to mate choice, difficult to uncover, especially in natural populations (Reynolds and Gross 1990). Alternatively, the mate choice we see may be driven by the Fisher process or by sexual conflict. However, to what extent is the lek paradox a genuine paradox?

There are two aspects of the paradox of the lek that we need to consider. Empirically, the data suggest that generally additive genetic variation for fitness is alive and well in natural populations (Burt 1995). Likewise, sexually selected traits themselves appear to retain additive genetic variation in the face of selection (Pomiankowski and Møller 1995). Theoretically, how is this variation maintained in the face of natural and sexual selection? The answer to this question is still one of the main puzzles of population genetics (e.g. Charlesworth and Charlesworth 2010), but the answers probably include mutation–selection balance and environmental variation in selection. In terms of mutation–selection balance, the concept of genic capture introduced earlier (Rowe and Houle 1996) argues that as traits become increasingly elaborated, so more genes are involved in the expression of the trait (i.e. become 'captured'). This creates a larger mutational target, with the greater input of mutations balancing the erosion of genetic variation by selection.

In terms of changing environments, another extremely influential idea has been the Hamilton–Zuk hypothesis (Hamilton and Zuk 1982). They argued that parasites and pathogens exert never-ending selection pressures on organisms to evolve mechanisms (such as immune systems) that provide resistance or tolerance to parasites and disease. As most parasites and pathogens have considerably shorter life-cycles than their hosts, they can readily evolve in response to their hosts' counter-measures, maintaining strong selection on the hosts and thus genetic variation in fitness. Given this scenario, if traits arise that can accurately reveal parasite or pathogen burden, mate choice for those traits can be favoured through indirect selection for the production of healthier offspring. Ten years later, Folstad and Karter (1992) introduced a related idea, the 'immunocompetence handicap', adding some mechanistic details (such as the potential for feedback between ornament display and immune function mediated by the hormone testosterone). These ideas have stimulated much interest, including in insects (where the role of testosterone is perhaps played by melanin or juvenile hormone), and the evidence for the role of parasites and pathogens in sexual selection is reviewed in detail in Chapter 13. For now, we note that while there is some evidence for the role of parasites and pathogens in mate choice, they do not appear to be a universal explanation for the maintenance of mate choice. However, the Hamilton–Zuk hypothesis forces us to remember that ecological systems are dynamic and prone to rapid change, and that such dynamism will maintain genetic variation in fitness in populations.

But perhaps the strongest critique of indirect sexual selection, and good-genes sexual selection in particular, came in a paper by Kirkpatrick and Barton (1997). They provided equations for the expected strength of indirect sexual selection and showed that it would likely be swamped by direct benefits to choice. Although these theoretical results do not themselves negate the action of indirect sexual selection, they do question its importance, for instance in explaining elaborate traits such as sexual ornaments or courtship displays. This paper has had a salutatory effect on the field. For a review of the empirical side of this question, see Chapter 8.

2.9.1 *A null model for the evolution of mate choice?*

In many ways, the Fisher process is uncontroversial. Rebranded as the LK (Lande–Kirkpatick) model of sexual selection, Prum (2010) has recently suggested that the LK model should be considered as the 'null model' for the evolution of mate choice, as it is an inevitable consequence of genetic variances in preference and ornament. Prum cites the hugely influential review by Kirkpatrick and Ryan (1991) as an early example (indeed the origin) of this thinking, and makes a strong case for the Fisher process as the default process of indirect sexual selection. The motivation for this position is clearly one of frustration with the apparent emphasis on good-genes sexual selection.

Here we briefly consider two points. First, on a more epistemological level, it might seem strange to position the Fisher process as the null model: one could argue that the most appropriate null model is actually no mate choice at all. Although the Fisher process will often commence when mate choice arises, this perhaps still seems a weak argument for positioning an active process as a null hypothesis. Second, perfectly valid forms of indirect selection are theoretically possible, and indeed have some empirical support, that are not expected to include the typical Fisher process, including genetic compatibility models whereby a genetic correlation between the trait that leads to non-random mating and the characteristic that signals genotype is not expected to arise. This could also argue against the prima facie explanatory primacy—or necessity—of the Fisher process in indirect sexual selection and so question placing it as a null hypothesis. Instead, perhaps the important point that needs to be made continually is that a Fisher process is central to indirect sexual selection when the benefit to choice is gained either by producing attractive offspring (only Fisher) or by producing offspring that are sexy and fit (Fisher and 'good genes').

2.10 Concluding remarks

With the focus on the conceptual basis of sexual selection theory, there is much that has not been covered in this introductory chapter. For instance, in recent years there has been growing interest in how sexual selection interacts with processes such as kin selection (Pizzari and Gardner 2012; Chapter 14) and sex ratio evolution (Fawcett et al. 2011). There has also been continuing work developing further the links between sex roles, mating systems and sexual selection: such links will be considered in more detail in Chapter 3.

Speciation research remains closely tied to sexual selection, and indeed some of the main developments in understanding the genetics of traits associated with sexual selection are coming from the study of traits associated with behavioural reproductive isolation (Chapter 4). Hybridization, and any resulting selection against hybrids (reinforcement: e.g. Kirkpatrick and Servedio 1999), may prove to be the most potent form of good-genes sexual selection we find in nature, although what that means for good-genes sexual selection as an agent driving the evolution of many of the elaborate traits seen in males remains to be seen. Modelling frameworks themselves are evolving too, for instance with theoretical models of indirect genetic effects growing in popularity (e.g. Moore and Pizzari 2005; Miller and Moore 2007; Chenoweth and McGuigan 2010; Bailey and Moore 2012). Finally, we have not been able to touch upon many other aspects of sexual selection, including the fascinating question of the evolution of multiple ornaments or displays (e.g. Iwasa and Pomiankowski 1994; Johnstone 1995), or on the role of mate sampling in the evolution of mate choice (e.g. Reid and Stamps 1997).

However, further developments in sexual selection theory will only flourish with conceptual clarity. Although definitions are never likely to be hard and fast, clearly delimiting pattern and process is important. For instance, there are general patterns in terms of male and female sexual behaviour across insects, but the processes of sexual selection that help generate those patterns do not need to be defined in terms of sex roles, or even in terms of gender. Our theory needs to be more explicit on that point. Furthermore, wrapping the downstream consequences of competition for mates and their gametes into definitions of sexual selection (for instance whether there are direct or indirect benefits associated with mate choice) can also lead to confusion. That sexual selection and other components of natural selection are hard to separate should also not surprise us, as Darwin's mechanism of sexual selection does not act merely on traits disfavoured by natural selection, even though such traits inspired the concept.

CHAPTER 3

Mating systems

Hanna Kokko, Hope Klug, and Michael D. Jennions

3.1 What's in a name? How to classify mating systems

Fertilization mode varies greatly among taxa. In many species, including all plants and many marine invertebrates, there is no mating as sperm (or pollen), or both sperm and eggs, travel independently of the parent. However, in most terrestrial animals, including insects, at least one sex is mobile and fertilization involves a direct sexual encounter between a male and a female (i.e. mating). Although the term 'mating system' is sometimes used to describe fertilization patterns in plants and sessile animals, behavioural ecologists usually restrict its use to taxa where actual mating occurs. Here we follow this convention.

The term 'mating system' is used to describe how mating and fertilization are achieved, and by whom, including reporting the numbers of individuals involved. A full population-level description of every interaction between all adults and their gametes is impossible. Researchers therefore summarize a population's mating system using a limited set of criteria. The most fundamental categorization uses the number of mates that each sex has within a defined time period (e.g. one breeding period, or over a lifetime). For instance, *monogamy* describes populations where both sexes have a single mate, whereas *multiple mating* describes cases in which individuals of at least one sex have several mates. Note, however, that these statements typically ignore individuals who never mate.

Monogamy or multiple mating can be defined relative to either a single breeding cycle or a lifetime. Monogamy within a single cycle does not preclude longer-term multiple mating ('sequential monogamy'). The female-specific terms *monandry* and *polyandry* refer to cases in which females mate either with one or with multiple males, respectively, during the defined time period. Similar terms *monogyny* and *polygyny* exist for males. If both sexes mate with multiple partners, the mating system is classified as *polygamous* (Emlen and Oring 1977; Thornhill and Alcock 1983; Davies 1991; Shuster and Wade 2003), with the occasional use of the terms *polygynandry* and *polyandrogyny* (respectively being greater variability in male than female mating success, or the reverse).

Although these categories provide a general sense of mating interactions between individuals, they have been criticized as overly simplistic (e.g. Chapter 6 in Shuster and Wade 2003). The challenge can be exemplified by lekking species (reviews: Höglund and Alatalo 1995; Shelly and Wittier 1997). Usually, only a small proportion of males obtain many matings. It is entirely possible that there are more males that mate only once than males who mate multiply, but the term monogyny is not used even if the modal mating success of a lekking male is 1. Instead, the term monogyny describes mating systems such as that

The Evolution of Insect Mating Systems. Edited by David M. Shuker and Leigh W. Simmons.
© The Royal Entomological Society 2014. Published 2014 by Oxford University Press.

of *Nephila* spiders where males no longer seek new females after inseminating one, and can become functionally sterile after breaking off parts of their copulatory organs (Fromhage and Schneider 2006). This highlights a link between theories for mating systems and sexual selection: our view of mating systems does not appear to be based solely on modal mate numbers, but subtly includes statements about the strength of selection on males (and females) to have multiple mates. Indeed, our central theme is that the distribution of matings depends greatly on how strongly individuals are selected to mate more (or less) often.

There are additional considerations at play when categorizing mating systems, including historical baggage. Much of the terminology was developed prior to paternity testing, which, once in place, revealed high levels of female multiple mating. This did not, as one might have expected, lead to increased use of the term *polygynandry* (Holman and Kokko 2013). Instead, the field often uses classifications based on questions that focus on selection on one sex and disregard that on the other. Such questions include when to expect *polyandry* (i.e. why females have multiple mates, Arnqvist and Nilsson 2000; Chapter 9) and under what conditions males benefit from *monogyny* (Fromhage et al. 2007).

Refinements to mating system terminology classify populations based on how individuals obtain mates (e.g. do males defend resources that attract females or search for receptive females) and characteristics of mating behaviour (e.g. mate guarding or the presence of social 'pair bonds'). Emlen and Oring (1977) introduced terms based on mechanisms of mate monopolization. In *resource-defence polygyny*, males attract females by defending territories containing resources that are essential for females; in *male dominance polygyny* aggregated males establish dominance hierarchies that determine their mating success; and in *scramble competition polygyny* a male's success relates to his ability to locate sexually receptive females (Emlen and Oring 1977, Thornhill and Alcock 1983).

Davies (1991) suggested that mating system terminology further incorporate details of the duration of the 'pair bond'. Although this term is suggestive of vertebrates, other taxa, including insects, can have pair bonds if these are operationally defined as lengthy associations before and/or after mating. Some researchers have proposed that mating system terminology should even reflect how much parental care each sex provides (Vehrencamp and Bradbury 1984; Reynolds 1996; Shuster and Wade 2003). Yet others have highlighted the distinction between patterns of mating and of paternity. Even if females mate multiply, strong first or last male sperm precedence might make the system genetically equivalent to monandry. Some authors therefore refer to 'genetic mating systems' (e.g. Jones et al. 2000).

The evolving nature of terminology is no surprise. Researchers will naturally emphasize different aspects of mating patterns when attempting to classify diversity using a limited number of categories. More interestingly, the last decades have seen advances in theoretical explanations for sexual differences in mate searching, mate choice, investment into sexually selected traits, optimal allocation of paternity, and the optimal number of mates and parental investment (Queller 1997; Gavrilets et al. 2001; Wade and Shuster 2002; Fromhage et al. 2007; Kokko and Wong 2007; Royle et al. 2012). This has shifted the key questions being asked.

Early work on mating systems emphasized how ecological parameters affect the spatio-temporal distribution of females, which then alter males' behaviour. Females were tacitly assumed neither to suffer from sperm limitation nor to gain significant fitness by mating with a second male (unless they could extract additional parental care). In contrast, males were assumed to benefit from each successive mating. The term *environmental potential for polygyny* (Emlen and Oring 1977) encapsulates this view: a quest to identify circumstances that permitted some males to acquire multiple mates, not whether it is beneficial to attempt this nor how females should respond.

3.2 Mating systems are the outcome of selection on individuals

In this chapter we briefly review the history of mating system theory, covering three landmark contributions (Bateman 1948; Trivers 1972; Emlen and Oring 1977). Our main focus is, however, to provide an update by reconciling explanations for why a population has a given mating system with recent theory about the evolution of sex differences. We place special emphasis on the evolution of sex differences in: mate searching, mate choice, and investment in sexually selected traits. To our knowledge, understanding which sex searches has not been explicitly discussed in the context of mating system theory. It should be, though, because traditional mating system theory simply assumes that males will 'seek out' females by moving to where females are or will soon be.

The mating system is a descriptive summary of interactions between individuals. To explain why a given population has a certain mating system we must understand selection on individuals. There is, however, feedback because selection operates on individuals within a mating system, which is itself contingent on how behavioural adaptations evolve. For example, if polyandry increases, there is stronger post-copulatory sexual selection on males to ensure paternity (Chapter 10). If there is a trade-off between pre- and post-copulatory performance, traits enhancing mating success might receive less investment. This could, in turn, affect the rate at which females encounter potential mates.

In this sense, mating systems are both cause and consequence of the sexually selected adaptations that evolutionary ecologists study. This insight can be useful to understand persistent patterns across populations or species. For instance, a reduction in the number of mates is often attributable to coercive traits that prevent a mate from acquiring a new partner. For example, in mosquitos and various *Drosophila*, males transfer seminal chemicals that reduce a female's propensity to remate (see Hosken et al. 2009; Chapter 10). Females can similarly impose monandry on males. For example, female burying beetles interfere with their mate's efforts to attract new females (Eggert and Sakaluk 1995).

3.3 A brief history of mating system theory

In a prescient study, Bateman (1948) graphed reproductive success against the number of mates for males and for female *Drosophila*. Male reproductive success increased with each successive mating, whereas, depending on the feeding regime, female success increased less strongly or even plateaued after a single mating. The experimental populations were, however, such that the findings do not directly illuminate what factors determine the distribution of mates acquired by each sex (i.e. mate availability). In the 1960s and 1970s the study of mating systems re-emerged, this time focusing on identifying correlations between environmental factors and how many mates individuals of each sex obtained. For example, John Crook showed that human societies ranged from monogamous to polygynous and polyandrous partly due to agricultural land tenure practices (Crook and Crook 1988).

This budding evolutionary perspective was solidified by Trivers (1972), who sought to understand why individuals of each sex differ in the benefits gained by investing in traits elevating mate acquisition abilities. He noted that differences in parental investment affect the availability of potential mates. Leaving aside effects of deviation from a 1:1 adult sex ratio, the sex with greater parental investment is a limiting resource for the other sex. The mate-limited sex then potentially experiences stronger sexual selection. In a general survey Trivers pointed out that parental investment per breeding event is usually greater for females than males. Consequently, he inferred that sexual selection is generally stronger on males. To emphasize this logic he highlighted that in species where male parental investment is higher than that of females, there is often 'role-reversal' with females competing for males.

Trivers (1972) was aware of, and highlighted, Bateman's (1948) work. More recently, the relationship between reproductive success and number of mates has become defined as the Bateman gradient (BG), originally called the sexual selection gradient (Arnold 1994; Arnold and Duvall 1994). It is an essential component of the strength of selection to improve mating rates (Jones 2009) (Figure 3.1). Females are argued to gain little by competing for mates if failure to mate and obtain sperm is improbable, such that most matings are nearly or completely superfluous, generating a shallow BG (see Section 3.4.2). Finally, Trivers (1972) noted that the spatial and temporal distribution of females should affect the intensity of mating competition, hence sexual selection. For example, when females are more tightly spatially clumped, the winner of a male–male contest gains access to far more females.

Five years later, Emlen and Oring (1977) famously formalized two aspects of mate availability. First, they defined the *operational sex ratio* (OSR) as the ratio of males to females available as mates at a given point in time and in a given location (these individuals form the 'mating pool', although Emlen and Oring did not use that term). Greater female parental investment shifts the OSR towards males. The OSR is a key parameter in most theoretical models of the evolution of sex differences, partly because it is easy to produce models where parental investment is represented by the time taken to return to the mating pool (so-called 'time in/time out' models: Clutton-Brock and Parker 1992; Parker and Simmons 1996). The sex with the shorter time out is typically more abundant in the mating pool, thereby experiencing mate limitation.

Emlen and Oring (1977) also proposed another term, the *environmental potential for polygamy* (EPP), to predict whether polygyny or polyandry will occur. The underlying idea is to identify ecological factors that determine the distribution of the mate-limiting sex (usually females). The resultant spatio-temporal spread then influences the economic feasibility for the mate-limited sex to monopolize access to multiple mates. Monopolization

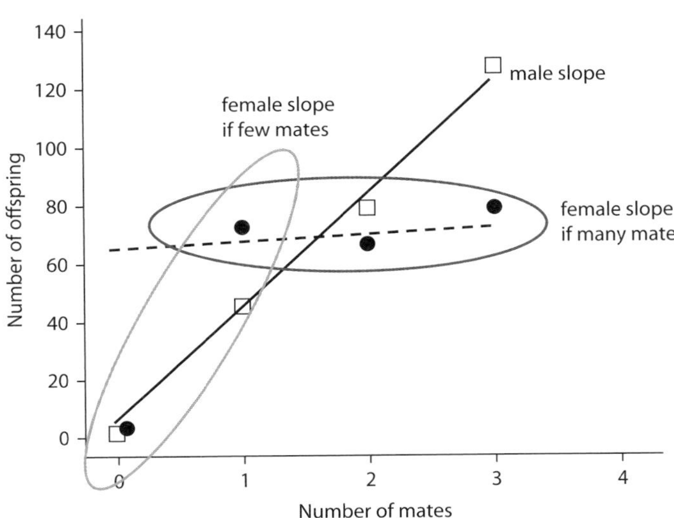

Figure 3.1 Interpreting the Bateman gradient (plot of reproductive success against number of mates) makes implicit assumptions about the likely number of mates encountered. Any statement that the female slope is 'shallow' has a hidden assumption that females do not often remain unmated.

was argued to be more likely to occur when females are spatially clumped, so that males only have to defend a small area to monopolize several females, or when female sexual receptivity is asynchronous so that high-quality males can sequentially control access to females.

Today, the EPP is rarely invoked. This hints at low explanatory power. If females are clumped, a male might find it either easier or more difficult to fend off competitors. The latter possibility is apparent once one notices that the number of males 'per clump', interested in mating, will increase when females form fewer clumps. Whether one male can monopolize many females, or paternity is shared more equally, depends on proximate species-specific determinants of the outcome of competitive interactions. The EPP therefore does not readily lend itself to theoretical modelling, and there has been no concerted effort to find widely applicable parameters that determine the EPP.

3.3.1 *Progress means identifying problematic points*

Bateman (1948), Trivers (1972) and Emlen and Oring (1977) led to a widely accepted view that females usually distribute themselves through time and space to maximize access to resources that enhance their reproductive success. Males then distribute themselves accordingly to maximize the number of females with whom they can mate. The OSR, being seen as a good predictor of mating competition (sexual selection), plays a central role in explaining sex roles. For example, a female-biased OSR can reverse sex roles, as occurs when males are the limiting sex because they produce large spermatophores or provide extended parental care that delays their return to the mating pool (Simmons 1992).

In the last two decades, researchers slowly identified two main problems with the above description of the evolution of mating systems (Figure 3.2). First, mate monopolization need not increase as mates and resources become more clumped, nor as the OSR becomes more skewed (Klug et al. 2010). The OSR is a less straightforward predictor of the mating system than once thought (Arnold and Duvall 1994; Owens and Thompson 1994; Shuster and Wade 2003; Klug et al. 2010). It is clearly still a relevant parameter, however, as it measures sex differences in mate availability. In Section 3.4 we describe how the OSR and Bateman gradient interact to affect sexual selection.

Strictly speaking, sexual selection predicts how mate acquisition effort evolves, but this does not directly translate into a statement about the expected *variation* in the number of matings per individual. Since such information about mating number distributions is necessary to classify mating systems, current theory is arguably incomplete. Nevertheless, selection to acquire more (or fewer) mates is clearly a necessary component of any complete theory, which may justify our focus on sexual selection.

The second problem recognized concerns the statement that males will seek out females. This is implicit in the assumption that males 'map' themselves on to the female distribution. Whether this involves males actively moving to encounter females or moving to locations where females are or will be present, these behaviours are broadly forms of 'mate searching'. They increase the encounter rate with potential mates, typically at some cost to other fitness components (e.g. defending a contested area can be costly). Initially it seems a reasonable assumption that males are more willing than females to expend effort mate searching, given the textbook account of how mate-limitation acts. The problem, however, is that early theoretical models of mate searching failed to confirm this prediction (Hammerstein and Parker 1987). Despite models including mechanisms that create a male-biased OSR, there was no asymmetry in the likelihood that males rather than females will invest more into mate searching. Although Hammerstein and Parker (1987) did not

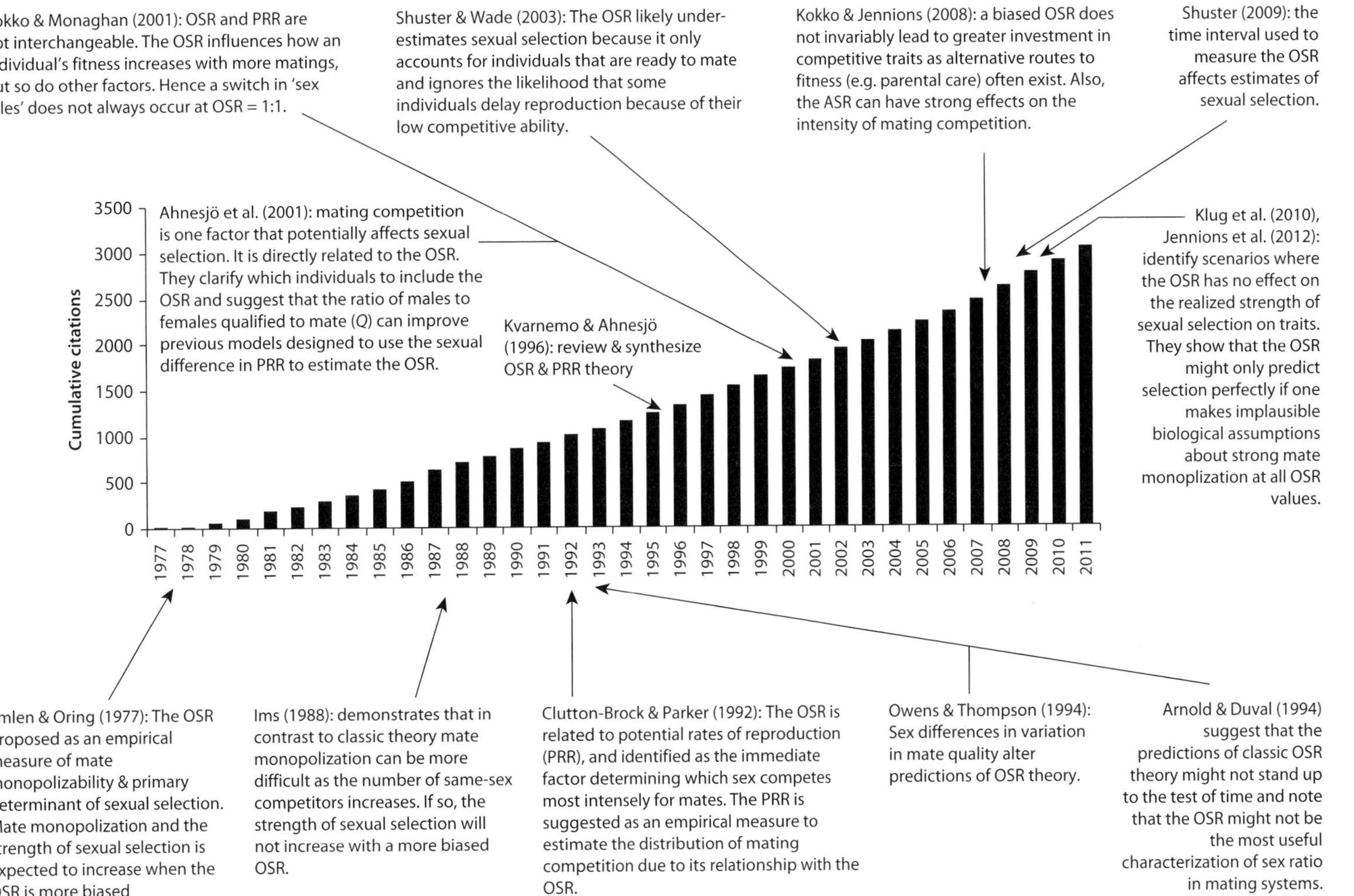

Kokko & Monaghan (2001): OSR and PRR are not interchangeable. The OSR influences how an individual's fitness increases with more matings, but so do other factors. Hence a switch in 'sex roles' does not always occur at OSR = 1:1.
Shuster & Wade (2003): The OSR likely under-estimates sexual selection because it only accounts for individuals that are ready to mate and ignores the likelihood that some individuals delay reproduction because of their low competitive ability.
Kokko & Jennions (2008): a biased OSR does not invariably lead to greater investment in competitive traits as alternative routes to fitness (e.g. parental care) often exist. Also, the ASR can have strong effects on the intensity of mating competition.
Shuster (2009): the time interval used to measure the OSR affects estimates of sexual selection.
Ahnesjö et al. (2001): mating competition is one factor that potentially affects sexual selection. It is directly related to the OSR. They clarify which individuals to include the OSR and suggest that the ratio of males to females qualified to mate (Q) can improve previous models designed to use the sexual difference in PRR to estimate the OSR.
Kvarnemo & Ahnesjö (1996): review & synthesize OSR & PRR theory
Klug et al. (2010), Jennions et al. (2012): identify scenarios where the OSR has no effect on the realized strength of sexual selection on traits. They show that the OSR might only predict selection perfectly if one makes implausible biological assumptions about strong mate monoplization at all OSR values.
Cumulative citations
3500
3000
2500
2000
1500
1000
500
0
1977
1978
1979
1980
1981
1982
1983
1984
1985
1986
1987
1988
1989
1990
1991
1992
1993
1994
1995
1996
1997
1998
1999
2000
2001
2002
2003
2004
2005
2006
2007
2008
2009
2010
2011
Emlen & Oring (1977): The OSR proposed as an empirical measure of mate monopolizability & primary determinant of sexual selection. Mate monopolization and the strength of sexual selection is expected to increase when the OSR is more biased
Ims (1988): demonstrates that in contrast to classic theory mate monopolization can be more difficult as the number of same-sex competitors increases. If so, the strength of sexual selection will not increase with a more biased OSR.
Clutton-Brock & Parker (1992): The OSR is related to potential rates of reproduction (PRR), and identified as the immediate factor determining which sex competes most intensely for mates. The PRR is suggested as an empirical measure to estimate the distribution of mating competition due to its relationship with the OSR.
Owens & Thompson (1994): Sex differences in variation in mate quality alter predictions of OSR theory.
Arnold & Duval (1994) suggest that the predictions of classic OSR theory might not stand up to the test of time and note that the OSR might not be the most useful characterization of sex ratio in mating systems.

Figure 3.2 Brief history of statements about the operational sex ratio (OSR), and the cumulative number of citations to Emlen and Oring (1977). PRR, potential rate of reproduction; ASR, adult sex ratio.

present their findings as a critique of Emlen and Oring (they only cited their work to introduce the OSR), it remains clear that to understand mating systems we must identify those conditions that favour mate searching by one or both sexes. We cover this topic in Section 3.5.

In Section 3.6 we synthesize our arguments based on selection on individuals of each sex to promote an approach that derives mate numbers from first principles based on fundamental considerations. These are sex specificity in the encounter rates with potential mates, sex differences in factors determining whether an encounter will lead to mating and, if so, what determines how long it is until an individual is then ready to mate again.

3.4 Reasons for sex-specific strength of sexual selection

3.4.1 *Traditionally used approaches to quantifying sexual selection*

Three key population-level parameters have been used to describe variation in the strength of sexual selection among species and/or between the sexes. Two of these, the BGs and the opportunity for sexual selection (described below), are traditionally more closely linked to formal selection theory (Jones et al. 2002) than is the OSR. Until recently, with a few notable exceptions (e.g. Jones 2009), there had been few attempts to reconcile the OSR with selection theory.

3.4.1.1 *Bateman gradients and the opportunity for sexual selection*

The BG concept (Figure 3.1) is closely associated with 'three principles' (Arnold 1994) that emerge from empirical studies showing: (i) greater variance in male than female reproductive success, (ii) greater variance in male than female mating success, and (iii) a stronger relationship between reproductive success and mating success (i.e. the BG) in males than females. According to Fisher's fundamental theorem, variance in fitness (measured here as reproductive success) places an upper limit on the magnitude of selection. The implication is that selection to increase mating success is potentially stronger on males whenever they experience higher variance in fitness due to a combination of high variance in mating success and a steep BG. Male traits that predict mating success are then expected to be under strong sexual selection.

The 'opportunity for sexual selection' (I_s or I_{mates}) is the variance in the number of mates scaled to account for the mean number of mates across all individuals. It equals the maximum possible selection on mating success. Unfortunately, any inference that the role of sexual selection in a mating system is readily measured using I_s (e.g. Shuster and Wade 2003) is perilous (Klug et al. 2010; Jennions et al. 2012). There is no guarantee that selection is stronger on males in a population with a higher male value for I_s. Some variation in male mating success is causally related to more successful males possessing traits that elevate mating success, but the remainder arises because of chance events (Sutherland 1985). The problem is far worse than chance simply adding noise to an otherwise correct estimate. There is a systemic bias associated with I_s because chance events tend to become more important determinants of success when there are, on average, few matings per male. This means that I_s values increase rapidly as the OSR becomes more male-biased, regardless of whether or not there is stronger selection on mate-acquisition traits (Jennions et al. 2012).

Despite concerns about I_s, the BG itself remains an invaluable parameter when determining the likelihood that sexual selected traits will evolve. Selection to acquire mates, as well as to reject a potential mating (i.e. be choosy), is obviously related to the BG. All else being equal, the benefits of mating are greater when the BG is steeper. In practice, the BG has been used to predict successfully which sex experiences more intense mating competition. For example, in sex role-reversed species the BG appears to be steeper for females than males (e.g. Jones et al. 2000).

3.4.1.2 *Operational sex ratio*

The OSR is a concise summary of mate availability, and underpins a second common approach to quantifying sexual selection (Emlen and Oring 1977). The argument is made that mate availability determines the intensity of competition for mates, hence the likelihood that individuals invest in sexual selected traits. Following earlier criticism that formal models linking the OSR to the strength of sexual selection were absent (e.g. Arnold and Duvall 1994), several theoretical models have now quantified how the OSR affects mating systems (e.g. Clutton-Brock and Parker 1992; Parker and Simmons 1996; Kokko and Monaghan 2001) (Figure 3.2).

It seems intuitive that competitive investment is more rewarding when competition is intense, but this belief does not withstand scrutiny. If survival and mating success are the only fitness components, theory predicts that any trade-off between elevating mating success and reducing survival (e.g. investing in costly ornaments) yields the same optimal investment strategy regardless of whether males compete for few or many females (or, equivalently, face many or few male competitors) (Kokko et al. 2012). If, on the other hand, fitness is a composite of several components, then a strongly biased OSR can select against investing in competitive traits. For example, the relative fitness returns from paternal care will increase compared to the profitability of competing for mates, if securing new matings is difficult so that it only offers, on average, meagre fitness prospects (Kokko and Jennions 2008).

3.4.2. *How to unify Bateman gradients and the OSR*

Costly ornaments and competitive weaponry are more common on males than on females. Males also reject potential mates less often than females (Jennions and Kokko 2010). Because empirical studies often document male-biased OSR values and steeper male than female BGs, both measures appear to predict reality equally well (this applies to I_s too, although one can argue this is merely a result of its strong tendency to co-vary with the OSR). It is, however, incorrect to infer that the BG and OSR are equivalent measures of the same biological process. Their relationship is more subtle. The OSR and BG provide complementary information about the value of investing in sexually selected traits. The BG reveals how fitness rises with each additional mating, but neither how difficult it is to obtain a mating, nor the level of investment required to mate sooner. The OSR quantifies how difficult it is to mate, but not the extent to which fitness improves when this happens (Kokko et al. 2012). Individuals are only expected to invest in mate acquisition traits when they have difficulty acquiring a mating *and* would benefit if they overcame this difficulty. We need to work out under what conditions both factors co-occur.

Predictions based solely on the BG or OSR implicitly make assumptions about the value of the other parameter. For example, the claim that females have a shallow slope in Bateman's classic dataset assumes that females typically mate once, twice, or three times; if they often remained unmated, the relevant slope should actually be measured where it is steep (Figure 3.1). This point has been made before (Arnold 1994; Jones and Ratterman 2009), but the consequent challenge to understand the relationship between the origin of mating difficulties (OSR) and the benefits of mating (BG) was not taken up until recently.

To claim that the BG is flat requires that we state the likely number of matings per individual. Consequently, quantifying and understanding the origins of changes in the OSR is a necessity.

We need to be explicit about one problem with the BG: the use of 'numbers of mates' (*x*-axis of Figure 3.1) leaves undefined the temporal aspect of how frequently mating occurs. For this reason, it is often useful to derive the Bateman *differential*, which describes selection to shorten the time spent in the mating pool. When mate limitation extends the time that elapses until a mating occurs, the Bateman differential measures the increase in fitness if this time were to be shortened (Kokko et al. 2012). If the OSR is biased towards individuals of one sex, they spend longer on average in the mating pool. This tends to increase the Bateman differential, which is ultimately the causal link between the OSR and BG.

What, then, is the origin of a biased OSR? The number of individuals in the mating pool (Figure 3.3) is clearly influenced by how many are alive. The primary sex ratio and sex-specific mortality rates determine the adult sex ratio (ASR) and will therefore alter the OSR (Clutton-Brock and Parker 1992). Mortality is particularly important because sex differences in trait expression often affect juvenile mortality and adult lifespan, which can cause feedback that either amplifies or diminishes sex differences (Lehtonen and Kokko 2012).

The OSR is, however, not synonymous with the ASR because the OSR is additionally influenced by sex differences in parental investment. Whenever individuals providing parental care cannot simultaneously be in the mating pool (sometimes they can, e.g. nest-guarding male fish: Stiver and Alonzo 2009), sex differences in care affect the OSR. Large nuptial gifts such as an edible spermatophylax, or limitations on the rate of gamete production, can likewise delay the rate of return to the mating pool (Simmons 1992; Kvarnemo and Simmons 2013). These influences are usually conceptualized with an idealized life cycle where individuals are either 'in' or 'out' of the mating pool. The latter state is often called 'time out' (Clutton-Brock and Parker 1992) but perhaps more easily remembered as 'dry time' (out of the pool; Kokko et al. 2012). 'Dry' individuals count towards the ASR but not the OSR, as they are alive but not yet ready to mate.

Factors that make one sex provide more parental care than another are well understood (Queller 1997; Kokko and Jennions 2008, 2012; Chapter 12), including females' greater certainty of parentage than that of males (due to sperm competition; see Chapter 10). In insects, however, 'dry time' is rarely spent providing parental care (for exceptions see

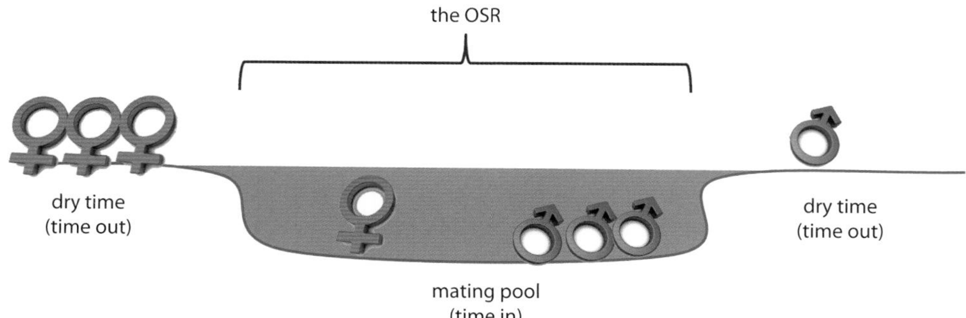

Figure 3.3 A conceptual view of mating systems. Individuals can be in the mating pool or outside spending 'dry time'. If the length of the dry time differs between the sexes, a sex bias in the number of individuals in the pool is likely. Males, in this example, have to wait far longer (on average) in the pool than do females.

Chapter 12). When compared with a typical vertebrate, the life history of many insects is more aligned with the concept of 'capital breeding' than 'income breeding'. Resources for gamete production are largely acquired during larval feeding, growth ceases at adulthood and the total budget available can be relatively fixed at eclosion/metamorphosis. This means that the 'dry time/pool time' framework is less insect-oriented than an entomologist might wish.

Nonetheless, the models' central features are robust to much taxon-specific detail. Consider capital breeding where females that have run out of eggs never return to the mating pool. The availability of females with fertilizable eggs then easily becomes lower than the availability of males ready to mate. The average time that an individual spends in the mating pool before it next mates *must* then be longer for males than females (assuming that each mating involves two opposite-sex individuals). There are potential ways to reduce the waiting time, including mate searching, moving to specific locations where potential mates meet (hill-topping, Alcock 1987), pheromone production (Cardé and Baker 1984), less choosiness upon mate encounter, and a greater ability to convince the other sex to mate (either through attractiveness or coercion). Such traits shorten the time to the next mating, but usually trade off with other fitness components. This trade-off is usually, but not necessarily, modelled as a reduction in lifespan. Net selection for a trait subject to this trade-off is more likely to be positive for the sex whose mean wait until the next mating was longer to start with (i.e. the majority sex in the OSR). A mathematical way to express the robustness of selection to shorten the time spent in the pool, against naturally selected costs, is the *scope for competitive investment* (SCI, Kokko et al. 2012).

Predictions from this mathematical approach highlight why neither the OSR nor BG alone are sufficient. Consider two scenarios A and B where male dry time is very short but they differ in the OSR (Figure 3.4). The wait time is shorter in case A than in case B, indicating a less biased OSR in A. In both cases A and B, a trait that greatly shortens the lifespan (by 30%) and slashes the wait time until the next mating will spread. The OSR difference is unrelated to the strength of sexual selection on males, which is the same in A and B. This is confirmed by a mathematical analysis: each additional mating is equally valuable whether males compete for few or many females (Kokko et al. 2012). Whether the wait is long or short, the only way to elevate fitness is to shorten the wait time even further. The importance of the OSR only reappears when an individual's dry time is longer. This is illustrated by case C (Figure 3.4). Here, shortening the wait time from an already insignificant part of the life cycle to something even shorter cannot compensate for a 30% shorter lifespan. In such circumstances, increased lifetime reproductive success will not select for greater effort directed towards mate acquisition. The link between sexual selection and the OSR emerges because a short wait time must be associated with an OSR value that predicts little difficulty in mating. It can now even become beneficial to reject some matings (Section 3.6).

3.5 Which sex searches, and why? Encounter rates with potential mates

3.5.1 *A counter-intuitive finding*

One factor that influences the frequency of mating encounters is the extent to which individuals actively search for mates. Much of mating system theory assumes that one sex need not search. Davies (1991) suggests that: 'For males, therefore, reproductive success is limited by access to females; the more females a male can mate with the greater his

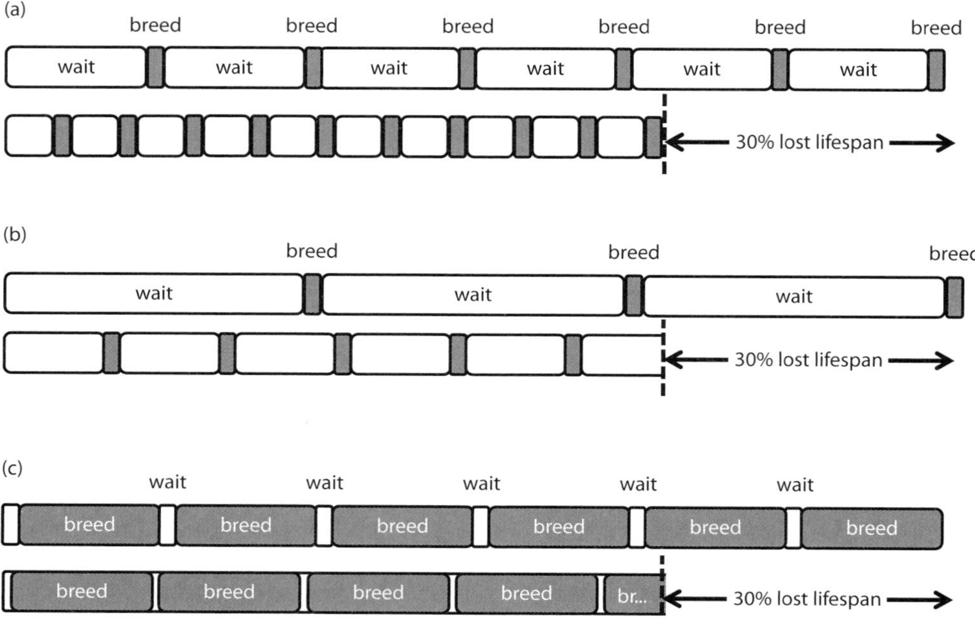

Figure 3.4 The operational sex ratio (OSR) matters, but not always. If a trait increases mating success greatly (visualized as shortening of the wait time) but causes a 30% reduction in lifespan, it will spread in cases (a) and (b) but not (c) where the dry time is long. In (c), having the trait leads to fewer completed breeding cycles than not having the trait.

reproductive success . . . thus it is males who tend to search for females'. Single-sex search effort is not necessarily true for insects, not only because sperm availability can be an issue (e.g. if sperm is packaged into large spermatophores which limits their availability; Simmons 1992) but because locating conspecifics can be unusually demanding. Consider wingless female moths. Despite the effort that males put into mate searching, females still invest in mate-attracting pheromones (Bengtsson and Löfstedt 2007). Even in species with milder sexual dimorphism, it can still be true that selection on males alone does not ensure that all females acquire a mate (Rhainds 2010).

Intuition might suggest that the mate-limiting sex should invest little and the mate-limited sex invests much into mate searching. Surprisingly, a major early model of mate searching did not reach this conclusion (Hammerstein and Parker 1987). As soon as selection favoured greater investment into mate searching by one sex, the other 'relaxed' and searched less. This led to mate searching by only one sex. Crucially, however, there was no bias towards males being more likely than females to become the mate-searching sex. This contrasts with Davies' quote that males are more likely to search because their fitness depends more heavily than that of females on how many mates they acquire (a BG asymmetry, Section 3.4.1.1).

The discrepancy has been resolved by more recent models that include additional biological details (Kokko and Wong 2007). There are several causal routes to sex-biased searching. First, male searching is predicted if the absolute cost of searching is greater for females (e.g. mobile females are preyed upon more often than mobile males due to the impediment of transporting a large clutch of eggs). There is, however, no obvious reason for this to be generally true. One could equally well argue that elaborate sexually selected traits make locomotion more costly for males.

A more convincing explanation for greater male searching is that polyandry leads to sperm competition (Chapter 10). Mate encounters are intrinsically random events: a female might never encounter a mate, but the probability of failure diminishes if females or males search actively. This will increase the mean number of mates that she encounters (*N*) during her mating window, and lead to sperm competition if more than one mating occurs. An increase in *N* has different consequences for each sex's reproductive success. It shifts the female BG towards the 'flat' part of the graph in Figure 3.1. Females usually gain only modest direct benefits by mating multiply, and the genetic benefits are even smaller (Slatyer et al. 2012). For males, too, each mating brings about smaller expected benefits if *N* is large. This is regardless of whether there is a strong pattern of sperm precedence or whether sperm are simply diluted in competition with other males' sperm. Males on average must sire a smaller number of offspring per mating if *N* is large. For males, however, the near linearity of the BG persists even if the fitness gain per mating is systematically reduced. The net result is a steeper male BG with a sex bias towards males searching more than females (Kokko and Wong 2007).

Females are, however, sometimes predicted to invest more than males into mate searching. The BG can be steeper for females than males when the risk of never mating is high (de Jong and Sabelis 1991; Rhainds 2010). Another exception to the prevalence of male searching occurs when the female BG remains steep because multiple mating is beneficial (e.g. due to nutritious nuptial gifts that enhance egg production; McCartney et al. 2012; Chapter 9). For example, in some bushcrickets (genus *Poecilimon*) males rely on females finding them. Males call to attract females, and do so rather than actively searching even though males benefit from multiple mating. Interestingly, male nuptial gifts are significantly larger in species where females rather than males actively mate search. This provides some evidence that the searching role switches to females when polyandry offers substantial direct benefits (McCartney et al. 2012).

3.5.2 *Mate searching: insect-specific themes*

Studies of insect mating systems have been especially important in the development of mate search theory for two reasons. First, insects offer excellent study systems for cases where both sexes must 'do something' for mating to occur. Hill-topping, for example, is quite a common way for the sexes to find each other, and it obviously involves movement of both sexes for the sole purpose of mating. The costs paid are presumably higher for the sex that remains at the mating site, as this precludes other activities such as feeding. The arguments developed above then predict that males will tend to be the majority sex in these swarms. Pheromone signalling by the less vagile sex to attract mobile mate searchers is another commonplace solution. If it is near impossible for mates to encounter each other unless they actively signal their presence, the prediction is a signalling system where only a tiny (perhaps unmeasurable) signalling cost paid by the less vagile sex combines with extensive searching by the other (Kokko and Wong 2007). Indeed, it has been difficult to document naturally selected costs of female pheromone production (e.g. Harari et al. 2011), although eavesdropping predators or parasitoids might cue in on their signals (Zuk and Kolluru 1998; Chapter 13).

Second, insect mating systems often feature rapid temporal shifts in mate availability. This can favour one sex emerging before the other sex at the start of the breeding season (Fagerström and Wiklund 1982; Iwasa et al. 1983; Morbey and Ydenberg 2001). Males usually emerge first (protandry), as often observed in solitary bees (Thornhill and Alcock 1983). This can be considered a 'temporal' mate-searching strategy. It occurs because of an asymmetry in mate availability that arises whenever females disappear from the breeding population sooner than males. Consider a situation where the distributions of male and female emergence times overlap perfectly. The ratio of males to females will increase over time, if males tend to remain alive and are available to mate for longer than females (i.e. females exit the mating pool once they mate). This lowers the mean male mating rate for males emerging later in the season, and favours males that emerge earlier. Each day-length shift towards earlier male emergence can offer an extra day of mating opportunities early in the season accompanied only by a *reduction* (not total loss) of opportunities later (Wiklund and Fagerström 1977; Bulmer 1983; Zonneveld and Metz 1991).

Intriguingly, the mean female and male emergence times can become so disparate that some late-emerging females fail to mate (Calabrese and Fagan 2004; Calabrese et al. 2008). This is an unexpected population outcome of selection on each male to find mates efficiently! Of course, there is then scope for a co-evolutionary response for earlier female emergence. This is seen in grasshoppers *Sphenarium purpurascens* where mating probability is highest for females with intermediate maturation time (stabilizing selection; del Castillo and Núnez-Farfán 2002). There is, however, a limit to how early females can emerge. The presence of a breeding season implies that there are naturally selected costs to emerging too early.

Finally, temporal shifts in the rules of mate searching might also occur over shorter time-scales. Ide and Kondoh (2000) showed that the best strategy for a male might be a switch from a dangerous activity (e.g. searching for females at emergence sites) to a less risky activity (e.g. lekking and waiting for females) when mate availability decreases and/or the reproductive value of potential mate falls below a critical threshold. Insect mating systems also offer many other exciting examples of alternative mating strategies (Chapter 7).

3.6 Encountering mates, deciding to mate and willingness to remate

We have already explained why males often (but not always) invest more than females into mate acquisition traits, including mate searching. What the theory in this field has not yet achieved, however, is a process-based approach that derives the distribution of mate numbers from encounter rates (which result from the search effort expended) together with a consideration of whether potential mates actually mate when they encounter each other. We will now outline current progress towards such a theory.

3.6.1 *A null model for sex differences in mating rates: encounter rate*

Even if one knows how individuals of each sex search for mates, we still have to understand why, and when, any mating system deviates from a null model. We follow Kokko and Mappes (2013) in suggesting that a useful null model is: *regardless of sex, behave identically in every mate encounter, and mate whenever the opportunity arises*. Even simple alternatives, such as monandry, or polyandry in which females mate in a certain proportion of encounters, invoke many more assumptions. Our minimal model avoids the need to specify which of many different possible proportions of mate encounters should lead to a mating (a necessary task if the null model included such an option). It also avoids any a-priori sex-specific assumptions. Monandry specifies that females, but not males, reject new matings once they have mated. This, of course, occurs in some species, but we want to explain and not simply assume evolved sex differences. If females show a clear behavioural change after the first mating ('reject new mates') we are faced with a likely adaptation that requires an evolutionary explanation. Our emphasis is different from the approach that has been prevalent for decades. Specifically, the oft-phrased question 'why do females mate multiply?' implies that female unreceptivity beyond the bare minimum is tacitly assumed to be the default, which in itself requires no explanation.

Our null model focuses attention on the question of whether one sex more often deviates from the null behaviour described in the model. Males or females often reject, or attempt to reject, some of the potential mates they encounter. Intuition suggests that individuals of the sex with the steeper BG (or stronger Bateman differential) should less often be rewarded by rejecting a potential mate. This is generally true but it is important to realize that the BG (Figure 3.1) excludes key aspects of mate choice. Choosing typically entails individual variation in the fitness gained from a given mating opportunity. For example, potential mates can vary in the size and quality of their nuptial gift, their relatedness to the potential partner, and choosy individuals can vary in how long they have been mate-deprived (Moore and Moore 2001). It is useful to compile a list of such factors that emphasizes key conceptual issues. We do this in Section 3.6.2.

3.6.2 *Reasons to disagree over whether to mate*

Sex differences are often expressed as sexual conflict over mating rates (e.g. Gavrilets 2000; Arnqvist and Rowe 2005; Maklakov et al. 2005; Gavrilets and Hayashi 2006). We consider it more illuminating to identify situations in which one sex but not the other benefits if a mating occurs ('conflict zone', Parker 1979, 2006; Hammerstein and Parker 1987). This is because the decision to mate—or to initiate or accept mate guarding—is rarely dependent solely on the time since the last mating. Considering other factors that affect mate acceptance and rejection will ultimately provide a more comprehensive theory than a focus on mating rates as evolving traits. This is not to say that mating rates are irrelevant. The mating rate is an emergent property of the mate-searching effort of each sex and their behaviours at each mate encounter depends on whether an individual's fitness is improved by mating (or pairing), or avoiding mating when a potential mate is encountered.

We can group the relevant factors affecting mate choice/rejection by asking five questions.

A. *Have I mated yet?* For females that store sperm, the fitness gains from mating are greatly diminished if she has already mated and can fertilize eggs. It is also possible that virgin and experienced males make different mating decisions (e.g. because their effect on fecundity affects their attractiveness as mates: Torres-Vila and Jennions 2005).

B. *How long until a potential mate will release gametes?* It might be in the interest of one or neither sex to initiate mate guarding if the female will not shortly fertilize her eggs (Parker 1974), or if the male will not release sperm. Whether pairing is beneficial for one or both sexes is also sensitive to costs. The cost of being guarded can include energetic costs imposed by the opposite sex (Watson et al. 1998) and greater predation risk (Arnqvist and Rowe 2005). Conflict over guarding is not inevitable, though, as it sometimes improves the guarded individual's survival (Cothran et al. 2012).

C. *Do I accept this particular mate?* Mate choice theory is well developed. It explains why an individual accepts one potential mate and rejects another (Chapters 2 and 8). The main reason is that potential mates vary in quality (due to the direct and/or genetic benefits they provide). Male quality affects the reproductive value of the set of offspring produced after a given mating. Kin-selected gains are relevant in some cases (Kokko and Ots 2006; Szulkin et al. 2013). Female mate choice is more likely to occur than male choice, as males often face a longer wait in the mating pool (see Section 3.4.2); thus, rejecting one mating would prolong males' absolute wait time more. Male mate choice is still possible if an elevated output after mating compensates for a lower mating (hence breeding) rate (Bonduriansky 2001; Jennions and Kokko 2010). Male choice is readily favoured if mating rates are the same whether a male is choosy or mates randomly, as occurs when several females are simultaneously available. Indeed, male choice is actually common if cryptic decisions to strategically vary ejaculate size based on female quality are classified as choice (Kelly and Jennions 2011).

D. *What are the costs of actually mating? Are there any compensatory gains?* Mating is often costly, but the magnitude and origin of these costs differ between the two sexes. For a male a major cost is sperm depletion if that reduces his ability to engage in future encounters with females (hence strategic adjustment of sperm allocation; Kelly and Jennions 2011; Chapter 10). For females, a major cost is the lost opportunity to mate with a better male; this might only become possible again after a prolonged spell away from the mating pool (e.g. when ovipositing). More generally, there are often immediate costs of mating. For example, mating can attract predators (e.g. Han and Jablonski 2010), and a winged insect's take-off ability can become severely compromised (Almbro and Kullberg 2009). In addition, there are also longer-term costs such as male seminal chemicals that lower female fitness (Arnqvist and Rowe 2005). As we shall see, these costs may be mitigated by a range of factors including replenishment of sperm supplies, material benefits, and the ability to use post-copulatory mechanisms to bias paternity to elevate offspring fitness (Chapter 9).

E. *How costly is it to resist a mating?* This is a factor with much potential to shape mating decisions. If females can cheaply resist a mating (e.g. by simply curling her abdomen away from a male) they should reject unwanted matings. If the costs of resisting are high, however, it might be cheaper to pay the actual costs of mating ('convenience polyandry': Thornhill and Alcock 1983). The cost of resistance can affect co-evolution between the sexes. For example, selection on male traits greatly depends on whether forced matings are successful (Eberhard 2009).

In principle, all the considerations listed above, even if we only provided a single-sex example, can apply to both sexes. However, the magnitude of their influence is rarely identical for both sexes. There is potential for sexual conflict whenever it is only in the interest of one individual to mate. These resultant systematic sex differences need to be explained. For example, when there is a risk of hybridization, males often appear to be less discriminatory than females about ensuring that a potential mate is a conspecific (e.g. Parker and Partridge 1998; Friberg et al. 2008; but see Espinedo et al. 2010). Finally, sexual conflict does not end with disagreement over mating. Even if both individuals agree to mate, conflict can persist over copulation duration, how much sperm is transferred, and many other traits (Arnqvist and Rowe 2005).

How do the factors discussed in A–E above combine to create a mating system? Despite sex differences, it can still be in the interest of both sexes to mate when they encounter each other. For example, when population density is low, one expects a mating system where neither sex attempts to evade the other. Polyandry and polygyny might then co-occur, as individuals who fortuitously happen to meet several potential mates should not reject any. When population density is higher, polyandry will intensify sperm competition. This can select for males competing with each other in ways that increase female mating costs, leading to greater sexual conflict during mate encounters (Härdling and Kaitala 2005). Such conflict can, in turn, lead to the evolution of a diverse set of traits from female mating resistance to mating plugs that prevent females from remating (Fromhage 2012; Chapter 10), even if she would benefit from polyandry (Baer et al. 2001).

The quest to identify ecological factors that affect mate numbers initiated by Emlen and Oring (1977) has not yet been finished. Ecological factors, for example, population density (its mean, and variation) do influence the rate of mate encounters and therefore selection on individuals. However, it appears that we currently have a better understanding of sex differences per se than we do of the precise ecological circumstances that affect the costs and benefits of each behaviour relevant for generating the various mating systems we observe.

3.7 Conclusions

The issues we have covered provide insights into the sex differences that underlie mating systems. How often individuals of each sex encounter potential mates requires information on population density, on which sex searches (Section 3.5), and on the adult and operational sex ratio (Section 3.4). The encounter rate with mates then has to be augmented with an understanding of why an individual might not mate every time it encounters a potential mate (Section 3.6), which can involve 'conflict zones'. The value of elevating the mating rate can therefore differ between the sexes.

We make two recommendations. First, start with a null hypothesis involving no sex bias. We believe a null model where every encounter results in a mating is appropriate (Section 3.6.1). It reduces the risk of easy pitfalls. For example, one might be tempted to argue that variation in female fecundity should always select for male mate choice. This is false. The null model reminds us that the simplest solution for males could be to mate with everyone. If matings are rapid affairs (i.e. no opportunity costs of missing another mating) and sperm depletion is not imminent, a male's fitness is maximized if he accepts every female he encounters. At the extreme this can even be true even when females vary in their propensity to be sexually cannibalistic, if a male's prospects of encountering another willing mate are sufficiently meagre (Barry and Kokko 2010). The null model also reminds us that variation in the number of mates might be an emergent property of selection on females to ensure at least one successful mating.

Second, it is necessary to integrate two widely used approaches to quantifying the strength of sexual selection. Bateman gradients tell us how much fitness is gained by each successive mating for each sex. The caveat is that the relevant slope depends on the likely range of mates, so we have to think about probable encounter rates. These depend on the OSR. Together the OSR and Bateman gradient help us understand the likelihood that an individual can afford to be choosy or the likelihood that selection favours greater investment into sexually selected traits even if they impose substantial fitness costs (e.g. a reduced lifespan).

As with any simplifying approach, we have omitted many system-specific details. For example, the OSR is only indirectly connected to post-copulatory sexual selection. Female multiple mating is likely to depend on male availability (which increases with a male-biased OSR), but an exactly linear relationship between the number of mates and selection on post-copulatory traits is unlikely. Consequently, how the number of mates per female alters selection on ejaculatory traits will depend on details of the mating system (Zeh and Zeh 1994), including trade-offs (Simmons and Emlen 2006) between male investments into traits that are favoured by pre-copulatory and post-copulatory selection (Shuster 2009).

We have mainly focused on factors that determine, in principle, whether individuals end up with few or many mates over their lifetime. Male and female mating strategies interact with ecology and contingent life-history trade-offs (i.e. changes in the cost and benefits of specific behaviours) to produce this number. For us, the major intellectual challenge when studying mating systems is found less in deriving the actual distribution of the number of matings than in exploring the evolutionary strategies that develop due to eco-evolutionary feedback between males, females and the ecological and social circumstances they experience that select for different traits in each sex.

CHAPTER 4

The genetics of insect mating systems

Michael G. Ritchie and Roger K. Butlin

4.1 Introduction

Thornhill and Alcock (1983) had a great impact on studies of the evolution of insect mating behaviour, raising many questions about the function, ecology, and mechanisms involved. However, a genetic perspective on insect mating systems was more or less lacking, though clearly many of the questions lent themselves to a genetic approach. Probably the only explicit discussion of genetics in Thornhill and Alcock (1983) concerned speculation on 'good-genes' models of sexual selection. Now, insect mating behaviour is very much subject to the full panoply of 'post-genomic' analyses of its causation, development, and function. For instance, many aspects of *Drosophila* mating systems are used as classic phenotypes for genetic studies, providing model behaviours, especially for neurogenetic studies of behaviour. For many insect systems, we now have genome-wide association studies of behavioural variation, gene expression analysis, numerous studies of neural organization, as well as an increasing appreciation of the multivariate and interaction-driven nature of much of the causation of complex behaviour. Single genes with large effects on insect mating behaviour have been identified and are among the best-studied loci in non-human animals, and genes with large effects on learning and the social organization of behaviour are being identified, which might give insight into between-species patterns of variation in social organization. Finally, gene families are being identified that now allow comparative analyses of genomic correlates of insect mating system variation.

It is interesting to reflect on the extent to which genetic analyses have addressed the questions that stimulated the imagination of the original readers of Thornhill and Alcock (1983) or whether they have simply provided new questions to explore. Adopting a genetic approach to the analysis of behaviour (or any 'complex' phenotype) has often been questioned, perhaps most infamously around the same time-period as the original Thornhill and Alcock book (Grafen 1984). If the aim of a genetics approach is seen solely or even primarily in terms of identifying the loci responsible for most phenotypic variation in a trait (or 'mapping' the relationship between genotype and phenotype), there is legitimacy to these criticisms. The resolution available for mapping this relationship is poor, certainly for non-model species but probably also model species (e.g. Rockman 2012, Slate 2013, Travisano and Shaw 2013) (an eye-watering statistic is that the global population of humans is probably too small to accurately map many loci of small effect for human traits

The Evolution of Insect Mating Systems. Edited by David M. Shuker and Leigh W. Simmons.
© The Royal Entomological Society 2014. Published 2014 by Oxford University Press.

even using genome-wide approaches). However, evolutionary genetics is not only about finding genes, but also for identifying or understanding processes. For example, gene expression studies are telling us about physiological interactions and the extent of sexual antagonism. Comparative quantitative genetics can tell us whether genes of large effect are under selection and if there is continuity in their influence from within-species selection to differences between species (which would suggest that these genes do not represent 'fool's gold' in the search for evolutionarily important loci; Rockman 2012). In this chapter, we will not attempt to provide a 'how to' guide to behaviour genetics for insects, but rather choose some areas of research and relevant study systems that illustrate the opportunities and possibilities for behaviour genetics of insects in the modern era, and how these are addressing old questions and raising new ones. By describing these examples, we hope to illustrate the potential for a genetic perspective to enrich the understanding of mating systems in insects (and other organisms).

4.1.1 *Original questions*

Over the last 30 years the most consistent debates about the genetics of insect mating behaviour have concerned the extent to which behavioural traits are heritable and how genetic variability in natural populations is maintained. More specifically, focal questions have been why heritability should exist for male courtship traits, despite apparent directional selection from female preferences, and whether signals, preferences, and fitness should co-vary (see also Chapter 8). It is probably fair to say that whether insect behaviour is heritable has been answered; heritable genetic variation can be found for most things, including behaviour (Lynch and Walsh 1998), though evidence for a genetic correlation between behaviour and measures of fitness is less clear (Prokop et al. 2012). The question as to why such variation should be plentiful probably has multiple answers, although not all necessarily make good-genes models of sexual selection more likely. For example, genotype-by-environment (G × E) interactions, whereby signal expression alters with the environment, may be widespread (Ingleby et al. 2010a). To give two examples, wax moth song variation is influenced by rearing temperature (Greenfield et al. 2012), and the genetic architecture of both song and contact pheromones of cactophilic flies depends upon the cactus in which the larval flies develop (Etges et al. 2010). As such, global fixation of advantageous alleles in mixed populations is unlikely. However, by the same token this phenotypic variation does not necessarily provide reliable information to females about genetic quality (Higginson and Reader 2009, Kokko and Heubel 2008); how does a female 'know' whether a male has an advantageous genotype or whether he just happened to develop in a particular environment? From an anthropomorphic good-genes perspective, we might say that all she can tell is that he is in good condition, and hopes that some of that reflects genetic rather than environmental variation. That a female's preferences might be finely tuned to the developmental experiences of signalling males is perhaps asking a lot of her perceptual ability, but is certainly not impossible. As discussed in Chapter 8, the complementary nature of G × E interactions on the evolution of male traits and female mate preferences is certainly an important area for future research.

One potential resolution to the 'lek paradox' which has achieved wide acceptance since Thornhill and Alcock (1983) is condition dependence of male mating traits (Chapters 2 and 8). If overall condition influences the expression of male traits or performance, the genetic variation for such traits might 'capture' genetic variation for condition (Rowe and Houle 1996). This is potentially a powerful genetic process, which might provide the raw fuel for adaptive sexual selection to flourish, and there is supporting evidence that condition dependence may contribute to the heritability of insect sexual signals (Tomkins et al. 2004). Examples considered below will show how new genomic approaches to insect mating systems suggest that condition dependence may be important and that we can begin to understand some of the regulatory processes involved.

The extent of any genetic covariance between male signalling traits and female preferences is another area of considerable historical importance in the study of insect mating systems. Initially this was highlighted as an important empirical issue by theoretical studies of Fisher's 'runaway process'; Fisher's process is more likely if strong covariance exists (Lande 1981) and measuring the covariance became a major focus of research effort (e.g. Bakker 1993, Qvarnstrom et al. 2006). More recently it has been appreciated that this covariance can arise under multiple forms of sexual selection (and even that runaway sexual selection does not strictly require a genetic covariance, e.g. Bailey and Moore 2012). However, it is still important to understand the extent of the covariance because it can contribute to faster co-evolution of signals and preferences, and could influence sexual-selection-driven speciation in the absence of an additional influence of environmental selection on traits (Prum 2010). A more specific, related question is the genetic nature of any such covariance; the relative importance of linkage disequilibrium, physical linkage or pleiotropy in generating co-evolution is an important area with tantalizing new data (Sections 4.2 and 4.5).

The contribution of variation in insect mating systems to insect biodiversity is also an important area. Thornhill and Alcock (1983) clearly saw an intimate association between sexual selection and speciation. Indeed, their book was published when people were beginning to question the importance of 'species recognition' as an evolutionary process, in favour of sexual selection inadvertently driving speciation—a debate which continues today (Mendelson and Shaw 2012). That sexual selection contributes to speciation seems an obvious conclusion given that so many insect systems differ primarily in sexually selected traits such as song, pheromones, genitalia, or extravagant male morphologies. However, empirical evidence of greater speciation rates in, say, more polyandrous clades of insect, is ambiguous (e.g. Gage et al. 2002, Kraaijeveld et al. 2011). Comparative genomics is an area which is beginning to contribute much to this debate, and may provide decisive data: if we can identify the most rapidly evolving genes or gene families in insects, we can examine whether they are under natural or sexual selection and their contributions to reproductive isolation.

The ability to manipulate gene expression in many systems, especially *Drosophila*, allows experimental testing of the role of specific loci in behaviour, and the relative ease with which genome-wide expression and sequence data can be obtained in many organisms means that comparative analysis of mating system genomics and evolution across insects is now quite feasible. In the last 30 years the genetics of insect mating systems has gone from a relatively minor footnote to one of the most vibrant areas of research in the field, and we hope the examples described here give a flavour of this excitement.

4.2 Quantitative trait loci, linkage, and pleiotropy

Quantitative trait locus (QTL) mapping provides an ability to detect the chromosomal distribution and effect sizes of genes influencing complex traits, provided that crosses can be made between divergent (preferably inbred) individuals and that sufficient genetic markers are available for informative co-segregation analysis (Mackay 2001). Many early studies provided useful results despite limited marker availability, but, in the era of next-generation sequencing, markers are much more easily generated in most insects. For instance, sexual isolation in flies and the behavioural traits involved are now relatively well studied (Mackay et al. 2005, Arbuthnott 2009). While most suggest that polygenic inheritance of insect behaviour is common, genes of large effect can be implicated (Boake et al. 2002, Orr 2001). In some areas of insect mating behaviour, studies more consistently identify genes of large effect—e.g. for pheromonal variation—so we highlight such studies later. Some robust patterns are becoming clear, including a disproportionate role of sex chromosomes (Qvarnstrom and Bailey 2009), at least for some groups of insects. Importantly, QTLs identified for sexual isolation or associated traits between pairs of *Drosophila* usually have different chromosomal locations, even between closely related species (Gleason and Ritchie 2004, Moehring et al. 2006). Although few studies allow really good comparisons, one pattern emerging (at least for studies of courtship song and isolation in flies) is that the chromosomal locations implicated for within-species variation are different from those influencing between-species differences (reviewed in Arbuthnott 2009), supporting the idea that segregating genetic variation within populations may be different in kind from genes fixed between species, perhaps under positive selection. More generally, studies suggest that different types of genetic mutation may be involved in short- and long-term adaptive changes in populations and species, meaning that different types of genetic substitutions may be favoured (Stern and Orgogozo 2008).

Although many studies do not support a common chromosomal location for genes influencing male traits and female preferences, some of the more intriguing studies are those suggesting that strong linkage or potentially pleiotropy (i.e. one gene influencing more than one trait) may underlie rapid recent divergence. For example, *Laupala* is an endemic cricket that has undergone an extremely rapid adaptive radiation in Hawaii, with numerous species on each island (Mendelson and Shaw 2005). The geological history of the islands allowed Mendelson and Shaw (2005) to calculate that the radiation is one of the fastest known, rivalling or exceeding more celebrated radiations such as that of Hawaiian *Drosophila*. Sexual communication lies at the heart of this radiation as species are characterized by distinctive songs and mate preferences but are not ecologically distinct (see also Chapter 8). A series of studies has used QTL approaches to examine the co-segregation of song and preference in crosses between species (e.g. Shaw and Lesnick 2009). A particularly interesting observation is that experimental selective introgression of song QTLs into the genetic background of the other species is accompanied by changes in preference, suggesting that these regions contain genes influencing preference for the same song traits (Wiley et al. 2012). What does this mean? The genes could be pleiotropic, for example the QTL regions could contain genes exerting a common effect on neurological pattern generation and recognition. This would be an example of genetic coupling, a concept introduced in the 1970s (Hoy et al. 1977) for which empirical evidence is generally weak. Alternatively, physical genetic linkage could exist between genes with independent effects on traits and preferences, perhaps in a region of reduced genetic recombination (Butlin and Ritchie 1989, Boake 1991).

Laupala provides an exceptional example of QTL analysis of traits intimately involved in the sexual communication system of a non-model insect. Another example of such a study is provided by the butterfly *Heliconius*, in which the aposematic warning coloration of female wings is a target of mate choice by males. QTL analysis places genes for both the warning coloration of females and male mate choice of those females in the same QTL regions (Kronforst et al. 2006, Merrill et al. 2011). This implies an intimate genetic association between traits, preferences, and natural selection in this species-rich group. However, whether either of these examples represents pleiotropy or clustering of loci is currently impossible to tell. Precise testing requires a system in which individual loci can be identified and manipulated, which is not yet possible for many non-model organisms. However, *desat1* in *D. melanogaster* does provide an example of such a study (Marcillac et al. 2005). This desaturase enzyme influences the profiles of cuticular hydrocarbons (CHCs), some of which function as pheromones. Inserting a transposon into *desat1* altered the CHC profiles of both sexes, changing the ratio of saturated to unsaturated compounds, including the main pheromones 7,11-heptacosadiene and 7-tricosene. Male flies normally direct courtship to appropriate (same species) females based largely on their CHCs. However, the *desat1* mutant males cannot discriminate the sexes and so court females inappropriately, at least when visual cues are unavailable. Detailed expression analyses (Bousquet et al. 2012) show that alternative transcripts of the gene are expressed extensively throughout a range of tissues. One responsible for the CHC changes is expressed in the CHC-producing oenocytes, with another causing changes in male gender recognition in the antennal lobes (including neurons implicated in pheromone perception), other neural tissues and sensilla of the antennae. Hence the ability to produce a pheromone and the ability to detect the pheromone are influenced by a single gene. It would be extremely interesting to test whether this gene underlies co-ordinated changes in the signal and preference components of the communication system by making comparisons across species. There is an exciting possibility that the gene's effects on both components might facilitate rapid signal-preference co-evolution because a single mutation might allow coupled rather than successive changes, with no need for the establishment of linkage disequilibrium.

QTL studies are a painstaking approach to examining the genetics of both model and non-model insects but the potential rewards for such studies are high. Marker development is increasingly easy and an important advance in QTL analysis will be to look at the standing genetic variation within natural populations. Greater progress is being made here outside of the insects, perhaps because long-term cross-generational monitoring is more straightforward (although they are possible, as in the exciting study of field crickets, *Gryllus campestris*, by Bretman et al. 2011). However, the statistical issues are challenging and studies with insufficient loci may overestimate the effect of individual QTL loci due to integrating across large areas of the genome (Slate 2013).

4.3 Major genes

Many genes have been identified which cause major effects on insect mating behaviour (Boake et al. 2002), and some of these show consistent effects between species, implying that they provide good candidate genes for further study (and are consistently important in trait evolution). The gene *fruitless* is arguably the best-known example (see also Chapter 8). *fru* is a transcription factor which, in interaction with *doublesex* and *transformer*, establishes many of the sexually dimorphic behaviours of *Drosophila* fruit flies. The mechanism is not well understood yet, but it is expressed in neurons of the central nervous system and manipulation of sex-specific splicing patterns can change gender-typical behaviours, for example producing a female who courts like a male (Demir and Dickson 2005). Many genes showing differential expression between the sexes, or in response to sexual interactions, interact with the sex-determining pathway to varying extents. *fru* in particular seems to establish the gender identity of the central nervous system in *Drosophila* (Demir and Dickson 2005, Billeter et al. 2006, Rideout et al. 2007) and many sexually dimorphic behaviours are influenced by these loci (Neville and Goodwin 2012). Several studies of other insects now demonstrate that this role is probably conserved, and *fru* has been implicated in behaviour of divergent species, for example knocking down *fru* reduces male mating success in locusts (Boerjan et al. 2011) and houseflies (Meier et al. 2013). Hence these sex-determining loci are prime examples of candidate genes identified in *Drosophila* that are now being used to understand differences in sexual behaviour in a wider range of insects (Fitzpatrick et al. 2005).

Clock genes provide another classic example of well-studied individual loci of large effect in a range of insects. Clock genes also interact with many 'downstream' loci to influence a wide range of behaviours which rely on temporal patterning at a variety of levels, and are discussed in Section 4.5.

4.3.1 *Major gene effects in pheromone systems*

Pheromonal signalling is a key part of the mating system of many insects. It is particularly highly developed in the Lepidoptera where release of pheromones by females for the long-range attraction of males is widespread. Attraction is highly species specific and this is achieved through tightly controlled mixtures of a relatively small number of compounds, mostly unsaturated long-chain fatty alcohols, acetates, and aldehydes. On the face of it, tight co-ordination between the pheromone blend released by females and the sensitivity and preference of males is required for this system to work. Thus pheromone evolution provides a particularly clear example of the apparent conflict between the expectation of stabilizing selection within species and the evidence for divergent signal–response co-evolution in a highly diverse group of insects (Smadja and Butlin 2009). Unravelling the genetic basis of pheromone and response variation within and between species will reveal the evolutionary steps involved in this divergence and so may help to resolve this paradox.

The European corn borer, *Ostrinia nubilalis*, provides a very informative example. Two races of this species use pheromones comprising contrasting mixtures of two isomers: E and Z forms of doubly unsaturated 14-carbon acetate in 98:2 (E race) or 3:97 (Z race) proportions (Cardé et al. 1978). Males respond specifically to these blends although their response windows tend to be wider than the variation in blend. Hybridization is rare in nature but crosses can be made easily in the laboratory. Classic work in the 1980s showed that the blend difference is controlled by a single autosomal locus. In males, antennal sensitivity is controlled by a second autosomal locus and male behavioural response is determined by a sex-linked locus (Roelofs et al. 1987). Since this finding, the search has been on to identify the loci.

Pheromone production starts from fatty-acyl-CoA precursors and involves a small number of steps catalysed by a defined array of enzymes, often specifically expressed in pheromone glands. Changes in chain length depend on β-oxidases; introduction of double bonds is carried out by desaturases; and then conversion of the active group is catalysed by fatty-acyl-CoA-reductases to produce alcohols, which may be further modified to acetates or aldehydes. Sequencing of the *Bombyx mori* genome made access to these enzymes much more straightforward, making a candidate gene approach very attractive. A key goal was to find the step in the process that is responsible for the control of the proportions of compounds in the pheromone blend. For *Ostrinia*, this question has been answered in a very elegant study (Lassance et al. 2010). Lassance and co-workers showed that the gene for a fatty-acyl-CoA-reductase that is expressed in the pheromone gland (*pgFAR*) maps to the autosomal locus responsible for the E/Z blend difference. This is consistent with the observation that race-specific ratios of isomers are not seen after desaturation but do appear after the reduction step. Moreover, they were able to express the two alleles (*pgFAR-E* and *pgFAR-Z*) in yeast and show that they differed strongly in affinity for the E and Z isomers of the precursors: *pgFAR-E* produced almost exclusively the E alcohol and *pgFAR-Z* the Z alcohol.

This important achievement has been followed up with an interspecific study (Lassance et al. 2013) using the European corn borer (ECB) and its close relatives, including the Asian corn borer (ACB), *Ostrinia furnacalis*. ACB mainly uses acetates with double bonds in the 12 and 14 positions, rather than 11 and 14 as in ECB. This difference is also mediated by *pgFAR*. Phylogenetic analysis identified strong signals of positive selection on multiple non-synonymous substitutions separating the two species. This leads to the very interesting question of whether the large, single-gene effect has been built up by multiple substitutions of small effect or whether one substitution is crucial, with others perhaps acting as modifiers. This issue was tackled by site-directed mutagenesis of ACB *pgFAR* at eight different non-synonymous sites, one at a time. The modified enzymes were then expressed in yeast so that their substrate specificity could be monitored. Remarkably, just one of these substitutions caused a change in specificity to a pattern very similar to ECB, with all other substitutions having relatively minor effects (Figure 4.1). The reverse modification of this position makes an ECB enzyme behave very like an ACB one (Figure 4.1 inset).

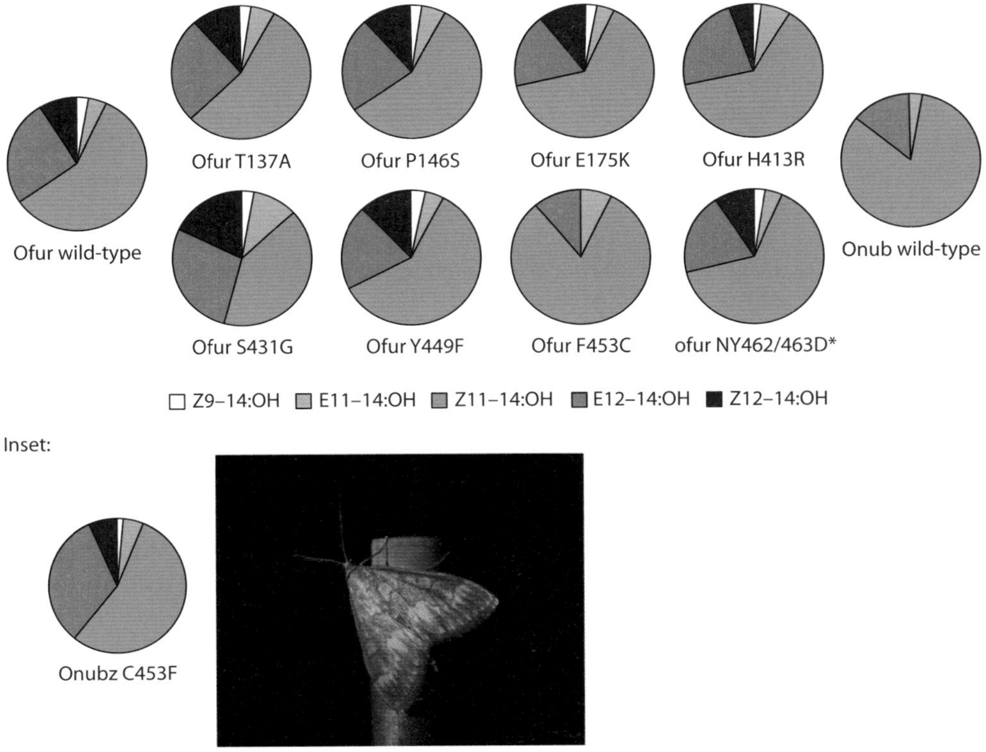

Figure 4.1 Ratios of alcohols produced by wild-type and modified *pgFAR* when expressed in yeast. The F453C substitution alters the Asian corn borer (Ofur) pattern to be close to that of European corn borer (OnubZ). Inset: The C453F modification of the ECB enzyme causes the reverse transformation. Reproduced with permission from Lassance et al. (2013). Photo: European corn borer on a pheromone-releasing rubber septum during response testing (courtesy of Callie Musto and Charlie Linn).

Adding additional modifications at other sites, along with the major effect substitution, made only slight differences to the ratios of alcohols produced. The historical sequence of events in the moths is unknown at present but these data strongly suggest that one major substitution was fixed by positive selection. Understanding the source of that selection remains an open question. Knowing the genes involved on the male side would certainly help to complete this fascinating story (see Section 4.6 for some progress in this direction).

One possible evolutionary explanation is known as 'asymmetrical tracking' (Bengtsson and Lofstedt 2007). The essence of this idea lies in the different strengths of selection on males and females. For example, weaker selection on females might allow new components to be added to the blend that initially have little impact on male response. Later, males might evolve the ability to use these components as cues, thus bringing them under stronger selection. In the parasitoid wasp, *Nasonia vitripennis*, males attract females using a simple, three-component pheromone blend but related species use only two of these compounds. The difference is explained by a single QTL which overlies a small group of related candidate genes (Niehuis et al. 2013). Behavioural tests suggest that female response was not influenced by the additional compound when it first appeared, even though it now enhances mate attraction in *N. vitripennis*. Thus, it seems that response—in this case by the female—evolved after a new signalling element became available. The new element—in this case an additional compound—may have spread by genetic drift or perhaps under selection for a function unrelated to mate attraction. The same principles could apply to any signal–response system.

Variation in pheromone blend and response does not always have a simple genetic basis. For example, continuous variation in the amount of an acetate component, which is crucial for interspecific discrimination, occurs within and among populations of the moth *Heliothis subflexa*. A QTL of large effect explains about 40% of this variation and also contributes to the interspecific blend difference (Groot et al. 2013). In this case, candidate loci do not map to the QTL position and the authors speculate that a transcription factor may be present at the QTL location that is involved in regulation of biosynthetic enzymes. Another unexpected feature of pheromone variation in this species is adaptive plasticity. Females of *H. subflexa* exposed early in life to pheromones of *H. virescens* subsequently produce a greater amount of the acetate component that inhibits response of *H. virescens* males (Groot et al. 2010). It is possible that such plastic responses precede the evolution of fixed genetic differences in blend.

Where loci of large effect are involved, pheromones provide excellent confirmation of the potential to go from genetic mapping, via good candidate gene arguments, to identification of loci at the molecular level and then functional studies. Understanding the selection pressures involved typically lags behind this genetic dissection, but it is a challenge that has been taken up in several groups and so it is an area where exciting results are likely in the near future.

4.4 Expression analyses

There has been a recent expansion of studies in genome-wide gene expression analysis, at least for *Drosophila*. Whole genome microarrays have allowed fairly straightforward analysis of gene expression changes involved with mating behaviour (though these are now being superseded by tissue-specific RNA sequencing). Because behaviour is a very flexible phenotype, analyses of gene expression changes during behaviour seem particularly appropriate for identifying key genes or gene networks involved in the control of mating behaviour. One would expect these to be upregulated in preparation for or during courtship, for example. Following some key studies (e.g. Chapman et al. 1995, Rice 1996), it has been widely appreciated that mating interactions represent the major arena within which many aspects of sexually antagonistic gene action may be played out. Although physical damage inflicted on partners during copulation is an important aspect of sexual conflict (e.g. sexual cannibalism or male genital damage to female reproductive tracts; see Arnqvist and Rowe 2005), the interplay between males and females is also due to biomolecular as well as behavioural interactions (Figure 4.2). Studies of detrimental effects of mating on female fitness led to the identification of a range of accessory proteins (ACPs) passed to females in semen, including the well-studied sex peptide of *D. melanogaster* (Liu and Kubli 2003, Wolfner 1997). Seminal proteins are among the fastest evolving genes in the fly genome and have a range of physiological effects on females, which sometimes seem to have evolved to favour males through functions in sperm competition or the manipulation of female fecundity, sexual attractiveness or libido (Chapman 2001, Wolfner 2002) (see also Chapter 10). Indeed, if females are experimentally prevented from co-evolving with males with whom they interact (by the use of non-recombining, 'balancer' chromosomes that allow the inheritance of individual chromosomes to be manipulated), male damage to females can increase (Rice, 1996). Normally females must evolve a response to sexually antagonistic gene expression in males, so the precise details of genetic interactions between males and females during and after copulation are potentially exquisitely complex. Gene expression studies are starting to reveal the nature of these interactions in *Drosophila*.

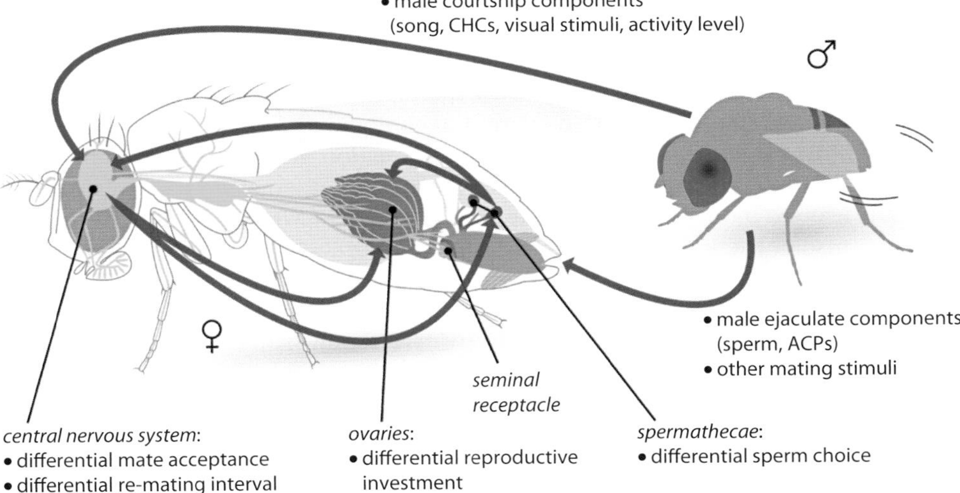

Figure 4.2 Courtship and copulatory interactions between male and female flies occur on many levels, and include behavioural effects, effects of copulation, and sperm transfer and storage. Bailey et al. defined mate choice as any bias in male reproductive success caused by female responses (active or passive) to phenotypic differences between males. Hence any labelled female tissues are possible locations for mechanisms of mate choice (bullet points). The arrows represent routes by which mate choice may occur. Active female choice is represented by arrows starting or finishing at the female brain; passive female choice by any arrows that do not involve the female brain. Arrows between tissues within the female represent neuronal and/or hormonal responses. ACPs, accessory gland proteins; CHCs, cuticular hydrocarbons. Used with permission from Bailey et al. (2011).

4.4.1. *Females*

As soon as females are courted by males, some genes change expression levels. Lawniczak and Begun (2004) used whole body analysis of females and found that up to 20 genes with a range of functions showed rapid expression changes following courtship (independent of mating). Functional analysis showed that upregulation of immune function genes was involved, including genes expressed in the sperm storage organs. This could be to protect against sexually transmitted infection, or to facilitate long-term sperm storage. Immonen and Ritchie (2012) found that a range of genes were upregulated in female heads in response to courtship song (using playback of artificial song, so independent of the presence of any males); song perception influenced genes involved in cognition and neuropeptide signalling activity. Interestingly, genes involved in the antenna and in olfaction were upregulated, and the strongest effects were in a gene family involved in immunity. It seems likely that females upregulate the expression of genes important in detecting and possibly assessing males during courtship and anticipate important mating responses, such as the immune effects.

Responses to mating in females are better studied, with changes in many more genes identified and a wide range of data now available. A precise comparison of numbers of genes detected in different studies is probably meaningless due to variation in resolution, statistical criteria, and experimental methodologies, but some trends are probably reliable. Immune function genes are clearly important, and Lawniczak and Begun (2004) point out how similar the genomic responses are to mating and to infection, as are genes influencing female reproductive output. In an attempt to distinguish female gene expression changes due to male effects, McGraw et al. (2004) compared virgin females to those mated to normal males or mutant males who lack either sperm or sperm and accessory gland proteins. About 13% of the genome (~1,500 genes) changed expression in response to mating, though the levels of change were modest. Surprisingly, perhaps, only around 160 loci were influenced by ACPs, around 500 by sperm, and the majority, around 1,000, by other aspects of mating (courtship, copulation or compounds from the ejaculatory bulb). Those responding to ACPs included the immune function genes (these were the largest changes in expression detected). This possibly suggests that female counter-responses to ACPs were detected more than male manipulation effects. McGraw et al. (2008) examined the time course of these changes over 24 h—more than 2,000 transcripts showed rapid changes (1–3 h). With increasing time post-mating, expression-level changes increased, but the number of genes showing changes declined. ACP-induced changes are perhaps greater after about 8 h. Individual ACPs varied widely in how many genes they influenced soon after mating, and intriguingly some of the common targets were affected in different ways, so even individual ACPs might have antagonistic interactions.

Mack et al. (2006) concentrated on gene and proteomic changes in the lower female reproductive tract, where gene expression associated with sperm storage and sperm competition might be expected to be concentrated. Around only 6 h post mating seemed to be a key period, when greatest upregulation in gene expression changes occur (Figure 4.3). They suggest that this demonstrates a two-stage response to mating, from early male-induced changes (via ACPs) to longer-term sperm storage and female reproduction effects. Gioti et al. (2012) examined responses in the female head or body post mating to males with or without sex peptide (see also Chapman et al. 2003b)—one of the best-studied male ejaculatory components with clear antagonistic effects—and identified an extraordinarily wide range of physiological effects. Dalton et al. (2010) also focused on temporal changes in the head of female flies and concluded that male accessory proteins are responsible for most

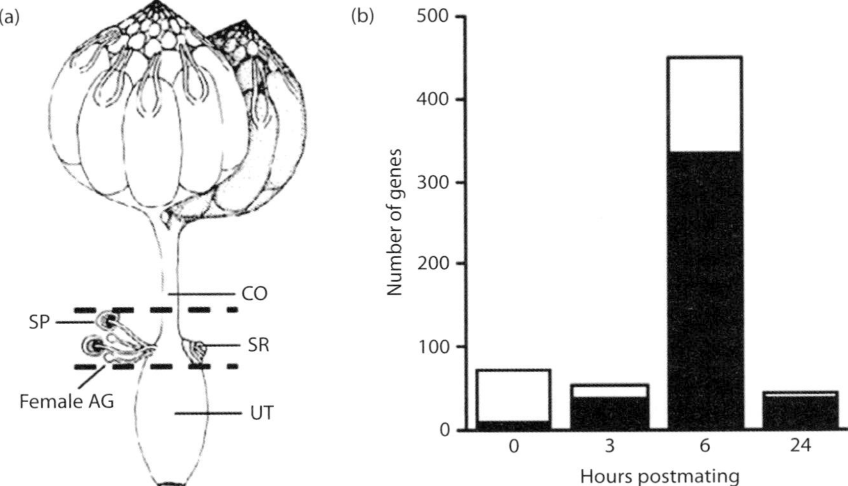

Figure 4.3 Post-mating gene expression changes in the lower reproductive tract of *Drosophila melanogaster* females. (a) UT, uterus; SR, seminal receptacle; SP, spermatheca; female AG, female accessory gland; CO, common oviduct. (b) Whole bars indicate the number of differentially expressed genes ($P < 0.0001$) at each time point in mated three-day-old females compared with unmated three-day-old females. Black shading in each bar represents the proportion of genes upregulated in mated females; white portion of bar represents the proportion of genes downregulated in mated females. Used with permission from Mack et al. (2006).

of the changes in female brains, which primarily involved changes in metabolic pathways. Interestingly, an immediate response involved suppression of some immune functions. Studies focused on the head have revealed (perhaps surprisingly) that gene expression changes in the fat body surrounding the brain (including juvenile hormone esterase and *female-specific independent of transformer*) may be as important as changes in the neural tissue itself, perhaps suggesting that hormonal changes are as significant as gene expression within the brain. This is seen in both sexes (Dalton et al. 2010, Ellis and Carney 2011). Some further aspects of the hormonal control of behaviour are introduced in Chapter 5.

4.4.2 *Males*

It is clear that the social environment (including sexual competitors) has a large effect on the expression and heritability of behavioural traits (Moore et al. 1997, 2002) and that apparently simple organisms such as male flies show subtle but behaviourally significant responses to the social environment (e.g. Krupp et al. 2008). Chapter 10 discusses how social conditions influence male investment into various components of their ejaculates, but here we consider some of the many studies that have concentrated on the effects of social interactions on gene expression in males.

Carney (2007) showed that virgin *D. melanogaster* males that court females rapidly upregulate a modest number of loci, including genes associated with sperm function and olfactory behaviour. Surprisingly, the majority of genes downregulated have an immune function, perhaps indicating that males are diverting resources from expensive immune function to other areas. Olfactory loci could be contributing to female detection, though such loci have a wide range of functions. Comparisons of expression changes of males exposed either to males or to females potentially allow identification of genes involved in courtship versus other social interactions such as male–male competition. Ellis and Carney (2011) (examining only male *Drosophila* heads) showed that twice as many genes changed expression when males interacted with each other as when males encountered females, indicating that intrasexual interactions had a greater effect on gene expression. Only 16 genes were non-overlapping and responded only to exposure to females rather than to any other social interactions, and these did not implicate particular functions. One gene, *egghead*, was shown to disrupt male courtship ability when knocked down. Perhaps a curiosity of such studies is that some of the loci we might expect to show changes are not among those detected. For example, we know that pheromone production in male flies is very sensitive to social interactions, including the number and type of males encountered, and that the production of some pheromones is ramped up when males are exposed to females. Focused studies of candidate genes involved in pheromone biosynthesis such as *desat1* show changes in expression, but these are often not implicated in the genome-wide studies. There are a number of interpretations of this: perhaps the studies of key genes can detect smaller changes, or, probably more importantly, such studies are often tissue specific. Pheromone-influencing genes expressed in oenocytes are drowned out in whole body analyses or are undetectable in studies of heads (where they are not expressed). More generally, we still know rather little about the mechanics of expression variation and its regulation, and the potential role of epigenetics in behavioural flexibility. An RNA-seq study (Smith et al. 2013) of cactus-dependent male mating success in the cactophilic fruit fly *D. mojavensis* revealed genes usually associated with epigenetic effects such as methylation (implicated in regulation of gene expression).

4.4.3 *Different species*

Potentially a major opportunity to explore the role of insect mating behaviour in speciation is to examine differential gene expression responses to different species. When encountering or mating with a heterospecific male, we might predict different types of responses to occur in a female. Ellis and Carney (2009) examined gene expression in *D. melanogaster* males who courted either conspecific or *D. simulans* females. They confirmed that males rapidly downregulated immune function genes, but there were virtually no differences when courting heterospecific versus conspecific females. Similarly, Immonen and Ritchie (2012) found that conspecific or heterospecific courtship song had similar effects on female gene expression in the head. We know that females are stimulated by heterospecific song, but to a lesser extent, so perhaps this is not surprising. It would be interesting to expose females to males they will always refuse to mate with, but this experiment has not yet been done.

A particularly interesting study in this area is that of Bailey et al. (2011) who examined the post-mating gene expression changes when *D. melanogaster* females from Zimbabwe were mated to their preferred Zimbabwe males or non-preferred Cosmopolitan males. They found that about 1,500 loci showed differential mating responses by male type. These were more likely to be expressed in the central nervous system (both female head and probably thoracic abdominal ganglia) and the ovaries. Perhaps surprisingly, genes expressed in sperm storage organs—thought to be likely targets of sexually antagonistic gene expression—were not differentially affected. Interestingly, differentially expressed genes were physically clustered in the genome, especially on the sex chromosome. They interpret these loci as contributing to mate choice; for example, differential investment in eggs indicated by expression differences in ovaries could represent a form of cryptic female choice (Figure 4.2). This is very interesting, but not entirely convincing. Outcomes of sexual interactions are a classic 'interacting phenotype' (Moore et al. 2002), and it is difficult, if not impossible, to say whether responses have evolved primarily due to female benefits or male exploitation. Males with which females have not co-evolved might be better at inducing changes in females (similar to the classic Rice (1996) experiment) or, if the female response is 'ahead' in the sexual arms race (Partridge and Parker 1999), males may be less able to manipulate females. Nevertheless, it is certainly true that sexually antagonistic interactions have the potential to contribute to reproductive isolation between diverging forms of insect (Gavrilets 2000, Martin and Hosken 2003), and their role in speciation may be underappreciated.

4.4.4 *Gene expression: conclusions*

Reviewing these studies, it seems clear that gene expression may be particularly tissue specific and many studies will need to be much more targeted than the initial whole organism—whole genome studies, which may miss subtle effects (Mank et al. 2013, Neville and Goodwin 2012). Another issue of great importance is the time course over which gene expression is studied. For post-mating responses, the studies examining different time periods have clarified the nature of the responses. But we still usually do not know when genes of the central nervous system or hormone-producing tissues need to be expressed to determine adult behaviour: for example, finding genes for female choice this way seems fraught with difficulty due to our poor understanding of the development of the nervous system. Are genes that predispose females to mate with attractive males or males with appropriate species-specific signals likely to be expressed when females meet males, or is the key expression period for such genes when the central nervous system is developing, to set up biases in receptor systems, perhaps long before the animal is adult? In which neurons do such loci need to be expressed, and are there many or only a few? The current wave of expression analyses seems to have succeeded more in detecting mechanistic effects, such as physiological and immunological loci, than examples of important neurogenetic loci that influence behavioural flexibility.

4.5 Systems approaches

Determining that a gene influences a trait can be a big step forward in understanding that then opens up interesting mechanistic or evolutionary questions. However, the connection between genotype and phenotype is not so simple: many genes have pleiotropic effects on multiple traits, and most traits are influenced by multiple loci, often with epistatic interactions. Furthermore, none of this is independent of the environment. Where genes and their products operate in complex networks, it is often not easy to see how a change in one gene will interact with others to modulate the response of the network to external stimuli. These problems can be tackled by adopting a 'systems approach' and this way of thinking has great potential for tackling evolutionary genetics questions (Loewe 2009).

Systems approaches have so far received little attention in the context of insect mating systems, though some behavioural traits including mating speed (Ayroles et al. 2009) and olfactory behaviour (Swarup et al. 2013) have been studied. One can see how they might be used in understanding the trade-off between mating effort and immune function (see Chapters 8 and 13) as the network of genes underlying insect immunity becomes better understood (Rolff and Reynolds 2010). However, one of the best-characterized networks of interacting genes in insects is the system that underlies the circadian clock. Since timing of reproductive activity is a key component of any mating system, and potentially critical for reproductive isolation between species, there is a major opportunity here for genetic analysis.

The operation of the circadian clock in *Drosophila* has been worked out in exquisite detail (Figure 4.4). The critical components are a series of feedback loops that generate approximate 24 h rhythmicity in the absence of external stimuli, and an entrainment system that senses the environment and modulates the rhythm accordingly. The major feedback loop depends on the *period* and *timeless* loci: transcription of these genes is activated by products of the *clock* and *cycle* genes (CLK and CYC). As the gene products PER and TIM increase they form a heterodimer which inhibits CLK–CYC and so represses further PER and TIM production. This cycle can be entrained to the external day–night cycle through the action of cryptochrome, CRY, which is produced when light is detected and degrades TIM. This machinery is described in greater detail by Hardin (2011), for example. Many features of the clock mechanism are shared across insects but it is becoming clear that there is also much variation among insect orders (Tomioka and Matsumoto 2010). There is also a continuing debate about the extent to which the circadian clock underlies seasonal, as opposed to daily, behavioural and other rhythms in insects (e.g. Bradshaw and Holzapfel 2010).

Interest in the role of timing in insect mating behaviour, particularly in allochronic reproductive isolation (where asynchronous mating activity leads to isolation), goes back to classic studies on crickets by Alexander and Bigelow (1960) and on lacewings by Tauber and Tauber (1977). From the speciation point of view, this is an aspect of mating behaviour for which there is a clear link to the environment. This is important because direct divergent selection on timing of reproduction may generate reproductive isolation in the face of gene flow more easily than indirect selection, for example on a mating signal or mate preference. This is because there is no need to build up an association between adaptive and mating traits, because the trait causing assortative mating is, itself, a locally adaptive trait (Smadja and Butlin 2011). The potential of clock genes to influence allochronic isolation

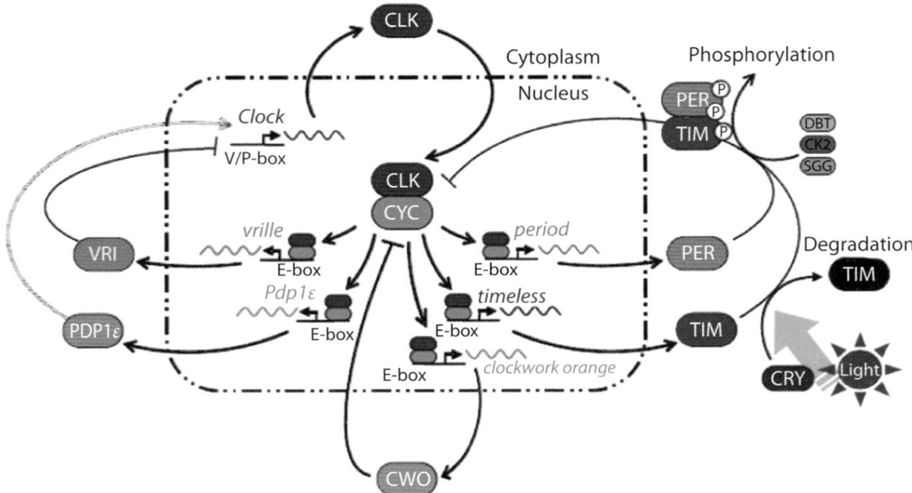

Figure 4.4 Core machinery of the *Drosophila* circadian clock, consisting of interlocked feedback loops with modulation from external light stimuli via cryptochrome (CRY). Reproduced with permission from Tomioka and Matsumoto (2010).

has been elegantly demonstrated by genetic transformation experiments in *Drosophila*, where moving the *period* gene of *D. pseudoobscura* into a *period* null strain of *D. melanogaster* produced flies with a daily activity cycle of *D. pseudoobscura* (Peterson et al. 1988). Detailed studies have shown that such transformed flies show assortative mating, and that this is not purely due to overall activity cycles (Wyman et al. 2010). These studies illustrate nicely the potential of a pleiotropic genetic influence on allochronic effects, though the role of temporal isolation in speciation between these divergent species is probably slight.

A more ecologically relevant example is provided by the melon fly, *Bactrocera cucurbitae*, which is a major agricultural pest. Strains of this species selected for short or long development time have been found to mate at different times of day, resulting in strong assortative mating. This mating barrier can be removed by using light cues to shift the photoperiodic behaviour of the flies: when the timing difference is removed, flies from the different strains mate at random (Miyatake et al. 2002). Involvement of the circadian clock was confirmed by cloning the melon fly homologue of *period* and showing that its expression cycled differently in the two strains. Following this pioneering work, the sequencing of the *period* gene established that in this case this gene was unlikely to contain the substitutions responsible for strain differences, or differences in timing between related species. However, knowledge of the clock system made it possible to identify other candidate loci and it now appears that substitutions in *cryptochrome* may be crucial (Fuchikawa et al. 2010). Two amino acid substitutions in *cryptochrome* distinguish the short- and long-development-time strains, and the pattern of expression in heads also differs between strains. The frequency of these substitutions also correlates with the free-running periodicity of behaviour in other strains. The causative role of one or both of the substitutions has yet to be formally established but knowledge of the underlying network shows how changes in *cryptochrome* might alter the free-running rhythm of the clock or its responsiveness to day–night transitions. Genetic manipulation in *Drosophila* of the region of CRY containing one of the substitutions does lengthen the circadian rhythm. This is therefore a gene that may well turn out to have a major effect on the timing of mating and thus reproductive isolation, not just in these strains but also in other species of fruit fly that differ in peak timing of mating activity. This work also has an interesting applied aspect: sterile male release is an important element of melon fly control and it is critical that released males show the same diurnal pattern of mating as the populations they are intended to control (Miyatake 2011).

As well as influencing diurnal cycles, the *period* gene of *D. melanogaster* influences the periodicity of an important component of fruit fly courtship song, the interpulse interval. Genetic transformation experiments have shown that the gene controls species-specific variation in the song cycle between *D. melanogaster* and *D. simulans* (Wheeler et al. 1991). The song cycle also influences female song preferences in *D. melanogaster* and relatives (Ritchie et al. 1999). Because clock genes are so fundamental to rhythmic behaviours, it was thought that this might be an example of genetic coupling if *period* also influenced preferences, perhaps even contributing to the central pattern generator. However, so far studies have not suggested that an influence of *period* on song preference is coupled to the effect on song traits (Greenacre et al. 1993, Ritchie and Kyriacou 1994).

One of the more exciting applications of a systems approach to behaviour genetics concerns recent work on the importance of particular pathways in signalling. Condition dependence is central to sexual signalling, both because it will lead to signals that reliably indicate quality and also because it may lead to 'genic capture' of loci influencing condition into sexual trait expression, providing a possible resolution to the lek paradox (Rowe and Houle 1996, Tomkins et al. 2004). In an attempt to investigate potential networks that might be involved in condition dependence, Kuo et al. (2012) manipulated genes involved in the insulin/insulin-like signalling (IIS) pathway of *D. melanogaster*. This pathway influences resource allocation and may play a central role in converting the resources of individuals that are in good condition into extra longevity, fecundity, and development. When knocked down, several genes in the IIS pathway all led to changes in the CHCs that (as we have seen) are important pheromones in sexual signalling. Males found these females to be less sexually attractive (courtship occurs due to male preference for female CHCs), and Kuo et al. (2012) demonstrated that it was the altered CHCs that were responsible by experimentally 'perfuming' females (with genetically non-functional oenocytes) with the CHCs of manipulated females. Also, the expression of loci involved in pheromone production was sensitive to IIS signalling. Hence there is an intimate association between IIS signalling and an important sexual trait. It would be most interesting to know if other sexual traits, such as song parameters, were similarly affected in mutant males. Emlen et al. (2012) have demonstrated that insulin signalling might be a general mechanism for exaggerated sexual signals by genetic manipulation in developing rhinoceros beetles. Their hypothesis was that more exaggerated morphological traits may be more sensitive to IIS signalling during development, and therefore be particularly reliable signals of developmental condition. They used RNAi (a method of suppressing gene expression) to interfere with the insulin receptor gene, and predicted that there would be a hierarchy of effects on genitalia (relatively insensitive), wings (moderately sensitive), and the elaborate antlers of male stag beetles (highly sensitive). The results conformed precisely to these predictions (Figure 4.5)

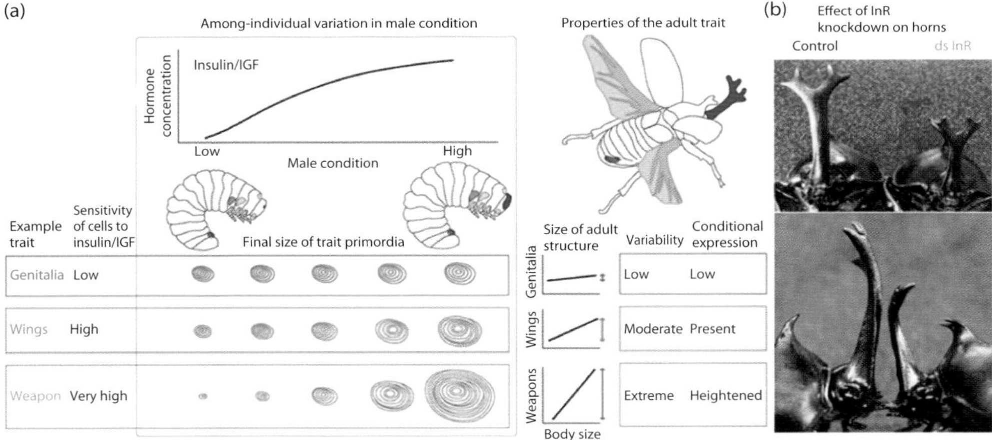

Figure 4.5 (a) Mechanism for the evolution of trait exaggeration through increased cellular sensitivity to insulin/insulin-like growth factor (IGF) signalling (shown for the disc-like appendage primordia of insects). Individual nutritional state and physiological condition are reflected in circulating concentrations of insulin-like peptides and IGFs, which modulate the rate of growth of each of the trait primordia. Traits whose cells are sensitive to these signals [e.g. wings (green)] exhibit greater nutrition-dependent phenotypic plasticity and among-individual variability than other traits whose cells are less sensitive to these signals [e.g. genitalia (red)]. An increase in the sensitivity of cells within a particular trait [e.g. horns (blue); see text] would lead to disproportionately rapid growth of that trait in the largest, best-condition individuals (i.e. exaggerated trait size) and smaller trait sizes in low-condition individuals. (b) Head and thorax shown in two orientations (top and bottom) for same-sized control (left) and insulin-receptor-suppressed beetles (right). Reproduced with permission from Emlen et al. (2012).

with the antlers being eight times more sensitive to the manipulation than wings. They suggest that numerous exaggerated traits in animals may show such heightened sensitivity to condition-dependent signalling during development. As Kuo et al. (2012) conclude: '. . . many sexually attractive characteristics, including those unique to individual species, may convey a universal aspect of beauty by accurately reflecting the molecular activity of a small number of highly conserved pathways that influence longevity and reproductive output across taxa.'

4.6 Comparative genomics

From the perspective of insect mating systems, comparative genomics is in its infancy, probably because so few genes have been clearly identified to play a major role in behaviour. However, indirect evidence suggests that sexual selection or mating system variation is a major driver of divergence within the genome. One of the fastest evolving gene families in comparative studies of *Drosophila* is that of accessory gland proteins or ACPs, probably driven by sexual selection (Haerty et al. 2007) (see Chapter 10). Genes that are sexually dimorphic in expression evolve quickly, both in coding sequence and expression variation (Drosophila 12 Genomes Consortium 2007, Ellegren and Parsch 2007). Sexual dimorphism of gene expression is condition dependent in *D. melanogaster*, with increases in both the number of loci showing sex-biased expression and the level of expression differences in individuals in good condition, especially for male-biased genes (Wyman et al. 2010). Genes which are male-biased in expression evolve more quickly between species, with greater signatures of adaptive substitution, than non-dimorphic loci, perhaps as males are subject to stronger combined intra- and intersexual selection. However, exploration of the role of mating system variation and sex-specific selection in genome and transcriptome evolution is just beginning (Mank et al. 2013).

4.6.1 *Comparative genomics and chemical communication*

The *Bombyx mori* genome sequence has been a great help in the identification of candidate loci for moth pheromone production, as discussed above. Now, more and more genome sequences and transcriptome analyses are being conducted and they offer fascinating opportunities for comparative approaches to the genetics of insect mating systems. For example, the desaturase gene family of *Drosophila* is well-characterized, allowing novel duplications and the role of sex-specific selection in divergence to be explored (Fang et al. 2009, Shirangi et al. 2009, Keays et al. 2011).

In the context of chemical communication, gene families involved in chemoreception are clearly interesting targets, complementing the desaturases and reductases implicated in the production of chemical signals. In insects, the major gene families code for the odorant binding proteins (OBP) and the odorant (OR) and gustatory (GR) receptors. Characterization of these gene families has been a major achievement of comparative genomics. The OR and GR loci in *D. melanogaster* were initially identified bioinformatically (Clyne et al. 1999) using genome sequence data and search algorithms based on the structure of odorant receptors originally identified in rats, although it has since been established that mammalian and insect gene families are not homologous (Bargmann 2006). This starting point enabled the identification of gene families in other insects, as genome sequences became available (e.g. honeybee: Robertson and Wanner 2006). The problem is that these are large gene families (totalling >100 loci, sometimes >500; Sanchez-Gracia et al. 2009) in which substrate specificity is known for few, if any, loci in most species. Work on *Drosophila* has identified ligands for some receptors, mainly ORs but including a GR that is a putative pheromone receptor (Bray and Amrein 2003). Odorant recognition works in a combinatorial fashion: most receptors respond to a few related compounds, although the precision of tuning varies, and most compounds elicit responses from more than one receptor (Bargmann 2006). In the search for receptors important in mating behaviour in other organisms, comparative approaches are needed that aim to narrow down the set of candidates within these large gene families.

The search for the male antennal receptor of the female sex pheromone in the silk moth, *Bombyx mori*, shows the sort of strategy that is possible. The silk moth has a very simple pheromone with one major component, bombykol, and a secondary component, bombykal, which is released by females but is not essential for male orientation. The high sensitivity and specificity of male response suggested that a finely tuned receptor was involved, probably expressed only in males and present in the long sensilla on the antenna known to be responsible for pheromone detection. Sakurai et al. (2004) used a differential cloning strategy to identify an OR gene, *BmOR1*, that is expressed exclusively in male antennae. Of 29 OR loci detected in the silk moth genome sequence, only two were expressed in male but not female antennae. Expression of *BmOR1* was concentrated in the epithelium underlying the pheromone-sensitive long sensilla. Finally, heterologous expression of *BmOR1* in *Xenopus* oocytes rendered them specifically sensitive to bombykol and ectopic expression of the gene in female antennae rendered them responsive in electroantennogram tests to bombykol but not bombykal. Subsequent work (Grosse-Wilde et al. 2006) has shown that the pheromone-binding protein, an OBP specific to pheromone-sensitive sensilla, not only functions to transport bombykol across the sensillar lymph but also enhances the specificity of response. This ground-breaking work clearly opens the way for the discovery of other pheromone receptors and so for understanding the evolutionary changes associated with signal–response co-evolution.

For the European corn borer, Lassance et al. (2011) argued that odorant receptor loci are candidates for the male behavioural response gene that was known to be sex-linked. Therefore, they tested the ECB homologues of OR genes that were known to be sex-linked in *Bombyx* or *Heliothis* and that were also male-biased in their expression, using a comparative genealogical approach together with genetic mapping. Three loci were strongly differentiated between races and showed evidence of selection, but none mapped to the same position as the behavioural locus, suggesting that the differences between races are not caused directly by variation in odorant receptors.

The chemosensory gene families evolve by a process of gene duplication and loss. Clade-specific expansions within the gene families show evidence of positive selection (e.g. Smadja et al. 2009) and it is tempting to suggest that these expanded subfamilies are associated with the specific way of life of the insects involved. One way to test this is to extend, to whole gene families, the principle that loci critical for behavioural differences between species or populations are expected to show higher levels of divergence than the genomic background. This is the 'genome scan' approach (Stinchcombe and Hoekstra 2007). An analysis of this type has been carried out in the pea aphid, *Acyrthosiphon pisum*, which has many host-specific races in Europe. Because mating occurs on the host, chemosensory choice of the host on which to feed is also a critical determinant of mating pattern and so reproductive isolation among races. Smadja et al. (2012) use targeted gene capture technology to compare divergence levels across 150 chemosensory genes in three host races, finding a small number of loci with unusually high divergence. These outlier loci were mainly odorant receptor genes. This is an efficient approach to reduce the set of candidates to a manageable number of loci, but much work remains to be done to demonstrate a functional role in host-plant choice for any one locus.

Comparative genomics will become increasingly important as more genome sequences are published, as more bioinformatics tools become available, and as studies of gene expression or genomic patterns of divergence become more accessible in non-model organisms (e.g. Ekblom and Galindo 2011). Combined with the sorts of predictions used to narrow down candidate genes within the chemosensory gene families, the study of sex-biased gene expression and patterns of genetic differentiation among populations offer huge potential for the successful identification of genes underlying mating behaviour in insects.

4.7 Conclusion

The application of both advanced statistical methodologies in quantitative genetics and selection analysis (Chapter 8), and the wide range of techniques being developed in genomics, is bringing huge advances in our understanding of the genetics and evolution of insect behaviour and mating systems. However, many of the questions addressed indirectly by Thornhill and Alcock (1983) about sexual selection, sexual communication, and speciation remain to be resolved in detail. But there are good reasons to be optimistic, given the many new genetic and genomic technologies developed in the last 30 years now being applied to these problems. For example, it is now within reach to determine the role of pleiotropy in co-ordinating sexual signalling and the relative rates of evolution of genes involved with communication, such as loci influencing pheromone blends and their perception. It is particularly exciting that genetic manipulations are allowing us to explore the effect of fundamental signalling pathways on behaviour, possibly allowing us to see inside the black box that has previously been called 'condition-dependent genic capture'. How this might translate into indicator models of sexual selection is still not clear and requires a combination of advanced molecular manipulation combined with studies of heritability of behaviour and fitness variation. Understanding the variation at individual genetic loci and in gene networks within natural populations is clearly vital, yet remains challenging. Comparative genomics is leading to greater progress in analysing genes and gene families under divergent selection between species. Finally, correlating mating system variation (between species and under laboratory manipulation) with its effects on the evolution of the genome and transcriptome is likely to provide the most immediate breakthroughs in our understanding of the genetics of insect mating systems. We look forward to increasing emphasis on this understanding of the complex interactions between genotype, phenotype, and environment as we move beyond simply finding genes.

CHAPTER 5

Reproductive physiology and behaviour

Patricia J. Moore

5.1 Introduction

Research on the behavioural ecology of mating systems often revolves around plasticity in reproduction. Traditionally, many behavioural ecologists have focused on the role of 'reproductive life history' (Shuster and Wade 2003; Hodin 2009) in shaping mating systems. We ask about the decisions individuals make as to when to reproduce, how much to invest in a particular reproductive bout, and how often to reproduce. We typically estimate fitness by counting the number of offspring produced. We may examine variation in the number of gametes transferred during mating. We consider constraints on this variation in generic terms. For example, to say that 'sperm are cheap' is incorrect, as there are numerous studies examining the effect of sperm limitation on the evolution of mating strategies relating to both strategic ejaculation by males and avoidance of sperm-limited males by females (Chapter 10). Yet such heuristic shortcuts are commonplace, and when asked why males have limited capacity for sperm production we resort to energetic arguments without examining mechanisms (e.g. Olsson et al. 1997). What we really mean by 'sperm are cheap' is that generating sperm is different than generating eggs, and it is easier to make sperm than eggs. But we only know this if we know the mechanisms by which sperm and eggs are generated. The number and rate at which gametes can be produced directly impacts potential reproductive rates, and thus the evolution of mating systems (Chapter 3).

Given the importance of the number and quality of gametes both on fundamental fitness of an individual and on the ability to respond plastically to environmental conditions, both social and physical, it is sometimes surprising how infrequently behavioural ecologists consider mechanisms of gamete production. To truly understand both variation in reproductive capacity and the ability to respond plastically through the production of gametes, we need to understand the cell and developmental biology underlying gametogenesis. In this we are making the same argument that has been made in the life-history field (Zera and Harshman 2001; Flatt and Heyland 2011); we need to understand the mechanisms underlying these strategic decisions in order to understand the targets of selection, as well as any potential trade-offs that might constrain response to selection.

Here we consider some of the things a behavioural ecologist should know about reproductive physiology. The reproductive biology that currently informs much of the research

The Evolution of Insect Mating Systems. Edited by David M. Shuker and Leigh W. Simmons.
© The Royal Entomological Society 2014. Published 2014 by Oxford University Press.

on mating systems is focused on modes of reproduction (Chapters 1–3) and reproductive life history, defined by three characters (Shuster and Wade 2003): the number of times an individual mates, the number of reproductive events in a lifetime, and the duration of reproductive competence. Here, we will discuss some of the developmental and hormonal mechanisms underlying reproductive physiology and how these relate to plasticity in reproduction and behaviour. We are not able to review all of insect reproductive physiology, and there are several excellent textbooks that cover the amazing diversity of reproductive strategies represented in the insects (see Gullan and Cranston 2011; Chapman 2013). Rather, we would like to highlight some reproductive physiology research that we hope will not so much teach readers what they ought to know but rather why they should be curious about reproductive physiology and the mechanisms underlying gametogenesis. In particular we consider germline stem cells, including what is known about how the production of germ cells, and how the germ line responds to variation in the environment, including conditions known to be important in variation in mating behaviour (e.g. age and nutrition). In addition to considering germline stem cells, we discuss how reproductive capacity can respond both plastically to environmental conditions, but also evolutionarily, for instance through variation in ovariole number. We also briefly consider how the hormones essential to the production of gametes are often the same hormones involved in mating behaviour and secondary sexual characteristics, and thus represent a potential constraint on the ability of populations to respond to selection. Finally, we examine the trade-offs between reproductive and non-reproductive tasks, including the most extreme case, the social insects. In this group, the link between reproductive physiology, key reproductive genes, and behaviour has been elucidated.

5.2 Strategic investment in reproduction

Over the last thirty years it has become clear that individuals should, and do, strategically adjust expenditure on matings or gametes given the likelihood that a particular mate or mating interaction will result in fitness benefits. There is a myriad of examples that could be cited, providing empirical evidence to support the idea that there is strategic investment in mating and reproductive effort. Males differentially allocate sperm when there is a perceived risk of sperm competition and when mating with females of variable quality (reviewed in Wedell et al. 2002a; Cameron et al. 2007; Parker and Pizzari 2010; and Chapter 10 in this volume). For example, in *Drosophila melanogaster*, males transfer more sperm to mated females, where sperm competition risk is high, than to virgin females (Lüpold et al. 2011). Female mating behaviour can also be influenced by male mating experience, perhaps in a bid to avoid mating with sperm-limited males. In the cockroach *Nauphoeta cinerea*, females avoid mating with males that have consorted, but not necessarily mated with, multiple females (Harris and Moore 2005). In this species even sperm-depleted males can prevent female remating, so females have evolved mechanisms to avoid mating with males that may be of high quality socially, being dominant or attractive, but have low reproductive potential due to multiple mating opportunities. Chapter 10 contains more examples and more in-depth discussion on how sperm competition shapes ejaculate expenditure.

Reproductive potential of a mating partner will also influence reproductive effort. Males tend to prefer to mate with, or invest more in mating opportunities with, high quality females (Bonduriansky 2001). In addition to tailoring ejaculates in response to perceived risk of sperm competition, Lüpold et al. (2011) found that males also transfer more sperm to young or large females, with higher potential fecundity, than to old or small females. An individual's own condition and reproductive potential will also influence mating behaviour. Fecundity decreases with age, and in *N. cinerea*, females become less choosy as they age (Moore and Moore 2001). Thus, females are foregoing whatever benefits they derive from mating with preferred males in order not to lose reproductive opportunities as their reproductive potential declines over time. The external environment can also influence reproductive potential. In the milkweed bug *Oncopeltus fasciatus*, a laboratory population has been adapted to use sunflower seed as a food source but has retained the ability to use the ancestral diet of milkweed seed. Males from this population fed the two diets express different patterns of mating behaviour and life history (Attisano et al. 2012). Milkweed-fed males invest in both mating, as evidenced by the speed at which they initiate copulation and the rate at which they copulate, and fertility at the expense of survival as compared to sunflower-fed males.

Phenotypic and genetic plasticity in reproductive effort are familiar to most behavioural ecologists and these processes have been explored in some detail; we think most behavioural ecologists are familiar with the arguments and outcomes. Indeed, many of the chapters in this book review and discuss these issues. Instead, we will consider an area of research that is less familiar to behaviour ecologists, but plays an important role in our understanding of the targets of selection that underlie these amazing adaptations that allow insects, and indeed many taxa, to shape their reproductive strategies and investments.

5.3 Germline stem cells—the ultimate source of fitness

Germ cells are typically set aside early in embryogenesis, and give rise to sperm and eggs in adults (Ewen-Campen et al. 2010). The specification of the germ line is essential to reproductive success, and the genetic integrity of the germ line is critical and protected (Narbonne and Roy 2006). In many species, adults are able to produce gametes continuously throughout their reproductive lifespan. The source of these germ cells is a population of germline stem cells (GSCs) present in the gonads (Spradling et al. 2011). Thus the GSCs and resulting germ cells are ultimately responsible for fitness, as they represent the source of the alleles to be transmitted to the next generation. The key question is: does the germ line respond to social and physical environment, resulting in variation in reproductive effort? Or do individuals adjust reproductive effort downstream of the production of gametes?

5.3.1 *What are GSCs and where do they reside?*

The germ cells are housed in the somatic gonad. It is worthwhile briefly reviewing the reproductive structures in insects (for more in depth information, see Chapman et al. 2012). The reproductive system of female insects consists of paired ovaries opening into a Y-shaped oviduct that ends in a vaginal and genital opening (Figure 5.1). Each ovary consists of a number of elements, the ovarioles, in which eggs develop as they pass from the tip of the ovariole to the oviduct. Ovariole number varies, both within and among species (see Section 5.4). Within an ovariole, three regions are recognizable, the terminal filament, the germinarium containing the germ and prefollicular cells, and the vitellarium where yolk is deposited in the developing ooctyes. Ovarioles in the more primitive orders of insects, including Orthoptera, Isoptera and Odonata, are panoistic. That is, there are no specialized nurse cells. In panoistic species, all germ cells develop into oocytes and are surrounded by

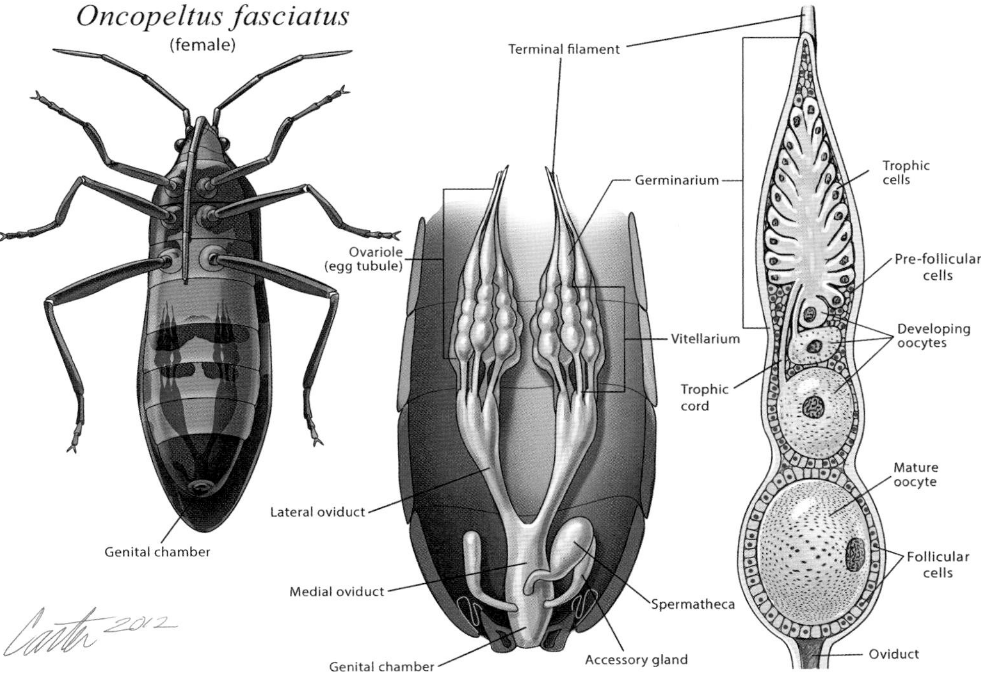

Figure 5.1 Diagram of the female reproductive system of *Oncopeltus fasciatus* (for placement only). The paired ovaries connect with a pair of lateral oviducts that join to form the medial oviduct. The medial oviduct opens posteriorly into a genital chamber. In some insects the genital chamber forms a tube, the vagina, which can form the bursa copulatrix. Accessory glands and the spermatheca for sperm storage open from the genital chamber. In *O. fasciatus*, each ovary consists of seven egg-tubes, or ovarioles. Each ovariole has three regions, the terminal filament, the germinarium, and the vitellarium. The ovarioles of *O. fasciatus* are telotrophic; the trophic cells remain in the germinarium at the distal tip but are connected to the developing oocyte though trophic cords.

a monolayer of somatic follicle cells. In meroistic ovarioles the germ cells divide to produce both the oocyte and its associated nurse cells that contribute to the cytoplasmic content of the developing oocyte. Meroistic ovarioles can be further subdivided depending on the spatial arrangement of the nurse cells relative to the oocyte. In telotrophic ovarioles, the nurse cells remain in the distal apex of the ovariole and are connected to the oocyte as it moves towards the oviduct through a nutritive cord. In contrast, in polytrophic ovarioles, such as those found in *Drosophila melanogaster*, the nurse cells move down the ovariole with their associated oocyte.

The male reproductive system consists of a pair of testes connecting to paired seminal vesicles and a median ejaculatory duct (Figure 5.2). It may also include various accessory glands. Typically the testis consists of a series of tubules, ranging in number from a single tubule, as seen in *D. melanogaster*, to more than 100 in grasshoppers. Like the ovarioles, each testis tubule has zones containing germ cells at various stages of development. An apical cell sits at the distal end of the tubule, followed by the germinarium, a growth zone where spermatocytes divide, a maturation zone where spermatocytes divide meiotically to form spermatids, and finally a transformation zone in which the spermatids develop into spermatozoa.

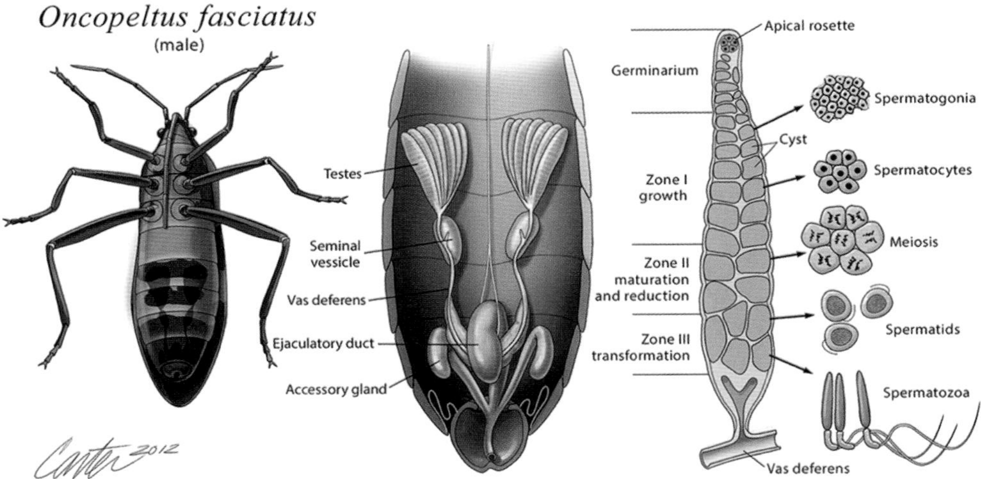

Figure 5.2 Diagram of the male reproductive system of *Oncopeltus fasciatus* (for placement only). As is typical for most insects, the male reproductive organs of *O. fasciatus* consist of a pair of testes and seminal vesicles. The seminal vesicles connect to a medial ejaculatory duct through the vas deferens. In most insects there are also several accessory glands. The testes can be further divided into testis follicles. In *O. fasciatus* each testis contains seven follicles. Sperm development proceeds from the distal tip of the germinarium. Zones of differentiation from spermatogonia through to spermatozoa are recognizable from the tip of the follicle down to the junction with the vas deferens. The germline stem cells lie at the distal tip, with each follicle having a hub, which serves as the niche for the germline stem cells (GSCs). As GSCs divide and daughter cells lose contact with the hub, differentiation occurs.

If we are interested in constraints to, or variation in, gamete production, we should look to their source, the germ line. In many species the continuous production of gametes in both males and females comes through the GSCs (Fuller and Spradling 2007; Spradling et al. 2011). GSCs, like all stem cells, reside in a micro-environment, the niche, provided by the somatic gonad (Kaczmarczyk and Kopp 2011; Spradling et al. 2011). GSCs depend on signals from the somatic niche for maintenance of their identity as stem cells. As with other types of stem cells, GSCs divide asymmetrically to produce one daughter cell that stays in the niche and remains a GSC and another daughter that leaves the niche and differentiates. In males and females the daughter cells that differentiate into sperm or oocytes are known as gonialblasts or cystoblasts, respectively.

Much of what we know about GSCs and their niches has been learned in *D. melanogaster*. In males, the niche is a hub of cells at the tip of the testis with space for 6 to 12 GSCs to remain in contact with the niche and thus remain GSCs (Figure 5.3a). Similar arrangements can be seen in other insects, including, for example, *Oncopeltus fasciatus* (Figure 5.3b; Schmidt and Dorn 2004). In the ovary of *D. melanogaster*, each ovariole contains a niche that supports two or three GSCs (Fuller and Spradling 2007). Male and female GSCs and niches have much in common, including the requirement of direct attachment to niche cells and the presence of somatic stem cells that enclose early germ cells. Furthermore, a number of common genes play a role in GSC dynamics in males and females. However, there are some differences. In particular, while in both males and females short-range signals from the niche are required for self-renewal, the particular molecular signals are not the same (Fuller and Spradling 2007).

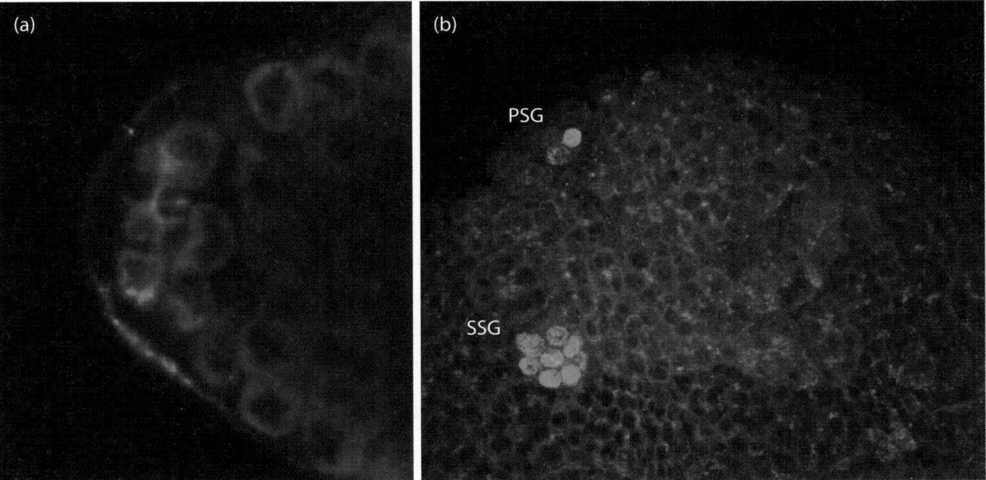

Figure 5.3 Testis tubules stained for the germ line. (a) Apical tip of the single testis tubule of *Drosophila melanogaster*. The hub is stained for the specific marker Fasciclin III (red) and the germline stem cells (GSCs) are stained with the germline-specific marker vasa (green). (b) Apical tip of one of the seven testis tubules in one testis of *Oncopeltus fasciatus*. The tubule was labelled ex vivo with 5-bromo-2-deoxyuridine incorporation (green) to label cells in the S-phase (preparing to divide). Two developing cysts of clonally derived spermatogonial cells are labelled as they divide mitotically before entering the spermatocyte stage of development. A primary spermatogonia (PSG) has only recently been produced from the GSCs. The secondary spermatogonia (SSG) is preparing to divide to produce a cyst of 16 spermatogonia. Mitotic divisions continue until a total of 64 primary spermatocytes. The tubule was also labelled with phalloidin that stains filamentous actin (red) present in cytoplasmic bridges between the daughter cells during mitotic divisions in the primary spermatogonia.

5.3.2 *Are GSCs responsive to the environment?*

Could variation in GSC maintenance under different social and physical environments influence reproductive effort and outcome? Again, most of the work examining how GSC dynamics change with age and environmental conditions has been carried out in *D. melanogaster*, but the evidence suggests that regulation of GSC populations does vary with conditions, particularly age and nutritional status. GSC populations are not fixed. In addition to the self-renewal asymmetric divisions in which the GSC mother replaces itself and also produces a daughter cell that will differentiate, GSCs can also divide symmetrically to produce either two GSCs, increasing the population, or two gonialblasts (in the testes) or cytoblasts (in the ovary) (Kaczmarczyk and Kopp 2011). In this way it is possible to increase the number of potential gametes at the expense of the GSC population. In addition, early germ cells can revert back to stem cells (White-Cooper and Bausek 2010; Kaczmarczyk and Kopp 2011). The decision either to differentiate or to remain undifferentiated depends on a balance between competing signals. While the molecular signals underlying this decision remain unknown in all but a very few species, this balance could control the number of gametes produced in response to the environment, either social or physical.

Reproductive capacity decreases with age, and in *D. melanogaster* it has been shown that GSCs are lost as individuals age in both males (Boyle et al. 2007; Jones 2007) and females (Margolis and Spradling 1995; Jones 2007). The loss of GSCs is primarily due to differentiation, rather than cell death (Kaczmarczyk and Kopp 2011). Kaczmarczyk and Kopp (2011) tested the hypothesis that early fecundity can be increased at the expense of late fecundity by increasing the number of cytoblasts (and thus ooctyes) produced by increasing the number of symmetrical divisions of GSCs to produce two cytoblasts. This was predicted to occur by removing the mother GSC from the population, thus sacrificing future reproductive potential. They found that whereas females from lines selected for early or late reproduction start out with the same number of GSCs, the early reproduction lines lose GSCs at a much faster rate. Thus, there appears to be heritable genetic variation in the control of GSC self-renewal versus differentiation with age that is available to selection.

GSCs are able to respond to the external environment as well as internal factors such as age. In particular, GSC numbers and rate of proliferation respond to nutrition. In *D. melanogaster*, egg production responds rapidly to changes in diet (Drummond-Barbosa and Spradling 2001). When females are exposed to a protein-restricted diet, the number of eggs laid decreases. The number of active GSCs is not different on the poor and rich diets, but the rate of proliferation falls when food is poor. GSCs sense nutritional quality or quantity through the insulin pathway and adjust the GSC proliferation accordingly (Drummond-Barbosa and Spradling 2001; Narbonne and Roy 2006). When nutritional conditions improve, and the insulin-signalling pathway is reactivated, the proliferation of the GSCs recovers. Interestingly, in *D. melanogaster* males, a decrease in nutrient availability leads to both a decrease in GSC proliferation and number. When males are starved, GSC populations decrease, although a small number are maintained and the population can recover quickly when the males are re-fed (McLeod et al. 2010). This change in GSC dynamics under the influence of diet in males also depends on an intact insulin-like signalling pathway. A similar response to food has been observed in *C. elegans* during adult reproductive diapause (Angelo and Gilst 2009). When late-stage larvae are starved, they transition into adults but cease reproduction. Entry into adult reproductive diapause depends on both nutritional and social environment (larval density). Animals entering adult reproductive diapause appear to be using oosorption as a strategy, exploiting differentiated oocytes for nutrients but preserving future reproductive potential by maintaining a small population of GSCs (Angelo and Gilst 2009). Thus, control of GSC maintenance and proliferation may be a common mechanism for responding to environmental conditions.

The research outlined above identifies molecular mechanisms underlying how GSCs respond to diet, but is there any evidence that this has ecological relevance? The preliminary data presented in Kaczmarczyk and Kopp (2011) suggest that GSC dynamics could play a role in the trade-off between early and late fecundity and lifespan in *D. melanogaster*. Angelo and van Gilst (2009)—cell and molecular biologists who can be forgiven for not being familiar with the term oosorption and for not referring to the ecological hypothesis by which females respond to unfavourable conditions by resorbing eggs and reallocating resources to survival until conditions improve—have clearly identified molecular mechanisms in *C. elegans* underlying both the resorption of eggs and protection of future reproduction in the GSCs. One study, although not looking at GSCs, suggests that life-history patterns of male fertility and survival under different dietary regimes could be due to GSC dynamics. In laboratory populations of the large milkweed bug, *Oncopeltus fasciatus*, even after many generations of adaptation to a diet of sunflower seeds, the ancestral milkweed seed diet maintains the signature of a high-quality food (Rion and Kawecki 2007), with males prioritizing reproduction, through both mating effort and fertilizing ability, at the expense of survival (Attisano et al. 2012). Milkweed-fed males have high fertility early in life compared to sunflower-fed males, but have shorter lifespans. One physiological difference between males fed the two diets is in lipid levels. Intriguingly, milkweed- and sunflower-fed males differ in the levels of oleic acid (Nation and Bowers 1982). This particular fatty acid has recently been shown to be a component of the pathway leading to extension of longevity through loss of the germ line (Goudeau et al. 2011), suggesting that GSCs may play a key role in life-history trade-offs.

Clearly a variety of open questions could be addressed through collaboration between scientists studying molecular mechanisms and those interested in evolutionary outcomes. For instance, why is there a difference between males and females in cell signalling pathways between GSCs and their niche (Fuller and Spradling 2007)? The authors themselves could only speculate about the selective forces that have led to this sexual dimorphism in cellular communication. Perhaps an investigator with more experience in the evolution of sex differences could refine and test this speculation. The universality of germline stem cells in insect reproduction is another question of interest to behavioural and evolutionary ecologists. As we have discussed, the data on GSCs and their response to environmental conditions comes almost exclusively from *D. melanogaster* and *C. elegans*. Whereas GSCs are widespread in males, adult ovarian GSCs have been identified in only a few species besides *D. melanogaster* and *C. elegans*, including medaka fish and zebrafish (Spradling et al. 2011). Although diverse orders of insects have been surveyed (reviewed in Büning 1994), it is not clear whether these cells are actually rare in females, or whether we simply lack the data. Perhaps the *D. melanogaster* ovary is an unusual case, allowing females to respond quickly to limited and ephemeral food and to oviposition sites (Spradling et al. 2011). If this is the case, we should look for female GSCs in other species that have similar reproductive life histories. Alternatively, are female GSCs actually rare and do other species use alternative mechanisms to respond plastically? If these species do not have GCSs, then reproductive potential is set at adult emergence by the number of germ cells present in the germinarium, and so variation in reproductive decisions must reflect how and when to use that potential.

5.4 Plasticity in reproductive capacity

The germ cells are housed in the ovarioles of females and testis tubules of males. Within the ovary or testis, each unit (ovariole or testis tubule) acts as an independent site of production of eggs and sperm. Thus we can think about the ovarioles and testis tubules as assembly lines for gametes (Hodin and Riddiford 2000). The number of gametes produced can be controlled by changing the number of assembly lines or by changing the rate of production along each line. Maximum reproductive output correlates positively with ovariole number because each ovariole can simultaneously mature an egg (Hodin 2009). But there are also trade-offs with ovariole numbers. Large ovaries can interfere with flight and rates of oogenesis may be inversely related to ovariole number (Hodin 2009). Presumably there are similar trade-offs associated with the testis tubules, although testes architecture has been given much less attention (Schärer et al. 2008). Thus, the potential exists for ovariole and testis tubule number to be shaped by natural selection.

As mentioned in Section 5.3, there may be variation in the number of these functional units, both within and among species. There has been a significant amount of work documenting the sources of variation in ovariole number in females. Hodin (2009) covers both variation in ovariole number and other sources of variation in female reproductive capacity in some detail. Here, we present the general picture of ovariole number variation and discuss some recent work investigating the molecular mechanisms underlying the control of ovariole number. In contrast to work on plasticity and variation in ovariole number, there seems to be remarkably little work done on variation in testis tubule number in males within and between species. Perhaps this is due to our bias that males are capable of producing more than sufficient sperm over their lifespan. However, given that we know that males can and do become sperm-depleted, it would be interesting to know whether the same factors that produce variation in female reproductive capacity, as well as the selection pressures that result in between-population and between-species variation, are also acting on males.

5.4.1 *Variation in ovariole number*

Ovariole number is variable both between and within species. Variation in ovariole number among populations can correlate with specific ecological niches. For instance, in both *D. melanogaster* and *D. simulans*, ovariole number varies along a latitudinal cline (Gibert et al. 2004). Interspecific variation in ovariole number also correlates with ecological niche specialization. For example, *D. seychellia* has about half as many ovarioles as its closest relatives. Given that *D. seychellia* has evolved the ability to use a fruit that is toxic to other sympatric drosophilids, thus reducing competition, the low ovariole numbers could indicate that there is a selective cost to producing large numbers of ovarioles (R'kha et al. 1997). Similarly, the Hawaiian drosophilids exhibit wide variation in ovariole number, with species that have specialized to use decaying leaves as an oviposition site depositing only one egg at a time, and having fewer ovarioles than those that lay many eggs under decaying bark (Kambysellis and Heed 1971; Hodin 2009).

There may also be intraspecific variation in ovariole number. In some taxa, such as the Lepidoptera, ovariole number is canalized, while in other taxa there is variation. In some cases, individual variation is due to allelic variation; in others, ovariole number is developmentally plastic (Hodin 2009; Chapman et al. 2012). Within the grasshoppers, species characterized by low ovariole number appear to have numbers genetically fixed, with low levels of variation. However, in species with large numbers of ovarioles, the number is highly variable and is set during development, with variation dependent both on allelic variation and larval diet (Taylor and Whitman 2010). Variation in ovariole number has been studied most intensively in *D. melanogaster*. Variation in ovariole number can be influenced by allelic variation (Bergland et al. 2008), but is strongly influenced by larval environment, particularly food quality, larval density, and temperature. In *D. melanogaster*, the intraspecific maximum number of ovarioles occurs under optimal larval nutrition and an intermediate temperature (Hodin 2009, Bergland 2011).

What mechanisms underlie control of ovariole number? In *D. melanogaster* larval nutrition affects ovariole number by modifying rates of differentiation of the specialized cells at the tip of the ovariole, the terminal filament cells, during the non-feeding stage of larval development (Hodin and Riddiford 2000). This suggests that endocrine signalling sets the rate of differentiation terminal filament cells and thus ovariole number (Bergland 2011). Quantitative trait locus (QTL) mapping studies of within- and between-species variation in ovariole number also suggest a role of endocrine signalling. Bergland et al. (2008) identified at least nine QTL underlying nutritional plasticity in ovariole number in *D. melanogaster*. Two of these loci contained genes to suggest that nutrient-dependent insulin signalling could act upstream to control terminal filament differentiation. Orgogozo et al. (2006) examined the genetic basis for the difference in ovariole number between *D. seychellia* and *D. simulans*. Fine-scale QTL mapping identified an autosomal region containing the insulin receptor gene, again suggesting that genetic variation in insulin signalling could contribute to variation in ovariole number. This is supported by the observation that insulin-signalling mutants have reduced ovariole numbers (Bergland 2011).

Recent work has also begun to examine how the developmental programme underlying ovariole development might influence ovariole number, and which cells and developmental pathways might be the target of selection on ovariole number. These studies investigate the mechanism by which the number of ovarioles is counted during development (Green et al. 2011). During larval development, terminal filament cells are specified beginning in the second instar (Sarikaya et al. 2012). Terminal filaments form during the third instar by intercalation of the terminal filament cells. As terminal filaments are the starting point of ovariole formation, the numbers of terminal filaments, rather than the number of germ cells, are the important determinant of ovariole number. Sarikaya et al. (2012) identified two mechanisms that can alter ovariole number. Between- and within-species variation due to nutritional stress appears to arise due to a change in the numbers of terminal filament cells. The number of terminal filament cells per terminal filament is constant; so fewer terminal filament cells give rise to fewer terminal filaments and thus fewer ovarioles. Within-species variation due to temperature, however, appears to be due to changes in the sorting of the terminal filament cells into terminal filaments. In this case, there is a difference in the number of terminal filament cells per terminal filament. In both cases, the terminal filament cell population is likely to be a target of selection, whether on mechanisms regulating the numbers produced or the mechanism by which they sort themselves into terminal filaments.

5.4.2 *Variation in testis tubule number*

Given the focus on sperm production, sperm usage, and sperm limitation in behavioural ecology, it is surprising how little attention has been paid to testis architecture (Schärer and Vizoso 2007). It could be argued that, just as ovariole number sets an upper limit to reproductive capacity in females, and just as variation in ovariole number has been a target of selection, testis tubule number may also constrain sperm numbers. Could variation in these 'sperm assembly lines' also be a target of selection for all these systems in which we know variation in sperm numbers is an important reproductive strategy? Why have we (including myself) been content to just examine testis 'size'? We know of only a single study examining intraspecific variation in testis tubule numbers. In a predatory stink bug, larval diet did not affect testis tubule number, suggesting that the number of tubules is species specific (Lemos et al. 2005). However, in this species, females are also invariant in ovariole number. It would be interesting to investigate the plasticity of testis tubule number, and its impact on fecundity, in a species where, unlike the drosophilids, there is more than one and where female ovariole number is known to be plastic.

5.4.3 *Limits to plasticity*

As we have seen, key reproductive traits can vary both within and among species. There is variation in the ability to produce unlimited numbers of gametes (via germline stem cells) versus finite numbers of germ cells. Variation in gonad structure can determine an upper limit to the rate at which eggs and sperm are produced. There is also variation in the timing of when gametes develop. In some species of insects, females emerge as adults with some eggs fully mature (Jervis et al. 2008). The proportion of mature eggs at emergence relative to a female's lifetime egg potential (ovigeny index) can range from strict pro-ovigeny (all eggs that a female will produce in her lifetime are mature at adult emergence) to synovigeny (all eggs will mature during her adult lifetime). A similar index exists for males (Boivin et al. 2005). Clearly pro-ovigenic and pro-spermatogeny species will be constrained in their ability to respond plastically during adult development, particularly if the species develops with a limited number of germ cells (i.e. does not contain germline stem cells). Given that we have limited knowledge, particularly in females, about where germline stem cells exist and how the rate of germ cell development is controlled, further studies into the interaction between these key reproductive traits will help us to define how reproductive potential shapes the evolution of mating systems.

5.5 Hormones, reproductive physiology, and trade-offs

No discussion of insect reproduction would be complete without a consideration of hormonal control of reproductive physiology and behaviour. Again, a review of this extensive body of literature is beyond the scope of this chapter. We would like to touch upon some of the issues that relate specifically to sexual selection and variation in reproductive success. A key regulator of reproduction in the insects is a hormone known as juvenile hormone (JH). JH is the primary co-ordinator of every major aspect of insect reproduction (Wyatt 1997; Flatt et al. 2005) and variation in JH signalling is proposed to underlie variation in fitness. JH regulates pheromone production, oocyte maturation, and production of accessory gland proteins, among other processes. Environmental influences on reproduction, such as nutrition and temperature, are mediated through the regulation of JH by insulin signalling and neuropeptides (Flatt et al. 2005). Mating experience can also influence JH synthesis and secretion. Mating often stimulates release of JH from the corpora allata in females, resulting in investment in oocyte maturation only when sperm are available to fertilize the eggs (e.g. Roth and Stay 1962). Mating experience can also stimulate JH release in males. In male Caribbean fruit flies, mating results in higher titres of JH, resulting in higher levels of sex pheromone release and higher mating rates (Teal et al. 2000), so that mated males have a competitive advantage over virgin males in acquiring mates. Thus the central role of JH on reproductive capacity and variation is clear.

We would like to focus on one aspect of JH that is particularly germane to a discussion of the evolution of mating systems: the role of JH in mediating trade-offs between primary and secondary sexual traits. Theory predicts that there should be some cost to the production of exaggerated secondary sexual traits (Andersson 1994); in insects, trade-offs between energetically expensive traits such as immunity and reproduction have been demonstrated (e.g. Hosken 2001; Simmons and Roberts 2005). Trade-offs between primary and secondary sexual traits, i.e. traits directly related to reproductive capacity and those related to the ability to secure a mate, have also been demonstrated (e.g. Moczek and Nijhout 2004; Simmons and Emlen 2006). But the patterns of the trade-offs are not always 'intuitive' (Simmons and Emlen 2006). In the horned beetle *Onthophagus nigriventris*, males manipulated to reduce horn growth responded with an increase in testis growth, as one might predict. However, across related species, there was no general relationship between horn size and testis size. There was a negative relationship between the allometric slope of horn size on body size and testis size on body size (Simmons and Emlen 2006), which could arise due to differential sensitivities to signals that co-ordinate growth in response to developmental conditions, such as nutrition. Since JH plays a key role in the pathway linking environmental conditions to internal physiology and development (Flatt et al. 2005), it seems worthwhile to explore the mechanisms by which JH could modulate male investment in fertility versus mating success, including perhaps those effects that could be mediated through insulin signalling. For example, recent evidence has shown that insulin signalling mediates female attractiveness in *D. melanogaster* (Kuo et al. 2012); a reduction in insulin signalling results in females that express a cuticular hydrocarbon profile that is not as attractive to males as females with an intact insulin signalling pathway. Another recently published example linking insulin signalling and sexually selected traits is the role of insulin signalling in the development of the sexually selected horn in rhinoceros beetles. Knockdown of expression of the insulin receptor by RNA interference results in a reduction in horn size, but in little or no change in size of wings or genitalia (Emlen et al. 2012). What is interesting in both these systems is that sensitivity of sexually selected traits to insulin levels may be a mechanism to enforce honest signalling relative to condition (Emlen et al. 2012; Kuo et al. 2012).

A few studies have attempted to examine the role of JH in trade-offs between male investment in primary sexual characteristics, such as sperm production, and physiological condition, such as immunity, with success at attracting females or in male–male competition. In stalk-eyed flies, females prefer males with exaggerated eye span, and eye span is correlated with success in male–male competition. Males treated with an analogue of JH during larval development develop increased eye span, thus increasing their potential mating success, but also have reduced testis size and transfer few sperm during copulations (Fry 2006). Thus, JH mediates a trade-off between mating success and fertility. In damselflies, application of JH also results in increased expression of sexually selected traits (Contreras-Garduño et al. 2009, 2011). Treated males had increased success in male–male competition, including increased aggression and territorial success (Contreras-Garduño et al. 2009) and wing ornamentation, and were preferred by females (Contreras-Garduño et al. 2011). In both these studies costs associated with the increase in sexually selected traits were demonstrated, either through a decrease in immune function or fat reserves. Both these studies suggest that JH could be a target of sexual selection as the mechanistic basis for secondary sexual character expression but also as a factor underlying the trade-off between mating success and fertility or survival (Flatt et al. 2005). A consideration not discussed in these studies, but one that deserves some thought, is that JH plays a primary role in reproductive capacity in both males and females. Thus selection in one sex may impact the other, and pathways involving JH may not be completely free to evolve in the sex on which selection is acting.

5.6 Reproductive physiology and trade-offs between reproductive and non-reproductive behaviour

It is clear that a knowledge of mechanisms underlying variation in fertility and fecundity will help shape our understanding of how selection can act on these fundamental fitness traits. But are there examples in which it is clear that selection has acted on these global reproductive regulatory networks that link genes, the germ line, physiology, and behaviour? One such example, in which there is extreme plasticity in reproductive potential, comes from the social insects that have castes of reproductive and non-reproductive individuals. And as we learn more about the mechanisms underlying this plasticity, we can begin to see that the evolution of the caste system in the social insects depends in part on selection on the pathways we have been discussing, linking the germ line and reproductive anatomy to development and nutrition.

One of the best-studied examples of a division of labour between reproductive and non-reproductive castes comes from honey bees. In honey bees we see extreme plasticity in ovariole number; queens develop with many ovarioles, and thus high reproductive potential, whereas workers develop with only a few ovarioles, and low reproductive potential (Rascón et al. 2011). The difference in ovariole number is set during larval development. In the fourth instar, both queens and workers have more than 100 incipient ovarioles, but, during the final larval instar in workers, most of these developing ovarioles undergo cell death and so workers emerge as adults with fewer than 10 ovarioles (Hodin 2009). The difference in development of queens and workers arises due to larval nutrition. Queens receive a nutrient-rich diet ('royal jelly') during development, whereas workers receive a more restricted diet. Recent work has begun to link this difference in diet to changes in the nutrient-sensing pathway. In the nutrient-rich diet conditions provided to the queen larvae, key components of the insulin-like signalling pathway are upregulated. And blocking a component of this pathway with RNA interference in young larvae results in larvae fed the queen diet developing as workers, suggesting that insulin-like signalling is necessary for the transition to a queen developmental pathway (Rascón et al. 2011; Wolschin et al. 2011). The increase in insulin-like signalling activity is accompanied by an increase in JH in the queen larvae as well as increased DNA methylation. Thus, insulin-like signalling and nutrient sensing may act upstream to regulate both DNA methylation and the endocrine system, resulting in differential development of the ovary and two castes, one reproductive and one non-reproductive.

Another example that may be less familiar—but of real interest to scientists interested in how development of the reproductive system may shape mating systems—comes from the polyembryonic wasps in the tribe Copidosomatini (family Encyrtidae). This group of parasitoids develops castes of larvae through clonal development of multiple offspring from a single fertilized egg (Strand 2009). One caste, the soldier larvae, lacks a reproductive system and never goes through metamorphosis. The role of the soldier larvae is to protect their siblings from intra- or interspecific competitors (Strand 2009). The other caste, the reproductive larvae, moults into reproductively competent adults. Reproductive and soldier larvae are clones, so the variation in reproductive development must come down to plasticity during development. The mechanism underlying this developmental plasticity has been best worked out in *Copidosoma floridanum*. The germ line is specified early during development by maternal determinants that become packaged into a group of primordial germ cells (PGCs). These cells are clearly marked as PGCs by expression of the conserved marker for the germ line in insects, vasa (Donnell et al. 2004). Ablation of the single vasa-expressing cell in the early embryo results in a brood composed entirely of soldier larvae (Donnell et al. 2004). During proliferation, most embryos inherit a few PGCs and then go on to develop into reproductive larvae. However, during proliferation some embryos fail to inherit PGCs, and these embryos develop as soldiers (Gordon and Strand 2009).

What makes this system particularly relevant to the evolution of mating systems is that the development of reproductive and soldier larvae is phenotypically plastic, both within and between species. For example, in *C. floridanum* the proportion of larvae developing as soldiers is highly variable and responds to the environment. When a single egg is laid in the host egg, the proportion of soldier larvae is about 4%. However, when a competitor egg is present, the proportion of larvae developing as soldiers increases (Harvey et al. 2000), so there is an adaptive shift in allocation to reproductive and non-reproductive larvae in response to competitors. The signals that stimulate this shift in allocation and mechanisms by which the proliferating morulae alter distribution of the PGCs remain to be elucidated.

A comparison of two species of polyembryonic wasps provides some insight into how selection on reproductive and non-reproductive castes can mediate a conflict over sex ratio (Smith et al. 2010). *Copidosoma floridanum* females lay one female and one male egg per host, producing mixed-sex broods. Males mate with their sisters, but also disperse and mate with unrelated females. The outbreeding of males produces sexual conflict between sisters and brothers (Grbic et al. 1992). In *C. floridanum*, female clones produce more soldiers and these emerge earlier than males and kill their brothers, manipulating the sex ratio to favour their sisters (Giron et al. 2007). Thus, the primary function of the soldier caste in *C. floridanum* is to resolve sexual conflict. A related species, *C. bakeri*, only oviposits a single egg and rarely produces mixed broods in the field. In *C. bakeri*, male and female clones produce similar numbers of soldiers (Smith et al. 2010), and the question was raised as to the function of the soldier caste in this species. In contrast to *C. floridanum*, *C. bakeri* soldiers express higher levels of aggression against heterospecific competitors than against intraspecific competitors. Given the variation in soldier behaviour and development between these two species, along with the available phylogenetic data, Smith et al. (2010) hypothesize that the soldier caste originally evolved to defend the host resource against heterospecific competitors, with males and females both investing in non-reproductive larvae. However, shifts in oviposition behaviour leading to sexual conflict resulted in a shift in the function of the soldier caste, in which the primary role became resolving conflict between brothers and sisters developing within the same host, resulting in sexual dimorphism in larval development and behaviour.

5.7 Conclusion

One reason that Thornhill and Alcock's (1983) volume was such a successful book was because they asked so many questions. In the book, they outlined directions in which we needed to go and, in so doing, stimulated 30 years of amazing research that has advanced our understanding of insect mating systems and beyond. We have attempted to update some of what we have learned about insect reproductive systems in the past 30 years, as well as introduce behavioural ecologists to reproductive physiology. But mainly, we hope we have updated the questions and outlined further areas for study in how the mechanisms underlying reproductive output can shape mating systems.

CHAPTER 6

Reproductive contests and the evolution of extreme weaponry

Douglas J. Emlen

6.1 Introduction

We saw in Chapters 2 and 3 that males very often face stiff competition for access to females. Competition can play out in many ways, from dashes across the landscape, as males race to be the first to find (and mate with) sparsely distributed females, to males competing with each other indirectly using elaborate courtship displays (Thornhill and Alcock 1983; Andersson 1994). It can even involve sperm competing for fertilization inside the reproductive tracts of females (Chapters 10 and 11; Simmons 2001). But in many species, male competition is overt, direct, and even fatal, as males fight outright with rival males in their struggle to achieve opportunities to mate.

Male contests are commonplace in an immense variety of species and habitats. In a few of these species, male contests have led to the evolution of spectacular weapons—structures such as legs or horns that have become so large that they appear exaggerated relative to other structures. Although such species are relatively rare, they hold a special place in biology because they include some of the most astonishing of all animal forms, displaying an extravagance and diversity of shapes that is breathtaking (Figure 6.1) (Emlen 2008).

What conditions give rise to such extremes in weapon form? And why is it that only a few species produce such elaborate structures, given that so many other species do not? It turns out that in each of these examples—indeed, in essentially every insect species with extreme male weapons—the answer is the same, and it comes from the logic of economics.

6.2 The economics of resource defence

Natural selection is the ultimate economizer, relentlessly culling individuals who allocate resources poorly. Over time, populations evolve to become efficient in their use of resources, investing in the growth of large structures such as weapons only when the benefits of having them outweigh any associated costs—that is, when they are cost-effective. By any measure, weapons are costly. Horns in beetles, for example, may comprise as much as 20% of the weight of a male, and horn growth can stunt growth of eyes, wings, and testes, depending on the species (Emlen 2001; Tomkins et al. 2005; Simmons and Emlen 2006). Weapons are expensive to produce, and they are expensive to wield. Males incur risks when

The Evolution of Insect Mating Systems. Edited by David M. Shuker and Leigh W. Simmons.
© The Royal Entomological Society 2014. Published 2014 by Oxford University Press.

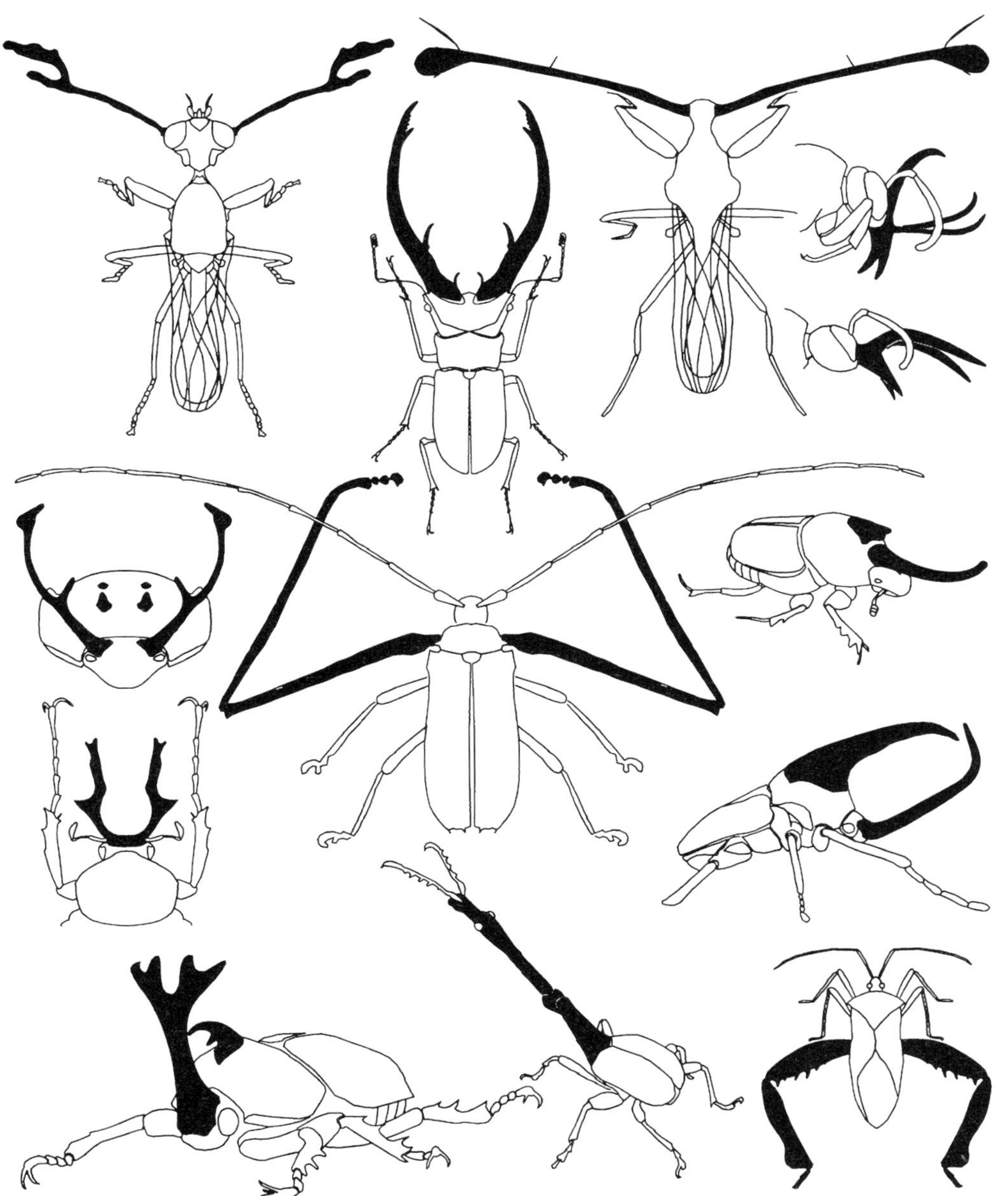

Figure 6.1 Examples of exaggerated male weapons in insects. Top row: Antlered fly (*Phytalmia antilocapra*), Lucanid 'stag' beetle (*Cyclommatus elaphus*), 'stalk-eyed' fly (*Cyrtodiopsis dalmanni*), 'tusked' wasps (head only) (*Synagris fulva* [upper], *S. cornuta* [lower]). Middle row: 'horned' beetles, dung beetle (head and thorax only) (*Onthophagus rangifer* [upper]), flower beetle (head only) (*Dicranocephalus bourgoini* [lower]); harlequin beetle (*Acrocinus longimanus*), dung beetle (*Onthophagus nigriventris* [upper]), flower beetle (*Theodosia viridiaurata* [lower]). Bottom row: 'rhinoceros' beetle (*Trypoxylus dichotomus*), 'giraffe' weevil (*Lasiorhyncus barbiornis*), 'leaf-footed' bug (*Acanthocephala declivis*).

they fight, and fighting over territories or resources requires time and energy that could be expended on other tasks like feeding. However, males with the largest weapons may derive significant reproductive benefits, if these weapons help them secure access to females and, at the same time, keep rival males away. Occasionally, the reproductive rewards can be huge, and, in these instances, the benefits of weapon production can outweigh the costs. Selection in these populations should favour males that allocate energy and resources to the production of large weapons.

Considering this problem like an economist involves weighing the relative costs and benefits of having a weapon. Costs include the material and energetic resources needed for growing and wielding the weapon, as well as the time and energy expended using that weapon to guard a resource. The benefits almost always involve increased access to females. Males benefit if they can mate with more females than rival males, and the greater the asymmetry in male reproductive success, the greater will be the pay-off to a male with the largest weapons (e.g. Shuster and Wade 2003). But under what circumstances will the benefits of investing in weaponry outweigh the associated costs? And when will the net benefits (benefits minus costs) of a weapon be the most profound? For many insects the answer depends on the kinds of resources they exploit, and how easy these resources are to defend.

Imagine a food resource spread out uniformly across a landscape (Figure 6.2a). Where would a male guard? Even if it were absolutely necessary for females to visit places with this resource to feed, and even if females were willing to mate with him if he happened to be there when they fed, where would that location be? For food resources broadly distributed in space, there is no obvious location that is better than any other. It would be impossible for a male to anticipate where females are likely to visit since they could find their food everywhere. A male could still invest in the production of weapons, and he

(a)

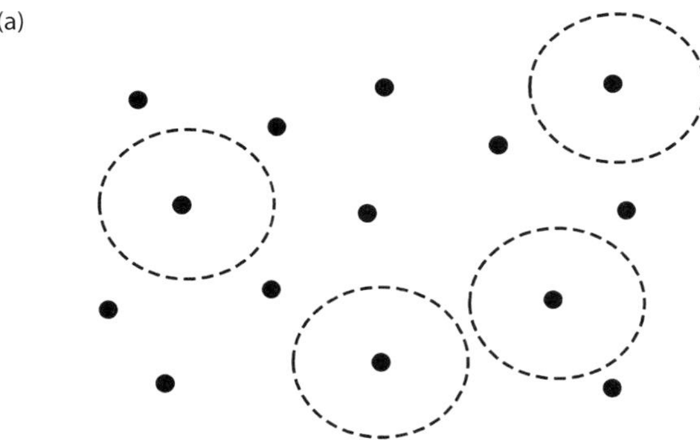

(b)

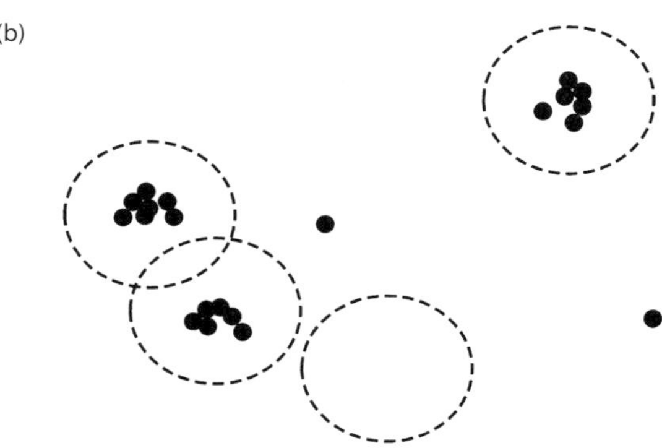

Figure 6.2 Economics of territory defence. All else being equal, guarding territories is cost-ineffective (a) when critical resources (filled black circles) are abundant and distributed uniformly across a landscape (territories indicated by dashed circles), and cost-effective (b) when resources are clumped into limiting and defensible patches. See text for explanation.

could use his weapons to deflect rivals from a piece of real estate. But why bother to guard one particular area if all of the surrounding areas are just as good? And why pay the price of producing a weapon and fighting, if other males without weapons or territories do just as well as him? Such behaviour would not be cost-effective.

If, instead, those same food resources were sparse, and especially if they were clumped into rare but concentrated patches, then the male would face a very different set of pay-offs for guarding a territory (Figure 6.2b). He would still pay a price for producing the weapon, and for expending time and energy fighting to keep rival males out of his territory. But now these territories would matter, and the benefits he could glean from guarding one may be significant (provided his territory included one of the rare resource patches). Females would be much more likely to visit him, since the resources they needed were few and far-between, and one of the only places where they could access them was inside his territory. Indeed, lots of females might visit his territory, and if the resources were localized enough to be readily defended from rival males, he might be able to mate with a disproportionate number of these females—more females than other males who were not able to guard a territory, and possibly even more than males who *were* guarding a territory, if his territory were bigger or better than theirs or contained more of the limiting resources.

What this exercise in the logic of economics reveals is a central tenet of animal behaviour: animals benefit most from fighting to guard territories when those territories contain valuable and limiting resources (J. L. Brown 1964; Emlen and Oring 1977; Thornhill and Alcock 1983). The more limiting they are, and the more economically defensible they are, the higher will be the pay-offs for successful guarding behaviour. However, whether or not a particular resource is valuable enough, or sufficiently localized as to make its defence cost-effective, depends entirely on the ecology and life history of the species in question, and what is defensible and valuable to one species may not be to another.

6.3 Two views of the same resource: harlequin beetles and pseudoscorpions

Harlequin beetles (*Acrocinus longimanus*) get their name from the angular streaks of orange, brown, and black that adorn their wing covers and body, but their most distinguishing features are their weapons: a pair of exaggerated forelegs which, in the largest males, can span almost 16 inches (Figure 6.1 and 6.3). Harlequin beetles are active during the rainy season in lowland tropical forests of Central and South America. Female beetles drill their eggs into trunks of fig trees after they have fallen to the forest floor. When the beetles arrive, they crowd on to the shaded undersides of the trunk—a fallen tree still sits on its branches, so that much of its length is actually perched several feet above the ground. Here, on the cool underside of the trunk, sap oozes from slashes nicked in the bark during the fall. Harlequin beetles feed on this sap and, more importantly, females use these parts of the trunk to insert their eggs under the bark of the tree.

Male beetles battle for ownership of these prized spots. There may be only one or two sap flows per tree, and the nearest other fallen fig may be miles away. From the perspective of a male beetle, sap flows are a critical resource needed by females. They are also rare, and highly defensible. Males battle vigorously with each other for possession of these territories and the consequent opportunities to mate. Males with the longest forelegs are most likely to win these battles, and males that win are the ones who mate with the females (Figure 6.4; Zeh et al. 1992). The extreme paucity of suitable egg-laying locations, combined with the restricted size and ease of defence of these resources, creates in this species an ecological situation where the benefits of fighting are enormous. For harlequin beetles,

Figure 6.3 Male harlequin beetles use their exaggerated forelegs as weapons in battles over sap flows visited by females. Photo: D. Emlen.

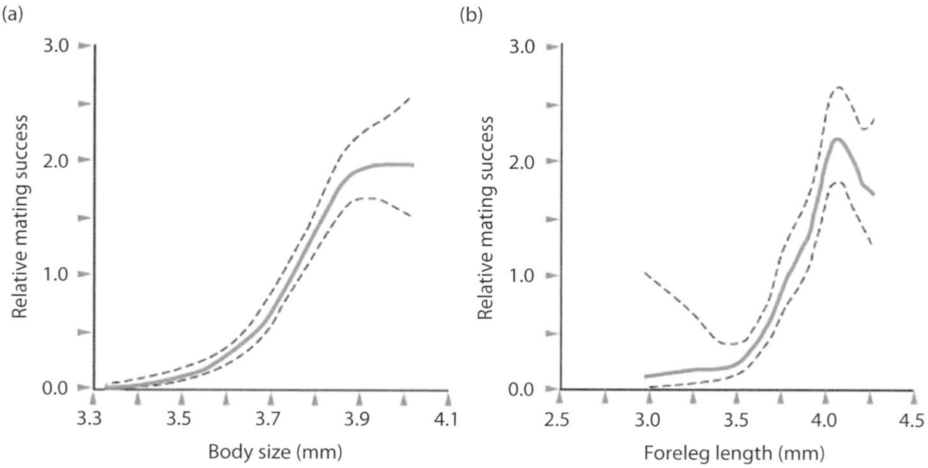

Figure 6.4 Battles between male harlequin beetles result in strong sexual selection for body and weapon size. Males with large bodies (a) and long forelegs (b) have higher mating success than males with smaller bodies and shorter forelegs. Redrawn by Echo Medical Media from Zeh et al. (1992).

males able to invest in large weapons fare much better than rival males who do not, so the weapons are cost-effective.

However, the most remarkable part of the harlequin beetle story does not actually involve the beetles. Instead, it involves an even stranger arthropod that hitches a ride on the beetles. Nestled on the abdomen, and tucked beneath the wing covers, are often a number of tiny pseudoscorpions (*Cordylochernes scorpioides*). This species also feeds on fallen fig trees, and their nymphs develop inside the rotting wood. Like their beetle counterparts, male pseudoscorpions have weapons—a pair of elongated, clasping pedipalps—and they use them in fights with rival males over opportunities to mate with females. Their weapons are impressive in their proportion too: much longer in males than in females, and extreme in the largest of males. But in absolute terms these animals are miniscule, and because of this, the context of the battles they wage is different. Indeed, because of their tiny size these males encounter a wholly different world from the beetles, and this illustrates why the defensibility and value of a resource depends entirely on the perspective of the species in question.

To a giant beetle with a reach of 16 inches, a wound on the belly of a fallen tree is both small and amenable to guarding. To a male pseudoscorpion, whose reach in the best of circumstances may span just a quarter of an inch, the same wound on that tree is enormous. An aggressive, territorial male may keep rivals away from one side but, while he does this, ten other males can approach females from other sides. The effort would be futile, and investing in weapons for such fights would be a waste.

However, female pseudoscorpions depend on something else that is much easier for males to defend. Fallen fig trees may be miles apart, and these little arthropods lack wings. So they climb on to the backs of the harlequin beetles and ride from tree to tree. It turns out that the back of a beetle makes a perfect mobile mating territory, and males can fight with each other to guard this resource. Males use their claw-like weapons in these fights, and male pseudoscorpions with the most extreme weapons win (Zeh and Zeh 1992). A male successful at guarding the back of a beetle may be able to mate with two-dozen or more females before the beetle lands, and once the beetle lands, the mated females hop off and a new round of females hops on! The result is that male weapon size is positively correlated with the number of offspring sired (Zeh and Zeh 1992). Thus, the high reproductive value of territories, and the premium it places on fighting ability, appears to have favoured the evolution of males who invest substantially in weaponry in both harlequin beetles and pseudoscorpions.

Consequently, here are two arthropods that feed on the same food, sap oozing from fallen fig trees. One can fly and the other cannot, and one is more than fifty times larger than the other. These simple differences have profound implications for territorial behaviour, causing sap flows to be economically defensible for one of the species, but not for the other. However, both species have a resource of critical importance to females that is rare and localized, and in both species males successful at guarding these resources are able to mate with many different females. That is, they can translate their success in fighting into success in reproduction. In both cases, this blend of ecological circumstances has led to a history of strong sexual selection for large male weapons.

6.4 Diversity of defendable resources: oozes, holes, burrows, and branches

Limiting, localized resources set the stage for the evolution of extreme weapons because in these situations the benefits of investing in armaments most dramatically exceed the costs.

Table 6.1 Examples of exaggerated morphological weapons in insects

Example	Weapon	Context of fights	Reference(s)
Thrips	Enlarged forelegs	Oviposition sites	[1]
Leaf-footed bugs	Enlarged hindlegs	Harems/branches	[2–4]
Wetas	Enlarged mandibles	Galleries/burrows	[5, 6]
Earwigs	Forceps	Crevices under stones	[7, 8]
Bees	Enlarged heads, mandibles	Burrows/nests	[9–11]
Wasps	Tusks	Mud burrows	[12]
Antlered flies	Antlers	Oviposition sites	[13–15]
Stalk-eyed flies	Eyestalks	Rootlets	[16–18]
Harlequin beetles	Enlarged forelegs	Sap sites	[19]
Fungus beetles	Horns	Bracket fungi	[20, 21]
Flower beetles	Horn	Branches	[22]
Stag beetles	Enlarged mandibles	Sap sites	[23]
Rhinoceros beetles	Horns	Burrows/branches/sap sites	[24–27]
Dung beetles	Horns	Burrows	[28–30]
Weevils	Horns	Oviposition holes	[31]

[1] Crespi (1988a), [2] Miyatake (1995), [3] Miyatake (1997), [4] Eberhard (1998), [5] Kelly (2006a), [6] Kelly (2008a), [7] Moore and Wilson (1992), [8] Tomkins and Brown, [9] Kukuk and Schwarz (1988), [10] Danforth (1991), [11] Houston and Maynard, [12] Longair (2004), [13] Moulds (1978), [14] Dodson (1997), [15] Schutze et al. (2007), [16] Burkhardt and de la Motte (1988), [17] Burkhardt et al. (1994), [18] Wilkinson and Dodson (1997), [19] Zeh et al. (1992), [20] Brown et al. (1985), [21] Conner (1988), [22] Holm (1993), [23] Hosoya and Araya (2005), [24] Daguerre (1931), [25] Eberhard (1978), [26] Eberhard (1982), [27] Hongo (2007), [28] Rasmussen (1994), [29] Emlen (1997), [30] Moczek and Emlen (2000), [31] Eberhard et al. (2000).

This simple observation turns out to be an almost universal feature of species with exaggerated weapons. Although the resources themselves differ greatly from species to species, their economic defensibility does not (Table 6.1).

Stag beetles (Lucanidae) battle over sap flows on the sides of standing trees. Like those of the figs, sap flows come from wounds to the tree. Sap flows are sparse and localized, and males use their elongated mandibles in fights over possession of these spots (Figure 6.1 and 6.5). Females visit sap sites to feed before they fly off to lay their eggs, and they mate with the resident male while they are feeding. This pattern occurs in many stag beetle species and, where there are exceptions to this rule—species that no longer feed and mate at localized sap flows—the species also have lost the enlarged male mandibles (Hosoya and Araya 2005).

New Guinean antlered flies (*Phytalmia* spp.) fight over tiny holes in the bark of fallen trees. Females must get through the bark to lay their eggs, and they can only do this by using an existing hole (the flies are not able to drill holes of their own). Males stand guard over a hole, moistening and marking it to attract females and fighting with all trespasser males who come by. Fights involve ritualized pushing matches, often with males stilting up on their hindlegs, and males in all seven *Phytalmia* species wield elaborate, rigid facial protrusions (antlers) that function in these male–male contests (Figure 6.1 and 6.6; Moulds 1978; Dodson 1997; Schutze et al. 2007).

Defence of burrows is probably the most widespread ecological situation that has led to the evolution of enlarged weapons. Shrimp and crabs that fight over burrows have huge claws (Knowlton and Keller 1982; Christy and Salmon 1984; Jennions and Backwell 1996). Wasps with long tusks fight over the tubular entrances to mud-pot nests (burrows) that they construct on the undersides of leaves (Figure 6.1; Longair 2004). Big-headed, big-mandibled male solitary bees fight over nest burrows excavated in soil (Kukuk and

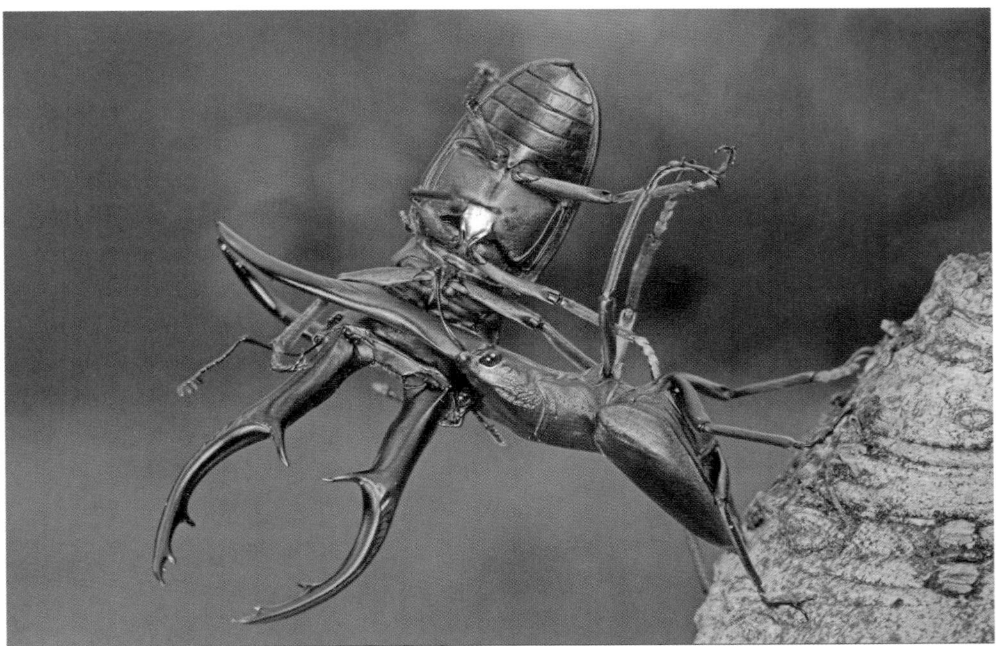

Figure 6.5 Male stag beetles fight over sap oozes on the trunks of trees. Photo: Igor Siwanowicz.

Figure 6.6 Antlered flies (*Phytalmia* sp.) battle over tiny holes in the trunks of fallen trees. Photo: Densey Clyne.

Schwarz 1988; Danforth 1991; Houston and Maynard 2012). Staphylinid beetles with exaggerated mandibles fight over burrows carved into fungi (Hanley 2001), and many species of rhinoceros beetle fight over burrows—either tunnels in the soil, or hollowed-out stems of plants such as sugar cane (Daguerre 1931; Eberhard 1979, 1987). Burrows, by their very nature, are restricted spaces where animals can brace themselves and twist or pry with rigid body protrusions, such as enlarged mandibles or a horn. The confines of a burrow appear to confer maximal leverage to many types of weapon, enabling males to efficiently utilize these structures to block the entrance and keep rival males away.

Branches work the same way. In essence, they are 'inverse tunnels' since a branch, like a tunnel, is a linear substrate that can be blocked, and along which a rival must pass. Thus, from the perspective of a guarding male, branches also can be considered localized and defensible. Insects as diverse as rhinoceros beetles (Eberhard 1978) and leaf-footed bugs (Miyatake 1995, 1997; Eberhard 1998) defend branches, blocking the passage of rival males and gaining access to females in the process. Many of these species invest in elaborate male weaponry (Figure 6.1).

6.5 Three prerequisites for an arms race

Contests between rival males can lead to the evolution of all sorts of male traits that improve a male's chances of victory, including bigger body sizes, aggression, faster flying or running speeds, and increased energy reserves for sustaining prolonged contests. Contests do not always lead to the evolution of exaggerated weapons, however, because in many species males with bigger weapons do not perform better than males with smaller ones. It turns out that in order for male weapons to evolve to extreme sizes, three conditions must be met.

First, males must face intense competition over access to females. This is the backdrop against which all sexually selected ornaments and weapons evolve, locally numerous reproductively available males forced to compete for a limited number of available females (Chapters 2 and 3). As a general rule, the more extreme the competition, the stronger the selection for traits such as weapons that improve a male's odds of securing access to females. Second, something about the biology or behaviour of a species must act like a 'bottleneck'—creating opportunities for males to mate with multiple females while keeping rival males away. Localized and economically defensible resources provide a substrate where males can plant themselves strategically, and in so doing gain disproportionate access to females. Only when reproductive competition is intense, *and* when resource defence is both practical and economical, will it pay for males to invest in battles over territory defence.

Third, the fights themselves musts be conducive to weapon evolution. Some fights, by their nature, are especially likely to confer an advantage to males with large weapons. Male battles can be roughly categorized into two basic types: scrambles, where many males attack simultaneously in a tangle, and duels, where rivals face each other one-on-one. Scrambles occur in the open, often in the air, and they tend to be wildly chaotic. Speed and agility may matter more than weapon size, and in these instances males with big weapons fare poorly if they are clumsy or awkward, or if the resources invested into weapon growth and maintenance detract from the pool available for stamina and speed. Duels, on the other hand, tend to be ritualized and predictable, head-on contests of strength or endurance, and in these fights the larger, stronger, or better-armed male almost invariably wins. When acrobatic duels occur in the air, as for example in many species of butterflies (reviewed in Kemp and Wiklund 2001) and damselflies (Grether 1996; Contreras-Garduño et al. 2006; Serrano-Meneses et al. 2007), victorious males tend to be those in better physiological condition, and, as in scramble contests, selection for agility and manoeuvrability appear to offset advantages of large weapons. However, when duels unfold in restricted places, such cavities, branches, or burrows, better-armed males consistently win, leading to strong selection for increasingly large weapons. All else being equal, we expect species in which males duel each other in confined spaces to be more likely to evolve extreme weapons than species where males fight acrobatically in the open, or chaotically in multimale scrambles.

6.6 Dung beetles with and without horns

Dung beetles face intense competition for their food resource (Hanski and Cambefort 1991; Risdill-Smith and Simmons 2011). Rich in nitrogen and other nutrients, dung is prized by thousands of species of flies and beetles that use it as provision for their offspring. In most habitats beetles must find dung fast, and, once they arrive, they have to contend with hordes of other insects that are attempting to steal the dung for themselves. The result is a race to stash dung away from the crowd, and one strategy that dung beetles employ is to carve a hunk from the pile, sculpt it into a smooth-sided ball, and roll it away from everybody else.

Ball-rolling is most often performed by males, and in extreme cases balls can be 50–80 times the size of the beetle. Females will join males as they leave the main dung pad and either cling to the ball and somersault along for the ride, or simply follow the male until the pair reaches an adequate patch of soft or moist soil. The beetles then co-operate to bury the ball, laying eggs beside or on top of the buried dung, depending on the species (Halffter and Edmonds 1982; Sato and Imamori 2008).

Females are not the only ones to approach males as they roll their balls away. Rival males constantly challenge each other over ownership of dung balls, and vigorous battles are commonplace. But these fights occur out in the open on the exposed surface of the soil, and males attack from all angles at once. Battles in ball-rollers often entail chaotic scrambles between three or four males. These species lack the restricted confines of a burrow or branch, and male fights are best described as scrambles rather than duels. Not one of the thousands of ball-rolling dung beetle species has horns.

Another strategy adopted by many dung beetles is tunnelling (Halffter and Edmonds 1982). Females of these species fly into dung and immediately begin to excavate burrows into the soil below. Once they have dug sufficiently deeply (a decimetre to a meter, depending on the species), they begin pulling pieces of dung down into these tunnels and stashing them away from the other dung-feeding insects above (Figure 7.1b). Females may make 50 or more trips to bury sufficient dung to provision just a single egg, and they will repeat this process for a string of successive eggs. Males fight among themselves to guard the entrances to tunnels, and victorious males mate repeatedly with the resident female. Males of tunnelling species often have horns (Figure 6.7).

When females reside inside tunnels, they are localized and readily defendable—exactly the situation in which we would expect to find a performance advantage to large weapons. Tunnels also physically restrict access in a way that aligns the interactions of opponents. A rival male has to enter the tunnel before he can challenge the guarding male, and there is space only for one rival to enter at a time. The restricted confines of tunnels align male battles so that they necessarily occur as a series of successive duels, rather than scrambles,

Figure 6.7 Many species of 'tunnelling' dung beetles have horns, which males use in fights inside burrows containing females. Clockwise from top left: *Onthophagus rangifer*, *O. nigriventris*, *Oxysternon conspicillatum*, *Phanaeus igneus*. Photos: D. Emlen (*O. conspicillatum*: D. Emlen and J. M. Rowland).

and males with longer horns are most likely to win these contests (Rasmussen 1994; Emlen 1997; Moczek and Emlen 2000; Pomfret and Knell 2006a). In dung beetles, species that fight inside tunnels and one-on-one often have elaborate horns; species that fight out in the open in chaotic scrambles do not. Simple changes in the way dung beetles hide their food resource therefore have had profound effects on the evolution of their weapons (Figure 6.8; Emlen and Philips 2006). Nonetheless, even within tunnelling species the shear density of beetles competing for resources can make tunnel defence uneconomical. In at least one guild of tunnelling dung beetles, the evolutionary origin of weapons has been found to depend strongly on the degree of crowding, with more crowded species less likely to have evolved exaggerated weapons (Pomfret and Knell 2008).

6.7 Stalk-eyed flies

Males in most species of Diopsidae have eyes protruding from their heads on the ends of long stalks (Figure 6.1), but males in a few of these species do not (Baker and Wilkinson 2001). Males use their extreme eyestalks to appraise and fight with rival males, and here, as with the dung beetles, we can begin to understand this variation by examining the details of their behaviour. Species such as *Teleopsis whitei* and *T. dalmanni* each have males with huge eyestalks, and spend their daytimes walking along the ground or on low vegetation near forest streams, feeding on fungi, molds, and yeast from decaying leaf litter

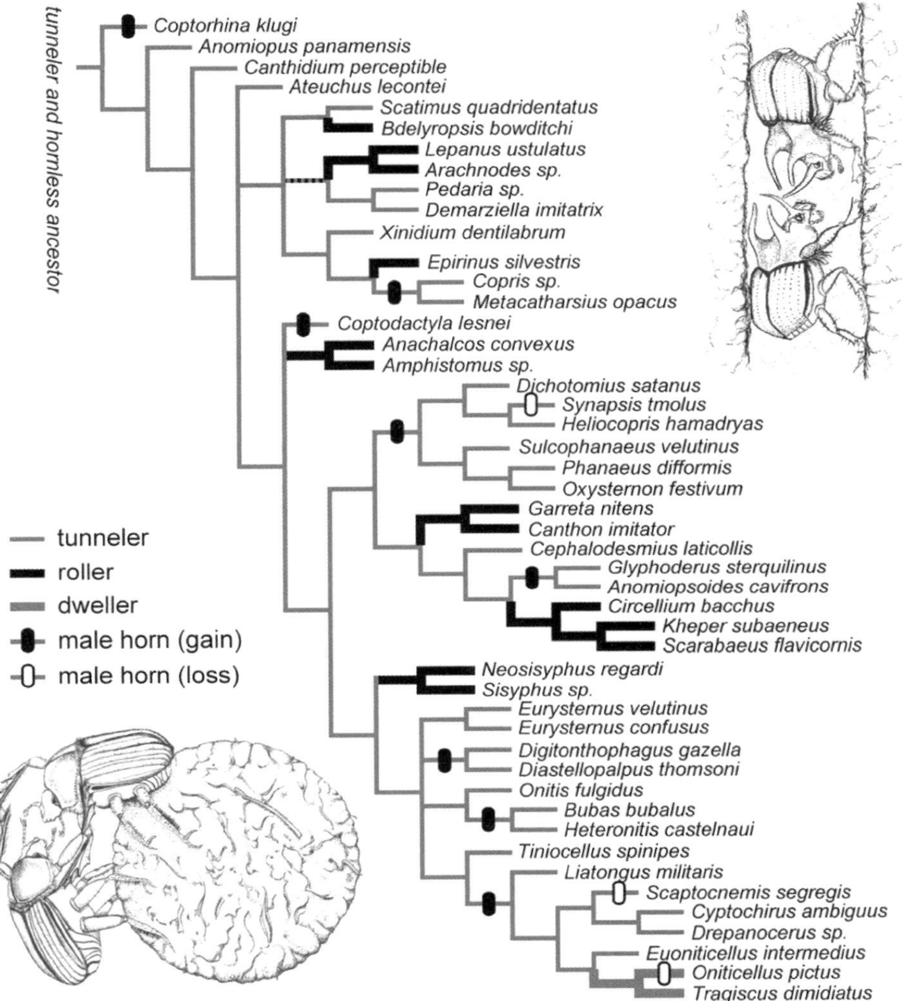

Figure 6.8 Phylogenetic hypothesis for the relationships among 46 species of dung beetle illustrating evolutionary transitions from tunnelling (thin grey branches) to non-tunnelling (ball-rolling or dwelling) behaviour, and gains (closed, vertical bubble) and losses (open, vertical bubble) of male horns. Non-tunnellers either roll balls (thick black branches) or form nests directly within dung (thick grey branches), both of which resulted in male contests occurring above ground. Gains of horns were significantly concentrated on branches of the tree that were also scored as tunnelling (8/8 gains of horns), and one of the three losses of horns occurred on a branch scored as non-tunnelling. Reprinted from Emlen and Philips (2006); phylogeny from Philips et al. (2004); illustrations by Melisa Beveridge.

or dead animals (Burkhardt and de la Motte 1988; Burkhardt et al. 1994; Wilkinson and Dodson 1997). During the day they forage alone, and they are aggressive to other flies that approach, whether they are male or female.

Each night, these same flies cluster together to roost in dense aggregations on dangling perches in the forest. In sheltered alcoves beneath the undercut banks of small streams, tiny rootlets hang down like threads. Some rootlets are longer than others, and long threads can hold more flies than small threads. Females often sidle in among as many as 20 or 30 other flies on a thread, forming a linear, hanging harem (Figure 8.3).

From a male fly's perspective, rootlets are critical resources routinely used by females, and long rootlets hold more females than small ones. For the few males that are able to do it, guarding a thread means securing disproportionate access to females, which, given the numbers of females per thread, translates into huge reproductive benefits to the males. Rootlets are also economically defensible: linear, localized, and easy to guard. Males battle to keep out rival males from their dangling territories.

When a rival male approaches, he hovers in front of the resident male, eyestalk to eyestalk. If the newcomer is smaller than the resident, he generally departs without incident. But if he is equal in size or larger, then a battle ensues. The intruding male lands on the thread and walks up to the guarding male. Face to face, forelegs outstretched, the males grapple and twist for control of the thread and, in virtually every instance, the male with the longer eyestalks wins (Panhuis and Wilkinson 1999). Successful males may mate with 20 or more females before they disperse the next morning to feed. Thus, for these species, the benefits of territory defence appear to vastly outweigh the costs of producing and bearing a weapon, even a truly enormous and awkward weapon. And the critical resource for which they fight is a localized, linear perch where females collect at night to roost. The resource is economically defensible, and, because of its linear shape and the way males must approach each other, confrontations between males occur face-to-face and one-on-one. These flies have all of the ingredients for arms races.

Remarkably, when people studied related flies that lacked large eyestalks, they found that the single biggest difference in their behaviour was that these flies did not roost communally at night (Burkhardt and de la Motte 1988; Burkhardt et al. 1994; Wilkinson and Dodson 1997). Flies such as *Teleopsis quinqueguttata*, which have only rudimentary eyestalks in males and females, never grouped into defensible clusters. They, like their relatives, fed in isolation during the daytime on fungi. But at night they roosted alone, dispersed among the vegetation. All mating occurred in the daytime, in happenstance and brief encounters between the sexes. There appears to be no critical resource or situation that concentrates females in these species, no opportunity for males to economically defend a territory or prevent rival males from mating with females, and no structure to the male confrontations that would cause them to occur in duels rather than scrambles. Consequently, related species of stalk-eyed fly occupy similar habitats, feeding on similar substances, but some of the species have bottlenecks—critical, economically defensible resources and one-on-one interactions—whereas others do not. Only the species with ecological bottlenecks have all the ingredients for an arms race, and these are the species with exaggerated weapons.

6.8 Conclusion

Many insect species face intense reproductive competition, with males battling rival males for opportunities to mate with females. Here and there, sprinkled among the multitudes of insect lineages, are species in which the intensity and nature of reproductive competition have led to the evolution of extravagant weapons—disproportionately large mandibles, antlers, horns, or legs—that aid males in battle. Although these species are relatively rare, they include some of the most fantastic and exaggerated examples of animal form. Examining the behaviours and natural histories of heavily armed species, and comparing these with related species lacking big weapons, hints at the ecological circumstances likely contributing to the evolution of extreme weapon sizes. Escalated weapon evolution appears most likely when insects depend on limiting resources that are localized and economically defensible, and when fights over these resources unfold on or in substrates that restrict access to guarding males, structuring male–male encounters so that they tend to occur as a succession of duels, rather than multi-male scrambles. This combination of (1) intense reproductive competition, (2) limiting, localized resources, and (3) one-on-one contests of strength, appears a potent formula for rapid evolution of extraordinary weapon sizes. In this chapter we have focused on how variation in the economics of resource defence among species can account for variation in sexual selection pressures that favour the evolution of exaggerated male weaponry. It is also the case that within species males may exhibit considerable variation in the phenotypic expression of weapons. In Chapter 7 we shall see how intense competition among males for access to females can favour the evolution of alternative phenotypes within species, such that males who lack the resources necessary to produce weapons or engage in aggressive interactions with rivals frequently adopt alternative means of gaining access to females.

CHAPTER 7

Alternative phenotypes within mating systems

Bruno A. Buzatto, Joseph L. Tomkins, and Leigh W. Simmons

7.1 Introduction

Insects were for a long time considered simple organisms with unvarying behavioural repertoires, incapable of complicated behavioural responses to changing environments and/or social conditions. However, nothing could be further from the truth; phenotypic plasticity is widespread in insect development, life history, physiology, and behaviour (Whitman and Ananthakrishnan 2009). Plastic responses to environmental and social conditions are actually central to the remarkable adaptability of insects, and have played a crucial role in their evolutionary histories (Moczek 2010; Simpson et al. 2011). Moreover, phenotypic plasticity in insects is not merely restricted to simple responses in metabolism or activity to abiotic factors such as temperature, but can be extremely elaborate, an illuminating example of which is the learning ability of honeybees (Menzel 1993; Hammer and Menzel 1995; Menzel and Muller 1996; Giurfa 2007).

Insect mating systems are no exception to this pattern. This chapter explores the intrasexual variation in behaviours and morphologies found in insect mating systems. More specifically, we focus on the evolution of the alternative means by which individuals obtain fertilizations, generally referred to as 'alternative mating tactics', or 'alternative mating phenotypes' (AMPs). The first studies to describe what we can interpret today as cases of AMPs in insects date back to at least the 1930s (Salt 1937), but it was only in the 1970s that the number of studies reporting this phenomenon started to accumulate. In 1983, when the classic examples of AMPs in digger bees and scorpionflies were reviewed by Thornhill and Alcock (1983), approximately 50 cases of AMPs in insects were already known, a number that has now surpassed the 200 mark. Here, we review the theoretical and empirical advances that have been made in this area since Thornhill and Alcock's volume.

We start by describing two illustrative systems in detail, gryllid field crickets and onthophagine dung beetles. These two groups were chosen because their reproductive biology is well known, and because the contrasting degrees of behavioural plasticity and morphological specialization between alternative phenotypes in these two groups illustrate the diversity of AMPs that has evolved in insects. We then discuss the genetic models that have been proposed to account for the evolution and maintenance of such dimorphisms, before reviewing the occurrence of AMPs in insects more generally. Finally, we

The Evolution of Insect Mating Systems. Edited by David M. Shuker and Leigh W. Simmons.
© The Royal Entomological Society 2014. Published 2014 by Oxford University Press.

discuss the relatively limited evidence for AMPs in female insects, a somewhat new and very promising area for future research.

7.1.1 *Behavioural plasticity in field crickets—calling versus searching for females*

Male gryllid crickets adopt an acoustic signalling behaviour in which they remain stationary within a protected burrow or crevice and produce a species-specific calling song to which females are attracted, often over considerable distances (Figure 7.1a; Loher and Rence 1978; Zuk and Simmons 1997). Once contact is established, the male ceases calling and begins to produce a second acoustic signal, the courtship song, which stimulates the female to mount, upon which the male inserts a spermatophore into her genital opening. Sperm drain from the spermatophore into the female's reproductive tract over a period of 40–60 min, during which time the male guards the female in an attempt to stop her from leaving and/or removing the spermatophore before sperm transfer is complete. During guarding the male also generates a fresh spermatophore, which he will offer to the female following a second bout of courtship song, or, if the female has escaped his attentions, he will begin to call again to attract another female (Zuk and Simmons 1997).

While calling behaviour is common within natural populations of field crickets, it is by no means the only behaviour that males adopt in their search for females. Males have also been shown to adopt a behaviour of silently searching for females, and a satellite behaviour whereby they remain close to a calling male and attempt to intercept females attracted to that male. The behaviour adopted by males does not appear to be fixed; any male can adopt any behaviour, depending on current environmental and social conditions. For example, in low-density populations most males call to attract females, and calling males achieve the majority of female encounters. However, when population density is high, males can have greater success in finding females by wandering silently through the habitat and encountering females by chance, rather than waiting for females to approach them (Simmons 1986a; Hissmann 1990).

Calling can attract unwanted attention. Male crickets are often subject to parasitism by acoustically orienting flies, *Ormia ochracea* (Cade 1984). The fly deposits larvae on the singing male, and these burrow into his body cavity to feed on his internal organs. Parasitism is lethal, with the infested male dying when mature fly larvae leave their host to pupate in the soil. Thus, although calling to attract females can be subject to sexual selection via increased male mating success, parasitism can represent a significant selection pressure against calling (Zuk et al. 1998). As a result, these opposing forces of sexual and natural selection can maintain genetic variation in the duration of nightly calling activity (Cade 1984). Finally, aggressive interactions among mate-searching males are intense, and calling can attract the attention of nearby males who aggressively attempt to monopolize the acoustic space for attracting females (Simmons 1988). Subordinate males who are unsuccessful in male contest competition are less likely to call and more likely to adopt the satellite and searching behaviours than are dominant males (Burk 1983; Simmons 1986a).

Calls are not the only important sexual signals, however. Male and female crickets also secrete long-chain cuticular hydrocarbons (CHCs) on to their cuticles that function in mate recognition and attractiveness (Thomas and Simmons 2009). Males that become subordinate after an aggressive interaction will upregulate their CHC secretion in order to make themselves more attractive to females via olfaction (Thomas and Simmons 2009; Thomas et al. 2011). It is clear from our example of these crickets that male insects can adopt a variety of means by which to find a mating partner, and that the behaviours an individual male adopts at any moment in time may depend on short-term changes in

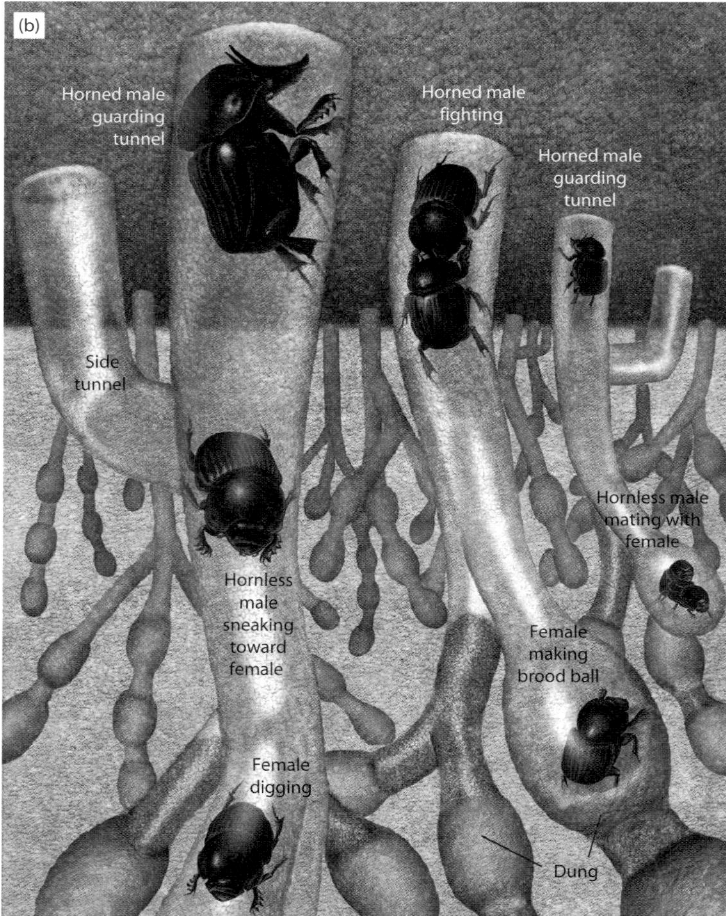

Figure 7.1 Examples of alternative mating phenotypes (AMPs) in insects. (a) Male field crickets call to attract females, silently search for females, or act as satellites of calling males. This male *Gryllus campestris* is calling at the entrance to its burrow. (b) Male dung beetles of the genus *Onthophagus* exhibit a suite of behavioural and morphological traits that characterize AMPs. Females dig tunnels in the soil beneath fresh dung, and provision brood masses for their offspring. Horned major males guard these tunnels from take-overs by other major males, and assist females in brood provisioning. Small hornless males dig side tunnels, sneak into breeding chambers and copulate with females while guards are defending tunnels or collecting brood provisions. (Illustration by Utako Kikutani.)

its environment and/or social pressures that affect his ability to find and compete for females either through acoustic or olfactory signals. Such behavioural plasticity can allow an individual cricket to make rapid adjustments to changing conditions. However, as we shall see in Section 7.1.2, for some species alternative behaviours can be associated with morphological differences that constrain an individual to adopt one mating behaviour for its entire lifespan.

7.1.2 *Morphological dimorphisms in dung beetles—guarding versus sneaking*

Onthophagine dung beetles are attracted to fresh animal droppings which they use as a resource for feeding and breeding (Simmons and Ridsdill-Smith 2011). Female dung beetles excavate vertical tunnels into the ground beneath the dung. They drag fragments of dung from the surface and pack it into the blind ends of side tunnels, producing a mass of dung or 'brood ball' into which they deposit a single egg. As with our example of field crickets, male dung beetles adopt alternative mate-securing behaviours but in this case they exhibit considerably less flexibility in terms of which behaviour they adopt (see also Chapter 6). Large males develop horns on the head and/or thorax, and compete for access to the tunnels within which females are provisioning offspring. These 'major' males will mate with resident females and guard the entrance to their breeding tunnels, fighting with other horned males for the sole access to the breeding tunnel (Figure 7.1b). Both body size and horn size are strong predictors of a male's competitive ability and ultimately his reproductive success (Hunt and Simmons 2001). Guarding males also assist females with brood provisioning, dragging fragments of dung down into the breeding tunnel and delivering it to the female to pack into her brood balls (Hunt and Simmons 2002b). Male parental care increases the size and/or number of offspring the pair can produce, and reduces the longevity costs of reproduction for females (Hunt et al. 2002).

A second class of males are smaller in body size with only rudimentary horns or no horns at all. These 'minor' males dig independent tunnels beneath the dung that intercept breeding tunnels, allowing them to sneak into the females' breeding chamber and copulate (Figure 7.1b). If discovered, sneaks are quickly evicted by the resident guard who will engage in retaliatory copulations with his mate (Hunt and Simmons 2002a). Thus majors specialize in fighting and defending females whereas minors specialize in sperm competition. Accordingly, minors invest more heavily in sperm production than do majors (see Chapter 10). The alternative morphologies of these beetles are set by the amount of dung that a beetle is provided with by its parents, with majors emerging from large brood balls and minors emerging from small brood balls. The behaviour of beetles also appears both fixed and dependent on morphology. Thus, minors never provide parental care. Majors do exhibit some behavioural plasticity insofar as they will adjust their parental care, reducing rates of brood provisioning and increasing time spent fighting and guarding when the frequency of sneaks in the population rises. However, they do not appear to adopt sneaking behaviour. Rather, guards will abandon the breeding tunnels earlier than sneaks and migrate to fresh droppings where there is less immediate competition for access to females (Hunt et al. 1999).

The mating systems of field crickets and dung beetles show us that male mating behaviours can vary both within individual males, and between males within a species. Moreover, they illustrate the range of traits that can vary; alternative phenotypes are not only characterized by phenotypic plasticity in behaviour, but can also involve specialization across different signalling modalities, body morphology and/or reproductive physiology. In the next section we will examine the evolutionary basis to the origin and maintenance of AMPs.

7.2 Modelling the origin and evolutionary maintenance of alternative mating phenotypes

The coexistence of discretely divergent phenotypes within a population raises an intriguing evolutionary question: how can these alternatives persist through evolutionary time, without one eventually driving the other to extinction? Evolutionary game theory, developed principally by Maynard Smith (1982), aimed to find solutions to this question of coexisting phenotypic variation. In the parlance of game theory, evolutionary stability (coexistence of alternative phenotypes) is sought from strategies that compete against one another. A complex terminology has been built around the game theoretic modelling of alternative reproductive strategies and to aid discussion we offer a glossary of these terms in Box 7.1. The evolutionarily stable strategy (ESS) is the strategy or set of strategies (an evolutionarily stable polymorphic state; Maynard Smith 1982, p. 11) that is resistant to invasion by new mutant strategies. At this point, it is worth rehearsing the terminology of game theory before continuing (Box 7.1): a 'strategy' or 'strategy set' is a 'decision rule', or set of rules, that has a genetic basis, whereas a *tactic* is the phenotypic expression of a

---

Box 7.1 GLOSSARY OF TERMS IN THE CONTEXT OF MODELLING ALTERNATIVE PHENOTYPES[a]

Strategy
A genetically based decision rule, such as 'fight if larger than opponent, sneak if smaller than opponent' (for a conditional strategy), or 'always fight' (for one of the strategies when alternative strategies coexist).

Evolutionarily stable strategy (ESS)
A strategy or combination of strategies that cannot be invaded by any other mutant strategy.

Pure strategy
When a population is composed of a single strategy (i.e. the conditional strategy and the mixed strategy are pure strategies).

Mixed strategy
A decision rule with a probabilistic basis (e.g. play 'sneak' with a probability of 0.25, play 'guard' with a probability of 0.75).

Alternative strategy
The occurrence in a population of a stable mixture of different, genetically based decision rules (e.g. strategy *a* = always sneak, strategy *b* = always guard).

Conditional strategy
A decision rule containing a conditional clause (e.g. fight if larger than opponent, sneak if smaller than opponent).

Genetic polymorphism
The occurrence in a population of alternative phenotypes reflecting genetic differences that are inherited as if they reflected differences at a single locus; genetic polymorphism is synonymous with alternative strategies.

Phenotype
morphology and/or behaviour produced by a strategy, synonymous with 'tactic' but without the baggage of being confused with 'strategy'.

---

continued

Box 7.1 *Continued*

Tactic

The phenotype generated by the decision rule (e.g. sneaking male behaviour). Tactic is synonymous with 'phenotype' and carries no information about the kind of strategy behind the tactic.

Switch point

The value of an environmental cue necessary to switch an individual's development from one phenotypic alternative to another.

Mean switch point

The value of an environmental cue necessary to switch the development of 50% of the population from one phenotypic alternative to another.

[a] After Tomkins and Hazel (2007).

strategy (Austad 1984). Hence, for the Onthophagine dung beetles discussed in Section 7.1.2 there is a single conditional strategy, giving rise to two tactics: majors and minors. In the damselfly *Mnais costalis* there are two genetically based alternative strategies each giving rise to a different tactic. Although 'tactic' is the game theoretic terminology, some authors prefer to use 'phenotype' because it is by definition different from a genotype (Tomkins and Hazel 2007). Thus, both tactic and phenotype can be used interchangeably, and we will use phenotype from now on, for the sake of consistency. It is also important to emphasize that phenotype might refer to a behaviour, or to a morphology. Game theory has advanced three solutions to the problem of the evolutionary stability of alternative phenotypes within a population: alternative strategies, the mixed strategy and the conditional strategy (Box 7.1; Gross 1996). Although game theory provides the basis for understanding how alternative phenotypes might coexist (Lively 1986; Gross 1996), quantitative genetic models explore the effects of selection and genetic variation on the coexistence of alternative phenotypes, as well as the ecological factors that affect which strategy sets will evolve (Hazel et al. 1990, 2004; Hazel and Smock 2000; Tomkins and Hazel 2007).

7.2.1 *Genetically based alternative mating strategies*

Where reproductive competition for females is intense, disruptive selection might favour individuals that circumvent competition, as well as those that engage directly in combat. In such cases it is conceivable that the fitness pay-offs for these combative or contrastingly sneaky males might be negatively frequency dependent (i.e. where a phenotype is fittest when it is rare). For example, a mutant male phenotype that is combative might invade a population of males that avoid fights; conversely a mutant sneaker male phenotype might invade a population composed only of aggressive guarding males. The fact that strong negative frequency-dependent selection can maintain genetic polymorphisms (Box 7.1) leads to the game theoretic premise for the existence of an evolutionarily stable state with two (or more) alternative genetic strategies (Maynard Smith 1982; Gross 1996). Where there are two alternative genetic strategies, their fitness is often thought to be equal; however, it is worth noting that the fitnesses of alternative strategies are *only* equal at the ESS frequency and that when there are three strategies frequency-dependent cycles of fitness may occur (Sinervo and Lively 1996). As a defining feature of alternative strategies, looking for equal fitness is more or less uninformative, particularly since demonstrating equality of fitness amounts to proving what is in general the null hypothesis of 'no difference'. Hence any conclusion of no difference can be criticized on the basis of statistical power. Furthermore, since neither equality of fitness nor negative frequency dependence separates alternative strategies from mixed or conditional strategies (Hazel et al. 1990; Tomkins and Hazel 2007, 2011), the research goal ought to be focused on what can separate these alternatives, that is, the genetic basis of the phenotypic variation in the population.

The distinguishing feature of alternative strategies is that their expression is insensitive to the environmental variation (Hazel et al. 2004). Hence the genetic architecture of these traits is traditionally thought of as a genetic polymorphism at one or a few loci, held in balance by negative frequency-dependent selection (Maynard Smith 1982; Gross 1996). However, since alternative strategies usually involve suites of traits, the notion that 'major genes' literally at a single locus are responsible for alternative strategies may in some cases be unrealistic. This is not a significant problem for the game theoretic modelling of alternative strategies, however, since it is the pattern of inheritance that is important, rather than the number of loci involved. For example, so-called 'supergenes' are tightly linked co-segregating (non-recombining) adaptive clusters of loci that tend to be inherited as one, and have been documented in the colour polymorphisms of Batesian mimetic butterflies (Joron et al. 2011; Jones et al. 2012). Supergenes seem to be a likely manner for many alternative strategies to be inherited. Hence, studies such as those of Tsubaki (2003) on the male-colour polymorphic damselfly *Mnais costalis*, although consistent with a single locus, may also indicate a supergene, since the polymorphism extends to size, behaviour and survival under parasite stress (Tsubaki et al. 1997; Tsubaki and Hooper 2004). Another possible genetic architecture for alternative strategies may arise from a chromosomal inversion (Gilburn and Day 1994a). Inversions have less recombination than other regions of the chromosomes (Kirkpatrick 2010) because recombination is rare to impossible in the heterozygote form. Furthermore inversions can have marked phenotypic effects (Tuttle 2003); these two features mean that inversions can become co-opted into the evolution of alternative strategies. A large chromosomal inversion system (three over-lapping inversions) occurs in the seaweed fly *Coelopa frigida*; this inversion system affects female mating preferences, male size and willingness to mate (Gilburn et al. 1992, 1993, 1996; Gilburn and Day 1994a), and it seems to be implicated in the determination of alternative male phenotypes in *Coelopa nebularum* (Dunn et al. 1999). Male reproductive success and the facultative polyandry of queens in the fire ant *Solenopsis invicta* was recently described as a supergene (Lawson et al. 2012). In this species, males of one genotype are unable to prevent queens from remating, while remating by queens mated to the alternative genotype is rare. Evidence now suggests that a genomic rearrangement in the form of an inversion is also behind this adaptive polymorphism (Wang et al. 2013).

7.2.2 *The mixed strategy*

Where strong negative frequency-dependent selection occurs there is a frequency of alternative phenotypes which is evolutionarily stable, and to which the population is expected to return following perturbation. Theoretically a female could follow a 'mixed strategy' and could partition her offspring into each phenotype at the ESS frequency (Alcock et al. 1977). Alternatively a male could enter a contest over a female and play each behavioural phenotype according to the ESS frequency (Maynard Smith 1982). The former was thought to be an important mechanism, particularly in the Hymenoptera where females have precise control over the sex and the size of their offspring (Alcock et al. 1977; Alcock 1995, 1996). Hence females could follow a mixed strategy and produce alternative male phenotypes in their offspring following the evolutionarily stable frequency. However, in Dawson's burrowing bee, for example, females are more likely to produce minor males if they are small or if it is late in the nesting season (poor foraging conditions), suggesting a conditional provisioning strategy, instead of a mixed strategy (Tomkins et al. 2001). There appears to be very little evidence for mixed reproductive strategies either generally (Gross 1996), or in the solitary Hymenoptera (Torchio and Tepedino 1980; Tomkins et al. 2001). The reason for this is likely to be that any adaptive tailoring (of the phenotypic alternative that is expressed) to an individual's circumstances (its size, age, or resource availability) will be favoured over a strict probabilistic decision rule, leading to the evolution of a conditional strategy.

7.2.3 *The conditional strategy*

Phenotypic plasticity occurs where changes in traits arise in response to environmental variability, and is an almost ubiquitous occurrence in living organisms (West-Eberhard 1989, 2003). In the context of reproductive competition, conditional strategies occur where there is plasticity in response to a cue, yielding divergent phenotypes aimed (evolutionarily speaking) at maximizing the organism's fitness given its circumstances. Exploring a conditional strategy (genetic decision rule) can make it clearer how conditional strategies operate. For example, frequently there are decision rules that involve competitive ability, which usually boil down to size; so the strategy might be: on the condition that the opponent is larger, 'play' (this is *game* theory) the phenotype *sneak*, and on the condition that the opponent is smaller, 'play' the phenotype *guard*. First, the strategy is conditional not because of differences in condition—although there frequently are—but because the strategy contains a conditional clause ('if *x* then do *a*; if *y* then do *b*'). Second, there is one strategy with two (or more) alternative phenotypic outcomes, but the two phenotypes are expressed conditionally on a cue. This contrasts with the mixed strategy in which phenotypes are expressed with a fixed probability and alternative strategies in which phenotypes are genetically determined. The conditionality in the decision rule reflects the expectation that individuals will adopt the phenotype from which they derive the highest fitness return for their circumstances (in this case whether they are larger or smaller than a rival). Frequency-dependent selection is not a necessary assumption for the evolution of conditional strategies; nevertheless it is a likely feature in many species simply because avoiding contests through sneaking is only viable when there are enough males in the guarding role to sneak on (as yet frequency dependence remains rarely quantified; Simmons et al. 2004). It follows that the average fitness of the different phenotypes arising from a conditional strategy need not be equal (Tomkins and Hazel 2007, 2011).

From this introduction to the conditional strategy, the game theoretic status-dependent selection (SDS) model of Gross (1996) follows most intuitively, even though this was actually pre-dated by the quantitative genetic model of the conditional strategy: the environmental threshold (ET) model (Hazel et al. 1990; Hazel and Smock 1993). Gross (1996) proposed that in most contexts where conditionally expressed phenotypes are selected for in males, male status is the factor that will determine the favoured phenotype. Hence low-status males tend to avoid aggressive interactions in favour of sneak matings, whereas high status males tend to fight for and guard females; this marries with the fact that in insects alternative phenotypes are frequently body size dependent (Hunt and Simmons 2001). Gross's SDS model assumes that fitness can be related to status in each morph, and that these fitness functions differ for the alternative phenotypes (Figure 7.2a). The intersection of the fitness functions is hypothesized to be at the ESS switch point (Box 7.1) under the SDS model; the status at which it is adaptive for individuals to switch phenotypes (Figure 7.3).

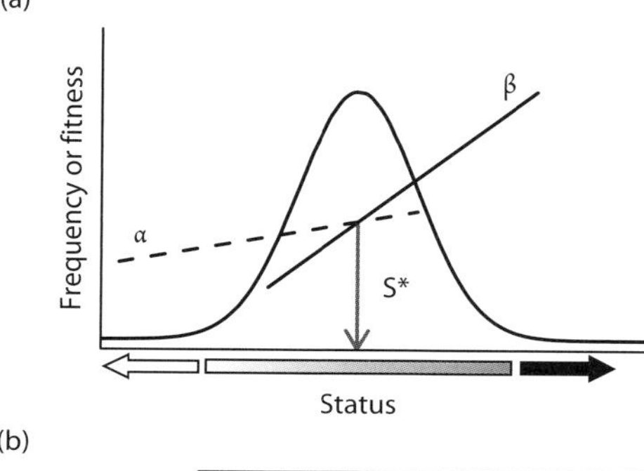

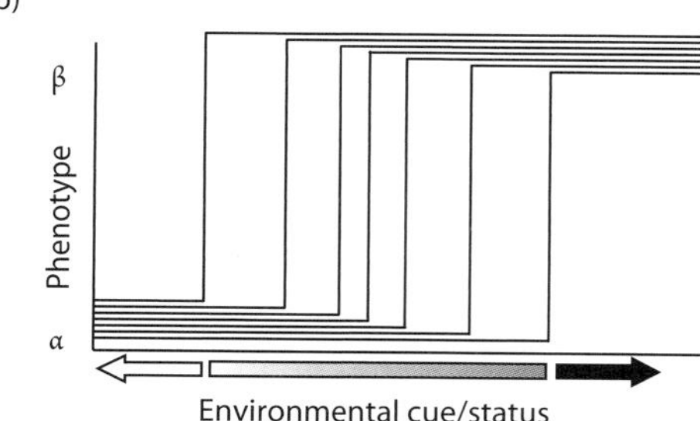

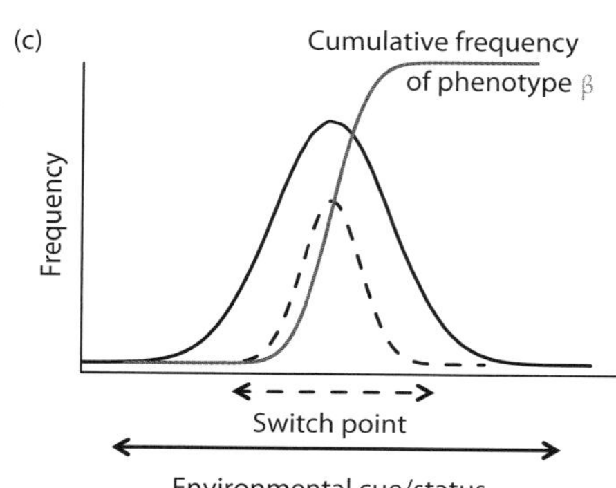

Figure 7.2 (a) The status-dependent selection (SDS) model proposed by Gross (1996) and (b) the distribution of switch points hypothesized under the environmental threshold (ET) model. (a) Under the SDS a normal distribution of status underlies phenotype fitness. Low-status individuals derive highest fitness from phenotype α whereas high-status individuals derive highest fitness from phenotype β. Under the SDS the ESS switch point S* is considered to be the point at which the fitness functions cross; note that the ET demonstrates that this is an oversimplification. (b) Under the ET model, genotypes vary in their response to the environmental cue. Here the sensitivities of seven genotypes are modelled in relation to variation in the environmental cue. (c) Here the distribution of switch points and the distribution of status (both assumed to be normally distributed) result in a cumulative frequency of males adopting phenotype β as status increases. Status is equivalent to the environmental cue. At the extremes of cue strength all individuals adopt the same phenotypes, whereas over a range of cue strengths some individuals will switch, but others not, depending on their individual switch points. The variance in switch points is a key factor in determining the effects of selection on switch points.

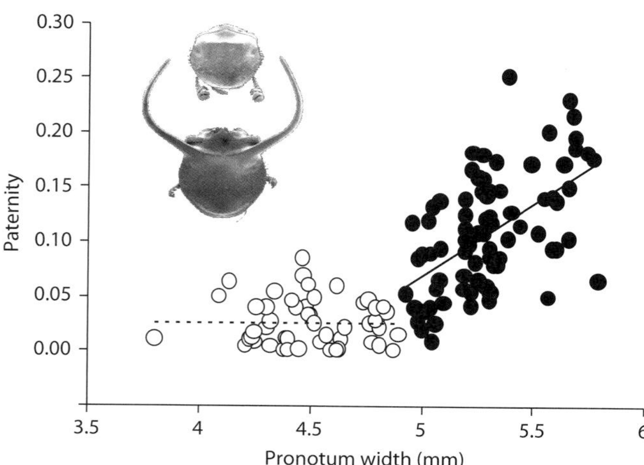

Figure 7.3 Status-dependent selection in the dung beetle *Onthophagus taurus*. Open circles, minors; closed circles, majors. The fitness functions are for fertile focal males competing against irradiated rival males or irradiated focal males competing against fertile rivals (Hunt and Simmons 2001).

The ET model (Hazel et al. 1990) provides a quantitative genetic model for the coexistence of alternative phenotypes. The genetic assumptions of this model are mathematically equivalent to the model proposed by Falconer (1965) to understand the heritability of dichotomous traits (Roff 1996). In Falconer's model, the population varies in a trait called 'liability', whereby individuals that exceed some threshold level of the liability go on to develop the alternative phenotype. The ET model uses the same principle but assumes that each individual in a population has a switch point. When the strength of the cue or cues (see Tomkins and Hazel 2011) exceeds an individual's switch point, that individual produces the alternative phenotype (Hazel et al. 1990; Tomkins and Hazel 2007, 2011). In a population, variation in switch points is modelled as a polygenic trait that is normally distributed and also has environmental variance; that is, the distribution of switch points can be modelled as being similar to any other quantitative trait (Figure 7.2b). The cue in the environmental threshold model is deemed to be environmental, hence it may not, at first sight, seem appropriate to model an individual's status (e.g. body size) as an environmental cue. However, body size and status in insects are often correlated with the amount of food ingested; such as in the case of the dung beetle *O. taurus*, where the amount of dung with which a beetle is provided by its parents determines its morphology: majors emerge from large brood balls, minors from small brood balls. Therefore, body size provides a good proxy for both status and environmental features such as resource availability. From a theoretical standpoint, as long as there is no genetic correlation between switch point and status, then status can be modelled in the same way as an environmental cue, despite its environmental and genetic basis (Hazel et al. 1990; Tomkins and Hazel 2007, 2011). In other cases it is clear that the status cue is wholly environmental, for example where behavioural phenotypes depend on relative size as in the rove beetle *Leistotrophus versicolor* (Forsyth and Alcock 1990). In the SDS model, the ESS switch point is hypothesized to be at the status that corresponds to the intersection of the fitness functions (Gross 1996). The ET model does not make this assumption, but rather uses information about the relationship between tactic fitness and the cue (Figure 7.2a), the distribution of the cue (i.e. status/body size/resource availability in many species), and the variation in the switch point distribution (Figure 7.2b, c), in order to estimate where the population's mean switch point should lie (Hazel et al. 1990; Tomkins and Hazel 2007, 2011). By quantifying these parameters researchers can use the ET to estimate how selection is acting on the switch point of a population, or whether there is a mismatch between the observed mean switch point and the ESS (Tomkins and Hazel 2011), i.e. whether the mean switch point is under selection.

7.3 Ecology and the genetic architecture of alternative mating phenotypes

When do we expect alternative genetic strategies to evolve and when do we expect a conditional strategy to evolve? This question goes beyond the architecture of the strategy itself and asks instead what ecological circumstances favour environmentally sensitive conditionality versus environmentally insensitive alternative genetic strategies. Insight into this question has come from studies of the morphologically plastic defence responses of barnacles to predation by a gastropod (Lively 1986; Hazel et al. 2004). If an organism is not phenotypically plastic it can be said to be 'canalized' such that a single strategy produces only a single phenotype, hence alternative strategies occur where mixtures of different canalized strategies coexist in a negative frequency-dependent manner. Alongside canalized strategies there are also conditional strategies that respond to environmental cues (Lively 1986; Hazel et al. 2004). It turns out that whether the ESS is a single canalized strategy (e.g. always guard), two alternative strategies (always guard for some individuals, always sneak for others) or a conditional strategy (e.g. guard or sneak dependent on some environmental cue) depends primarily on the reliability of the cue and the extent to which the individual's phenotypic 'choice' is tested by encounters with other individuals (Figure 7.4; Hazel et al. 2004). So, for example, if cues are unreliable (e.g. male size is a poor predictor of contest outcome), conditional strategies tend not to evolve but canalized strategies (e.g. always *fight* or always *sneak*) do. Where there is negative frequency-dependent selection there is also a small range of the parameter space where alternative strategies evolve to coexist as a genetic polymorphism (Figure 7.4; Hazel et al. 2004). Generalizing, where cues are reliable and an individual's choices are frequently tested by encounters with other individuals, the conditional strategy with two or more phenotypes evolves (Figure 7.4; Hazel et al. 2004).

This is perhaps the explanation for why, despite the diversity of reproductive competition and behaviour in the insects, about 95% of alternative male reproductive phenotypes represent conditional strategies (Table 7.1): status (size relative to an opponent) is a reliable cue of competitive ability. The only species for which alternative genetic strategies have been robustly documented are in the Odonata, where aerial pursuits are contests of stamina that are resolved according to energy reserves (Marden and Rollins 1994; Plaistow and Siva-Jothy 1996), rather than by physical grapples, and trials of size and strength. Much of the variation in success of territorial defence in damselflies comes down to age-related energy reserves, so while alternative behaviours in damselflies are common (e.g. Córdoba-Aguilar and Cordero-Rivera 2005), dimorphism *sensu stricto* based on size is unlikely, simply because size at the final nymphal instar is unlikely to be a reliable cue of success as an adult. In the rare case of *Mnais*, territory holding or non-territorial behaviour in males appears therefore to have been canalized genetically into two alternative strategies (Tsubaki 2003).

Finally, one way the change in genetic architecture from a conditional to a canalized strategy can be achieved is by mutation changing the sensitivity of the response to the cue. Genes with major effects can simply shift the switch point beyond the range of cues,

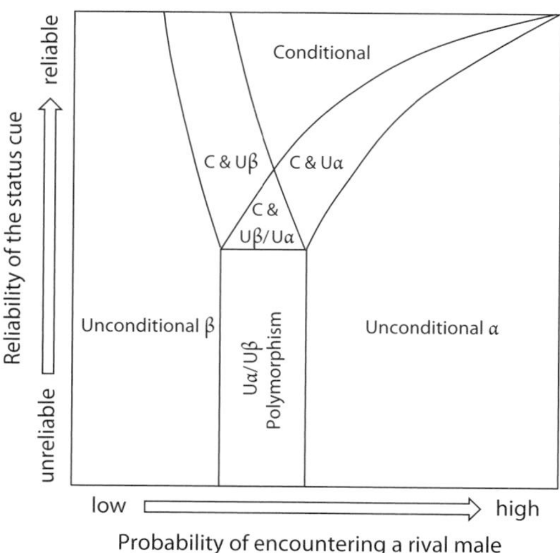

Figure 7.4 The ecological determinants of reproductive strategies (after Hazel et al. 2004). Hazel et al. (2004) showed that the conditional strategy evolves where cues are reliable and where encountering rival males is fairly common. In contrast, where cues are unreliable and encountering rival males is also common (and particularly where frequency-dependent selection is present) a genetic polymorphism can evolve. Depending on the variance in the switch point distribution, populations can evolve where unconditional and conditional strategists are present. The phenotype represented by α (minor phenotypes, such as 'sneaker' and 'satellite') and β (major phenotypes, such as 'fighter' and 'territorial') is likely to depend very much on the biology of the species.

Table 7.1 Cues that trigger the permanent or temporary adoption of conditional alternative mating phenotypes (AMPs) in insects[a]

Cue	No. of species	Extrinsic (E) / Intrinsic (I)
Body size or condition/diet	77	E × I
Density	36	E
Time of day/season	12	E
Age	10	I
Territory/female availability	9	E
Previous success	7	E × I
Food/host availability	6	E
Weather	6	E
Mate competition	5	E
Sex ratio	3	E
Body/wing colour	2	I (E × I?)
Female/copulation site	2	E
Oviposition site availability	2	E
Phenotypic frequency	1	E
Total	178	12 E, 4 I

[a] May be extrinsic (such as weather conditions), intrinsic (such as age), or an interaction between both.

so that the switch is always tripped, or never tripped (Hazel et al. 2004). Evidence for these kinds of major genes in insects has come from colour-polymorphic tobacco horn worms *Manduca sexta* (Suzuki and Nijhout 2008) where a mutation in the juvenile hormone regulatory pathway revealed a conditional alternative phenotype that was otherwise hidden. Clearly if these kinds of mutations occur, mixtures of conditional and unconditional strategies can result, and there is some evidence for this in male dimorphic acarid mites (Buzatto et al. 2012a), but not to our knowledge in the AMPs of insects.

Plaistow et al. (2004) developed a model in which the costs and limits of plasticity are considered as constraints on the extent to which populations of purely conditional strategists (Box 7.1) are likely to occur. Their modelling suggests that alternative genetic strategies will be more widespread than currently thought, or at least that purely conditional strategies should be very unusual. The data we present in Section 7.4 for insects suggests that conditionality, as has been the long-held view (West-Eberhard 1989, 2003), does in general predominate.

7.4 Alternative mating phenotypes are widespread in insects

When reviewing any aspect of insect biology, we must keep in mind that the diversity of the group has the habit of hampering generalizations. Insects can be found in nearly every terrestrial and freshwater habitat, occupying most ecological niches, feeding on any form of organic matter and interacting with all other organisms in the ecosystem (Grimaldi and Engel 2005; Schowalter 2006). It is therefore unsurprising that their mating systems are so diverse, and summarizing or categorizing the sexual behaviours and reproductive strategies across the whole class is a daunting challenge. We nevertheless attempted to review the occurrence and diversity of AMPs at the level of insect orders (Figures 7.5 and 7.6), as a small step towards understanding the evolution of intrasexual diversity in the mating systems of these organisms.

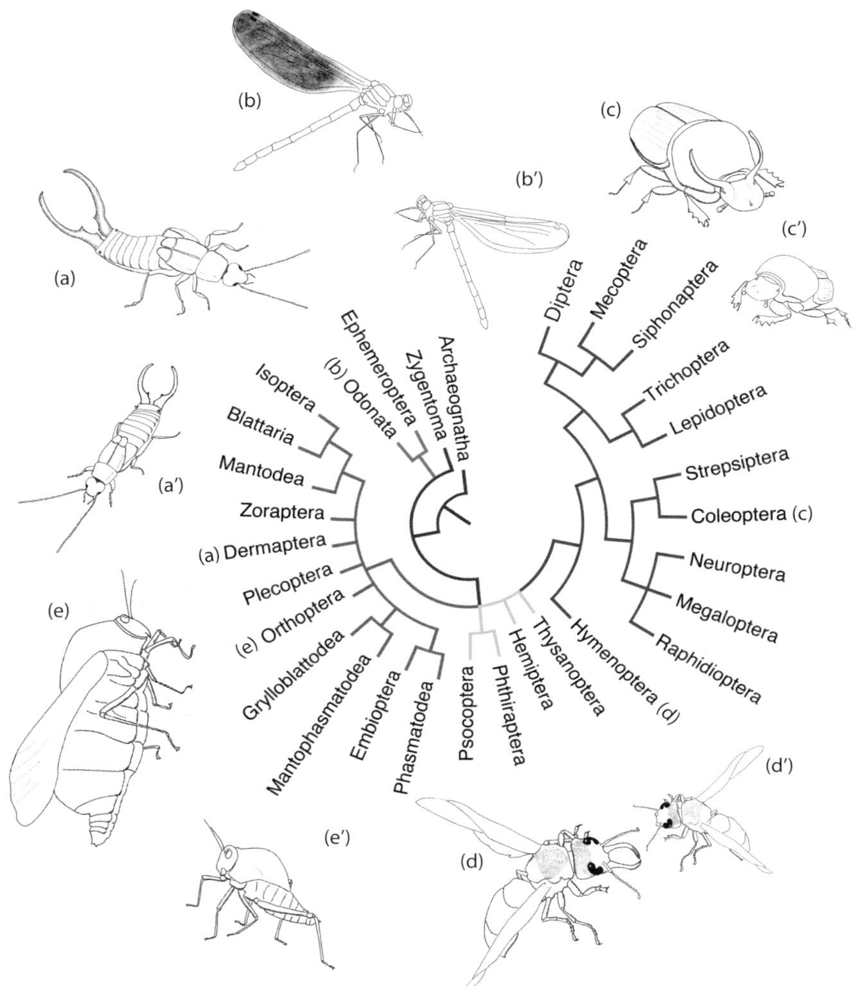

Figure 7.5 A recent phylogeny of Insecta (according to Ishiwata et al. 2011) showing the major insect clades (Holometabola in blue; Paraneoptera in yellow; Polyneoptera in brown; Palaeoptera in red), and wingless insects in grey. The drawings (by Rachel Werneck) illustrate five species in which male dimorphism is associated with the expression of AMPs: (a) a major and (a′) a minor of the European earwig *Forficula auricularia* (Dermaptera, Forficulidae); (b) a major and (b′) a minor of the damselfly *Mnais costalis* (Odonata, Calopterygidae); (c) a major and (c′) a minor of the dung beetle *Onthophagus taurus* (Coleoptera, Scarabaeidae); (d) a major and (d′) a minor of the potter wasp *Synagris cornuta* (Hymenoptera, Vespidae); and (e) a major and (e′) a minor of the bladder grasshopper *Bullacris membracioides* (Orthoptera, Pneumoridae).

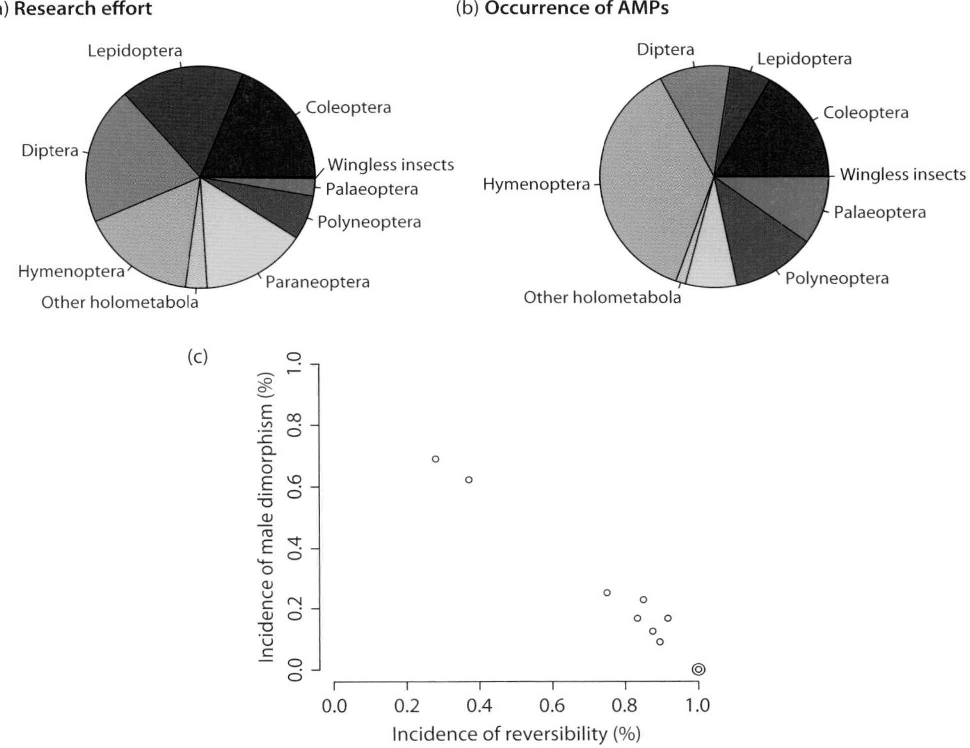

Figure 7.6 (a) The proportions of papers found in Web of Science by searching the name of each order (labelled 'research effort'); and (b) the proportion of species known to present alternative mating phenotypes (AMPs) (labelled 'occurrence of AMPs') in each major insect clade, with the extremely diverse Holometabola divided into five categories (its four largest orders and 'other holometabola'). (c) When AMPs are associated with male dimorphism, it is common for one, and sometimes both, male morphs to be specialized and therefore constrained to employ one of the possible phenotypes. The incidence of reversibility in insect AMPs is therefore usually significantly lower in orders with high frequency of male dimorphism ($r_s = -0.930$, $P < 0.001$, $n = 10$ orders). Although this correlation is very clear, there currently are no hypotheses to explain why some orders have higher frequencies of male dimorphism than others. Dermaptera (with a single species, *Forficula auricularia*) is removed from this analysis because, despite the well-known cases of male dimorphism, the degree of reversibility in AMPs is not well understood in the group. The two overlapping open circles represent the orders Blattodea and Mecoptera, which have the same combination of 100% reversible AMPs and 0% with male dimorphism.

AMPs have been recorded in more than 200 insect species, spread across 11 orders and all four major insect clades (Table 7.2). Within the Holometabola, AMPs can be found in a few species of butterflies and moths, flies and mosquitoes, scorpionflies, almost 40 species of beetles, and more than 80 species of wasps and bees. Within the Polyneoptera, AMPs have been described for crickets, bush-crickets, mole-crickets, grasshoppers, wetas, cockroaches, and earwigs. Within the Paraneoptera, AMPs are known for planthoppers, true bugs, water striders, and thrips. Finally, within the Palaeoptera, AMPs are expressed by males of several species of dragonflies and damselflies. These patterns strongly suggest that AMPs are ubiquitous. In fact, every new insect mating system that is investigated in

Table 7.2 Incidence of male alternative mating phenotypes (AMPs), described for more than 200 species spread across 11 orders and all four major insect clades

Clade	Order	No. of species[a]	Incidence of AMPs		Conditionality[b]	Fixed/reversible[c]	Morphology[d]	Dimorphism[e]
			Families	Species				
Holometabola	Coleoptera	386,500	12	37	29C (4?)	18F (3?), 7R	13D (1?), 12I (2?), 1Irr	9N, 20Y
	Diptera	155,477	12	22	17C (1?), 2M	2F, 17R	11I, 9Irr (2?)	20N, 2Y
	Hymenoptera	116,861	18	81	60C (28?), 1M?	49F (22?), 29R (10?)	47D (20?), 20I (2?), 11Irr (9?)	31N (8?), 51Y (9?)
	Lepidoptera	157,338	3	13	11C	1F, 11R (1?)	1D, 4I, 7Irr (2?)	11N, 2Y
	Mecoptera	757	2	3	3C	3R	3I	3N
	Megaloptera	354	0	0	–	–	–	–
	Neuroptera	5,868	0	0	–	–	–	–
	Raphidioptera	254	0	0	–	–	–	–
	Siphonaptera	2,075	0	0	–	–	–	–
	Strepsiptera	609	0	0	–	–	–	–
	Trichoptera	14,391	0	0	–	–	–	–
Palaeoptera	Ephemeroptera	3,240	0	0	–	–	–	–
	Odonata	5,899	5	22	16C (2?), 4M (1?)	3F (1?), 17R (5?)	3D, 5I, 11Irr (4?)	17N (8?), 5Y
Paraneoptera	Hemiptera	103,590	4	12	12C	2F, 10R	2D, 8I (4?), 2Irr	10N, 2Y
	Thysanoptera	5,864	2	4	3C	1F?, 3R	1D?, 3I	3N, 1Y?
	Phtiraptera	5,102	0	0	–	–	–	–
	Psocoptera	5,720	0	0	–	–	–	–
Polyneoptera	Blattodea	4,622	1	1	1C	1R	1Irr?	1N
	Dermaptera	1,978	1	1	1C	1R?	1I	1Y
	Embiodea	463	0	0	–	–	–	–
	Grylloblattodea	34	0	0	–	–	–	–
	Isoptera	2,692	0	0	–	–	–	–
	Mantodea	2,400	0	0	–	–	–	–

	Mantophasmatodea	15	0	0	–	–	–	–
	Orthoptera	23,855	6	24	23C (6?), 1M	3F (1?), 21R (8?)	3D (1?), 2I, 19Irr (12?)	21N (5?), 3Y
	Phasmatodea	3,014	0	0	–	–	–	–
	Plecoptera	3,743	0	0	–	–	–	–
	Zoraptera	37	0	0	–	–	–	–
Wingless insects	Archaeognatha	513	0	0	–	–	–	–
	Zygentoma	560	0	0	–	–	–	–
Total		1,013,825	66	207	176C (41?),8M (2?)	79F (28?), 118R (25?)	70D (23?), 69I (8?), 61Irr (30)?	126N (21?), 87Y (10?)

[a] From Zhang (2011).

[b] AMPs may be conditional (C) or Mendelian (M, pure genetic polymorphism),

[c] fixed (F) or reversible (R).

[d] We also included whether male morphology determines (D), influences (I) or is irrelevant (Irr) for the mating behaviour employed by individuals, and [e] whether there is morphological dimorphism (Y for yes, N for no) among males. '?' indicates uncertainty, whereas lack of information is omitted. This review was based on a search of the literature (through Web of Science) using the keywords 'alternative', 'mating', 'reproductive', 'tactic(s)', 'strategy(ies)', and combinations thereof. Even though the focus was only on mating phenotypes, and not on other features of insect reproduction (such as oviposition strategies), the term 'reproductive' was included among the keywords because the term 'alternative reproductive tactics/strategies' in the literature often encompasses AMPs. However, some studies describe the kind of intraspecific variation in mating phenotypes that falls under the umbrella of AMPs, without using any of the keywords mentioned above, and have likely been missed. Moreover, to restrict our literature search to studies on insects, the keywords 'insect(s)' and 'Insecta' were also used, and studies that only treated their focal species by a common name might have been missed. We therefore anticipate that our review is somewhat incomplete, and that AMPs are probably even more widespread than currently recognized. We would not be surprised if they were the rule, rather than the exception, in insect mating systems. Our review covered a total of 353 references, including journal articles and chapters in peer-reviewed books, published between 1934 and 2012 (data deposited at Dryad: doi: 10.5061/dryad.kp826).

detail reveals some sort of intrasexual variation in reproductive behaviour, and a broad definition of AMPs would probably encompass the great majority of well-studied species. We focus here on summarizing some studies in which the authors have clearly described males' reproductive behaviours as AMPs, or studies in which the description of the mating system hints at the presence of dichotomous means by which males obtain fertilizations. In the following sections we will illustrate the diversity of AMPs in insects, and elaborate on the continuum between plastic and fixed alternatives (sometimes involving morphological dimorphism) in the mating phenotypes of insects.

7.4.1 *Phenotypic plasticity and reversibility between alternative phenotypes*

Our examples of field crickets (Section 7.1.1) and dung beetles (Section 7.1.2) illustrate one of the axes of variation along which AMPs may be categorized: plastic (or reversible) versus fixed (or irreversible) throughout an insect's adult life. The behavioural plasticity of calling by male crickets illustrates how AMPs may be reversed or alternated during the adult life of a male, whereas the morphological dimorphism of male dung beetles represents AMPs that are irreversible after a male reaches adulthood. However, there seems to be a continuum between these categories in nature. An example of extremely reversible AMPs is found in the butterfly *Lycaena hippothoe*, where males use a flexible combination of territoriality (perching) and patrol flights, the frequencies of which are correlated with a fluctuating ecological feature—favourable weather conditions (Fischer and Fiedler 2001). Such highly plastic mating behaviours are to be expected whenever dynamic extrinsic factors play an important role in the relative fitness of the different alternatives (Tomkins and Hazel 2007). However, these factors need not only be related to climate or weather as the social environment can exert a similar effect. High population density usually triggers the switch from a calling to a satellite behaviour in crickets (Cade 1980, 1981; Cade and Wyatt 1984; French and Cade 1989), bushcrickets (Feaver 1983; Bailey and Field 2000), and also *Ligurotettix* grasshoppers (Greenfield and Shelly 1985; Shelly and Greenfield 1985, 1989). In all these cases, the relative pay-offs of calling and satellite behaviours seem to be so strongly influenced by fluctuating environmental conditions that, even if an intrinsic male trait (such as size or age) affects which behaviour is more likely to be adopted, either behaviour may be the optimal decision for any given male under the right conditions.

In contrast, in some mating systems only a limited number of switches between alternatives, or only switches in one direction, are possible during an adult's lifetime. This is illustrated by the damselfly *Calopteryx maculata*, whose males defend territories at the beginning of their adult life, switching to a sneaking behaviour later in life when their resource-holding potential declines (Forsyth and Montgomerie 1987). Similarly, in the scarlet dwarf dragonfly *Nannophya pygmaea*, territorial males also switch to a sneaking behaviour after being displaced from their territories by younger males (Tsubaki and Ono 1986, 1987). Constraints on the number or direction of switches are expected when the expression of one of the behaviours is clearly more costly than the alternative, and such costs can only be paid by males that are large, young, or in good condition (hereafter called majors). As a consequence, majors can often switch between behaviours, whereas smaller/older/poorer-condition males (hereafter called minors) are more commonly constrained to adopt the less costly mating behaviour. But this rule is not without exceptions, and the opposite pattern is found in the damselfly *Paraphlebia zoe*, whose majors (black-winged) seem fixed at performing a territorial behaviour, whereas minors (hyaline-winged, usually satellites) with relatively large body sizes can become territorials under certain social conditions (Romo-Beltran et al. 2009; Munguia-Steyer et al. 2010).

For heuristic purposes, in Table 7.2 we have categorized insect AMPs as 'reversible' or 'fixed' during an adult insect's life. Although we recognize that this is probably an oversimplification, it does allow us to speculate on general patterns. In particular, the category 'reversible' includes different degrees of reversibility. Accepting this criterion, reversibility seems extremely common in insect AMPs, occurring to some extent in at least 60% of the reported cases. This proportion is not general across orders, however; while full behavioural plasticity is the most common type of AMP in the Mecoptera (all three cases), Lepidoptera (11 out of 13 cases), Diptera (17 out of 22 cases), Orthoptera (21 out of 24 cases), Hemiptera (10 out of 12 cases), and Odonata (17 out of 22 cases), AMPs are usually fixed in the adult stages of Coleoptera and Hymenoptera, whose adults present reversibility of behaviours in only seven out of 37 cases and in 29 out of 81 cases, respectively.

7.4.2 *Costly secondary sexual traits and conditionality*

Secondary sexual traits in insects include male weapons, such as spines, spurs, horns, and the elongation and/or thickening of legs, mandibles, antennae, and forceps; as well as ornaments, such as the expansion and/or presence of colourful patches on wings, legs, and thoraces of males. As we have seen in Chapter 6, these traits often exhibit 'exaggerated' morphologies (*sensu* Emlen and Nijhout 2000), being much larger than other appendages, and can even be larger than the rest of the body. Producing and maintaining enlarged weapons and ornaments is inevitably costly (Emlen 2001; Jennions et al. 2001), and such costs cannot be paid to the same extent by all males in a population. Consequently, exaggerated traits typically display remarkable variation, with large males expressing them to a disproportionally greater extent than small males (Wilkinson and Taper 1999). Moreover, the degree of phenotypic variation in exaggerated structures is expected to be greater in species under more intense sexual selection, which has been demonstrated for the forceps of earwigs (Simmons and Tomkins 1996), the extremely long eye stalks of stalk-eyed flies (Wilkinson and Taper 1999), and the enlarged mandibles of stag beetles (Knell et al. 2004).

Heterogeneity in the ability of males to produce and maintain secondary sexual traits makes fighting and defending territories or females only profitable to the large males of a population (Chapter 6). In contrast, small males may benefit from avoiding the costs of producing weapons or engaging in fights altogether (Oliveira et al. 2008). Weapons and ornaments of small males are often much reduced or completely absent, which generates dichotomous morphological variation among males, known as male dimorphism (Gadgil 1972) or intrasexual dimorphism. In such cases, one or both male morphologies (or *morphs*) are usually specialized and constrained to employ one of the AMPs, and therefore the incidence of reversibility in insect AMPs is usually lower in orders with high frequencies of morphological dimorphism (Figure 7.6). In the Coleoptera and in the Hymenoptera, 54% and 63% of all cases of AMPs are accompanied by dimorphisms in a way that the morphology of males determines, or at least strongly influences, the mating behaviour employed by them. Meanwhile, in the other insect orders, far fewer male dimorphisms occur in species with AMPs. The exception here is the Dermaptera (earwigs), in which the nature of and flexibility in AMPs is not well understood despite well-known cases of male dimorphism.

When distinct male morphs are determined by Mendelian inheritance, the adoption of a mating phenotype is fixed, and completely insensitive to the environment. This phenomenon has traditionally been referred to as 'alternative strategies' (Section 7.2.1; Gross 1996), and seems very rare in insects, being perhaps applicable in one genus of seaweed flies (Dunn et al. 1999), but confirmed only in one genus of damselflies (Tsubaki 2003), and in one species of cricket (Tinghitella 2008). More commonly, fixed AMPs employed by distinct male morphs are conditionally expressed, such that environmental effects play the predominant role in phenotypic expression, even if a male's behaviour and morphology is irreversible after it reaches adulthood. Traditionally named 'conditional strategies' (7.2.3; Gross 1996), in these cases environmental cues determine which phenotype is expressed by a given male at a given time (for reversible AMPs) or throughout his adult life (for fixed AMPs). This sort of sensitivity to the environment is central to insect development, behaviour, and general life history (Whitman and Ananthakrishnan 2009), and it is no different with their mating systems.

There are more than 170 species of insects in which conditionally expressed AMPs have been detected (Table 7.2). The cues that trigger the permanent or temporary adoption of a mating phenotype can be almost as diverse as the mating phenotypes themselves (Table 7.1). Cues can be extrinsic factors, such as density of competing males or weather conditions, or intrinsic traits, such as age, body size, or colour. In many cases intrinsic and extrinsic factors interact, for example when intrinsic competitive ability interacts with the density of competing males to determine which mating phenotype to adopt. The most common cues used in the expression of conditional mating phenotypes in insects are body size and condition, both emerging as an interaction between intrinsic traits (such as foraging efficiency) and extrinsic traits (such as food availability). We refer to the cues that determine conditional phenotypes as 'environmental cues', because of the general influence of the environment on this type of AMP.

7.4.3 *Variation in AMPs and male dimorphisms*

Male dimorphisms have been documented in more than 80 different insect species so far. In the basal Palaeoptera, all cases of AMPs were described for the order Odonata, and in some damselflies these alternative behaviours are correlated with male dimorphisms. In the genus *Mnais*, for instance, majors have orange wings, are territorial, and guard females, whereas minors have pale wings, are non-territorial and sneak copulations (Tsubaki 2003). This dimorphism is linked to a genetic polymorphism at a single autosomal locus, and the orange wing pigmentation is inherited in a Mendelian fashion, constituting one of the rare cases of alternative strategies (*sensu* Gross 1996) in insects.

Within the Polyneoptera, the classic cases of AMPs consist of full behavioural plasticity in the calls of different species of field and bushcrickets (see Section 7.1.1), without any morphological correlates. However, there are some cases of male dimorphisms in this clade as well, most of which represent conditionally expressed dimorphisms. These conditional dimorphisms include the swollen abdomens of bladder grasshoppers (Donelson and van Staaden 2005), the enlarged mandibles of wetas (order Orthoptera; Koning and Jamieson 2001), and the elongated forceps of earwigs (order Dermaptera; Simmons and Tomkins 1996; Tomkins and Simmons 1996; Tomkins 1999). In all these cases, male morph expression is tightly linked to body size, which is greatly influenced by diet in the earwigs (Tomkins 1999) and in insects generally (Nylin and Gotthard 1998). But at least in bladder grasshoppers, the expression of the minor morph is also triggered by the crowding effect of high population density (Donelson and van Staaden 2005).

Within the Paraneoptera, the dichotomy between dispersing versus fighting seems to be common and wingless/winged dimorphism in both sexes has been reported for thrips (order Thysanoptera; Crespi 1988b), planthoppers (Langellotto et al. 2000), and chinch bugs (order Hemiptera; Fujisaki 1992). In all these cases, wingless males, or males with shorter wings, attain copulations more frequently at their place of birth than their winged counterparts, generally because they are more aggressive and repel other males. Winged males are capable of dispersing and seeking copulations in other places. In contrast, in the thrips *Elaphrothrips tuberculatus*, all males are winged, and, whereas large males are fighters and defend egg-guarding females, small males sneak copulations with females guarded by large males (Crespi 1988c). Other than male size, local sex ratio also influences the adoption and switching between fighting and sneaking behaviours in this species (Crespi 1988c).

Finally, male dimorphism has been extensively reported for the Holometabola, manifest mainly in the horns and mandibles of several groups of beetles (order Coleoptera; Emlen et al. 2007), the colours of butterfly wings (order Lepidoptera; van Dyck and Wiklund 2002), as well as the wings, heads, mandibles, and body size of several families of Hymenoptera (e.g. Danforth 1991; Simmons et al. 2000; Cook and Bean 2006). It seems that this group surpasses any other insect group in terms of the repetitive evolution of morphological dimorphism among males. However, the effect of research effort must not be overlooked, and the fact that these orders are by far the most studied (see Figure 7.6) might be the reason why male dimorphisms have more often been recorded for them.

It is also important to emphasize that the degree of male dimorphism of a given species can also vary in a continuous fashion, and interestingly the whole continuum can be found within the Hymenoptera alone. For instance, in some species of sweat bees from the genus *Lasioglossum* (Halictidae), males may be clearly assigned to one of two morphs, each of them with very distinct head width and mandible size (Houston 1970; Kukuk 1996). Large-headed males are flightless fighters that never leave their nest to mate, whereas small-headed males fly out of the nest and disperse before mating (Kukuk and Schwarz 1987). At the other end of the continuum, the AMPs adopted by a male can depend on its body size, but the distribution of body sizes or the distribution of secondary sexual trait sizes can be unimodal, with no evidence of morphological male dimorphism. In the wasp *Polistes dominula* (Vespidae), males can employ a resident mating behaviour that involves territoriality, aggressiveness, and site-faithfulness, or a transient mating behaviour that involves larger ranges, no aggressiveness, and little site tenacity (Beani and Turillazzi 1988). Larger males are residents more frequently than small males, but some individuals switch between behaviours, and the distribution of male sizes is not bimodal (Beani and Turillazzi 1988). Indeed, many species fall somewhere in the continuum that exists between monomorphism and dimorphism. In an undescribed species of *Philotrypesis* fig wasp, for instance, males from the 'aggressive' morph are much more likely to fight than males from the 'passive' morph, but male morphs have overlapping distributions of both body sizes and mandible sizes (Cook and Bean 2006). The dimorphism can only be detected from the ratio of mandible size to head size (which is higher in the aggressive morph), constituting a phenomenon described as 'cryptic male dimorphism' (Cook and Bean 2006).

7.4.4 *Insect development and the evolution of male dimorphism*

As in all arthropods, insect growth is constrained by their hardened exoskeletons, resulting in a punctuated growth pattern restricted to moulting events. The number of such events is usually fixed within species, and the last moulting event precedes the adult stage, when sexual maturity is reached and growth ceases in the great majority of insects (Triplehorn and Johnson 2005). This pattern of growth holds interesting implications for the evolution of fixed AMPs. The lack of growth in adult insects means that if the adoption of AMPs is tightly connected to body size, sexually mature males cannot switch between different mating behaviours, as reversibility is constrained by the fixed adult size. For example, in the bladder grasshopper *Bullacris membracioides*, AMPs are tightly connected to male size and morphology, which are strikingly divergent (van Staaden and Romer 1997). Large males have wings and inflated abdomens, which are used to emit high-intensity calls to females, whereas non-inflated small males are wingless and adopt a satellite strategy, parasitizing the sexual calls of large males (van Staaden and Romer 1997; Donelson and van Staaden 2005). When a male reaches sexual maturity, growth ceases, and his mating phenotype becomes fixed for the rest of his adult life (Donelson and van Staaden 2005). So when small male bladder grasshoppers moult into adulthood, they become permanent satellites, providing a classic example of fixed AMPs.

Another implication of insect growth patterns is that punctuated growth may underlie the origin of morphological polymorphisms. If different males are capable of arresting growth and reaching maturity after a different number of moults (or instars), adult size distributions may become multimodal. In the wellington tree weta, *Hemideina crassidens*, males can mature at the 8th, 9th, or 10th instar, generating a trimorphism in adult male head size (Kelly 2008b; Kelly and Adams 2010). This head polymorphism is linked to AMPs in the species, although only two behaviours (not three) exist: 10th instar males defend harems and engage in intense male–male fights, whereas 8th and 9th instar males seem to adopt a sneaker behaviour based on acquiring mates by either furtively invading galleries defended by 10th instar males (Kelly 2008b; Kelly and Adams 2010) or by mating with females in small galleries that are inaccessible to large males (Kelly 2006b).

A final implication of insect developmental patterns on the evolution of AMPs is the ability (or inability) to determine offspring sex, coupled to the importance of offspring provisioning. When parental provisioning decisions determine offspring body size—and consequently male behaviour—dimorphisms might be constrained by the parents' ability to determine or discriminate the sex of each individual offspring. This is the case of the dung beetle *O. taurus*, where parents gather all the dung on which the offspring will feed throughout their entire development, and the amount of parental provisions strongly influence offspring body size and thus male mating behaviour (Hunt and Simmons 2000). Horn length in this species is strongly dimorphic and associated with AMPs (see Section 7.1.2). But because dung beetle parents seem unable to allocate the sex of their offspring (Kishi and Nishida 2008), neither the brood masses built for male offspring nor the distribution of body sizes in those offspring are dimorphic (Buzatto et al. 2012b). It is hence possible that the inability of parents to discriminate the sex of their offspring is the reason why male offspring of intermediate body size and horn length are produced, even though these males are not optimally adapted to the fighter or sneaker mating behaviour. In contrast, it is well-known for females to adaptively allocate sex in the Hymenoptera, either in terms of the local mating conditions her offspring will face, or in terms of the resources available for offspring and the sex-dependent fitness consequences of those resources (so-called conditional sex allocation: Trivers and Willard 1973; see also West 2009 and Chapter 15). In Dawson's burrowing bee, the amount of pollen that mothers provide to a given brood cell determine the size and mating behaviour of their offspring, and hence these bees dig cells that are bimodal in size distribution even before they begin to provision those cells and allocate sex appropriately (Tomkins et al. 2001). Unsurprisingly, in this species no intermediate-sized males are produced, and male offspring body size is strongly dimorphic (Alcock 1997; Simmons et al. 2000).

7.5 Variation in female mating systems

At the time of the original publication of Thornhill and Alcock's (1983) volume, all cases of AMPs known were restricted to males. This is somewhat unsurprising considering that mating systems theory has tended to focus on the manner and the degree to which the limited sex (usually males) monopolizes mates and/or the resources that their mates depend on (Emlen and Oring 1977). This naturally leads to a male perspective on the study of mating systems evolution. But perspectives have shifted significantly, and intrasexual variation in the mating phenotypes of females is receiving increasing attention. So far, clear cases of female AMPs have only been recorded in damselflies and diving beetles, where divergent female phenotypes have evolved under sexual conflict over the optimum number of copulations for males and for females.

In damselflies, intrasexual colour dimorphisms in females seem to have evolved as a response to male harassment (Fincke et al. 2005). Sexual conflict arises when the optimal number of matings is much higher for males, who tend to harass females that are not sexually receptive, resulting in significant fitness costs to those females (Sirot and Brockmann 2001). In at least 115 species of damselflies and dragonflies, usually in groups where males search for mates rather than defend specific territories (Fincke 2004; Fincke et al. 2005), females may be 'heteromorphs' that are distinct (in behaviour and colour) from males, or 'andromorphs' that greatly resemble males in coloration and behaviour (Johnson 1975). Interestingly, in some species two heteromorphs (and no andromorph) exist, leading to the proposition of two hypotheses for the evolution of female dimorphism: the 'learned mate recognition' hypothesis argues that females avoid male harassment by confusing males due to variation in female signals, whereas the 'male mimicry' hypothesis argues that females avoid harassment by being similar to other males (reviewed in Fincke et al. 2005).

A similar phenomenon occurs in diving beetles from the family Dytiscidae, where males possess special structures on their forelegs that function as suction cups, allowing them to attach themselves to the females during the copulation (Miller 2003). Here again, the optimum number and/or duration of copulations are probably greater for males, and copulations can bring about significant costs to females because their access to the surface to breathe is restricted by the male while they are mating (Bergsten and Miller 2007). Female diving beetles have evolved an alternative phenotype consisting of a rough rather than the normal smooth elytral surface. The rough phenotype makes it more difficult for males to attach, and to remain attached, because the roughness hampers the function of males' suction cups (Bilton et al. 2008; Inoda et al. 2012). In both damselflies and diving beetles, these female dimorphisms and polymorphisms can be considered AMPs because they implicate different means to avoid male harassment and optimize the number (and potentially the quality) of copulations achieved.

In the Hymenoptera, female dimorphism may be found among ant queens, in which one queen morph is adapted to initiating new colonies by themselves, whereas the other is adapted to join colonies that are already established (Heinze and Keller 2000). Queens that start new colonies are usually heavier and possess large fat reserves and fully developed wings, whereas queens that join already established colonies, either by invading alien colonies or by simply returning to their own colony after mating, can be smaller and lighter, lacking large fat reserves, fully developed wings, and flight muscles. A similar case of female dimorphism is also found in three species of the parasitoid wasp genus *Melittobia* (Eulophidae; Gonzalez and Matthews 2008; Matthews et al. 2009). Here, short-winged females reproduce and oviposit in the same host from which they emerged, whereas long-winged females disperse to find a new host (Matthews et al. 2009). However, in all these cases of female dimorphic hymenopterans, alternative female phenotypes are not directly connected to alternative ways of obtaining fertilizations (and hence AMPs), but rather represent diverging adaptations to disperse or not.

It is clear that AMPs and dimorphic morphology are not as widely reported in female insects as they are in male insects. However, intrasexual variation in female mating systems must not be overlooked, and it is certainly a promising topic for future research. We emphasize that there are other sources of intrasexual variation in the reproductive biology of female insects that we have not covered in this chapter, as here we defined AMPs as *dichotomous* ways to obtain *fertilizations* (see Section 7.1). Variation in the choosiness of females of a given species has been described as 'alternative female choice tactics' (Thornhill 1984), for example. However, we refrained from covering these cases here in detail because variation in the degree of choosiness among conspecific females is a significant topic in its own right (see for instance Gray 1999; Lehmann 2007; Perry et al. 2009). Furthermore, variation in female choosiness is continuous more often than not, and does not constitute clearly dichotomous ways to obtain (or resist) fertilizations. We have also not covered alternative means to complete other stages of female reproduction, such as alternative ways of provisioning nests or finding oviposition sites, for instance. For a more complete review of female alternative *reproductive* phenotypes that involve all stages of reproduction, we refer the reader to Brockmann (2008).

7.6 Conclusion

Our poor knowledge of the astonishing insect diversity is surprising: it has been estimated that the nearly one million described species of insects represent only one fifth, and perhaps only one tenth, of the actual number of extant species (Grimaldi and Engel 2005). Nevertheless, the fraction of insects that have already been described reveals an enormous variety of natural histories, including a wealth of AMPs. Exploring the evolution of AMPs in insects can shed considerable light on the importance of phenotypic plasticity for the evolutionary origin of novel traits (West-Eberhard 2003). To illustrate this point we return to our example of behavioural plasticity in the calling behaviour of male crickets (Section 7.1.1). Males decide either to sing to attract females or to search silently for females, based on population density and/or recent success in aggressive interactions with other males (Simmons 1986a; Hissmann 1990). In the field cricket *Teleogryllus oceanicus*, parasitism from the acoustically orienting parasitoid *Ormia ochracea* selects against calling (Zuk et al. 1998). In at least two populations of this cricket—on the Hawaiian islands of Oahu and Kauai—there has been a rapid spread of a single locus mutation that affects the morphology of male wings, rendering them incapable of calling (Zuk et al. 2006; Tinghitella 2008). Because males harbouring this mutation cannot call, they always use the satellite mating phenotype and avoid parasitization by flies (Zuk et al. 2006; Tinghitella et al. 2009). Pre-existing phenotypic plasticity in male mating behaviours in these field crickets has thus predisposed the rapid fixation of a mutation that encodes male dimorphic morphology. Populations have shifted from adopting plastic AMPs to a mixture of plastic and fixed AMPs in just a few years (see Chapter 13). This fascinating system illustrates how promising the study of AMPs in insects can be, and we believe that investigating insects' AMPs from behavioural, developmental, and genetic perspectives offers great opportunities for expanding our understanding of evolutionary innovation, and subsequent speciation.

CHAPTER 8

Mate choice

John Hunt and Scott K. Sakaluk

8.1 Introduction

> If the foundation of this tower of adaptive logic is unsound, its fantastic and unwarranted elaboration is a stark comment on the present status of research in much of this area. (Wade 1984)

As the quote above reveals, Thornhill and Alcock (1983) was not as universally celebrated as the current retrospective might suggest. Wade (1984) was referring specifically to one of the broad themes advanced by Thornhill and Alcock's book: that female mate choice is adaptive. Thirty years later, there are few evolutionary biologists who would challenge the notion that females benefit either directly or indirectly by choosing certain kind of males over others. Wade's (1984) comments reflected a widespread disagreement at the time over whether female mating preferences are arbitrary or adaptive. This controversy was captured quite nicely by Kirkpatrick's (1987a) review of sexual selection in which he described two schools of thought: the 'good-genes' school (the adaptive version) and the Fisherian school (the non-adaptive version).

According to the good-genes school, females are selected to mate with those males of the highest viability, and the traits upon which they base their preferences are expected to reflect a male's underlying genetic quality. The Fisherian school, in contrast, posits that arbitrary mating preferences can become established because of a genetic correlation between the preferred trait and the preference that leads to a self-reinforcing process promoting the continued elaboration of both until opposed by natural selection. We now know that these processes are not mutually exclusive. Female mating preferences can function to secure genes that promote offspring viability, but the Fisherian process is undoubtedly a pervasive element of the coevolution of male sexual traits and female preferences. Indeed, Kokko et al. (2003) argue that the dichotomy is more apparent than real because, from the standpoint of a female's fitness, there is no difference between genes that enhance the viability of her offspring (e.g. Zahavi 1975) and those that promote the increased mating success of her sons (e.g. Fisher 1930). Thus, Kokko et al. (2003) advocate the use of the 'Fisher–Zahavi' model to capture the continuum between indirect benefits that are conferred via increased offspring survival at one extreme and increased offspring mating success at the other. We will return to the issue of direct and indirect benefits of female choice later, but first, it is worth reviewing the knowledge gaps that existed at the time that Thornhill and Alcock (1983) published their book.

The Evolution of Insect Mating Systems. Edited by David M. Shuker and Leigh W. Simmons.
© The Royal Entomological Society 2014. Published 2014 by Oxford University Press.

One of the major arguments advanced by Thornhill and Alcock (1983) was that female choice had played a fundamental role in influencing the evolution of male sexual traits. If this seems self-evident now, it must be taken in historical context. For much of the 20th century, female mate choice was discounted as the least compelling of the two primary forces that Darwin (1871) had envisioned driving the evolution of male secondary sexual characters (see Kirkpatrick 1987a), the other being male–male sexual competition. To be sure, Thornhill and Alcock's book was rife with speculation and untested hypotheses, but that was part of its appeal: the questions it generated helped inspire a whole generation of young scholars to devote their entire careers to the study of sexual selection in insects. The anniversary of this volume has seen an interesting turnaround in the relative emphasis placed on male–male competition and female choice: while mate choice has become a small cottage industry within the last 30 years, male–male competition might have been largely neglected were it not for the recognition that competition between males continues after mating within the reproductive tract of females in the form of sperm competition (Chapter 10).

In their overview of mate choice, Thornhill and Alcock (1983) identified three fundamental barriers to a more widespread acceptance of female mate choice as an important purveyor of evolutionary change in males: (1) a lack of direct evidence that females actually choose certain kinds of males on the basis of some phenotypic feature(s); (2) the traits that appear to influence female preferences often appear to be irrelevant to female fitness; and (3) female choice of males on the basis of traits that reflect underlying male genetic quality should impose strong directional selection on males, and ultimately erode the genetic variation in males upon which female preferences are designed to exploit, a problem widely known as the 'lek paradox' (Borgia 1979).

To what extent are these issues regarded as problematic 30 years later? If the evidence for female mate choice in insects was somewhat depauperate in 1983, this is certainly no longer the case. The literature is replete with empirical demonstrations of female choice, to the extent that an entire volume could be devoted to cataloguing the taxonomic breadth of the mating preferences of females, and the sheer variety of traits upon which they choose. In crickets and other acoustic insects, for example, females mate selectively with males producing songs exhibiting the most attractive temporal rhythms (Jang and Greenfield 1998; Shaw and Herlihy 2000; Brooks et al. 2005; Bentsen et al. 2006). The tempo of the substrate-transmitted vibrational signals produced by male treehoppers on plant stems greatly influences the choice of males to which females orient (Sullivan-Beckers and Cocroft 2010). Female fireflies are closely attuned to the bioluminescent flash signals of males, and their responsiveness to males is greatly influenced by the rate and duration of the flashes (Lewis and Cratsley 2008). Butterfly females attend closely to the size of eyespots of males (Robertson and Monteiro 2005), and their mating preferences can also be altered by the brightness, colour, hue and ultraviolet iridescence of males' wings (Kemp 2008; Morehouse and Rutowski 2010). The exaggerated eye stalks of male stalk-eyed flies, although employed in male–male combat for access to receptive females, also influence female mating preferences (Wilkinson and Dodson 1997). In addition to these sorts of static visual signals, behavioural displays per se, such as wing waving and complex flight manoeuvres of male long-legged flies can influence who gets mated and who does not (Zimmer et al. 2003). Although the majority of studies of mate choice in insects have focused on acoustic and visual displays, studies examining the role of mate choice based on chemical cues such as cuticular hydrocarbons are also beginning to emerge, particularly in *Drosophila* (Chenoweth and Blows 2003; Skroblin and Blows 2006), but also in crickets (Thomas and Simmons 2009; Weddle et al. 2013). The list goes on. Readers looking for convincing examples of mate choice in insects will have no difficulty finding them, regardless of the mode of communication.

What of the concern that the traits that appear to influence mate choice often appear to be irrelevant to female fitness? This is no longer viewed as troublesome because although the majority of male sexual traits lack practical utility (or even hinder male survival), females may still benefit from choosing males on the basis of these traits if their expression is correlated with male condition, and male condition is heritable (Rowe and Houle 1996). This is also directly related to the third concern raised by Thornhill and Alcock (1983), namely, that mate choice should invariably exhaust genetic variation in male quality, eliminating any benefits to females who continue to select among males on the basis of the male sexual trait (i.e. the lek paradox). Rowe and Houle's (1996) genic capture model offers a simple but elegant solution to the lek paradox because male condition is expected to be influenced by many loci, such as those involved in the acquisition and efficient use of resources, thereby sustaining ample genetic variation in condition. This is not the only possible solution to the lek paradox, and we will consider others below when we next consider the various models addressing the evolution of mate choice.

8.2 Theoretical models of mate choice

There has been a bewildering number of theoretical models developed to account for the evolution of mate choice, but they can basically be classified into two types, those in which females derive *direct benefits* by choosing certain males over others, and those in which females obtain *indirect benefits*.

8.2.1 *Direct benefits of mate choice*

Direct-benefits mate choice describes those situations in which males offer resources that directly enhance a female's survival or reproductive output, and females that mate selectively acquire more of these resources. The kinds of resources that males offer vary widely, and may include food, territories suited to rearing young, parental care, protection from predators, among others. This kind of mate choice requires only that males vary in the resources that they offer females, and that this variation is tied to the expression of the trait(s) upon which females can choose (Kokko et al. 2003; Jones and Ratterman 2009).

Direct-benefits choice has not engendered the controversy of indirect-benefits choice in part because it does not require variation in male genetic quality: some males may have more resources to offer merely by virtue of having developed under more favourable environmental circumstances (Kokko et al. 2003). Not surprisingly, then, there is good empirical support for this form of mate choice in insects. One of the most convincing examples comes from the ornate moth, *Utetheisa ornatrix*, the larvae of which feed on plants high in toxic pyrrolizidine alkaloids (PAs) (Figure 8.1). The larvae are able to safely sequester these compounds, which are retained through metamorphosis. PAs not only afford protection to adult moths, but females transfer a portion of their PAs to their eggs, thereby conferring a modicum of protection to their young. During copulation, females receive a substantial direct benefit from the male in the form of a large nutritious spermatophore that enhances female fecundity by up to 15% (LaMunyon 1997) and to which the male allocates a significant portion of his PAs, contributing directly to the female's own PA budget (Conner et al. 1990; Dussourd et al. 1991). Males vary both in the size of the spermatophore and in the amount of PAs transferred to females at copulation. Males use a portion of their PA

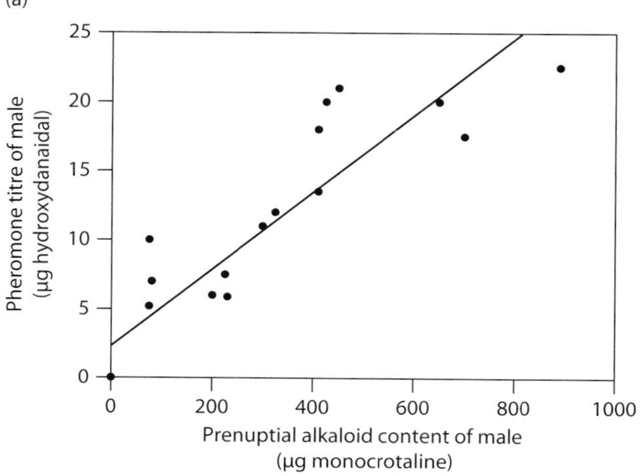

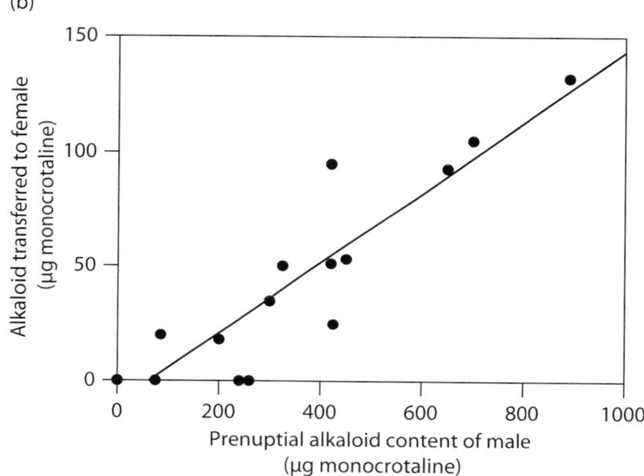

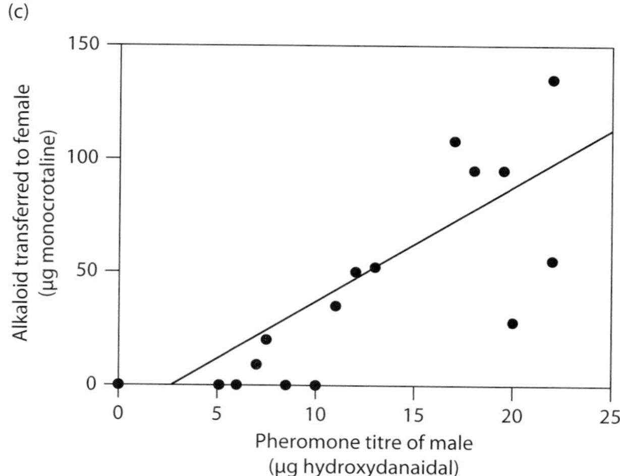

Figure 8.1 Associations between male adult alkaloid (monocrotaline) content, adult male pheromone (hydroxydanaidal) titre, and the amount of alkaloid transferred to the female during mating in the ornate moth, *Utetheisa ornatrix*. Figure redrawn after Dussourd et al. (1991).

budget in the manufacture of their courtship pheromone, hydroxydanaidal (HD), and females select among males on the basis of HD levels. Females preferentially mate with males producing more HDs, and because HD content is significantly correlated with male spermatophore size and higher PA levels in spermatophores, females receive a direct benefit by so doing (Iyengar et al. 2001).

8.2.2 *Indirect benefits of mate choice*

Indirect-benefits choice encompasses those situations in which males provide no material benefits to females, but females nonetheless distinguish among prospective mates on the basis of some phenotypic trait. The benefits in these cases are genes, and they are regarded as indirect because their effect is manifest in the fitness of the offspring produced by choosy females. Indirect-benefits models come in two forms: those involving additive genetic effects and those involving non-additive effects (Neff and Pitcher 2005). The first comprises those situations in which females discriminating among males on the basis of a sexual ornament produce more attractive offspring or offspring of higher viability (Hunt et al. 2004); essentially, this is the Fisher—Zahavi (i.e. 'good-genes') model. The second concerns those situations in which the genetic benefits of female choice arise as a consequence of an interaction between paternal and maternal genomes (Zeh and Zeh 1996, 1997). Here a female may choose a particular male because dominance effects, epistasis, or a heterozygous advantage promote increased offspring survival; such males are said to be genetically more 'compatible'. A primary difference between the two kinds of models is that a male with good genes should always produce superior offspring, whereas a male with compatible genes only produces superior young when mated to a particular maternal genotype (Neff and Pitcher 2005).

The controversy over models invoking genetic benefits has generated an upsurge of empirical work focused on this form of mate choice, much of it directed at insect mating systems. As a result, several studies have succeeded in showing that female mating preferences can enhance individual components of offspring fitness. For example, in *Drosophila montana*, male song characteristics are the focus of female mate choice, with females preferring male songs with short sound pulses and a high carrier frequency (Hoikkala et al. 1998). Females in this species gain indirect genetic benefits from their mate choice, with the frequency of the male song being positively correlated with the survival rate of the male's offspring from egg to adulthood (Hoikkala et al. 1998; Figure 8.2). However, the frequency of the male song was not correlated with the fecundity of his mating partner, suggesting that there are no direct benefits to the female for choosing males based on this song character.

The previous study constitutes an example in which additive genetic effects accrue to female mating preferences; we next consider an example involving non-additive effects. Female spotted cucumber beetles (*Diabrotica undecimpunctata howardi*) distinguish between males based partly on cuticular hydrocarbons (CHCs), lipid compounds forming the waxy surface of the insect cuticle (Brodt et al. 2005). Females signal their receptivity via pheromone emission, and males mount receptive females with little fanfare. Males are able to achieve partial intromission within a few seconds, but they are thwarted from transferring the spermatophore by the female's rigid vaginal duct muscles that prevent access to her bursa copulatrix. Male's stroke females with their antennae during copulation, after which, the female may relax her vaginal duct muscles allowing spermatophore transfer to occur. Ali and Tallamy (2010) conducted two experiments, one in which they employed gas chromatography to compare the CHC profiles of males that were rejected or

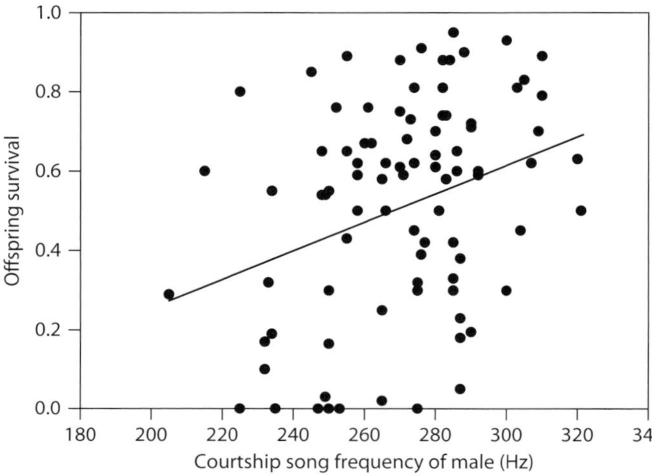

Figure 8.2 Correlation between the frequency of the male courtship song and the proportion of offspring surviving from egg to adulthood in *Drosophila montana*. Figure redrawn after Hoikkala et al. (1998).

accepted by a female, and another in which they compared the CHCs of each experimental male and two females, one that had voluntarily mated with him and another that was forced to accept his spermatophore. In both experiments, immune responses of males and their offspring were measured by inserting a nylon filament into the body cavity of the beetle and measuring the degree of melanization that ensued upon encapsulation of this foreign body by haemocytes. In the first experiment, there was greater divergence in the CHC profiles of females and accepted males than there was between females and rejected males. There was, however, no difference in the immune responses of accepted and rejected males. In the second experiment, the immune responses of offspring produced by females that were forced to copulate with males were significantly lower than those of the offspring of females that voluntarily mated with males. Ali and Tallamy (2010) concluded that females seek males with whom they are 'immunocompatible', and use a form of self-referencing to identify compatible and incompatible males on the basis of similarity in CHC profiles. In contrast, the lack of a difference in immune responses of accepted and rejected males would appear to rule out additive genetic effects.

The examples of indirect-benefits choice that we have considered thus far represent cases in which investigators have succeeded in showing that female mating preferences enhance specific components of offspring fitness (e.g. body size, survival, immunocompetence). However, Hunt et al. (2004) strongly caution against the use of single fitness components, noting that there is no a-priori reason why single fitness components should be strongly or even positively related to total fitness, and that total lifetime reproductive success is the best measure of a male's genetic quality. This is because life-history trade-offs or genotype × environment interactions (GEIs) may undermine correlations between single fitness traits and total fitness. In the case of body size, for example, although larger body size may confer a mating advantage, there may be a survival disadvantage if larger individuals require a longer period of development prior to adult eclosion. With respect to possible GEIs, it is entirely possible that large body size may confer a fitness advantage in one environment (e.g. one in which there are numerous predators), but confer a fitness disadvantage in another (e.g. one in which predators are largely absent).

This deficiency with respect to empirical studies of indirect genetic benefits led Kokko et al. (2003) to assert that: 'Despite the large amount of empirical work on indirect benefits of mate choice, an astonishing fact remains: the fundamental prediction that mating preferences increase net offspring fitness in species where direct benefits to mate choice can be excluded has not been empirically tested . . . after two decades of work there is still no study showing that mean offspring fitness is elevated.' Happily, this empirical barrier has since been hurdled. Head et al. (2005) mated attractive and unattractive male house crickets (*Acheta domesticus*) to randomly selected virgin females, reared their offspring to sexual maturity, and measured the attractiveness of the sons and fecundity of the daughters. They found that the sons of attractive males were nearly twice as likely to mate as those of unattractive males, and that this, along with the increased fecundity of daughters, was sufficient to confer a net fitness increase on the dams mated to attractive males. Thus, choosy females benefit by producing 'sexy sons', a possibility hypothesized many years earlier (Weatherhead and Robertson 1979).

8.2.3 *Resolving the lek paradox*

We saw earlier that one solution to the lek paradox is the genic capture model (Rowe and Houle 1996): if the expression of male sexual ornaments is condition dependent and condition is heritable, female mate choice will be favoured so long as genetic variation in male fitness is maintained in the population. In stalk-eyed flies (*Cyrtodiopsis dalmanni*), for example, males exhibit exaggerated eye spans and females prefer males with larger eyespans (Wilkinson and Dodson 1997; Figure 8.3). David et al. (2000) showed that the expression of male eyespan is condition dependent, specifically on diet, and, perhaps more importantly, that the extent to which male eyespan is influenced by diet is significantly heritable. Thus, females obtain genetic benefits by mating preferentially with males with the largest eyespans. We now briefly consider two other solutions to the lek paradox, one encompassed by the parasite-mediated model of sexual selection, and the other involving GEIs.

The parasite-mediated model of sexual selection originally was proposed to explain the evolution of the striking coloration of male birds (Hamilton and Zuk 1982), but has since

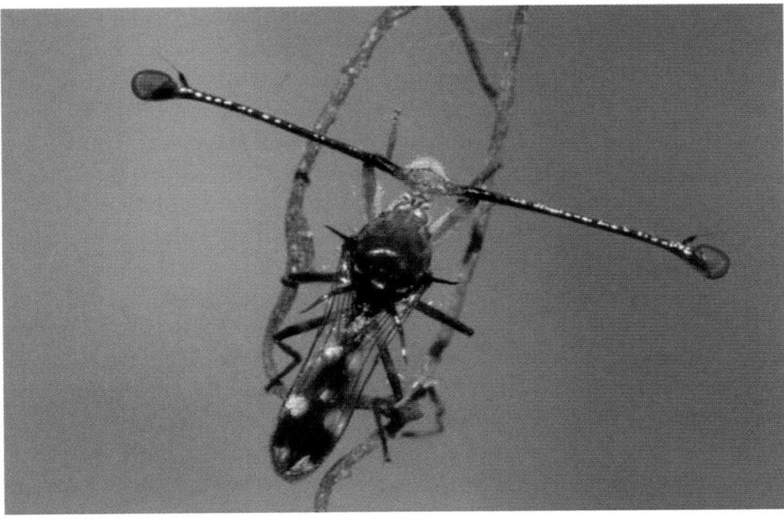

Figure 8.3 Male stalk-eyed fly, *Teleopsis pallifacies*. Females prefer males with larger eye-spans. Photo by Gerald Wilkinson.

been extended to other forms of male ornamentation across a variety of taxa, including insects (Chapter 13). The basic idea as it was originally conceived is that only males relatively free of disease or parasites should be able to develop the showiest plumage, and thus females selecting on the basis of a male's plumage should obtain genes that confer a measure of disease resistance to her offspring. But if males differ in their resistance to parasites, why wouldn't female choice rapidly deplete the genetic variation underlying that resistance? The solution to the paradox stems from the ongoing evolutionary arms race between hosts and their parasites. The spread of a resistant host genotype would favour parasite genotypes able to circumvent that resistance, which in turn would favour those host genotypes more adept at evading the most common parasite genotype. Hence, what constitutes the 'best' host genotype is continuously changing, so that females always obtain genetic benefits by choosing the healthiest males in the population (i.e. those with the greatest elaboration of the sexual ornament).

Much of the focus of parasite-mediated sexual selection has been on vertebrates because the immunosuppressive influence of testosterone has been seen as a pervasive mechanism enforcing the link between male trait expression and parasite resistance (Folstad and Karter 1992). However, as discussed in Chapter 13, a recent shift has seen the extension of the model to insect mating systems. In the damselfly, *Calopteryx splendens xanthostoma*, for example, males sport prominent melanized wing patches that influence the outcome of intrasexual contests, but also affect female mate-choice decisions (Siva-Jothy 1999). Both the production of this ornament and the expression of an important component of insect immunity (encapsulation and melanization of parasites) are influenced by the same enzyme, phenoloxidase. By experimentally manipulating levels of a gut parasite, Siva-Jothy (2000) showed that males with darker patches are unaffected by parasite load, whereas lighter males showed elevated phenoloxidase levels in response to increased parasite loads. He surmised that this difference emerged because the most resistant males (presumably the males with the blacker wing patches) were able to downregulate their immunity sooner, or that less resistant, light males produce melanin at a suboptimal rate so that they must speed up expression of the substrate upon which melanin production depends (i.e. phenoloxidase). In either case, the results are consistent with a version of the parasite-mediated model of sexual selection that holds that the honesty of the sexual signal is maintained by a trade-off between the resources needed for the expression of the ornament and the resources required for proper immune functioning (Sheldon and Verhulst 1996).

In addition to parasite-mediated sexual selection, GEIs have been viewed as a possible solution to the lek paradox (Hunt et al. 2004), the basic idea being that the genetic quality of a male may depend on the environment in which his genotype is expressed. Thus, what constitutes the best male genotype in one environment may not constitute the best in another, and differences in the strength and direction of sexual selection across environments may prevent the loss of genetic variation that normally results from female choice.

One of the most persuasive examples of how GEIs can maintain genetic variation in the face of female mate choice comes from studies of lesser waxmoths, *Achroia grisella*. Waxmoths breed in honeybee hives, where the larvae subsist on honey, beeswax and other resources. Males attract females through the production of ultrasonic signals that vary in their attractiveness to females. The sounds are produced through the downward and upward movements of the male's wings, which causes tymbals on the forewings to buckle. Females prefer males that produce sexual signals at higher sound rates (SR) and of greater peak amplitude (PA) (Jang and Greenfield 1998). Notwithstanding the strong directional sexual selection imposed by female choice, there is significant genetic variation in male signal characters, raising the spectre of the lek paradox. Jia et al. (2000) created high- and low-SR and high- and low-PA lines and, using a split-family design, reared offspring from these different lines under different environmental conditions that varied in the amount of food available, temperature, and photoperiod. In some environments, the offspring of low-SR genetic variants fared better as revealed by their developmental time and SR phenotype, whereas in other environments, the offspring of high-SR lines performed better. The resultant GEI thus provides a mechanism by which genetic variation in a sexual signal subject to strong sexual selection can be maintained indefinitely in a population.

8.2.4 *Other models*

In addition to the direct- and indirect benefits of mate choice, other models have been proposed to account for the evolution of female mating preferences, including the sensory exploitation model and the sexual conflict model.

The sensory exploitation hypothesis posits that male sexual ornaments can arise because they exploit pre-existing biases in females' sensory systems (Ryan 1998). One form of sensory exploitation is sensory traps, male signals that mimic stimuli to which females respond in other contexts and that elicit female behaviours that enhance male mating success (Christy 1995). What distinguishes sensory traps from other forms of sensory exploitation is that the biases they engage are currently advantageous to females outside the context of mate choice. Sensory traps may account for a male sexual trait that is particularly ubiquitous across insect mating systems: the provision of a nuptial food gift to the female at copulation. Such gifts occur in a bewildering array of forms, ranging from insect prey, various kinds of male bodily secretions, and even parts of the male's own body (Vahed 2007a). In the decorated cricket, *Gryllodes sigillatus*, the nuptial food gift takes the form of a spermatophylax, a large gelatinous mass forming part of the male's spermatophore, which remains attached outside the female genital opening after mating (Figure 8.4). The spermatophylax envelopes a small sperm-containing ampulla, and immediately upon dismounting the male after mating, the female rips the spermatophylax from the ampulla with her mandibles. While the female consumes this nuptial food gift, sperm are pumped into her reproductive tract from the ampulla. Within a few minutes of consuming the spermatophylax, the female removes and eats the ampulla. Because smaller gifts require less time to consume, males providing such gifts suffer premature ampulla removal and reduced sperm transfer (Sakaluk 1984, 1985). This, in turn, greatly reduces a male's fertilization success (Sakaluk and Eggert 1996).

The influence of the size of the gift on the number of sperm transferred by the male is not unique to *G. sigillatus*. In species as diverse as crickets, katydids, hangingflies, scorpionflies, and dance flies, females preferentially utilize the sperm of males providing the largest gifts. Why might this be so? Sakaluk (2000) adopted a novel approach to examine the possibility that nuptial gifts arise as a form of sensory trap that exploits pre-existing gustatory responses of females: he offered nuptial gifts synthesized by male *G. sigillatus* to females of several related cricket species whose mates offer no such inducements (Figure 8.5). Females of the non-gift-giving species not only fed on these 'foreign' food gifts, they selectively accepted more sperm from their mates than when no nuptial gifts were offered. These results support the hypothesis that nuptial food gifts and post-copulatory mate choice co-evolve through an unusual form of sensory exploitation.

What distinguishes the sensory exploitation model from the direct and indirect models of mate choice is that females need not benefit by responding to the sexual display of prospective mates. The sexual conflict model takes this one step further: females may suffer

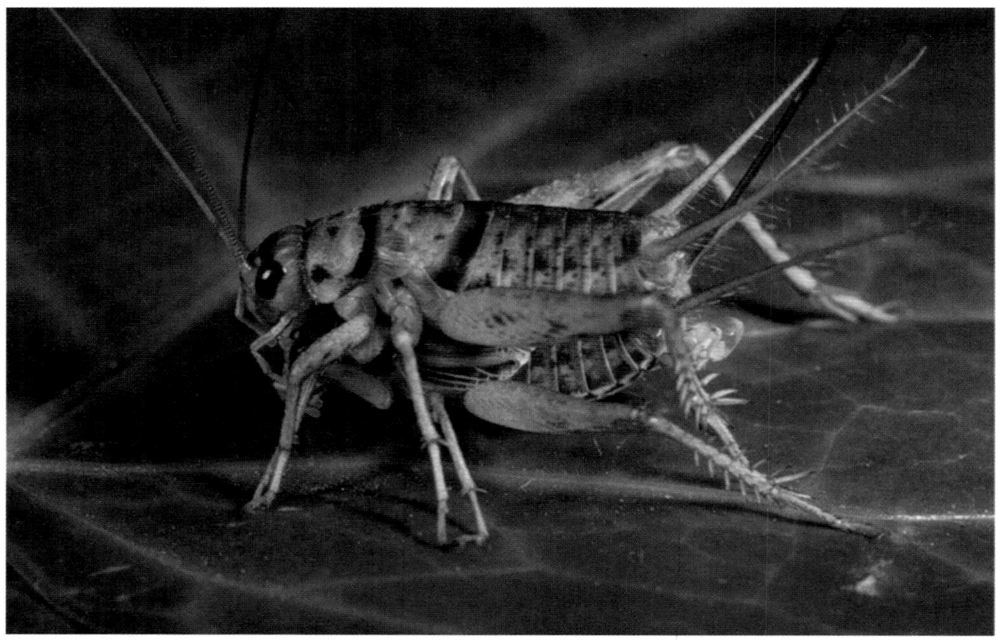

Figure 8.4 Copulation in decorated crickets, *Gryllodes sigillatus*. The spermatophore transferred by the male includes a large, gelatinous mass (visible as a translucent blob emerging from the tip of the male's abdomen) that the female consumes as a nuptial gift after mating. Photo by Scott Sakaluk.

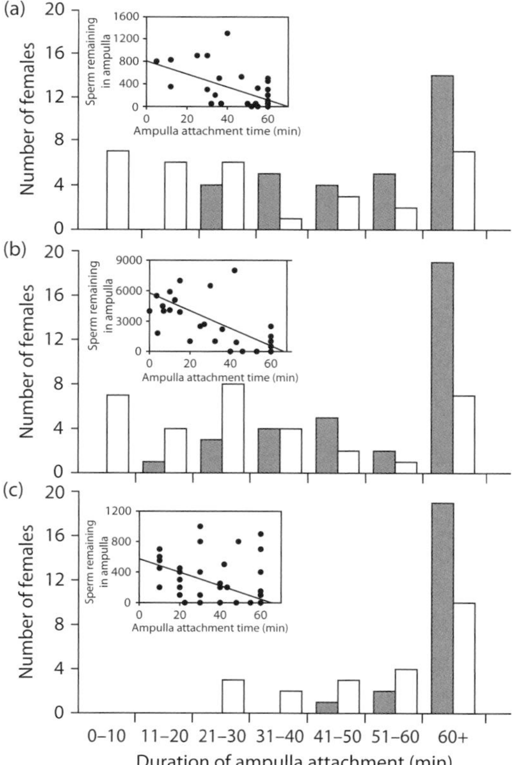

Figure 8.5 Frequency distribution of ampulla attachment duration of females mated twice to the same male and either given the opportunity to feed on a novel food gift, a *Gryllodes sigillatus* spermatophylax (grey bars), or not given this opportunity (white bars). Females of three non-gift-giving cricket species were tested (a, *Gryllus veletis*; b, *Acheta domesticus*; c, *Gryllus integer*) and in all cases females retained the sperm ampulla of their mate after consuming the gift. Inset in each frequency distribution is the relationship between ampulla attachment time and the amount of sperm remaining in the ampulla, illustrating that longer attachment time enables more sperm to be transferred to the female reproductive tract in each species. Figure redrawn after Sakaluk (2000).

decreased fitness by responding to male sexual displays (Arnqvist and Rowe 2005). The sexual conflict model begins with the recognition that the optimal outcome for many important fundamental reproductive decisions may differ between the sexes. For example, it may be in the best interests from the standpoint of a female to mate with many different males if she can secure additional direct and indirect benefits by so doing, but it is never in the best interests of each of her prospective partners, who may experience a decrease in fertilization success as a result of her behaviour. The ubiquity of this conflict is evidenced by mate guarding, a pervasive tactic employed by males to prevent females from remating with other males (Alcock 1994) (Chapter 10).

It should be immediately apparent that female choice invariably leads to sexual conflict because, while it might be in the female's best interest to avoid mating with a particular male, it is almost always in the best interest of males to consummate courtships. This conflict promotes the evolution of traits in males to impose unwanted matings on females. One problem in identifying such conflicts is that in practice it may be difficult to determine when female resistance to male mating attempts is being used as a screening device to choose the most vigorous males (i.e. mate choice), and when it is being used to avoid costly, unwanted matings (i.e. sexual conflict). A classic study on water striders, *Gerris odontogaster*, beautifully disentangles these competing explanations (Arnqvist 1992). Matings in water striders are brutish affairs: marauding males pounce on females and aggressively attempt to obtain genital contact. Females actively attempt to dislodge males using backward somersaults and other manoeuvres, while males cling to the females using abdominal claspers and spines (Figure 8.6). But is the female's resistance a form of

Figure 8.6 Pair of water striders, *Gerris lacustris*, engaged in a premating struggle. Photo by Ingela Danielsson and Jens Rydell. Inset: abdominal spines of male *G. odontogaster* used to grasp female abdomen during such struggles. Photo by Göran Arnqvist.

mate choice or an attempt to avoid the costs of superfluous copulations, which include an increased risk of predation, reduced foraging success, and energetic expenditures? To test these alternative hypotheses, Arnqvist (1992) manipulated both the operational sex ratio and population density of water striders. If female resistance is a form of mate choice, we would expect females to be more choosy as the number of available mates increases, and thus we would predict that the level of female resistance (as measured by the number of somersaults) should increase as the number of available mates increases, but that female mating frequency should be largely independent of population density and the number of available mates. In contrast, if female resistance functions to avoid costly matings, we would expect that the benefits of resistance would be diminished as males become more abundant and females are harassed more frequently, and hence we would predict that the level of female resistance should decrease with both an increase in population density and the abundance of males, whereas female mating frequency should increase under these circumstances. The results of Arnqvist's (1992) experiment were more closely aligned with the latter set of predictions, and thus he concluded that sexual conflict best accounts for the observed resistance of females.

8.3 Mechanisms of mate choice

Thus far, our focus has been on *why* females choose their mates, or more specifically, its adaptive significance. Indeed, this was Thornhill's and Alcock (1983) principal concern in their overview of insect mate choice. But a related and equally important question is: *how* do female insects choose their mates? Which sensory organs are involved in the receipt of the sexual signals that males transmit; which elements of a female's neural circuitry are involved in processing this information; how does the brain map this sensory information, the female's internal state and her previous experience in guiding her decision to mate with a particular male; and what genes control the female sexual response? A great deal of progress has been made in addressing each of these components in isolation. For example, numerous studies have measured female preference functions by presenting females with multiple stimuli or stimulus sets to determine precisely the features of males they find most appealing (Jang and Greenfield 1998; Wagner 1998; Shaw 2000). Decades of research have enabled neuroethologists to identify the individual neurons used in sound reception and phonotactic preferences of crickets, grasshoppers, and other acoustic insects (Kostarakos et al. 2008; Ronacher and Stange 2013). Single-gene mutations, gene manipulation, and genomics studies (Dickson 2008; Ferveur 2010; Immonen and Ritchie 2012) have identified many of the genes important in regulating the sexual responses of females (Chapter 4). We could not possibly review in this short space all that has been discovered about the neural underpinnings of female sexual behaviour. Despite these critical advances, however, we are still far removed from a comprehensive understanding of how these various proximate mechanisms work in concert to determine female mating decisions. To reveal the scope of the challenge, we consider one example in detail that offers, perhaps, the greatest promise of understanding female choice at the finest-grained mechanistic level, sexual decision-making in the model species *Drosophila melanogaster*.

At first glance, female mate choice in *D. melanogaster* might seem like it ought to be a perfunctory affair: the male produces his courtship display and the female need only decide whether or not to mate. This she does either by opening her vaginal plate for copulation, or by extruding her ovipositor and flying away (Dickson 2008). But in reality, the male bombards the female with a large number of cues in multiple sensory modalities that the female must process in a way that enables adaptive decision-making. Female mate choice is influenced by chemical cues and acoustic signals produced by the males. Males and females differ in their CHC compounds, and the primary male CHC, 7-tricosene, stimulates females to mate. CHCs enable sex and species recognition via taste receptors located on the mouthparts and legs. These sensory neurons are contained in specialized hair-like sensilla that are filled with fluid or lymph. CHCs enter the lymph, where they are bound to proteins that allow their detection by specialized receptor neurons (Billeter and Levine 2013). In addition to CHCs, males emit volatile olfactory pheromones that further stimulate female mating, the best known of which is the male pheromone, *cis*-vaccenyl acetate (cVA). cVA acts on female receptivity through the Or67d receptor, which is expressed in a subset of olfactory sensory neurons located on the third antennal segment. Axons from these olfactory sensory neurons extend to the antennal lobe, the primary olfactory centre, where the pattern of odour receptor activation is mapped. This map is relayed to higher brain centres via secondary projection neurons, and this entire neural circuitry has almost been fully characterized at the cellular level (Dickson 2008; Ferveur 2010).

Notwithstanding the importance of male chemical signals in stimulating female sexual receptivity, it is the acoustic signals produced by males that are of paramount importance in a female's decision to mate. Males unable to produce song because they have been experimentally muted, or those producing low-quality songs, rarely succeed in mating. However, the mating success of muted males can be restored upon the playback of high-quality song (Dickson 2008). Males sing by extending one of their wings and vibrating it. The song consists of two components, a 'humming' part (sine song) and a series of short, rapidly produced pulses (pulse song). The most important aspect of the song from the standpoint of female mating preferences is the interpulse interval (Dickson 2008; Ferveur 2010). Females detect song via the arista, specialized hair-like structures extending from the second antennal segment. Together these structures respond to the movement of air particles through rotational movements that activate stretch receptors in Johnston's organ, located on the second antennal segment (Dickson 2008; Immonen and Ritchie 2011). These receptors are acutely sensitive to such displacements, and different subsets of Johnston's organ neurons project to different regions of the brain, which may reflect different functions related to detection of courtship song, flight, or gravity-sensing (Dickson 2008).

Sex differences in behaviour in *D. melanogaster* appear to be influenced primarily by the expression of two genes, *fruitless* (*fru*) and *doublesex* (*dsx*). For example, females forced to express male *fru* court like males, but only sing upon simultaneous expression of male *fru* and *dsx* (Dickson 2008; Ferveur 2010). The precise role of these genes in regulating female receptivity is, however, unknown. Moreover, the neural circuitry within the female brain that integrates the sensory information coming from the olfactory, auditory, and reproductive systems in arriving at a decision to mate remains largely unstudied. Further complicating the issue is that learning and the transfer of accessory gland proteins in males' ejaculates also influences female receptivity. Thus, although much is known about the proximate basis of mate choice in *D. melanogaster*, future research employing powerful new methods for gene manipulation, cell-activity imaging and analytical biochemistry are needed to fully resolve female mate-choice decisions at a cellular level (Dickson 2008; Ferveur 2010).

8.4 Mate choice and the evolution of male sexual traits

As we have already outlined, Thornhill and Alcock's (1983) book largely centred on why females choose to mate with certain males over others, rather than how this mate choice influences the evolution of the male sexual trait(s) it targets. Interestingly, the same year as Thornhill and Alcock's book was published, Lande and Arnold (1983) published one of the most influential papers in evolutionary biology, detailing how to estimate the strength and form of sexual selection acting on a suite of correlated phenotypic traits. Yet, despite almost three decades of empirical research on mate choice, this approach still remains largely under-utilized (Hunt et al. 2008): most studies of female mate choice tend to focus on a single male trait at a time (e.g. horn length) using dichotomous experimental designs (e.g. males with short versus long horns). While this approach is sufficient to demonstrate that mate choice exists, it ignores the fact that male sexual traits are seldom independent of other traits possessed by the male (i.e. traits are often highly correlated) and therefore can *only* focus on the linear selection imposed by females on male sexual traits. Thus, important forms of non-linear sexual selection (i.e. stabilizing, disruptive, and correlational selection) are ignored in these studies. More recent work using the Lande and Arnold (1983) approach, much of which has been conducted on insects, has shown that this traditional approach to studying mate choice can seriously underestimate the complexity of female mate choice and the effects it has on the evolution of male sexual traits.

8.4.1 *The strength and form of sexual selection that mate choice imposes on male sexual traits*

Since Thornhill and Alcock (1983), there has been a growing appreciation that phenotypic evolution is a multivariate process. That is, male sexual traits in many species are complex and consist of many individual components (e.g. the spectral and temporal components of a cricket call) and sexual selection targets each of these components simultaneously (Lande and Arnold 1983). Moreover, these components are often genetically correlated, meaning that the evolutionary response of a given component results not only from direct selection operating on it but also from indirect selection operating on genetically correlated components (Lande 1979). Lande and Arnold (1983) addressed this first issue by developing a multivariate statistical approach that enables the strength and form of selection on a given trait to be measured independently (of other correlated traits), as well as sexual selection acting on the covariance between traits to be examined. Phillips and Arnold (1989) extended this approach to show how the canonical analysis of the matrix of standardized non-linear selection gradients ($\boldsymbol{\gamma}$) can be used to locate the major axes (i.e. eigenvectors) of non-linear selection. $\boldsymbol{\gamma}$ is a symmetric matrix that contains the standardized quadratic selection gradients along the diagonal (indicative of stabilizing ($-\gamma$) and disruptive ($+\gamma$) selection) and the standardized correlational selection gradients in the off-diagonal positions. Interpreting individual terms in $\boldsymbol{\gamma}$ can be difficult, particularly as the number of traits being examined becomes large, and may seriously underestimate the strength of non-linear selection (Blows and Brooks 2003). This is because selection rarely operates on individual traits but instead targets the combinations of traits associated with these eigenvectors (Phillips and Arnold 1989; Blows and Brooks 2003). Thus, a formal characterization of the strength and form of linear and non-linear selection requires both the estimation of the standardized linear and non-linear selection gradients, as well as the canonical analysis of $\boldsymbol{\gamma}$. We direct the reader to several recent reviews for a detailed overview of these statistical approaches (Hunt et al. 2009; Chenoweth et al. 2012).

Table 8.1 provides a summary of mate choice studies in insects that have applied the multivariate selection approach outlined above. The most discernible pattern emerging from Table 8.1 is that very few insect studies have conducted a complete multivariate selection analysis (including canonical analysis) to examine the strength and form of sexual

Table 8.1 Empirical studies illustrating the complexity of sexual selection that female mate choice exerts on male sexual traits

Species	Common name	Female sexual trait	Strongest selection gradient	Dominant eigenvector of non-linear selection	Curvature of fitness surface	References
Orthoptera						
Teleogryllus commodus	Field cricket	Structure of the advertisement call	Linear	Stabilizing	Peak	[1]
		Structure of the advertisement call and calling effort	Linear	Disruptive	Saddle	[2]
		Structure of the courtship call[a]	Stabilizing	Stabilizing	Saddle	[3]
		Structure of the courtship call[b]	Correlational	Disruptive	Saddle	[3]
Gryllodes sigillatus	Decorated cricket	Amino acid composition of the spermatophylax	Linear	Disruptive	Saddle	[4]
Teleogryllus oceanicus	Field cricket	Cuticular hydrocarbons[c]	Stabilizing	Stabilizing	Saddle	[5]
Gryllus pennsylvanicus	Field cricket	Morphology[d]	Linear	Stabilizing[e]	Saddle	[6]
		Morphology[f]	Correlational	Disruptive	Saddle	[6]
Diptera						
Rhamphomyia longicauda	Dance fly	Morphology	Linear	Disruptive[g]	Saddle	[7]
Drosophila serrata	Fruit fly	Cuticular hydrocarbons	Linear	Stabilizing[e]	Saddle	[8]
Hymenoptera						
Bombus terrestris	Bumblebee	Morphology	Linear	Disruptive	Saddle	[9]
Hemiptera						
Phymata americana	Ambush bug	Size and coloration[h]	Stabilizing	Stabilizing	Saddle	[10]
		Size and coloration[i]	Linear	Disruptive[e]	Saddle	[10]

[1] Brooks et al. (2005), [2] Bentsen et al. (2006), [3] Hall et al. (2008), [4] Gershman et al. (2012), [5] Thomas and Simmons (2009), [6] Judge (2010), [7] Bussièrre et al. (2008), [8] Chenoweth and Blows (2005), [9] Amin et al. (2012), [10] Punzalan et al. (2008).

[a] Female-only treatment.

[b] Male present treatment.

[c] Sexual selection based on mating success.

[d] Experienced females.

[e] Sexual selection on eigenvector is not statistically significant.

[f] Inexperienced females.

[g] Non-linear eigenvectors derived by chapter authors and therefore no statistical significance tests could be conducted.

[h] Sexual selection measured early in the season.

[i] Sexual selection measured late in the season.

Only studies in which a full multivariate selection analysis was conducted were included. In some studies, canonical analysis of γ was not performed. However, in these cases, $\boldsymbol{\gamma}$ was provided so we conducted the canonical analysis but were unable to test the statistical significance of sexual selection acting along the eigenvectors derived. The 'strongest selection gradient' refers to the strongest standardized linear or non-linear selection gradient (stabilizing, disruptive, or correlational) from the Lande and Arnold (1983) selection analysis. The 'dominant eigenvector of non-linear selection' refers to the eigenvector with the strongest non-linear sexual selection acting on it (i.e. highest $|\boldsymbol{\lambda}|$). The 'curvature of the fitness surface' refers to the overall shape of the fitness surface (Phillips and Arnold 1989). If all of the eigenvalues are negative, this indicates multivariate peak (i.e. stabilizing selection), if they are all positive this indicates multivariate bowl (i.e. disruptive selection), and if there is a mixture of positive and negative eigenvalues this indicates that the fitness surface is best described as a multivariate saddle.

selection that female mate choice imposes on male sexual traits. Some studies report the standardized linear and quadratic gradients but do not present the correlational gradients that are needed to construct the γ matrix (e.g. Lebas et al. 2004). Others provide all the linear and non-linear standardized selection gradients but do not perform a canonical analysis of γ (e.g. Bussière et al. 2008). Thus, while the major eigenvectors of selection may be extracted, the significance of linear and non-linear selection operating on them cannot be tested statistically. This may underestimate the strength of non-linear selection because the selection acting on trait combinations (i.e. eigenvectors) is often very different from that acting on individual traits (Blows and Brooks 2003). In fact, the studies we include in Table 8.1 illustrate this point, as, for most studies, the strongest standardized selection gradient is very different in form to selection acting along the dominant eigenvector of γ after canonical analysis. It is therefore not surprising that studies employing both multiple regression and canonical analysis have revealed that female choice can exert a complex pattern of sexual selection on male sexual traits. The studies contained in Table 8.1 demonstrate this point in two main ways. First, with the exception of a single study (Brooks et al. 2005), the fitness surface for male sexual traits is best described as a multivariate saddle. That is, at least one of the major dimensions of non-linear selection is disruptive and the other stabilizing in form (Figure 8.7). While it has been argued that the least-squares regression approach of Lande and Arnold (1983) may predispose the fitness surface to take this form (Shaw and Geyer 2010), signalling theory (Getty 1998) predicts that a saddle-shaped fitness surface (where a male signal attracts females at a greater-than-linear rate per unit investment) may be a general feature of honest signals. More work is

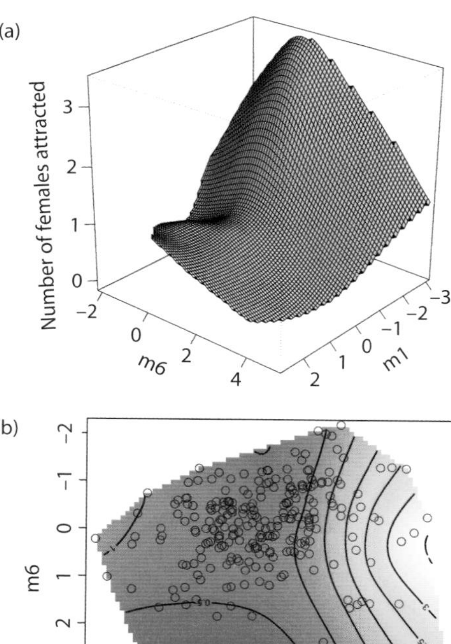

Figure 8.7 Thin-plate spline visualization (a, perspective view; b, contour view) of the fitness surface demonstrating the two major axes of non-linear sexual selection (m1 and m6) acting on male call structure and calling effort in the Australia field cricket, *Teleogryllus commodus*. In (b), lighter colours (yellow) represent regions of higher fitness. Redrawn after Bentsen et al. (2006).

needed to distinguish between these alternatives. Second, several of the studies contained in Table 8.1 show that the sexual selection imposed by female mate choice may vary over time (Punzalan et al. 2008) and with different social contexts (Hall et al. 2008, Judge 2010). For example, Punzalan et al. (2008) found that the sexual selection imposed by mate choice for size and body colouration in male ambush bugs (*Phymata americana*) changed from stabilizing to disruptive over the breeding season, as the study population became progressively more female-biased (Punzalan et al. 2010). Moreover, in the black field cricket (*Teleogryllus commodus*), female post-copulatory mate choice (measured as spermatophore attachment time after mating) exerts selection on the structure of the male courtship call, and this differs depending on whether or not the male physically guards the female after mating (Hall et al. 2008). Hall et al. (2008) found that when males were prevented from mate guarding, females exerted multivariate stabilizing selection on call structure, whereas when the male guarded, the fitness surface was saddle-shaped. As discussed further in Sections 8.4.3 and 8.4.4, this increased complexity of female mate choice decisions is likely to have important consequences for how male sexual traits evolve.

There has also been increased awareness over the last 30 years that female mate choice is a layered process, occurring both before and after copulation has taken place (e.g. Eberhard 1996) (Chapter 11). Surprisingly little attention, however, has been given to how these different 'episodes' of selection interact to determine the total strength and form of sexual selection acting on male sexual traits (Hunt et al. 2008). Whenever the selection imposed in each of these episodes is not identical, total sexual selection is likely to be very different to that operating in either episode (Hunt et al. 2008). The reason for this is that when episodes of mate choice occur sequentially, the first episode of selection (i.e. pre-copulatory mate choice) alters the phenotypic distribution of the male sexual trait available for selection to target during the second episode of selection (i.e. post-copulatory mate choice) (Arnold and Wade 1984a,b). The total strength and direction of linear selection ($\boldsymbol{\beta}_{\mathbf{total}}$) across sequential episodes can be calculated easily, however, by accommodating this change in the phenotypic distribution of the male sexual trait as $\beta_{\text{total}} = \mathbf{P}_0^{-1} \Sigma \mathbf{P}_k \beta_k$, where $\mathbf{P}_0^{-1}$ is the phenotypic (co)variance matrix for the male sexual trait before any selection, P_k is the same matrix before selection in episode k, and $\boldsymbol{\beta}_k$ is the linear selection gradient for the male sexual trait in episode k (Wade and Kalisz 1989). Estimating the total strength and form of non-linear sexual selection ($\boldsymbol{\gamma}_{\text{total}}$) across episodes is more complicated, however, as the non-linear selection gradients across each episode cannot simply be added after taking into account any changes in P across episodes (McGlothlin 2010). This is because linear selection in each episode also contributes to the total curvature of the fitness surface and therefore must also be accounted for as

$$\gamma_{\text{total}} = \mathrm{P}_0^{-1} \sum (\mathrm{P}_k \gamma_k \mathrm{P}_k) + \sum_{k \neq l} \mathrm{P}_k \boldsymbol{\beta}_k \boldsymbol{\beta}_l^{\mathrm{T}} \mathrm{P}_l \ \mathrm{P}_0^{-1}$$

where $\boldsymbol{\gamma}_k$ is the quadratic selection gradient for the male sexual trait in episode k and $\boldsymbol{\beta}_l^{\mathrm{T}}$ is transpose of the linear selection gradient in episode l (McGlothlin 2010).

Studies on insects have provided some of the best evidence confirming that female mate choice is a layered process. In both the black field cricket (*Teleogryllus commodus*) and the decorated cricket (*Gryllodes sigillatus*), females that prefer males in pre-copulatory mate choice also retain their spermatophore for longer after mating (thereby allowing more sperm to be transferred to the female), suggesting that pre- and post-copulatory sexual selection is reinforcing in these cricket species (Bussière et al. 2006; Ivy and Sakaluk 2007). Likewise, in the house cricket (*Acheta domesticus*) and the red flour beetle (*Tribolium *castaneum*), females mating with males preferred in pre-copulatory mate choice produce more eggs after mating (Edvardsson and Arnqvist 2005; Head et al. 2006). However, in the water strider *(Gerris lacustris*), larger males have a significant pre-copulatory mating advantage but smaller males gained higher fertilization success from each mating, suggesting that pre- and post-copulatory episodes of sexual selection on male body size act antagonistically in this species (Danielsson 2001). While these studies highlight that pre- and post-copulatory mate choice may exert considerable (and often different) sexual selection on males, none has formally combined episodes of selection to estimate the total strength and form of sexual selection on male sexual traits. Indeed, to our knowledge, only a single study on the dragonfly (*Libellula luctuosa*) has attempted to combine episodes of sexual selection using the above multivariate framework and found that total sexual selection on male morphology was very different in strength and form to that operating in each episode of male competition (i.e. territory success) and female choice (mating and fertilization success) (Moore 1990). Consequently, a large gap currently exists in this important area of research (Hunt et al. 2008).

8.4.2 *Genetic basis of mate choice*

The genetic analysis of mate choice is fraught with difficulties (Chenoweth and Blows 2006). This is because males frequently produce complex signals and displays (Sections 8.4.1 and 8.4.3) and female preferences for these traits are notoriously difficult to quantify (Wagner 1998). Consequently, relative to our understanding of the genetic architecture of male sexual traits, we still know very little about the genetic basis of female preferences for these traits. However, due to the ease of manipulation in the laboratory, work on insects leads this field of research.

Table 8.2 summarizes the studies on insects that have examined the genetic basis of mate choice, as well as the genetic covariation between mate choice and the preferred sexual trait. A range of different experimental designs has been used to examine the genetics of mate choice, ranging from studies examining the repeatability of mate choice (which has been argued to provide an upper limit to heritability) to artificial selection and experimental evolution studies (Table 8.2). It is clear from the studies listed in Table 8.2 that mate choice has a genetic basis in the insect species thus far examined. This is not restricted to female mate choice, as there is also a genetic basis to male mate choice in several lepidopteran species in which females produce the sex pheromone to attract males to mate (Table 8.2). Two particularly powerful approaches used in insects to demonstrate the genetic basis of mate choice are artificial selection and experimental evolution (Table 8.2). Studies using artificial selection have taken one of two approaches: either directly selecting on the mate choice behaviour (e.g. Collins and Carde 1990) or selecting on sexual trait expression and examining a correlated response in mate choice behaviour (i.e. indirect selection) (e.g. Collins and Carde 1989a). In contrast, experimental evolution studies manipulate the mating system (either directly through the number of males and females or indirectly through resources) and examine the evolutionary divergence in mate choice and sexual trait expression (Rundle et al. 2005). While the latter approach has been underused in the study of mate choice, the advantage of both approaches is that evolutionary change is 'realized': that is, evolutionary change is observed directly rather than being inferred from genetic estimates. As discussed in Section 8.4.3, this is important because a significant level of genetic variance does not necessarily mean that a trait (such as mate choice) will evolve.

It is important to point out that not all of the species contained in Table 8.2 show genetic variance for mate choice. Although some of these instances are likely attributable to the

Table 8.2 Empirical studies examining genetic variance in mate choice and the genetic covariance between choice and the preferred sexual trait

Species	Common name	Sexual trait	Trait manipulated?	Design	Genetic variance	h^2	Genetic covariance	r_A	Reference
Orthoptera									
Gryllus integer	Field cricket	Pulses per trill in call	Yes	Full-sib	Yes	0.32 ± 0.15	Yes (+)	0.51 ± 0.17	[1]
Gryllodes sigillatus	Decorated cricket	Amino acid composition of spermatophylax	No	Isofemale lines	Yes	0.93 ± 0.04	Yes (+)	0.62 ± 0.05	[2]
Teleogryllus oceanicus	Field cricket	Percentage long chirp in call	Yes	Full-sib	No	–	No	–	[3]
		Percentage long chirp in call	Yes	Common garden	Yes	–	No	–	[3]
Ephippiger ephippiger	Bushcricket	Syllable number in call	Yes	Crosses	Yes	–	No	–	[4]
Nauphoeta cinerea	Cockroach	Pheromones	No	PO regression	Yes	–	–	–	[5]
Laupala kohalensis/ L. paramigra	Swordtail crickets	Pulse rate of call	Yes	Cross	Yes	–	Yes (+)	–	[6]
Chorthippus brunneus	Field grasshopper	Syllable length of call	Yes	Artificial selection	Yes	–	No	–	[7]
Lepidoptera									
Achroia grisella	Lesser wax moth	Pulse rate and asynchrony interval of calls	Yes	Half-sib	Yes	0.21 ± 0.13	–	–	[8]
		Pulse rate of call	Yes	Full-sib	Yes	0.40 ± 0.17	–	–	[9]
Pectinophora gossypiella[a]	Pink bollworm moth	Pheromone composition	Yes	PO regression	Yes	0.38 ± 0.11	–	–	[10]
		Pheromone composition	Yes	Artificial selection	Yes	–	Yes (+)	–	[11]

continued

Table 8.2 *continued*

Species	Common name	Sexual trait	Trait manipulated?	Design	Genetic variance	h^2	Genetic covariance	r_A	Reference
		Pheromone composition	Yes	Artificial selection	Yes	0.16 ± 0.02	–	–	[12]
		Pheromone composition	Yes	Artificial selection	No	–	No	–	[13]
Utetheisa ornatrix	Arctiid moth	Body size	Yes	PO regression[b]	Yes	0.51 ± 0.11	Yes (+)	–	[14]
Ostrinia nubilalis[a]	European corn borer	Pheromone composition	Yes	Crosses	Yes	–	–	–	[15]
			No	Crosses	Yes	–	–	–	[16]
Agrotis segetum[a]	Turnip moth	Pheromone composition	No	Common garden	Yes	–	Yes (+)	–	[17]
			No	Common garden	No	–	No	–	[18]
Colias eurytheme	Sulphur butterfly	Pheromone composition	No	Correlation	Yes	–	–	–	[19]
Argyrotaenia velutinana[a]	Redbanded leafroller	Pheromone composition	No	PO regression	Yes	0.41	–	–	[20]
Diptera									
Cyrtodiopsis dalmanni	Stalk-eyed fly	Eye-span	No	Artificial selection	Yes	–	Yes (+)	–	[21]
Coelopa frigida	Seaweed fly	Body size associated with **β** inversion	No	Correlation	Yes	–	–	–	[22]
		Body size associated with **β** inversion	No	Correlation	Yes	–	Yes (+)	–	[23]
		Traits associated with *Adh* locus	No	Correlation	Yes	–	–	–	[24]
Drosophila mercatorum	Fruit-fly	Interpulse interval of call	No	Artificial selection	Yes	–	Yes (+)	–	[25]
D. melanogaster	Fruit-fly	*Yellow* mutation	No	Crosses	Yes	–	–	–	[26]

		Yellow mutation	No	Artificial selection	Yes	–	–	–	[27]
		Eye colour mutation	No	Crosses	Yes	–	–	–	[28]
		Cuticular hydrocarbons	No	Crosses	Yes	–	–	–	[29]
		Call	No	Artificial selection	Yes	–	Yes (+)	–	[30]
		Body colour and wing mutants	No	Artificial selection	Yes	–	–	–	[31]
		Interpulse interval of *per* mutants	Yes	Correlation	Yes	–	–	–	[32]
		Unknown	No	Isofemale lines[c]	Yes	0.58 ± 0.12	–	–	[33]
		Unknown	No	Isofemale lines[d]	Yes	0.86 ± 0.06	–	–	[33]
D. mojavensis	Fruit-fly	Unknown	No	Artificial selection	Yes	–	–	–	[34]
D. montana	Fruit-fly	Carrier frequency of call	No	Isofemale lines	No	–	No	–	[35]
	Fruit-fly	Call structure	No	Artificial selection	Yes	0.52 ± 0.24	–	–	[36]
D. pseudoobscura	Fruit-fly	Unknown	No	Isofemale lines	Yes	–	Yes (+)	–	[37]
D. serrata	Fruit-fly	Cuticular hydrocarbons	No	Half-sib	Yes	0.65	–	–	[38]
		Cuticular hydrocarbons		Experimental evo-lution	Yes	–	Yes (+)	–	[39]
		Cuticular hydrocarbons	No	Common garden	Yes	–	No	–	[40]
Drosophila simulans	Fruit-fly	*Ebony* mutation	No	Artificial selection	Yes	0.26 ± 0.11	Yes (+)	–	[41]
Coleoptera									
Adalia bipunctata	Two-spot ladybird	Elytra colour	No	Artificial selection	Yes	–	–	–	[42]
			No	Artificial selection	Yes	–	–	–	[43]
			No	Isofemale lines	No	–	–	–	[44]
			No	Isofemale lines	Yes	–	Yes (+)	–	[45]
Ips pini	Pine engraver beetle	Pheromone composition	No	Common garden	Yes	–	Yes (+)	–	[46]

continued

Table 8.2 *continued*

Species	Common name	Sexual trait	Trait manipulated?	Design	Genetic variance	h^2	Genetic covariance	r_A	Reference
				Crosses	Yes	–	–	–	[47]
Tribolium castaneum	Red flour beetle	Pheromone blend	No	Repeatability	No	–	–	–	[48]
Homoptera									
Nilaparvata lugens	Brown planthopper	Pulse repetition frequency of call	Yes	Isofemale lines	Yes	–	–	–	[49]
Ribautodelphax imitans[a]	Planthopper	Interpulse interval of call	Yes	Artificial selection	Yes	–	Yes (+)	–	[50]

[1] Gray and Cade (1999), [2] Gershman et al. (2013), [3] Simmons (2004), [4] Ritchie (2000), [5] Moore (1989), [6] Shaw (2000), [7] Charalambous et al. (1994), [8] Jang and Greenfield (2000), [9] Rodriguez and Greenfield (2003), [10] Collins and Cardé (1989a), [11] Collins and Cardé (1989b), [14] Iyengar et al. (2002), [15] Roelofs et al. (1987), [16] Hansson et al. (1987), [17] Löfstedt et al. (1986), [18] Hansson et al. (1990), [19] Sappington and Taylor (1990), [20] Roelofs et al. (1986), [21] Wilkinson and Reillo (1994), [22] Gilburn and Day (1994b), [23] Gilburn et al. (1993), [24] Engelhard et al. (1989), [25] Ikeda and Maruo (1982), [26] Heisler (1984b), [27] Dow (1977), [28] Tebb and Thoday (1956), [29] Scott (1994), [30] Cook (1973), [31] Crossley (1974), [32] Greenacre et al. (1993), [33] Narraway et al. (2010), [34] Koepfer (1987), [35] Ritchie et al. (2005), [36] Aspi (1992), [37] Millar and Lambert (1986), [38] Delcourt et al. (2010), [39] Rundle et al. (2005), [40] Chenoweth et al. (2010), [41] Sharma et al. (2010), [42] Majerus et al. (1982), [43] Majerus et al. (1986), [44] Kearns et al. (1992), [45] O'Donald and Majerus (1992), [46] Lanier et al. (1972), [48] Mustaparta et al. (1985), [49] Boake (1989), [50] Butlin (1993), and [50] De Winter (1992).

[a] Male mate choice, estimated as the amount of variation explained by the leading eigenvector of **G** for female preference using factor analytic modelling.

[b] Genetic basis of mate choice estimated from grandparent offspring regression.

[c] Females cold-shocked during development.

[d] Females maintained at constant 25°C during development.

Various experimental designs have been used to examine the genetic basis of mate choice, including repeatability analysis (which has been argued to represent an upper limit to heritability), rearing discrete populations in a common environment (common garden), genetic crosses (between different species or distinct genotypes), parent–offspring (PO) regression, isofemale lines, full-sib or half-sib breeding designs, artificial selection, and experimental evolution. When estimated, heritability estimates (h^2) and genetic correlations (r_A) for these studies are provided. If significant covariance between mate choice and the preferred sexual trait is detected, but the genetic correlation is not estimated, the sign of the covariance in parentheses (+ or –) is provided.

small sample sizes used or to the relative insensitivity of experimental design (or a combination of both) (e.g. Hansson et al. 1990; Simmons 2004), in others it is likely to represent important biological complexity. The best example of this is the two-spot ladybird (*Adalia bipunctata*) in which Majerus et al. (1982, 1986) artificially selected on female mate choice for male elytra coloration using beetles derived from a natural population. This selection regime produced a clear increase in female preference for melanic males and consequently this work was heralded as providing the first definitive evidence of a genetic basis to mate choice. However, subsequent work on these lines, as well as beetles taken from the same population, failed to show non-random mating (Kearns et al. 1992) suggesting that variation in the gene(s) governing this mating preference has been rapidly lost in this population (and selection lines). Work using beetles from a second population, where there is a much stronger preference for melanic males, confirmed the original findings of Majerus et al. (1982, 1986) indicating that important population differences in mate choice also exist in this species (O'Donald and Majerus 1992).

It is also apparent in Table 8.2 that few studies provide formal estimates of the heritability of mate choice, and even fewer estimates of the genetic correlation between mate choice and the preferred sexual trait. Surprisingly, the heritability estimates provided are moderate to high (mean ± SE across studies = 0.48 ± 0.06), suggesting a strong genetic basis to mate choice, although it is likely that this value is inflated by particularly high estimates from studies using isofemale lines (0.79 ± 0.11) that are known to overestimate genetic estimates. Only two studies have estimated the genetic correlation between mate choice and the preferred sexual trait and both of these cricket studies show a strong and positive genetic covariance between these traits (Table 8.2). However, numerous studies have used experimental designs that enable the presence and sign of this genetic covariance to be assessed. In the majority of cases (16 out of 24 studies), the sign of the genetic covariance between mate choice and the expression of the preferred sexual trait was positive, providing important support for the 'good genes' and Fisherian mechanisms of mate choice (Section 8.2).

As discussed in Chapter 4, an exciting new research direction is the use of genomic tools to dissect the complex genetic basis of mate choice (Chenoweth and Blows 2006). While many of the studies included in Table 8.2 use crosses between discrete genotypes or species to detect the chromosomes on which mate choice genes are likely to reside (e.g. Scott 1994; Iyengar et al. 2002), various genomic tools can be used to locate the position of these genes, identify the actual genes, as well as ascertain whether they co-localize with gene(s) for the preferred sexual trait. Although still in its infancy, several insect studies have applied genomic approaches to the study of female mate choice. Bailey et al. (2011) used microarrays to examine gene expression patterns in female *Drosophila melanogaster* mating with more versus less preferred males. They found expression differences in a total of 1,498 genes, many with high expression in the female central nervous system and ovaries, suggesting an important role for both pre- and post-copulatory mate choice in this species. Importantly, Bailey et al. (2011) also found that disproportionately large numbers of the candidate genes for mate choice were located on the X chromosome, forming a number of gene clusters with low recombination distances. This pattern of sex linkage and gene clustering suggests that mate choice in this species has a high potential evolutionary rate and is also likely to protect mate choice genes from the homogenizing effect of gene exchange between populations, thereby facilitating the process of sexual isolation and speciation.

Genomic approaches are also being increasingly applied to non-model organisms. Shaw and Lesnick (2009) examined the genomic location of quantitative trait loci (QTL) for male call structure (pulse rate) and female acoustic preference for this male trait in two closely related species of Hawaiian crickets, *Laupala kohalensis* and *Laupala paranigra*. They found a single acoustic preference QTL on Linkage Group 1 (LG1) (Shaw and Lesnick 2009) and five QTL for male pulse rate (LGs 1, 3, 4, 5 and 8) in both species (Shaw et al. 2007). Composite interval mapping was used to show that this single preference QTL co-localizes with the major QTL for pulse rate (on LG1) in these species. Furthermore, both the call and preference QTL make small-to-moderate contributions to the behavioural differences between species, suggesting that the divergence in mating behaviour among *Laupala* species is due to the fixation of many genes of minor effect. More recent work using marker-assisted introgression to move 'slower' pulse rate alleles from *L. paranigra* into the 'faster' pulse rate genetic background of *L. kohalensis* showed that four of the five pulse rate QTL significantly predicted the acoustic preferences of females from fourth-generation backcrosses, providing further direct evidence for genetic linkage between female acoustic preference and male pulse rate (Wiley et al. 2012). Collectively, this genomic work provides the functional co-ordination between female mate choice and the preferred male sexual trait needed to fuel signal-preference evolution and may facilitate the characteristically rapid speciation observed in this genus of crickets. More work is needed, however, to determine whether this pattern of genetic linkage is unique to *Laupala* or is a more widespread characteristic of signaller-receiver systems.

8.4.3 *Evolutionary dynamics of male sexual traits subject to female mate choice*

Thus far in this chapter we have shown that female mate choice (i) is a prevalent feature of many insect mating systems, (ii) can exert a strong and often complex pattern of sexual selection on male sexual traits, (iii) has a genetic basis, and (iv) is frequently positively genetically correlated with the expression of the male sexual trait. It would therefore be reasonable to assume that female mate choice is a major driving force in the evolution of male sexual traits. There is, however, surprisingly little direct evidence showing that female mate choice drives the evolution of male sexual traits, especially in natural populations (Svensson and Gosden 2007). Some of the most convincing evidence comes from work on insects where several studies have examined whether variation in female preference functions for individual components of the male sexual trait explains divergence in these sexual traits across populations (Simmons et al. 2001; Svensson et al. 2006; Grace and Shaw 2011). For example, in the Hawaiian cricket (*Laupala cerasina*) females express variable preference functions for three components of the male call: pulse duration has an 'open-ended' (linear) preference function, whereas pulse rate and carrier frequency have 'unimodal' (stabilizing) preference functions (Shaw and Herlihy 2000). *L. cerasina* is distributed across 13 spatially disjunct populations on the Big Island of Hawaii, and males from these populations show significant differences in pulse rate and carrier frequency but not in pulse duration (Grace and Shaw 2011). Females from these populations also express different preference functions for pulse rate (pulse duration and carrier frequency were not tested in this study) and importantly this variation in preference functions is positively correlated with mean pulse rate across populations, suggesting that female mate choice has driven the evolutionary divergence of the male sexual trait across populations in this species (Grace and Shaw 2011). However, in the field cricket *Teleogryllus oceanicus*, males show large differences in call structure across 15 populations spanning Australia and Oceania (Zuk et al. 2001), particularly in the proportion of the call consisting of the long chirp (Simmons et al. 2001). Female preference functions for the proportion of long chirp in a call also differed across these populations but were found to be unrelated to mean call structure across populations, possibly due to differences in predation by acoustic predators across populations (Simmons et al. 2001). As these examples clearly illustrate, the results of such studies have largely been equivocal and therefore uncertainty still exists over exactly how important female mate choice is for the evolution of male sexual traits.

Part of the reason for this uncertainty may represent the limitations of taking a univariate approach (i.e. relating a single component of a male sexual trait to female preference for this component) to address this question, as well as ignoring the important role that the genetics of the male sexual traits will also play in this process. Many sexual traits may be considered as 'complex' traits. They are 'complex' because the overall sexual trait that is the target of female mate choice (i.e. a male cricket's call), is made up of numerous individual components (i.e. pulse duration or pulse rate) that are often correlated. For such traits, it is both the genetic variance in the individual components, as well as the genetic covariance between these components, that directs how the sexual trait will evolve (Lande 1979). This pattern of genetic (co)variation between the individual components of the sexual trait is described by the genetic variance–covariance (**G**) matrix: a symmetrical matrix with genetic variances along the diagonal and genetic covariances in the off-diagonal positions (Lande 1979). Thus, studies wishing to examine how female mate choice influences the evolution of male sexual traits must take into account both the pattern of sexual selection that female mate choice exerts on male sexual traits and **G** for the male sexual trait (Lande 1979).

To our knowledge, this approach has only been comprehensively examined in a single insect species, *Drosophila serrata* (Chenoweth et al. 2010). *D. serrata* is native to Australia and occupies a long, narrow distribution along the coastal strip of eastern Australia (Chenoweth et al. 2010). In this species, female mate choice exerts sexual selection on a suite of nine male CHCs and the strength and form of sexual selection are known to vary across nine geographic populations sampled over a 1,450 km range (Chenoweth et al. 2008). Male CHCs have also diverged across these populations (Chenoweth et al. 2008), as has **G** for male CHCs (Hine et al. 2009), and neutral divergence due to genetic drift has been excluded as the cause of this divergence (Chenoweth et al. 2008). Using the multivariate framework of Zeng (1988), Chenoweth et al. (2010) examined the relative contribution of sexual selection imposed by female mate choice and **G** to the observed evolutionary divergence in male sexual traits across populations. Differences across populations in the direction of sexual selection were only weakly associated with the observed divergence in male CHCs, but the inclusion of **G** in the model significantly improved the alignment between the predicted and observed divergence in male CHCs (Chenoweth et al. 2010). This suggests that the genetic architecture of male sexual traits is relatively more important than female mate choice in driving the divergence in male CHCs across populations in this species. However, male CHC expression is also known to be correlated with maximum temperature across these populations, suggesting that natural selection also contributes to the evolutionary divergence of male CHCs to reduce evaporative water loss (Frentiu and Chenoweth 2010), a finding that is supported by laboratory evolution experiments in this species (e.g. Hine et al. 2011).

Although it is intuitively appealing to conclude that female mate choice is a major driving force in the evolution of male sexual traits, the examples discussed suggest that this conclusion would be premature. There is clear evidence that female mate choice imposes significant sexual selection on male sexual traits (Section 8.4.1), but there are many reasons why this selection may not necessarily result in phenotypic evolution, including the genetic architecture of the male sexual trait and the opposing effects of natural selection. More empirical work is clearly needed before we fully understand the implications that female mate choice has for the evolution of male sexual traits.

8.4.4 *Multivariate lek paradox: does the lek paradox actually exist?*

In Section 8.2.3 we outlined the empirical support for the various mechanisms that provide a resolution to the lek paradox. It is important to point out that each of these proposed mechanisms takes a univariate view of evolution (i.e. females show a preference for a single male sexual trait) where it is easy to see how the lek paradox can operate. However, several recent studies on insects have shown that when evolution is viewed as a multivariate process, it is unlikely that the lek paradox actually exists (Blows et al. 2004; Hine et al. 2004; Hunt et al. 2007; Van Homrigh et al. 2007; Hall et al. 2010b). Consider a hypothetical male sexual trait, such as the advertisement call of a field cricket, that consists of three individual components (chirp number, chirp duration, and duty cycle). This hypothetical cricket call can be considered a complex sexual trait and the pattern of genetic (co) variation between the individual components of this trait is summarized by the genetic variance–covariance (**G**) matrix (Section 8.4.3; Lande 1979). However, interpreting the individual elements of **G** is unlikely to provide a meaningful picture of the genetic (co) variance in the sexual trait (Blows 2007). Rather, **G** can be diagonalized to locate a series of genetically independent dimensions that encompass the total genetic variance in the sexual trait in multivariate space. Different amounts of genetic variance reside along each of these dimensions (with each successive dimension harbouring less of the total genetic variance in G) and selection targets these dimensions differently. It is therefore possible for genetic variance to be depleted along some dimensions of **G** where selection is at its strongest but maintained in other dimensions where selection is weaker. Consequently, as selection is unlikely to be effective at depleting genetic variance in all possible dimensions, at least some genetic variance will always remain in the sexual trait, thereby challenging the very existence of the lek paradox.

In a direct test of this hypothesis, Hunt et al. (2007) examined the alignment between **G** and the fitness surface for call structure in the black field cricket, *Teleogryllus commodus*. Using acoustic software, Hunt et al. (2007) phenotypically engineered calls differing in five call parameters (number of pulses in the chirp, chirp interpulse duration, trill number, intercall duration, and dominant frequency) and tested their attractiveness to females in dyadic playback trials in competition against the average call structure in the population. Formal multivariate selection analysis revealed that call structure is under multivariate stabilizing selection in this species (Brooks et al. 2005) and a paternal half-sib breeding design was used to estimate **G** for these individual components of call structure (Hunt et al. 2007). Diagonalization of **G** revealed five independent genetic dimensions, and factor analytical modelling provided statistical support for the first three of these dimensions that together explained 90% of the total genetic variance in **G**, and moderate support for the fourth dimension, explaining a further 7% of the genetic variance (Hunt et al. 2007). As predicted, if genetic variation is depleted along some dimensions where selection is strong, and preserved in others where selection is weaker, Hunt et al. (2007) found that the strength of stabilizing selection was inversely proportional to the amount of genetic variance across the first four dimensions of **G**.

Sexual selection may not always be effective at reducing levels of available genetic variation in complex traits if, for example, the major axis of sexual selection is poorly aligned with **G** (Blows 2007). The degree of this alignment is typically measured as the angle between the vector of linear sexual selection ($\boldsymbol{\beta}$) and the dominant eigenvector of **G** (g_{max}): the greater this angle the less effective sexual selection will be at depleting genetic variance in the male sexual trait (and also in driving phenotypic evolution) (Blows 2007). Several studies on *Drosophila* have estimated this angle for male CHCs that are known to play a key role in female mate choice in this genus (Ferveur 2010). In *D. serrata*, the angle between β and the dominant eigenvector of **G** is remarkably consistent in laboratory (74.9°; Blows et al. 2004) and field (73.5°; Hine et al. 2004) populations, and in *D. bunnanda* this angle is even more extreme (88°; Van Homrigh et al. 2007). Together, these studies highlight that sexual selection may often be poorly aligned with available levels of genetic variance and therefore ineffectual as a mechanism that depletes genetic variance.

8.5 Mate choice and speciation

Given the power of mate choice to effect evolutionary change in sexual traits, it will come as no surprise that the role of mate choice in promoting speciation has received increasing attention. Available evidence suggests that if sexual selection plays a role, it most likely does so through divergence of sexual traits that drive reproductive isolation in allopatry, that is, after populations have become geographically isolated (Ritchie 2007). Much of the evidence in support of this process comes from comparative studies, in which investigators seek to establish an association between the number of species (or rate of speciation) in clades that differ in the apparent intensity of sexual selection, using proxies such as the degree of sexual dimorphism or variation in mating systems as markers of sexual selection. Such studies, though informative, can only provide indirect evidence of an influence of sexual selection on speciation, and in insects have provided evidence both in favour of and against such an effect (Ritchie 2007). Studies involving experimental evolution offer considerable promise in providing more direct tests of the hypothesis that sexual selection promotes reproductive isolation, but such studies are still in their infancy. The most convincing evidence is most likely to come from detailed case studies, and we can think of no more convincing example than the work by Kerry Shaw and her colleagues on the Hawaiian crickets of the genus *Laupala* (Figure 8.8).

Figure 8.8 Male Hawaiian cricket, *Laupala pruna*. This genus shows the highest rate of speciation of any insect, which appears to be driven primarily by female mating preferences for male song. Photo by Kerry Shaw.

Laupala comprises a group of forest-dwelling Hawaiian crickets that boasts the highest known rate of speciation of any insect, approximately 4.2 new species per million years. Males produce simple songs consisting of trains of pulses to attract females, and among species, songs vary primarily in their pulse rate. Females exhibit strong preferences for the pulse rate of males of their own species. Moreover, variation across species in male pulse rate and female preferences are predicated on small additive genetic effects, and the genes controlling both songs and preferences seem to be tightly linked genetically (Shaw 2000; Mendelson and Shaw 2005). It is difficult to avoid the inference that sexual selection arising from female preferences for male pulse rate has promoted the reproductive isolation that underlies the explosive speciation in this group of insects (Mendelson and Shaw 2005). Of course, it is possible that ecological divergence accounts for speciation in this group, and that divergence in call traits only arose after species had become adapted to new environments. Arguing against this possibility, however, is the fact that closely related *Laupala* species are morphologically indistinguishable, exhibit no obvious differences in ecological requirements, and appear to be dietary generalists (Mendelson and Shaw 2005). Another alternative hypothesis—that divergence in male call traits comes from interspecies interactions that generate reinforcing selection—also appears to have been ruled out by intensive study of population differentiation in male mating signals and female preferences within a single species, *Laupala cerasina* (Grace and Shaw 2011). If population divergence in male calls is a result of species interactions, there should be evidence of character displacement between local populations of *L. cerasina* and the congeners with which they coexist sympatrically: there is, however, little evidence of this. Instead, *L. cerasina* exhibits marked population differentiation in both male pulse rate and female preferences, with female preferences being closely attuned to male pulse rates within populations, suggesting that sexual selection has driven sexual isolation in these populations (Grace and Shaw 2011).

8.6 Conclusions and future directions

The beauty of Thornhill and Alcock's (1983) book is that it poses more questions than it provides answers and this is particularly true with regard to its coverage of female mate choice. Consequently, this work has been a major influence in stimulating research over the last 30 years and our understanding of female mate choice has improved considerably since the book was first published. As we have sought to demonstrate in this chapter, research on insects has played a central role in this progression. We now have a more solid theoretical foundation and empirical support for why females choose mates and the mechanisms they use to differentiate between potential mating partners. We also know that female mate choice can impose a strong and complex pattern of sexual selection on male sexual traits and that there is a genetic basis to mate choice. Moreover, the genes controlling mate choice co-vary positively with the genes regulating the expression of the preferred sexual trait, which should facilitate the co-evolution of these traits. Despite this, variation in female mate choice does not always predict the evolution of the preferred male sexual trait, suggesting a complex co-evolutionary dynamic between these traits. Likewise, although female mate choice is also known to play an important role in reproductive isolation and eventual speciation, this process is unlikely to be simple.

Despite the considerable progress that has been made since the original publication of Thornhill and Alcock (1983), our chapter highlights several areas in the field of mate choice where more empirical work is needed and therefore future research would be fruitful. It is clear that females preferentially mate with some males over others, but the benefits of this mate choice are not always clear. Although we have outlined numerous examples where females gain direct and indirect (genetic) benefits, there are also countless examples where these benefits have not been detected. If the benefits of mate choice can only be realized in stressful environments, the benign conditions provided in laboratory studies (which predominate in insect research on this topic) may mask any potential benefits. Thus, more studies are needed that examine the benefits of mate choice in multiple environments (Hunt et al. 2004). Even though male mate choice appears widespread in insects (Bonduriansky 2001), most research focuses on the direct (fecundity) benefits received by choosy males. However, just as the genes provided by a male can be important to the mate choice decisions of a female, the reverse is likely also to be true for choosy males. Thus, more empirical work is needed on the indirect benefits of male mate choice, particularly whether males choose females of high genetic quality or based on compatible genes (Hunt et al. 2004; Neff and Pitcher 2005).

While few evolutionary biologists would question that mate choice exerts strong selection on the sexual traits it targets, very few studies have examined the full complexity of this selection even though a strong theoretical framework for this purpose was published in the same year as Thornhill and Alcock's (1983) book. Lande and Arnold's (1983) multivariate regression analysis enables both linear and non-linear forms of sexual selection to be estimated in a common currency: the standardized selection gradient. Moreover, canonical analysis of $\boldsymbol{\gamma}$ can be used to gain a true estimate of the strength and form of non-linear selection, as well as to help the biological interpretation of this selection (Phillips and Arnold 1989). Work on insects using these approaches has revealed that female choice exerts a complex pattern of sexual selection on male sexual traits, particularly that the fitness landscape most commonly represents a 'rising ridge' (Table 8.1). A landscape with this form is important as it suggests that male sexual traits may represent an honest signal of quality (Getty 1998). However, more studies applying these statistical approaches are needed to determine whether this is a general feature of male sexual traits. It is also known that mate choice is a layered process, occurring pre and post copulation. These discrete 'episodes' of selection can have important implications for how male sexual traits evolve by altering the variance in male sexual traits that each episode has to act upon. Again, the theoretical basis exists to statistically combine episodes of selection (Arnold and Wade 1984a) but is rarely used. More work is needed, therefore, to estimate selection across all episodes of selection and to combine them to estimate the total strength and form of sexual selection imposed by female mate choice. Likewise, female mate choice is often claimed to be a major force driving the evolutionary diversification of male sexual traits across populations, as well as significantly contributing to the process of reproductive isolation and even promoting eventual speciation. While some of the most compelling research on these topics comes from insects, most of our understanding comes from work on a few, extremely well studied, insect species (e.g. *Laupala*, *Drosophila serrata*) (Mendelson and Shaw 2005; Chenoweth et al. 2010). Extending this research to other insect systems with the same degree of thoroughness would greatly improve our understanding of the wider evolutionary implications of mate choice.

Finally, we provide strong evidence that there is a genetic basis to mate choice in insects and that there is also likely to be substantial genetic covariance between mate choice and the expression of the preferred sexual trait (Table 8.2). However, due to the experimental designs commonly used to assess the genetic architecture of mate choice in insects, there are very few estimates of heritability for mate choice and even fewer estimates for the genetic covariance between mate choice and the preferred male trait. These estimates are crucial for making quantitative predictions about the evolution of mate choice and the sexual trait(s) they target. The estimates that do exist suggest that heritabilities are high and the genetic correlations are strong and positive, although more quantitative genetic studies are needed before this may be accepted with any certainty. There is also genomic evidence demonstrating considerable linkage between male sexual signals and female mate choice, as well as specific genes that are involved in pre- and post-copulatory mate choice. While this work is also currently limited to a few species (e.g. *Laupala*, *Drosophila melanogaster*), the ever-increasing list of insect species with sequenced genomes is likely to make this line of research more feasible in the future. We believe that this will provide an even greater understanding of how female mate choice evolves and the implications this process has for the evolution of male sexual traits.

CHAPTER 9

The evolution of polyandry

Rhonda R. Snook

9.1 Introduction

We saw in Chapter 3 how insect mating systems encompass monogamy through to polyandry. Historically, monogamy was the presumed predominant mating pattern in females, but close observations and the subsequent advent of molecular techniques to allocate parentage has demonstrated that polyandry is rampant (Birkhead and Møller 1998). This is perhaps the single most important empirical advance since Thornhill and Alcock's (1983) volume. Classifying mating systems is rife with terminology, including several definitions for polyandry (Snook 2013; Chapter 3). However, the definition most widely used today is from Thornhill and Alcock (1983); females have more than one male as a mate during a breeding season (pp. 81–82). While polyandry is now recognized as being widespread, its origin and maintenance remain enigmatic due to variation between the sexes in the costs and benefits of reproduction. Central to the problem are theories on the evolution of anisogamy, Bateman's principles (Bateman 1948) and parental investment (Trivers 1972). Anisogamy is the defining feature of males and females, wherein males produce small but numerous gametes and females produce large, nutritious but generally rather few gametes. This fundamental difference between the sexes tends to limit female reproduction more strongly than male reproduction, with further delineation (or indeed reversal) of these 'sex roles' arising through subsequent variation in parental investment associated with ecological circumstances (Trivers 1972). Bateman's principles derive from observations, including Bateman's original experiments on *Drosophila*, which show that male fitness increases with an increasing number of mates whereas female fitness does not. These sex differences predict that male reproductive success will exhibit greater variance than female reproductive success. As a consequence, sexual selection will act more strongly on males than females. These patterns will be opposite in sex-role-reversed species, where males become the limiting sex and there is greater variance in success in competition for mates among females than males. Additionally, whereas sex is costly for both sexes, females appear to bear the sharp end of this expense. Aside from the energy and time expenditure required to engage in copulation (Thornhill and Alcock 1983), polyandrous females may, for example, experience greater predation (Arnqvist 1989), be injured (e.g. Eberhard 1996), and have reduced longevity as a consequence of the receipt of caustic male seminal fluids (e.g. Chapman et al. 1995). Sexual conflict over mate number can have profound influences on the behaviour, morphology, and physiology of insects (Arnqvist and Rowe 2005). Together, these data indicate that typically males should benefit most from

The Evolution of Insect Mating Systems. Edited by David M. Shuker and Leigh W. Simmons.
© The Royal Entomological Society 2014. Published 2014 by Oxford University Press.

multiple mating whereas females should benefit most from, and thus are predicted to exhibit, monogamy. However, females are rarely monogamous.

When predictions (females should be monogamous) do not match empirical data (females mate with more than one male), alternative explanations need to be sought. One such alternative is that the entire edifice (the trinity of anisogamy—Bateman's principles—parental investment) on which the predictions are made is false. Recently, several researchers have suggested this alternative, criticizing the data and interpretation of Bateman's own experiment (Tang-Martinez 2010; Gowaty et al. 2012). Some of these criticisms are justified—for example, problems with the genetic strains used—but others, such as a biblical-like literal interpretation that females do not benefit at all from multiple mating, are not. Whereas male fitness generally increases linearly with increasing number of mates, female fitness can increase asymptotically with number of mates, an empirical result found by Bateman (1948). Regardless of the experimental sins of Bateman, sufficient additional studies across different taxa by different researchers have supported the underlying principles—the non-limiting sex tends to gain more from multiple mating than the limiting sex—such that rejecting the entire idea is unwarranted.

However, it is likewise true that not all systems studied have conformed to Bateman's principles (Snyder and Gowaty 2013), that males can be choosy and females compete for males (Clutton-Brock 2009), and that polyandry can change the relative strength of sexual selection on each sex. For example, if females benefit from polyandry, then females may compete for access to multiple males resulting in increased sexual selection on females while polyandry simultaneously limits the male's ability to monopolize females, potentially reducing the strength of sexual selection acting on males (for review, see Kvarnemo and Simmons 2013). Thus, sex differences in competition and preference are more nuanced than the traditional Bateman view accommodates.

Regardless of the underlying evolutionary rationale for polyandry, the consequences of female multiple mating are indisputable. As discussed in Chapters 10 and 11, polyandry gives rise to post-copulatory sexual selection in both sexes and sexual conflict between the sexes, resulting in both competition within and co-evolution between the sexes. Indeed, the *raison d'etre* for Thornhill and Alcock's book was to '. . . attempt to explain characteristics bordering on the bizarre (the fantastic horns of certain male beetles, the complex penis structure of damselflies, the presentation of food gifts to females by male scorpionflies, and week-long copulations in walking sticks) as the products of social [sexual] selection for traits useful in an intensely competitive and coevolving sexual environment.' (Thornhill and Alcock, 1983, p. 54). Thirty years later, we understand that insect polyandry has additional impacts on, and is affected by, a variety of different evolutionary processes: the maintenance of genetic variability (Jennions and Petrie 2000 for review); sociality (Hughes et al. 2008; see Chapter 14); sex allocation (Ratnieks and Boomsma 1995); the spread of selfish genetic elements (Price et al. 2008a; see Chapter 13); speciation (Martin and Hosken 2003; Bacigalupe et al. 2007); and inbreeding (Michalczyk et al. 2011).

The widespread influence of polyandry on a variety of different evolutionary phenomena highlights just how important it is for us to understand this fundamental phenomenon. Studies on the evolutionary causes and consequences of polyandry have accumulated rapidly over the past 30 years, revealing a tremendous amount about the *raison d'etre* for those bizarre behaviours and structures that so fascinated Thornhill and Alcock and their early students. The goal of this chapter is to discuss a variety of recent developments regarding our understanding of polyandry, including perspectives that were nascent in Thornhill and Alcock's book, and to draw attention to areas of growing interest to which a new student of polyandry can contribute. To that end, this chapter includes (i) a brief consideration of the rise of studies of sexual conflict over mating decisions between the sexes that affect the economics of polyandry, (ii) how adoption of a quantitative genetics framework can elucidate the relative contribution of different indirect benefits and why this is important, (iii) the use of modern techniques, from experimental evolution through to genomics, for understanding the underlying phenotypic and genetic responses and potential constraints of polyandry, and (iv) how the economics of polyandry within and between species is affected by a variety of context-dependent factors. Consideration of these topics may open up new research areas for elucidating the evolutionary significance of polyandry. Such understanding hinges on the relatively poorly studied underlying genetic causes and consequences of female multiple mating.

9.2 Quantifying polyandry

Three different aspects of female reproduction can contribute to the effective levels of polyandry: the proportion of females in a population that remate, the number of males that an individual female mates with (or at the population level, the average number of males), and the (average) number of sires. Various different techniques can assess one or more of these components and may be applied to studies of both natural and laboratory populations. For example, in wild populations of lepidopterans, the number of times a female has mated can be determined by dissecting the female reproductive tract and counting the number of spermatophore remnants, as these remain in the female for her lifetime (for review, see Simmons 2001). However, this technique cannot determine whether these multiple matings were with the same individual or different males or whether, and the extent to which, sperm from each mating are used.

A second method applied in both wild and laboratory populations is to subject focal individuals to a continuous supply of the opposite sex to estimate the maximum number of mates within a given time. For example, in the seaweed fly, *Coelopa frigida*, whose mating system is characterized as convenience polyandry (see Section 9.3.2), Blyth and Gilburn (2006) collected wild males and females and measured in the field the number of pairs in which a male mounted a female within 5 min. These data suggested that, on average, males will mount a female approximately every 9 min; taking this into account together with rates of both male and female rejection of the mounts, an individual female was estimated to mate more than 30 times a day.

Various genetic techniques have been used to estimate the number of males a female has mated with. These originally included allozymes but have now moved on to microsatellite markers, including competitive polymerase chain reaction (PCR) techniques applied to sperm stored by females (Bretman and Tregenza 2005). The latter technique allows for an extensive evaluation of not only whether females are polyandrous but also whether there is any sperm sorting across sperm storage organs. For example, in the yellow dung fly, *Scathophaga stercoraria*, wild caught females were collected across a spring season and frozen upon collection, with the three spermathecae later dissected individually and the stored sperm extracted. Competitive PCR was then used to assign the number of males that females had mated with and whether sperm from these males were distributed equally between the three spermathecae. The researchers found that the proportion of females multiply mated increased sharply at the beginning of the season, remaining high, until shortly before the last sampling day. The average number of ejaculates ranged from 2.47 to 3.33 based on assumptions underlying the estimation process. Additionally, the number of ejaculates stored differed between the spermathecae, with the singlet spermatheca storing generally fewer sperm than either the inner or outer doublet (Demont et al. 2011). Results from the wild population mainly conformed to previous results based on laboratory populations (Demont et al. 2011).

A new technique to examine polyandry and its consequences in the wild is the combined use of DNA profiling and video monitoring of wild populations (Rodríguez-Muñoz et al. 2010). Two generations of a Spanish population of the flightless field cricket, *Gryllus campestris*, were observed with videos used to assess such factors as dominance, number of mates, and calling behaviour, combined with DNA profiling of the subsequent generation to assign parentage. Males had significantly greater variance in offspring number relative to females, confirming Bateman's principle in a wild population, but there was no difference between the sexes in the variance in the number of mates, contradicting Bateman's principle. The study also found that both males and females benefit from multiple mating by increasing the number of offspring, although what generates this increase is unclear. Females could benefit directly either through ejaculate donations or sperm replenishment, or indirectly either via good genes or via compatibility genes (see Sections 9.3.4 and 9.3.5).

As the ability to use more sophisticated techniques in the field improves, our understanding of what drives polyandry can only gain. Currently, many studies examining the evolutionary significance of female multiple mating rely on laboratory populations. While these are useful in allowing controlled conditions under which a researcher can isolate and assess the relative importance of one aspect of reproductive performance, these results may not always reflect what happens in the multivariate space of natural populations where both natural and sexual selection operate. For example, in *S. stercoraria*, patterns of polyandry and sperm storage in the field matched what was found in laboratory populations, although fertilization patterns and costs and benefits of polyandry have only been estimated in the laboratory (Demont et al. 2011). Hence, the relationships between polyandry, sperm storage and fitness cannot be compared between the field and laboratory populations. In contrast, the positive effect of dominance and male singing on male reproductive success in laboratory populations of *Gryllus* was not seen in the wild populations of *G. campestris*, for various reasons (Rodríguez-Muñoz et al. 2010) which can be followed up through experimental manipulations in a laboratory setting. A more integrated approach between field and laboratory studies of the evolutionary significance of polyandry is a crucial endeavour to generate far-reaching insights into sexual selection acting on natural populations of insects (e.g. Bretman and Tregenza 2005; Demont et al. 2011).

9.3 The evolutionary causes of polyandry

9.3.1 *Origin versus maintenance of polyandry*

A number of hypotheses have been proposed to explain the origin of polyandry: that is, why selection should favour the evolution of females multiply mating with different males (Table 9.1). However, most research has aimed primarily at studying the current utility of the phenomenon. While the origin and maintenance of polyandry may be closely linked, contemporary selection may not reflect historical selection, shifting the functional significance of polyandry across the evolutionary history of a species.

The origin of traits can be addressed using comparative methods, but, given the near ubiquity of polyandry, historical analyses may not be exceptionally informative. An

Table 9.1 Some direct and indirect benefits associated with the origin and maintenance of polyandry

Type of benefit	Evolutionary rationale for polyandry	Description	Prediction
Direct	Fertility assurance	Ensure lifetime sperm supply	Increased number of fertilized eggs compared to singly mated females or females mated multiply to same male
	Paternal donations	Transfer of nutrients	Increased fecundity and/or longevity as a result of donated nutrients
	Paternal donations	Transfer of chemicals	Decreased predation risk of remating females
	Convenience polyandry	Females acquiesce to remating	Increased longevity compared to females that are courted by males but do not remate
Indirect	Trading up	Females remate with partners of better genetic quality	Offspring with increased viability or increased reproductive success
	Genetic diversity/ genetic bet-hedging	Females guard against future environmental change by increasing genetic diversity of offspring	Increased probability of some offspring surviving
	Good sperm/ intrinsic male quality	Females remate with males to select superior fertilizing sperm which reflects male genetic quality	Increased offspring viability
	Sexy sperm	Females remate with males to promote increased fertilization efficiency (?) arising from sperm competition. Fertilization success?	Sons with increased fertilization success and daughters that promote sperm competition
	Genetic compatibility	Females multiply mate to increase genetic heterozygosity and avoid inbreeding through beneficial maternal and paternal genome combinations	Increased offspring fitness

alternative is to study species that are either monogamous or in which polyandry has evolved relatively recently, in order to understand what forces promote the evolution of a particular mating system. Studying these cases may shed light on the extent to which the origin and maintenance of the phenomenon are aligned, although currently this insight is limited.

For example, in the jewel wasp, *Nasonia vitripennis*, wild females are generally monogamous and laboratory females are reluctant to remate (Burton-Chellew et al. 2007). However, the longer the period of time in which strains are kept in the laboratory, the more frequently females mate with multiple males (Burton-Chellew et al. 2007). Additional work has shown that this response is heritable, although largely driven by non-additive effects (Shuker et al. 2007). The costs and benefits to polyandry in this system have not yet been studied, so the reason why polyandrous behaviour increases in laboratory culture remains unknown.

In *Drosophila subobscura*, females are resolutely monogamous (Smith 1956; Holman et al. 2008), yet this species shows nuptial feeding—a hallmark of direct benefits (Section 9.3.4) for mating multiply. Indeed, whereas direct benefits are dependent on nutritional status of both the male and female, females can increase fecundity from the nuptial gift and experience no identified mating cost (Immonen et al. 2009). Thus, in *D. subobscura*, females should benefit from multiple mating, yet they do not mate multiply.

Mating systems, despite often being described categorically, are actually a continuum encompassing spatio-temporal variation in ecology within and among populations of the same species. Thus, the general rule is likely to be that neither exclusive monandry nor polyandry is observed within most species. Indeed, in many insect groups, populations differ in the extent to which females engage in polyandry; and this may be heritable, as in *N. vitripennis* (Shuker et al. 2007). Thus, many species may have a plastic mating system in which the costs and benefits of mating are altered according to the particular number or quality of those partners. This understanding requires studying the conditions under which such plasticity is manifest, keeping in mind that the origin and maintenance of polyandry can be conflated.

9.3.2 *Non-adaptive explanations for polyandry: the rise of sexual conflict*

One early explanation for polyandry was that female mating rate was genetically correlated with male mating rate and was thus non-adaptive for females (Halliday and Arnold 1987). Although plausible, artificial selection experiments in which proxies for mating frequency were manipulated in *Drosophila melanogaster* (e.g. Gromko and Newport 1988) found no genetic correlation between male and female mating rate. Two subsequent selection experiments on the stalk-eyed fly *Cyrtodiopsis dalmanni* directly selecting for mating frequency (Grant et al. 2005), and on the adzuki bean beetle *Callosobruchus chinensis* selecting for female receptivity (Harano and Miyatake 2007a), also found no genetic correlation in mating frequency between the sexes. Thus, overall there is currently no support for polyandry arising as a correlated response to selection on increased male mating frequency, albeit with relatively limited data.

Thornhill and Alcock (1983) recognized the phenomenon of convenience polyandry in which females acquiesce to matings to alleviate constant and costly male harassment. In this case, females are making the 'best of a bad situation' and do not gain any benefits from remating. For example, in the seaweed fly, *C. frigida*, males do not perform any courtship behaviour and simply attempt to mount any female they come into contact with (for references see Blyth and Gilburn 2006). Females resist mating by shaking, kicking, and curling their abdomen, an energetically costly rejection response. Mating decreases longevity and wild females may experience more than 30 matings per day (Blyth and Gilburn 2006).

Convenience polyandry represents one manifestation of sexual conflict—an area of evolutionary biology that has been integral to recent studies of polyandry but not well-incorporated at the time of Thornhill and Alcock because its importance was not widely appreciated at that time (see Arnqvist and Rowe 2005). As discussed in Chapter 2, sexual conflict occurs when the fitness optima of the sexes differ (Parker 1979). A vast literature is now accumulating that documents the existence of sexual conflict over mating, including evidence from both comparative and experimental data (for review, see Arnqvist and Rowe 2005). Such evidence includes costs of mating to females, conflicts between the sexes over mating frequency, and the reproductive tactics employed by the sexes either to persuade or to resist mating. For example, mating itself may be differentially costly to females by increasing predation probability, as in *Gerris* water striders (Arnqvist 1989), or decreasing longevity due to physical damage to the female during mating, as seen in the bean weevil *Callosobruchus maculatus* (Crudgington and Siva-Jothy 2000), or chemical damage to the female after mating as seen in *Drosophila melanogaster* (e.g. Chapman et al. 1995). It is likely that many of these costs are collateral consequences of male–male competition; in other words a negative side-effect on females due to a conflict-driven adaptation in males, such as persistent courtship of females as seen in the drowning of female yellow dung flies, *Scathophaga stercoraria*, in the dung pats during mating struggles between males (see Chapter 10; reviewed by Arnqvist and Rowe 2005). Females are expected to evolve counter-adaptations to limit fitness costs, following which males may evolve a more powerful form of the costly trait, or selection may favour a different manipulative trait (Parker 1979, Holland and Rice 1998).

One recent example of such sexually antagonistic co-evolution can be found in the red-backed water strider, *Gerris gracilicornis*. Males of many *Gerris* species forcibly mate females whose genitalia are exposed. In *G. gracilicornis*, females have evolved concealed genitalia preventing male intromission (Figure 9.1; Han and Jablonski 2009). Females allow males access for mating only after the male performs tapping behaviour with his mid-legs, producing ripples on the water surface, which the authors interpret as a 'courtship signal'. Thus, females in this species have evolved a structure that prevents male manipulation. Subsequent work, however, found that the male courtship signal was actually a form of intimidation of the female. The tapping attracted predatory backswimmers, *Notonecta triguttata*

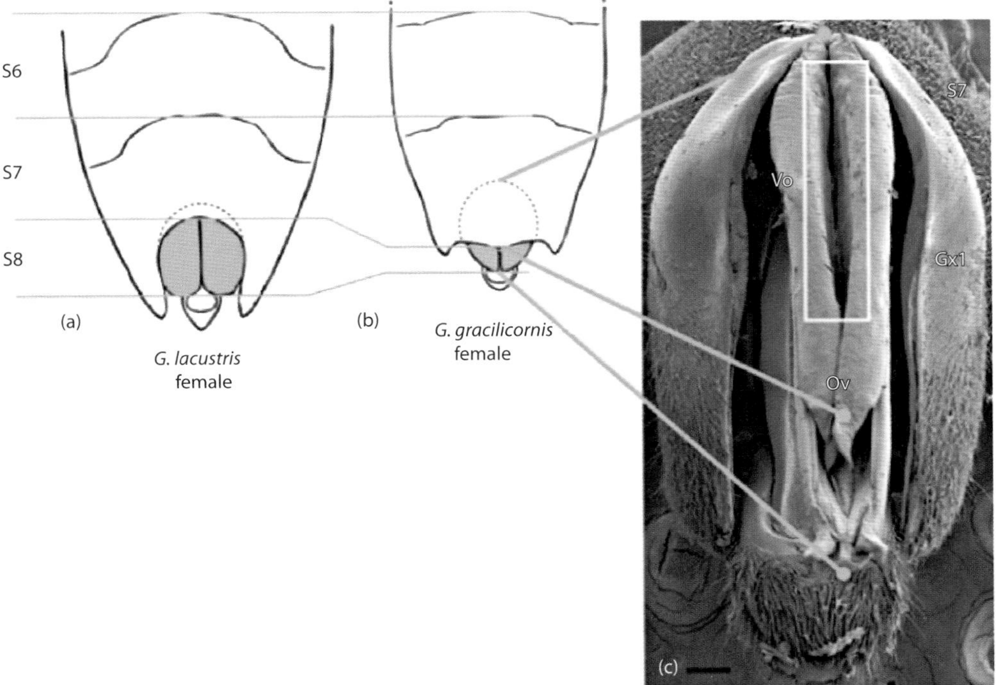

Figure 9.1 Genital segments of *Gerris lacustris* (a), a typical *Gerris*, and in *G. gracilicornis* (b), in which females have hidden genitalia from Segment 7 (S7). The vulvar opening (Vo) is normally hidden by the gonocoxa 1 (Gx1) which is partially inflated here (c). Scale bar: 0.1 mm. Modified from Han and Jablonski (2009).

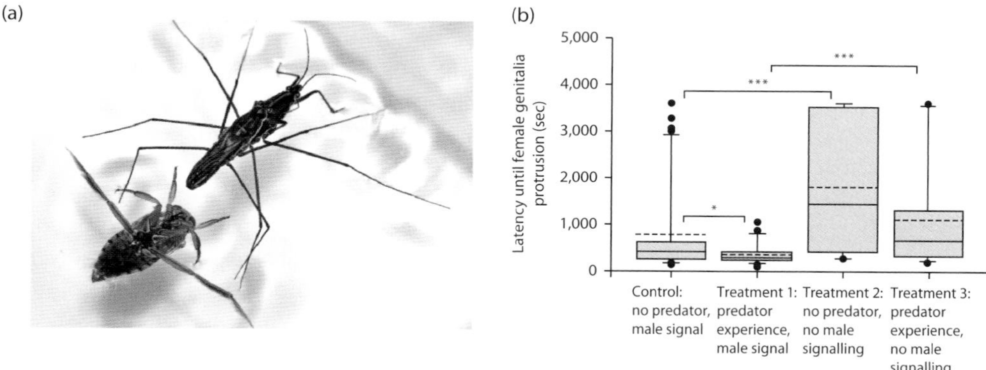

Figure 9.2 Female response to predator experience and male signals in *Gerris gracilicornis*. (a) Mating pair of water striders, female on the bottom, with a backswimming notonectid predator (from http://blogs.discovermagazine.com/notrocketscience/2010/08/10/male-water-striders-summon-predators-to-blackmail-females-into-having-sex/#.UXUX76LCZ8G). (b) Females were exposed to four fixed-order treatments and their latency to protrude their genitalia was measured. Both predator experience and male signalling influenced latency to protrude genitalia. Genitalia protrusion latency decreased when females had experienced a predator (Control vs Treatment 1; $^{*}P < 0.05$) but latency substantially increased if males were experimentally manipulated so that they could not tap their legs (Treatment 1 vs Treatment 3; $^{***}P < 0.001$). Boxes indicate the 25% and 75% quartiles with the solid line indicating the median and the dotted line the mean. Black dots represent outliers. Modified from Han and Jablonski (2010). Reproduced with permission from Nature Publishing Group.

(Figure 9.2a), which targeted female but not male water striders as prey. Females reduce predation risk by allowing males to copulate (Figure 9.2b; Han and Jablonski 2010).

Conflict over mating rate is a form of interlocus conflict (see Chapter 2). However, when examining the functional significance of polyandry, intralocus conflict may also affect the economics of female multiple mating. This is because the measures of reproductive success and hence fitness will reflect net selection acting on both sons and daughters, which obviously share some loci associated with offspring fitness, such as longevity and performance. Indeed, studies in insects have shown that there may be negative genetic correlations between son and daughter fitness with high-performing sires giving rise to high-performing sons but low-fitness daughters (e.g. Chippindale et al. 2001). In *D. melanogaster*, there was a positive correlation between male and female juvenile survival, but adult reproductive success was negatively correlated. While overall total fitness was not correlated between the sexes, substantial genetic × gender interactions were found (Figure 9.3). These results suggest pervasive sexual antagonism between the sexes at the genetic level; negative genetic correlations between the sexes, especially for shared fitness components, are fairly common (Poissant et al. 2010).

Intralocus sexual conflict poses a significant challenge for indirect-benefit models to explain the evolution of polyandry since mating with high-quality males may generate the production of low-quality daughters, cancelling any indirect benefit. Indeed, theory predicts that indirect benefits are likely to be smaller than direct costs of multiple mating (Kirkpatrick and Barton 1997; Cameron et al. 2003). In *D. melanogaster*, a number of studies from the Rice laboratory have found no evidence for indirect benefits but rather substantial mating costs (e.g. Stewart et al. 2005, 2008).

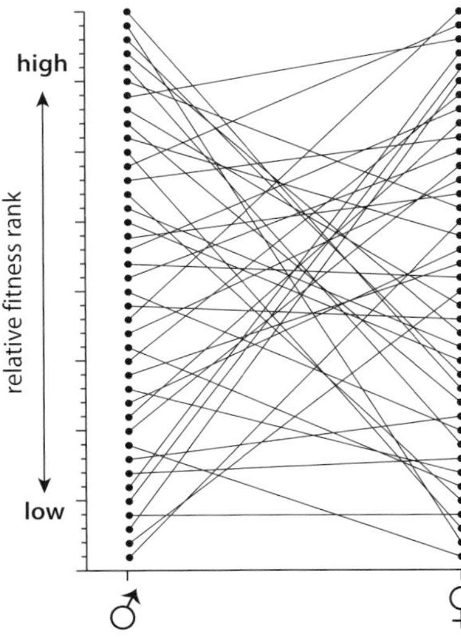

Figure 9.3 Significant interaction between relative total fitness rank, combining juvenile survival and adult reproduction, of male and female genomes in *Drosophila melanogaster*. Relative fitness between the genders frequently reversed, indicating that genomes producing high fitness in one sex can produce low fitness in the other sex. From Chippindale et al. (2001). Reproduced with permission from the US National Academy of Sciences.

9.3.3 *Overview of adaptive hypotheses for polyandry*

A number of adaptive hypotheses that focus on the currency of the fitness advantage have been proposed to explain why selection should favour the evolution of polyandry (Table 9.1). Thornhill and Alcock recognized that polyandry was easily explained if females gained directly from multiple mating either through increased longevity or increased production of offspring. The mechanisms of such benefits are often quite evident (i.e. nuptial feeding in bushcrickets) and relatively easy to study and manipulate (Section 9.3.4).

The conundrum over the evolutionary rationale for polyandry lies in species in which males make no material contribution to female reproduction. In many species, males apparently donate nothing more than the ejaculate, such that material benefits (beyond sperm and seminal fluid) are unlikely. In these cases, alternative indirect ('genetic') benefits have been suggested, such that females gain fitness from mating with different partners through increased performance of progeny. This general class of hypotheses has been more difficult to study than material benefits, as both the action and effects of indirect benefits are less conspicuous. Very large, longitudinal studies across generations are required, often only feasible in the laboratory on a subset of amenable species. Moreover, if such benefits accrue via mechanisms of post-copulatory sexual selection such as sperm competition (Chapter 10) and cryptic female choice (Chapter 11), these mechanisms, let alone their outcomes, are themselves typically difficult to study and disentangle. Indirect genetic-benefit models are also controversial because direct costs of mating multiply are predicted to outweigh indirect benefits to females and, whereas some studies in *D. melanogaster* have supported this prediction (Section 9.3.2), other studies in this species have found that indirect benefits outweigh direct mating costs to females (e.g. Rundle et al. 2007; Priest et al. 2008), and that indirect benefits can sometimes be even greater than direct benefits (Tuni et al. 2013). There is a substantial number of reviews on direct and indirect benefits for polyandry (e.g. Jennions and Petrie 2000; Zeh and Zeh 2001a; Simmons 2005; Puurtinen et al. 2009; Slatyer et al. 2012), so here we briefly cover only the major points.

9.3.4 *Direct benefits*

In many tettigoniids (bushcrickets), the male transfers a spermatophore containing a nutritious spermatophylax that is attached to the sperm-containing ampulla. After transfer of the spermatophore, females consume the spermatophylax while the sperm drain from the ampulla and are stored in her spematheca. In the Australian bushcricket, *Kawanaphila nartee*, spermatophylax feeding increases the number and weight of egg (Simmons 1990).

Such material benefits have a direct positive effect on female fitness, through a variety of fitness-related traits, including assuring fertility, increased fecundity, increased offspring number, increased longevity, predator avoidance, and paternal care. Many of these were discussed extensively by Thornhill and Alcock (1983, e.g. pp. 374–390). Recent meta-analyses have found strong support for direct benefits explaining polyandry generally, and in insects in particular (Arnqvist and Nilsson 2000; Slatyer et al. 2012). Many of these studies address the role of nuptial gifts and other forms of male donation in offspring production and female longevity. Here we shall focus on a second less-well-documented direct benefit, fertility assurance, and whether it can explain the origin of polyandry or its maintenance.

In some insect species, such as parasitoid wasps with local mate competition, males may mate with a large number of females in quick succession such that fertility assurance may explain female multiple mating (Boivin 2013). Likewise, females may remate to replenish sperm in the fire ant, *Solenopsis invicta*. While several papers suggest that *S. invicta* is predominantly monogamous, approximately 20% of queens from some areas can be polyandrous. Such polyandrous behavior is associated with whether their first mate carried a particular haplotype (Lawson et al. 2012). Males carrying a *Gp-9*^{b} haplotype produce fewer sperm than males carrying the *Gp-9*^{B} haplotype. Females mated to *Gp-9*^{b} males remate more frequently than those mated to *Gp-9*^{B} males. Thus, females may remate to replenish sperm stores—a direct benefit—although this does not entirely explain paternity skew when females mate with both male types (Lawson et al. 2012), leaving the potential for remating also to increase indirect fitness through selection against selfish genetic elements (Chapter 13).

However, in many insect species, fertility assurance is thought to be an unlikely explanation for the origin of polyandry because many researchers state that males in most (but not all) insect species transfer sufficient sperm for the female's lifetime in one copulation. One issue is that we generally don't know what a 'lifetime' is. For example, in a laboratory population of *D. pseudoobscura*, females that had mated just once had reduced average number of offspring per day but no differences in average number of eggs per day compared to females mated multiply with the same male, suggesting sperm limitation (Gowaty et al. 2010). However, this experiment lasted well beyond 45 days and it is unclear how long females live in the wild, or have opportunities to oviposit, as extrinsic mortality present in natural populations is not reflected in laboratory populations. If females on average do not live very long in the wild, or have few opportunities to find oviposition sites, then fertility assurance may be unlikely to explain the origin of polyandry.

Fertility assurance may, however, explain the maintenance of polyandry. While sperm are relatively inexpensive compared to eggs, the ejaculates themselves can be expensive and, to avoid sperm depletion, males of some species have evolved strategic allocation (see Chapter 10) (Wedell et al. 2002). As a consequence of strategic sperm allocation, females may not receive enough sperm per mating. Taken at face value, this then suggests that polyandry is less likely to have evolved originally in order to counteract sperm limitation (as males had plenty of sperm when females only mated once). However, once females become polyandrous, strategic allocation and other factors influencing male ejaculate sizes may generate sperm limitation which may reinforce whatever other selection has favoured polyandry. Sperm limitation may then be a factor maintaining polyandry in a population, rather than the original cause of polyandry. Even in sex-role-reversed species where males (and their sperm) are the limiting resource, sperm limitation may still not be the evolutionary origin of polyandry. For example, in some *Drosophila* species, males produce exceptionally long sperm and relatively few of them, such that females tend to remate often. While this result supports the idea of multiple mating evolving in response to fertility assurance, both theory and empirical results show that long sperm likely evolved in response to the post-copulatory process of cryptic female choice (Miller and Pitnick 2002). Thus, sperm limitation in this case may not be the origin of female multiple mating, but arises as a consequence of females preferring investment in costly long sperm.

9.3.5 *Indirect benefits*

The functional significance of polyandry in the absence of direct benefits was not readily apparent when Thornhill and Alcock's book was published. When direct benefits can be ruled out, only indirect ('genetic') benefits can offset the costs of polyandry. But whether and to what extent genetic benefits could explain polyandry was very unclear, as good-genes models of sexual selection (and indeed formal models of mate choice more generally; see Chapter 2) were still relatively new and the controversies over good-genes sexual selection only just warming up. Since then, many studies of genetic benefits, and the ongoing attendant controversies surrounding these, have been published.

Selection favours polyandry via genetic benefits when the mean offspring fitness of polyandrous females rises above the average value obtained via monandry, as a consequence of certain genes from the male (intrinsic male quality) or combinations of male and female genes (genetic compatibility). In quantitative genetics terms, these benefits are either additive, that is the effects on reproduction that are due to the male's breeding value for fitness, or non-additive, that is the effects on reproduction that are due to the genetic interactions between the parents' haplotypes. We have seen how additive and non-additive genetic benefits influence the evolution of mate choice in Chapter 8. In this section, we briefly discuss how these two dominant hypotheses for indirect benefits might influence the evolution of polyandry.

Polyandry incites sperm competition (Chapter 10) and facilitates cryptic female choice (Chapter 11). These post-copulatory mechanisms can bias paternity, such that males with greater breeding values for fitness are selected. Females may benefit from mating with multiple males either through ejaculate components that are correlated with intrinsic male quality (e.g. 'good sperm'; Sivinski 1984) or success in sperm competition ('sexy sperm'; Keller and Reeve 1995). One simple prediction is that if females gain from polyandry through either good or sexy sperm, then female mate choice should be congruent within a population; that is, females should prefer the same male phenotype. These additive good-genes benefits result in directional selection on the male trait which is expected to deplete additive genetic variation in associated fitness-related traits, implying that good-genes benefits should be small. Nonetheless, additive genetic variation for these traits may be quite high, resulting in the so-called lek paradox (see Chapter 8).

Genetic benefits through post-copulatory mechanisms can also be obtained through genetic compatibility which brings favourable parental genetic combinations together and/or avoids unfavourable gene combinations (i.e. the benefits are non-additive genetic benefits). In this case, if females gain compatible alleles from polyandry, then different females should prefer different males from the same population and no directional selection on male phenotype occurs. However, it may be difficult to envisage how females can optimize mating for genetic compatibility as the fitness outcomes for offspring may be hard to estimate. If such benefits are associated with a few linked loci, or reflect a relatively simple mechanism to avoid inbreeding, then polyandry may be beneficial. In contrast, if extensive pleiotropy occurs, then polyandry for genetic compatibility may be more difficult to evolve (Puurtinen et al. 2009). Few studies on the genetic architecture of polyandry exist, but of those we have, pleiotropy appears to be common (Section 9.4.3).

Because of these predictions, good-gene and compatible-gene models for the evolution of polyandry are often treated as mutually exclusive. However, the same alleles may contribute to variation in both additive and non-additive indirect benefits (Puurtinen et al. 2009). As such, females should make their mating decisions taking both into account. Depending on the extent to which intrinsic genetic quality versus genetic compatibility dominate the fitness consequences, we should expect some agreement among females as to the best male partners (perhaps all avoiding avowedly poor or sick males), but there should also be some disagreement among females associated with genetic compatibility. It will be an empirical challenge to tease these effects apart, especially given the sampling error that inevitably arises in behavioural studies and the large sample sizes needed to disentangle additive and non-additive effects in the required quantitative genetics framework. This line of enquiry is still too new for any studies to have accumulated.

9.4 Experimental approaches to understanding polyandry

9.4.1 *Measuring the economics of polyandry*

Much of the debate over the evolution of polyandry hinges on properly estimating the economics of polyandry. The relative magnitude of direct and indirect costs and benefits is likely not only to be taxon specific but also environmentally (context) specific, and thus could be highly labile (Section 9.5). This means that polyandry may vary across populations of the same species, or that benefits to polyandry may be found in some populations but not others (including populations living in the laboratory), rendering conclusions about the functional significance of polyandry challenging. Additionally, measuring the fitness consequences of multiple mating should include multiple measures of offspring fitness. For example, in *C. maculatus*, a quantitative genetics approach found variation in the contribution of additive and non-additive genetic variance depending on what fitness-related trait was measured (Bilde et al. 2008). Moreover, a recent study suggests that indirect benefits may be transient across multiple generations; *D. melanogaster* females exposed to high levels of sexual conflict produced sons with increased fitness but grandsons with decreased fitness assayed as competitive fertilization success (Brommer et al. 2012). These studies indicate that, while it may be easier to pick one reasonable fitness proxy and constrain measurements to the F_1 generation, these choices will impact our understanding of the evolutionary significance of polyandry.

Another concern with current studies is that most do not consider whether multiple factors could contribute to the evolution of polyandry. For example, when direct benefits are found, few studies examine whether indirect benefits may also be acting, likely because the gains through indirect benefits generally are thought to be relatively weak. One recent study, however, found that females in the nuptial-gift-giving spider, *Pisaura mirabilis*, benefited marginally from mating multiply via earlier reproduction—a direct benefit—but also acquired substantial indirect benefits through increased egg-hatching success (Tuni et al. 2013). Such indirect benefits may represent the origin of polyandry for this species or could be derived from male manipulation of the female motivation to forage (see Tuni et al. 2013). Likewise, both good-genes and compatible-genes benefits may be occurring simultaneously and their relative magnitudes have rarely been examined in insects (but see Section 9.4.3).

Some powerful techniques have been employed since Thornhill and Alcock that have helped to uncover the functional significance of polyandry. These include experimental protocols to clarify differences in mating number versus polyandry per se, genetic approaches including quantitative genetic models to partition additive and non-additive variance, and genomic techniques to identify genes and gene networks associated with mating.

9.4.2 *Controlling for multiple mating*

One issue with earlier studies on the evolution of polyandry was that experiments did not control for differences in multiple mating per se and mating with multiple partners (polyandry). The first experimental approach to combat this potentially confounding issue was that of Tregenza and Wedell (1998) who allowed females to mate multiple times with the same male or multiple times with different males. Using the cricket *Gryllus bimaculatus*, they found that polyandrous females had increased offspring production compared to females who mated multiply with the same male. Subsequent experiments suggested that the benefits of polyandry were associated with inbreeding avoidance (Tregenza and Wedell 2002). This protocol, along with including singly mated female controls (Ivy and Sakaluk 2005), is now widely adopted for discriminating the potential effects of polyandry on female fitness.

9.4.3 *Quantitative genetics*

Given that distinguishing the effects of intrinsic male quality and genetic compatibility hinge on whether such benefits are additive or non-additive (Puurtinen et al. 2009), quantitative genetics will play a significant role in our understanding of the evolution of polyandry. Yet only a few studies have examined the genetic architecture of the fitness benefits of polyandry in insects.

In the decorated cricket, *Gryllodes sigillatus*, diallel crosses in which a set of genotypes is crossed in all possible combinations revealed that whether intrinsic male quality or genetic compatibility explained polyandry depended on the fitness trait being measured (Ivy 2007). For development time and female adult offspring mass, non-additive effects strongly influenced phenotypic variance, whereas offspring survival to adulthood was primarily influenced by additive effects (Figure 9.4). A large proportion of phenotypic variance in hatching success was influenced by sire effects (Ivy 2007). Whether these were direct or indirect paternal effects was not determined. In the Australian field cricket, *Teleogryllus oceanicus*, a quantitative genetics design indicated additive benefits to polyandry arising from sires via increased survival of embryos (García-González and Simmons 2005a). This benefit could be mediated through good genes per se—that is, increased viability by direct transmission of paternal genes—or through indirect paternal effects mediated by ejaculate components, which themselves are heritable. Subsequent work supported the role of ejaculate components promoting embryo survival (García-González and Simmons 2007a).

Combined additive and non-additive genetic benefits may make the evolution of polyandry more likely in the face of direct costs to female multiple mating. However, a major

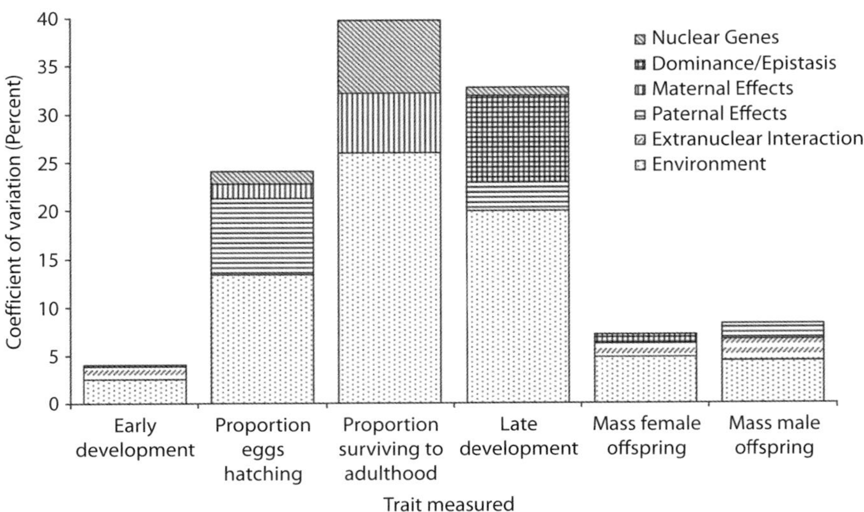

Figure 9.4 The relative impact of additive and non-additive genetic effects on phenotypic variation of six fitness-related traits in *Gryllodes sigillatus*. Six variance components, representing additive (nuclear genes and paternal effects) and non-additive (dominance/epistasis and extranuclear interactions), were estimated using a diallel cross design. From Ivy (2007). Reproduced with permission from Wiley-Blackwell.

caveat is that net selection for female choice is critical in determining the potential for good genes and genetic compatibility to drive polyandry. As previously mentioned, negative genetic correlations between son and daughter performance may occur (Figure 9.3), limiting indirect benefits even if both additive and non-additive benefits act. Additionally, While quantitative genetic designs are powerful, they cannot show that paternity is biased towards either intrinsically superior genotypes or genetically compatible genotypes. As Simmons (2005) points out, this is because paternity is assigned by the investigator at a relatively late stage of development, such that any earlier effects of genetic incompatibility or intrinsic sire effects on embryo mortality are unknown. Thus, early development paternal or sire-by-dam effects may confound estimates of the genetic effects of polyandry. Maternal effects via differential allocation may similarly confound genetic estimates. Hence, studies that employ rather sophisticated quantitative genetics designs are necessary to delineate additive, non-additive, and maternal effects while controlling for mate number and variation in early embryo mortality. The sample sizes required for such studies are large, and laboratory-friendly species are needed, potentially biasing the insects we choose for such studies.

9.4.4 *Experimental evolution*

Early artificial selection experiments contributed to our understanding of the functional significance of polyandry by showing that male and female remating rate were not genetically correlated (see Section 9.3.2). A relatively recent technique is experimental evolution in which the mating system, rather than a target trait, is manipulated over an evolutionary time frame to uncover (for instance) the effects of removing and/or elevating the strength of sexual selection and sexual conflict (Kawecki et al. 2012). In insects, such experimental sexual selection has shown that a variety of pre- and post-copulatory traits in males evolves under different mating systems (for example see Chapter 10). With relevance to polyandry, two of these experimental evolution approaches have revealed the adaptive significance of polyandry.

Selfish genetic elements in the form of meiotic drivers may result in sperm limitation since 50% of sperm—in sex-ratio distorters those bearing either the X or the Y chromozomes—are rendered sterile as a consequence of the driving element. Many natural populations harbour such drivers, and Chapter 13 discusses how in *D. pseudoobscura*, after just 10 generations of evolution, females in female-biased populations harbouring the X driver had evolved greater remating rates and a greater likelihood of remating on the first opportunity (i.e. decreased female choosiness) compared with populations lacking the X driver (Price et al. 2008b). Thus, selfish genetic elements may promote polyandry by allowing females to mitigate against mating only with males with depleted sperm and/or with sperm carrying a sex-ratio distorter. Interestingly, monogamous populations carrying the sex-ratio distortion gene were also more likely to go extinct, suggesting that polyandry can protect against population extinct risk (Price et al. 2010a).

Experimental evolution has also identified the role of inbreeding avoidance in generating polyandry. If polyandry is beneficial to avoid genetic incompatibilities arising from inbreeding, then such behaviour should be found in spatio-temporal conditions subject to high inbreeding risk, such as species in which females have to leave crowded resource patches, colonize an empty patch, and then whose offspring will be restricted to siblings as potential mates until further colonization. Theoretical modelling examining whether such conditions could explain polyandry via indirect benefits indicates that this depends on whether inbreeding depression occurs through deleterious recessive alleles or overdominance (Cornell and Tregenza 2007). While the latter can maintain inbreeding depression, thus fostering continual potential indirect genetic benefits for polyandry, deleterious recessive alleles would be purged from the system and benefits from polyandry would be small and of transient importance.

An experimental evolution study in which populations of the red flour beetle, *Tribolium castaneum*, a species that meets the spatio-temporal conditions of the Cornell and Tregenza (2007) model, has been used to show that inbreeding can promote the evolution of polyandry (Michalczyk et al. 2011). In this experiment, *T. castaneum* populations were subjected to eight generations of genetic bottlenecking via sib–sib matings and the fitness consequences for inbred and outbred populations were compared under evolutionary scenarios of either monogamy or polyandry. Monogamous inbred females had reduced composite fitness (offspring number and offspring survival) compared with non-inbred controls, but polyandrous inbred females regained this loss; these results indicate that inbreeding depression occurs and that polyandry may combat inbreeding depression (Figure 9.5a). Because polyandry benefitted inbred populations, female remating behaviour was compared with non-inbred control lines 15 generations after genetic bottlenecking. Previously inbred females had higher levels of polyandry than non-inbred controls (Figure 9.5b, c; Michalczyk et al. 2011). These results suggest that past inbreeding may have current effects on mating system structure. However, whether inbreeding depression in *T. castaneum* is mediated via deleterious recessive alleles or overdominance is currently unknown.

Thus, experimental evolution has been used to great effect to test aspects of sexual selection and sexual conflict in general and polyandry in particular. This technique is particularly well suited to insects as they tend to have relatively large effective population sizes, short generation times, are relatively easy to rear in the laboratory, and thus relatively easy

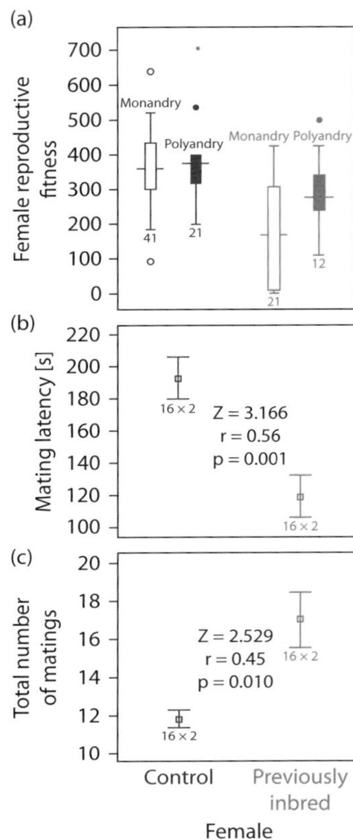

Figure 9.5 Female fitness and mating behaviour response to experimental inbreeding in *T. castaneum*. Inbred females regain fitness to control, non-inbred, levels if they mate polyandrously compared to a single mating (upper panel). Females from the previously inbred lines mate quicker (middle panel) and have more matings (lower panel) than control, non-inbred females. Modified from Michalczyk et al. (2011). Reproduced with permission from The American Association for the Advancement of Science.

to replicate. This approach will remain informative, especially in combination with quantitative genetics and genomics techniques (see Section 9.4.5). However, such laboratory studies must always be tempered with the realization that they serve as hypotheses to test in wild populations (see Section 9.3.1).

9.4.5 *Genomics*

One considerable difference between the current emphasis on understanding polyandry compared to when Thornhill and Alcock's book was published are the genetic techniques that have not only demonstrated that polyandry is pervasive (Section 9.2), but also now allow the study of the genes themselves that are associated with polyandry (see also Chapter 4). Microarray and next-generation sequencing permit the identification of genes that control polyandry. While the use of these techniques to study polyandry is still in its infancy, such studies represent a fast-growing area for understanding evolutionary consequences of polyandry at the genetic level.

Drosophila melanogaster has served as a model laboratory system in which to examine how specific genes influence the outcome of mating (see Chapter 4). Perhaps the best understood of these genes is Sex Peptide (SP), a seminal fluid protein that alters female behaviour and physiology and mediates sexual conflict. Relevant for studies of polyandry, SP reduces female receptivity and decreases female survival. Using microarrays, one study has focused on how the sole transfer of SP influenced female mRNA expression (Gioti et al. 2012). A large number of genes altered expression in response to the receipt of SP, indicating that it has extensive pleiotropic effects and that SP may be a 'global regulator' of female reproductive processes. As a consequence of large-scale pleiotropy, females may be constrained in evolving any response that mediates the toxic effect of receiving SP upon mating, rendering the mitigation of the negative fitness consequences of multiple mating difficult.

Another study of *D. melanogaster* used artificial selection to alter mating speed, also referred to as copulation latency, defined as the time it takes for males and females once together to begin copulation. Artificial selection was used to effectively increase ('Fast') or reduce ('Slow') female receptivity (Mackay et al. 2005), essentially serving as a proxy for populations to be either more polygamous or less polygamous respectively. After 29 generations of selection, microarrays were used to assess whole genome transcriptional response to changes in the mating system. Twenty-one per cent of the transcriptome changed as a consequence of selection. Of the genes that were differentially expressed between selection lines, approximately 38% were altered uniquely (i.e. up- or downregulated in only one line) with 68% of these upregulated in Fast females. A variety of different types of genes was altered in response to selection for mating system, suggesting extensive pleiotropy. Intriguingly, some genes with altered expression are involved in the basal sex determination pathway (*doublesex*, *transformer*, *transformer 2*, and *fruitless*), with Fast females exhibiting upregulation of female sex determination genes and Slow females exhibiting upregulation of male sex determination genes (Mackay et al. 2005). Perhaps the mating system either influences or is influenced by genes associated with sex determination and downstream effects on sex-specific patterning of neural architecture, and potentially impacting the extent of maleness and femaleness.

Genomics studies on the evolution of mating systems are still in their infancy (Chapter 4). Two studies in *D. melanogaster* suggest pervasive and pleiotropic effects on the genome in response to sexual conflict and changes in female mating behaviour (Mackay et al. 2005; Gioti et al. 2012). The extent to which this constrains or facilitates responses to sexual conflict, and affects direct and indirect benefits to polyandry, has not yet been examined. Future work might capitalize on the decreasing costs and increasing ease of genomic interrogation in non-model organisms. When combined with experimental evolution, such studies have the potential to substantially increase our understanding of the evolution and maintenance of polyandry. Although powerful, these studies can only suggest candidate genes that may be important in mediating functional responses to polyandry. Subsequent work is necessary to determine the evolutionary relevance of such changes regarding the economics of polyandry.

9.5 The ecology of polyandry

9.5.1 *Mating system plasticity and ecology*

An explosion of data on the evolutionary significance of polyandry has accumulated since Thornhill and Alcock. There is now a large and ever-expanding body of literature on the consequences of polyandry for specific species; somewhat of a directory of costs and benefits for each species. Likewise, as the chapters in this volume attest, there is a growing appreciation for the extensive effects of polyandry on other evolutionary processes. Just within the past few years, researchers have found new evolutionary factors that promote polyandry, such as selfish genetic elements, population extinction, and inbreeding.

Yet, despite the tremendous advance, explanations for the underlying interspecific patterns that generate variation in remating and the economics of polyandry, particularly when material benefits are absent, are lacking. Individual studies provide the building blocks for supporting or rejecting particular hypotheses regarding the evolution of polyandry that can then be tested generally using meta-analyses, which have found substantial and uncontroversial evidence for material benefits but little evidence for genetic benefits, despite some taxa clearly benefitting indirectly from polyandry (Arnqvist and Nilsson 2000; Slatyer et al. 2012). How can we reconcile the individual cases with the overall pattern? What are the abiotic and biotic conditions, if any, that *generally* promote genetic benefits arising from polyandry? What ecological conditions limit such benefits?

For more than 30 years, behavioural ecologists have recognized that mating systems are influenced by ecology; indeed mating system theory is predicated on ecology (e.g. Emlen and Oring 1977). That is, mating systems are context dependent. For example, in the burying beetle genus *Nicrophorus*, individuals may experience monogamy, polyandry, polygyny, or polygynandry depending on the number of males and females that show up at the carcass, which is at least partially determined by carcass size (Muller et al. 2006). In bushcrickets, in which males produce expensive spermatophores that directly benefit females, the availability of resources in the environment determines the limiting sex (Gwynne and Simmons 1990; for review, see Lehmann 2012). It is therefore odd that, despite this strong history of recognizing the role of ecology in mating systems, this appreciation has not informed our understanding of the economics of polyandry to the extent that it perhaps should have done. If we assume that polyandry is a variable life-history strategy, then studying parameters that influence such life-history decisions may provide a more general understanding of what types of benefits females may obtain from mating and whether these benefits outweigh the costs of multiple mating.

The burying beetle and bushcricket examples are clear cases in which females benefit directly, and changes in ecology have clear implications for the extent to which females gain from multiple mating. Not all examples, even for direct benefits, are so clear cut. For example, in the butterfly, *Pieris napi*, females should be polyandrous as males transfer a large, nutritious, ejaculate (Figure 9.6a) which increases female fitness with little cost, as polyandrous females live longer than monandrous females (Wiklund et al. 1993). Yet, while polyandry is under genetic control (Wedell et al. 2002), populations vary in the proportion of females that are polyandrous, with more northern populations being more monandrous than southern populations (Figure 9.6b; Välimäki and Kaitala 2010). The association between latitude and female mating pattern suggests that the costs and benefits of multiple mating vary based on environmental and associated life-history parameters that are linked to latitude. In particular, northern populations will experience a truncated summer and lower average temperatures throughout the year, limiting reproduction and offspring development. Hence, populations may vary in the number of generations per year which could change the relative costs and benefits of polyandry within the same species. Intriguingly, whereas northern females have lower overall fecundity under standard conditions, they have greater early fecundity because they avoid the timing costs of additional matings (Välimäki et al. 2008), suggesting that in a shortened reproductive season northern females would not benefit from polyandry despite receiving direct benefits from multiple males. In the southern multivoltine populations, females from directly developing generations exhibit higher levels of polyandry than females from the pupal diapausing generation, suggesting that something about the economics of polyandry differs even in a species with material benefits. What this difference may be is currently unknown (Larsdotter Mellström and Wiklund 2010).

If different ecological conditions and life histories select for different levels of polyandry within a species, this may have downstream effects on other aspects of the mating system.

(a)

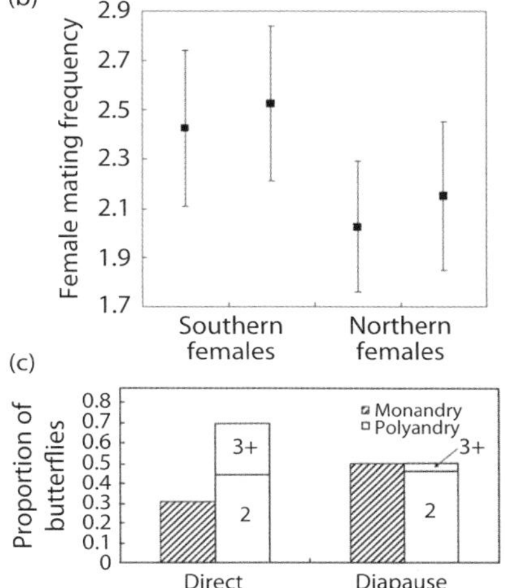

Figure 9.6 Variance in mating frequency between and within populations of *Pieris napi*. (a) Copulating *P. napi*. Reproduced with permission from P. Välimäki. (b) Females from southern populations are more promiscuous than females from northern populations. Modified from Välimäki and Kaitala (2010). Reproduced with permission from Elsevier. (c) Females from the directly developing generation are more promiscuous than diapausing females. Numbers on the open bars are lifetime number of matings. From Larsdotter Mellström and Wiklund (2010). Reproduced with permission from Elsevier.

For example, in *P. napi*, since females from the directly developing generation exhibit higher levels of polyandry, perhaps males may be sperm-depleted, which may impact on other aspects of sexual selection and the costs and benefits to polyandry (for review, see Kvarnemo and Simmons 2013). Thus, determining the causes of interspecific variation in the functional significance of polyandry, taking into account abiotic and biotic processes within a species that facilitate and limit benefits to female multiple mating, is challenging. It will require integrating studies of intraspecific plasticity of mating systems driven by identified ecological variables, with comparative studies across species incorporating those variables.

9.5.2 *Quantitative genetics and mating system plasticity*

In a quantitative genetics framework, the context dependency of mating systems is reflected by genotype × environment interactions, that is, the relative phenotypic performance of different genotypes is influenced by the environment in which genotypes are expressed. In quantitative genetic studies of indirect benefits to polyandry, sources of phenotypic variance may be significantly affected by the environment (Bilde et al. 2008; but see Watson and Simmons 2011). The effect of environmental heterogeneity on the economics of polyandry can arise from abiotic factors (e.g. latitude, *P. napi*; see Section 9.4.1) or from biotic factors (indirect genetic effects; Wolf et al. 1998) such as those arising between the interacting male(s) and female. A study of *D. melanogaster* which altered social heterogeneity by changing group composition found that female reproductive behaviour changed in a single generation with an increase in offspring diversity through alterations in either female preference or remating frequency (Billeter et al. 2012). While the study did not assess female costs to altering mating behaviour, or directly quantify these in a quantitative genetics framework, the results clearly argue for such studies. A caveat is that a quantitative genetics framework will be limited to species that lend themselves to such studies (see Section 9.4.3). Nevertheless, such integration will facilitate an understanding of the association between relevant abiotic and biotic parameters and how changes in these might influence the costs and benefits (both direct and indirect) of polyandry.

9.5.3 *Reproductive mode and indirect benefits*

Other researchers have proposed that focusing on life-history trait differences between species may help elucidate the evolutionary significance of polyandry (e.g. Hosken and Stockley 2003, Zeh and Zeh 2001a). For example, Zeh and Zeh (2001a) proposed that reproductive mode can influence the adaptive significance of polyandry. Reproductive mode is a catch-all phrase, used to describe asexual versus sexual reproduction, whether reproduction is terrestrial or aquatic, whether fertilization is internal or external, or—as considered by the Zehs—whether females are viviparous or ovi/ovoviviparous (give birth to live young or lay eggs). The hypothesis that reproductive mode influences the evolution of polyandry arises from a distinction between whether genetic benefits gained from polyandry are additive or non-additive, and the effect of physiology, immunology, and evolutionary conflict on maternal/zygote interactions during embryonic development. As argued by Zeh and Zeh (2001a), if species are viviparous, then maternal–zygote interactions during development are critical for female reproductive success and any evolutionary genetic conflict with the paternal genetic contribution may have a detrimental fitness effect. In this case, polyandry should be associated with benefits arising from genetic compatibility. In oviparous taxa, because the zygote is not reliant on continuous input of maternal nutrient contributions *during* development, polyandry should perhaps be associated with benefits arising from intrinsic male quality.

Even though all insects produce eggs and there is no placenta, some insects are nonetheless pseudoplacental, giving birth to live offspring and providing maternal nutrients during development (Chapter 12). Conflict over such investment may occur between the female and her mate(s), the female and developing embryos, and between developing siblings which may lead to spontaneous abortions (Zeh and Zeh 2001b). Whether such conflict is manifested in insects has not been determined. Viviparity in dipterans may have evolved independently more than 60 times (Meier et al. 1999). Given the prediction about reproductive mode and benefits to polyandry, comparative analyses of oviparous/ovoviviparous with viviparous dipterans would predict that the former should exhibit intrinsic benefits and the later non-additive benefits, all other things being equal. Such studies have not been done, likely because the benefits of polyandry have not been well studied in viviparous systems.

Whereas reproductive mode is one specific life-history trait, the general point is that very little progress has been made moving from the catalogue of intraspecific studies to understanding the broader interspecific patterns. As argued in Sections 9.5.1 and 9.5.2, such integration is important in helping to understanding polyandry and its fundamental consequences for evolutionary change.

9.4.5 *The potential applied relevance of polyandry*

Female multiple mating behaviour may be influenced by anthropogenic environmental modifications, such as climate change, habitat fragmentation, pollution and selective harvesting, which changes the species' ecology and will affect the underlying genetic variation for sexually selected traits (Lane et al. 2011). For example, one of the first biological responses to climate change is an alteration in the temporal distribution of life-history traits (Parmesan 2006). Since variation in the life histories of males and females determines the relative costs and benefits of multiple mating, any phenological response to climate change also has the potential to change the frequency and functional significance of polyandry. Likewise, if habitat fragmentation results in the loss of lekking or nesting sites, then smaller areas will support fewer males and females. How is the mating system affected? Sexual selection could be intensified on males to compete for fewer females that are patchily distributed, or the system may move to a more monogamous one. If females still mate multiply, then whether they do so to gain primarily additive (good- or sexy-sons benefits) or non-additive benefits could depend on the extent to which genetic factors are altered as a consequence of habitat fragmentation. Does habitat fragmentation increase the chances of inbreeding and subsequent inbreeding depression? If so, then females may mate multiply to avoid costs of inbreeding, assuming that inbreeding depression is mediated through overdominance. Selective harvesting can eliminate dominant males from the population and have a variety of genetic consequences on the population but can also change life-history traits and demographics (for review, see Coltman 2008), which may influence the mating system. Understanding whether anthropogenic influences on the environment alters the mating system and, if so, what the likely consequences are for the evolutionary trajectories of affected populations has only recently been considered as an important research paradigm in behavioural ecology, and thus few studies have been published. However, given the effect of ecology on mating systems and the rapidity of anthropogenic changes in ecology, our relative ignorance of the long-term consequences of such changes should make such studies highly relevant.

9.6 Conclusion

In the 30 years since Thornhill and Alcock's book, our understanding of the putative drivers of polyandry has increased substantially. The study of indirect benefits, in its infancy at that time, has now moved into a quantitative genetics framework. Perhaps the largest change in emphasis has been on the direct costs of polyandry as a consequence of sexual conflict. Although individual empirical studies have demonstrated indirect benefits to female multiple mating, meta-analyses have shown overall weak support for indirect benefits. The use of experimental evolution has documented the effect of polyandry on a variety of putatively sexually selected traits, has demonstrated the action of sexual conflict, and has identified the benefits of polyandry in different evolutionary contexts. Further development of genomics techniques has identified pervasive and pleiotropic genetic effects of polyandry. And the consequences of polyandry on a variety of different evolutionary phenomena has been discovered, many of which are discussed throughout this volume.

Yet much work remains for the new student of polyandry. A continuing key challenge is to determine the functional significance of polyandry, particularly in species with no direct benefits. Many researchers suggest, and find, that direct mating costs, mediated primarily through sexual conflict, are too large relative to the predicted weak genetic benefits. However, this is not always true; recent work has found evidence for indirect benefits that can even be larger than both direct benefits and direct costs. Assuming that genetic models of sexual selection hold true, what is the relative contribution of additive and non-additive effects? Quantitative genetics studies can help pinpoint this, but such efforts must include multiple fitness-related traits and integrate between laboratory and natural populations. Integration is complicated by the fact that mating systems are still typically studied as a static phenomenon, despite the recognition that the economy of polyandry is context dependent, influenced by both abiotic and biotic interactions, and potentially across more than one generation. Environmental influences may change the economics of polyandry but empirical studies, particularly in wild populations, remain scarce. To understand the origin and maintenance of polyandry, researchers should move beyond examining individual static case studies to investigating how spatiotemporal variability in ecology and life history within a species changes the costs and benefits of female multiple mating. Such studies should be done in controlled laboratory settings but also attempted in the wild. We currently have a catalogue of intraspecific studies that have helped tremendously in our understanding of polyandry, but we remain largely ignorant of how we can scale up our understanding of what drives adaptive significance of polyandry between species—that is, how we might predict the cost/benefit ratio for any given system. The increasing affordability of genomic data can be used to understand how polyandry influences and is influenced by genetic architecture, and what the evolutionary consequences of such interactions may be. Rapid ecological change and the context-dependent nature of the economics of polyandry strongly argue for studies that link these two to foster an understanding of how anthropological effects may influence mating systems.

Over the past 30 years, studies of the impact of polyandry have demonstrated its evolutionary power, such that polyandry recently has been argued to be the most fundamental agent of evolutionary change (Kvarnemo and Simmons 2013). Given the realization that net selection operating on males and females is more nuanced than previously appreciated, the study of polyandry remains at least as vibrant and as important as it was 30 years ago. There is no indication that we are at the limit of our understanding and new students of polyandry have much to offer.

CHAPTER 10

Sperm competition

Leigh W. Simmons

10.1 Introduction

We have seen in Chapter 9 how female insects can obtain a variety of benefits from copulating with several different males. The storage of sperm by females following copulations with multiple males can generate intense selection for male adaptations that promote an individual's competitive fertilization success. Indeed, Parker (1970a) noted that sperm competition would promote opposing adaptations in males that, on the one hand allowed a male to pre-empt the sperm stored by females from her previous mates, while on the other, ensure that future rivals are unable to pre-empt a male's own sperm. Sperm competition is thus the post-copulatory equivalent of male mating competition discussed in Chapter 6; once successful under pre-copulatory sexual selection for mating opportunities, males must again compete post copulation to fertilize the limited supply of ova that a female will produce.

This chapter explores the evolutionary consequences of sperm competition for male reproductive biology. Sexual selection will also act on males via the post-copulatory equivalent of female choice—cryptic female choice whereby females employ mechanisms that determine which male's sperm are stored and/or used to fertilize their eggs (Thornhill and Alcock 1983; Eberhard 1996). Cryptic female choice is discussed in Chapter 11. In this chapter we explore the consequences of sperm competition for the evolution of behavioural and morphological adaptations that allow males to avoid and/or succeed in competitive fertilization. These areas were already well developed in 1983 and synthesized by Thornhill and Alcock (1983), and more recently by Simmons (2001). We then focus on the advances that have been made since Thornhill and Alcock's volume, in how sperm competition can affect male expenditure on the ejaculate. We begin with a tale of two flies, the yellow dung fly, *Scatophaga stercoraria*, and the fruit fly *Drosophila melanogaster*. We do so, both because work on these flies has helped to lay the foundations of this research field, and because their contrasting mating systems illustrate well the varied adaptations that can arise in response to selection from sperm competition.

10.2 A tale of two flies

Males of the yellow dung fly are conspicuous occupants of summer meadows in northern Europe and North America. The males can be found on and around freshly deposited cow

The Evolution of Insect Mating Systems. Edited by David M. Shuker and Leigh W. Simmons.
© The Royal Entomological Society 2014. Published 2014 by Oxford University Press.

pats, often in their hundreds to thousands, and are easily recognized from their golden yellow pelage. They await the arrival of the drab-olive-coloured females. Gravid females typically arrive from upwind, landing in the grass and walking back to the dung pat where they lay a clutch of around 40 eggs. Males search on and upwind of the pat, and rather unceremoniously seize arriving females and immediately engage them in copula. Those females captured on the pat are carried downwind into the surrounding grass for copulation. Gravid females rarely resist copulation, which continues for around 35 min, before the male returns the female to the dung surface to lay her clutch of eggs. During oviposition the male remains mounted on the female for a further 16 min before the pair go their separate ways (Figure 10.1).

The reproductive behaviour of these flies led Parker to ask three questions. First, why do males remove females from the dung surface to copulate? The surface of the pat provides a warm microclimate, so that copulation could be accomplished some 30% faster if flies remained, thereby reducing the cost of mating for males and increasing their potential reproductive rate. Second, why do males remain with females for so long after copulation is completed? After all, time spent with females after copulation should be costly in terms of a male's ability to find and copulate with additional females. Third, why does copulation take so long, when, in many other species of insect, insemination is accomplished in a few seconds? Again, time spent copulating with the current female should come at a cost to a male's ability to find and copulate with additional females. The answers to all of these questions become clear when one considers that females will mate repeatedly so that sperm from different males compete to fertilize the clutch off eggs that she eventually deposits in the dung.

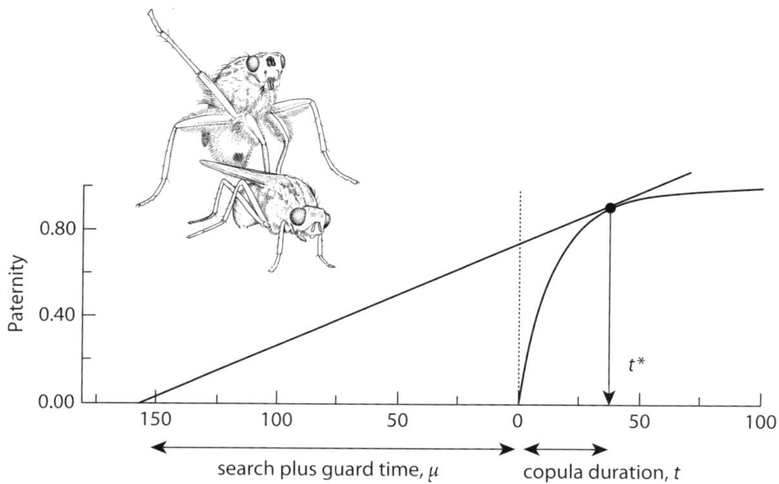

Figure 10.1 Sperm displacement by male yellow dung flies *Scatophaga stercoraria*. During copulation a male displaces sperm from the female's sperm stores so that his paternity (proportion of eggs fertilized) rises with diminishing returns. Drawing the tangent to the fertilization gain curve from the time cost associated with finding, copulating, and guarding a new female yields the predicted optimal copula duration that should maximize male fitness (Parker 1970d). Inset, a male yellow dung fly guards his mate for 15 min following copulation, while she deposits eggs into the dung surface. When encountered by a mate-searching male, the guard will push the female close to the dung surface, raising his body and mid-legs, so as to maximize the distance between mate and rival.

Parker (1971) noted that the probability of a pair emigrating downwind of the pat depended on the density of males searching for mates on the dung surface. The higher the density of males, the higher the encounter rate between searching males and copulating pairs. When a searching male encounters a copulating pair, intense struggles ensue. In 36% of cases, struggles result in take-overs, often by larger and competitively superior males. Males successful in a take-over will copulate with the female before allowing her to lay her clutch of eggs.

Parker (1970b) used a technique to assign paternity to a female's clutch that relied on sterilizing one of two males so that the paternity of each could be determined from whether an egg hatched or not. He found that the copulating males' paternity increased with time spent in copula. Thus, males are able to pre-empt the sperm stored by females from her previous mating. Males deliver sperm to the bursa copulatrix, from which they are transferred by the female into the spermathecae (Simmons et al. 1999). A single copulation is sufficient to fill the spermathecae so that new sperm entering do so by displacing resident sperm. Hence, as copulation proceeds, a male begins to displace his own sperm at an increasing rate, generating diminishing returns for male time investment in copulation (Figure 10.1). Taking into account the cost of finding a new female, marginal value theorem predicts that males should have an optimal copula duration of 42 min and obtain 80% of fertilizations, a prediction that is close to the observed copula duration of 36 min (Parker 1970b) (Figure 10.1). Subsequent work has refined these models, taking into account the effects of male and female body size on the probability of successful take-over and on the rate of paternity gain, and predicted optimal copula durations conform very closely to the distribution of copula durations seen in natural populations of flies (Parker and Simmons 1994).

The finding that males can pre-empt rival sperm explains why males remain with females during oviposition. Parker's (1970c) models predicted that, while a mutant male that lacked the guarding phase would have a higher fitness in the absence of take-overs, non-guarding males would suffer an extremely rapid decline in fitness as male density increased, so that guarding males would experience a 400% advantage at the average densities of flies recorded in natural populations. Emigration from the site of mate searching, prolonged copulation, and mate guarding during oviposition make perfect sense in the light of the risk of remating by females and offensive sperm competition via sperm displacement. Parker's research on dung flies thus led to the general realization that sperm competition could be a potent force of post-copulatory sexual selection shaping male behaviour, physiology, and morphology (Parker 1970a).

The fruit fly *Drosophila melanogaster* is perhaps the most widely used animal model in evolutionary biology and genetics. We therefore know more about *Drosophila* reproductive biology than perhaps that of any other animal. The full DNA sequence of its genome was reported in 2000, leading to the identification of genes that control the insect's development, behaviour, and physiology (Misra et al. 2002). Our ability to manipulate gene expression in this fly has allowed us to dissect the genetic architecture of the fly's reproductive physiology, and provide an unprecedented insight into the evolutionary consequences of sperm competition. The reproductive behaviour of *D. melanogaster* differs markedly from that of the yellow dung fly. Flies congregate on rotting fruit where they meet, mate, and females lay their clutch of eggs (Markow 1988). Mating peaks occur in the early morning at dawn and in the late afternoon at dusk. After encountering a female, males engage in a courtship dance, in which they deliver both acoustic and olfactory signals that are essential for eliciting copulation (Bastock and Manning 1955; Dickson 2008; detailed discussions of *Drosophila* courtship and mating can be found in Chapters 4 and 8).

Early studies of *D. melanogaster* found that, like yellow dung flies, females frequently produced offspring sired by more than one male, with the last male to copulate generally siring the majority (Gromko et al. 1984). Copulations last for around 20 min; however, unlike yellow dung flies, males abandon females immediately and females lay their clutch of offspring alone. Given the high density of mate-searching males at rotting fruit, why do males not guard their females during oviposition? Another difference between these two species of flies is that copula duration has no simple relationship with fertilization success. Sperm transfer appears to occur within the first 8 min of copulation so that interruptions in copula after this time have little impact on a male's paternity (Gilchrist and Partridge 2000). However, copula duration does influence whether or not a female will remate; matings in excess of 8 min generate a period of non-receptivity that lasts 30% longer than matings interrupted within 8 min. Observations in the field suggest that, despite repeated courtship attempts by males, the majority of ovipositing females are unlikely to accept additional matings (Partridge et al. 1987), negating any fitness advantage that might accrue to post-copulatory mate guarding.

Flies that have been genetically modified to be spermless via a mutation in the gene *tudor* have been used to show that seminal fluid proteins are involved in the mobilization and displacement of sperm from the females' sperm stores (Harshman and Prout 1994), and flies genetically modified to selectively knock out specific seminal fluid proteins, including sex peptide and ovulin, have shown us that seminal fluids are responsible for the loss of sexual receptivity of females following mating and an increased rate of offspring production respectively (Wolfner 1997; Chapman 2001). In effect, male *D. melanogaster* biochemically mate guard the females with whom they have mated.

Sperm competition in these two species of fly appears to select for alternative mechanisms that affect sperm displacement and its avoidance. In dung flies, sperm numbers appear important in the physical displacement of sperm and behavioural mate-guarding in its avoidance. In fruit flies, both sperm displacement and its avoidance are mediated biochemically via products of the reproductive accessory glands. Accordingly, experimental evolution studies of these two flies, in which post-copulatory sexual selection is reduced in some lineages compared with others, have shown that for yellow dung flies post-copulatory sexual selection generates divergence in testes size and sperm production (Hosken and Ward 2001), whereas for *D. melanogaster* it generates divergence in accessory gland size and productivity (Linklater et al. 2007). More generally, it is now clear that the insects exhibit an extraordinary diversity of reproductive behaviours, morphologies, and physiologies that only make sense in the light of sperm competition (Simmons 2001).

10.3 The ubiquity of insect sperm competition

Multiple mating by female insects is common. There are now data available on the patterns of sperm utilization following double matings by females for 156 species of non-social insects from 12 orders (see Table 10.1 and Simmons 2001). Sperm utilization patterns are generally provided as the species' mean value of P_2, the proportion of offspring sired by the second of two males to mate with a doubly mated female, and the variation about that mean. It is clear from the data that among species, cases of first male sperm precedence are rare (Figure 10.2). More typically paternity is either shared equally among a female's mates (85% of species studied report P_2 values ranging from 0.4 to the median value of 0.68), or the second male to copulate experiences a significant advantage (25% of species report P_2 values above the median). Generally, there is considerable within-species variation about

Table 10.1 Patterns of sperm utilization in non-social insects[a]

Order/species	Mean P_2	Range	SD	Method	Source of variance	Reference
Dermaptera						
Euborellia bruneri	0.57	0.00–1.00		R	Relative male genital length; relative male age	[1]
Orthoptera						
Ephippiger ephippiger		0.00–1.00[b]		M		[2]
Mantodea						
Pseudomantis albofimbriata	0.58	0.00–1.00		R		[3]
Hemiptera						
Cimex lectularius	0.68		0.19	R		[4]
Phyllomorpha laciniata	0.43	0.10–0.78	0.29	M		[5]
Pyrrochoris apterus	0.59	0.38–0.98	0.29	P		[6]
Coleoptera						
Anomala orientalis	0.58		0.38	R	First male mass (+); first male genital size (–)	[7]
Callosobruchus subinnotatus	0.62		0.46	R		[8]
Carabus insulicola	0.30–0.56		0.09–0.11	M	Guarding duration of first male (–); fertilization order (–)	[9]
Diaprepes abbreviatus	0.76	0.47–0.97		R		[10]
Gatrophysa viridula	0.45		0.37	R	Male contest competition (–)	[11]
Gonipterus scutellatus	0.63	0.00–1.00		R		[12]
Photinus greeni	0.59	0.00–1.00		M	Male attractiveness (–)	[13]
Mecoptera						
Panorpa germanica	0.62	0.00–1.00	0.28	M	Relative copula duration	[14, 15]
Diptera						
Drosophila bifurca	0.98			M		[16]
Drosophila buzzatti	0.69	0.64–0.74		M		[17]
Drosophila santomea	0.99		0.01	P		[18]
Drosophila yakuba	0.93		0.12	P		[19]
Merosargus cingulatus	0.84	0.55–1.00		M		[20]
Teleopsis dalmanni	0.0	0.00–1.00	0.38	M		[21]

continued

Table 10.1 *continued*

Order/species	Mean P_2	Range	SD	Method	Source of variance	Reference
Hymenoptera						
Anisopteromalus calandrae	0.0			P		[22]
Lepidoptera						
Cadra cautella	0.49–0.63	0.00–1.00[b]	0.49–0.63	R	Relative spermatophore size	[23]
Ephestia kuehniella	0.86		0.21	C	Remating interval (–)	[24]
Pectinophora gossypiella	0.71	0.00–1.00[b]		M	Insecticide resistance (–)	[25]
Bicyclus anynana	0.26	0.00–1.00[b]		P	Male age and spermatophore size (+)	[26, 27]
Limenitis arthemis	0.71	0.00–1.00[b]		P		[28]

(+/–) indicates direction of change in P_2 value with described variable; R, irradiated male technique; M, molecular markers; P, phenotypic markers; C, chemosterilization.

[a] This table is an update to Table 2.3 in Simmons (2001). Sperm utilization expressed as the proportion of offspring sired by the second male to mate, P_2, when a female is mated by two males. Estimates of variation in the species-specific value of P_2 are provided in the form of the range and/or standard deviation, and known causes of variation in P_2 noted. Species grouped within orders by commonly recognized major divisions.

[b] Bimodally distributed with modes at zero and one.

[1] van Lieshout and Elgar (2011), [2] Hockham et al. (2004), [3] Barry et al. (2011), [4] Stutt and Siva-Jothy (2001), [5] García-González et al. (2003), [6] Schöfl and Taborsky (2002), [7] Wenninger and Averill (2006), [8] Rugman-Jones and Eady (2007), [9] Takami (2007), [10] Harari et al. (2003), [11] Kozlowski (2004), [12] Carbone and Rivera (1998, 2003), [13] Demary and Lewis (2007), [14] Kock et al. (2006), [15] Kock and Sauer (2007), [16] Luck et al. (2007), [17] Bundgaard et al. (2004), [18] Chang (2004), [19] Chang (2004), [20] Barbosa (2009), [21] Corley et al. (2006), [22] Do Thi Khanh et al. (2005), [23] McNamara et al. (2009), [24] Xu and Wang (2010a,b), [25] Higginson et al. (2005), [26] Brakefield et al. (2001), [27] Kehl et al. (2013), and [28] Platt and Allen (2001).

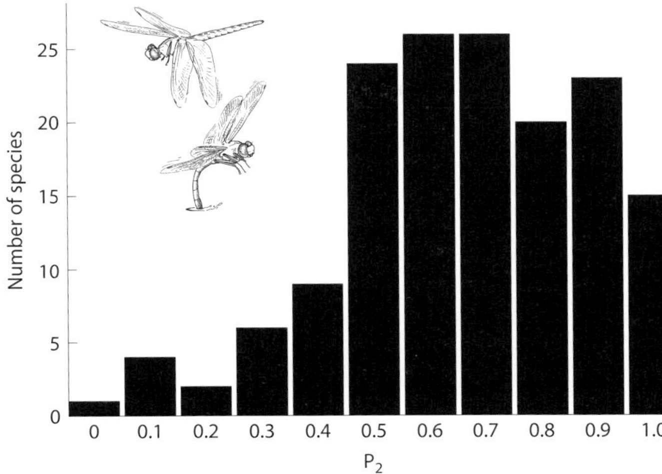

Figure 10.2 The distribution of species-specific mean values of P_2, the proportion of offspring sired by the second of two males to mate with a double-mated female, for 156 species of insect from 12 orders. Cases of first male sperm precedence are rare; sperm mixing appears to be the most common pattern of sperm utilization, with some species showing high last-male precedence. (Data from Table 10.1 and Simmons 2001.)

mean P_2 values. Species in which mechanisms have arisen for the displacement and/or removal of rival ejaculates, such as dragonflies and damselflies that use their secondary genitalia to remove rival sperm from the female's sperm storage organs, tend to be characterized by high last male sperm precedence with very low variance (Simmons and Siva-Jothy 1998; Simmons 2001). In contrast, where sperm mixing occurs, variance in last male paternity is large, and a male's competitive fertilization success is often associated with characteristics of the male and/or his ejaculate (Table 10.1).

Studies that have examined sperm utilization following matings with more than two males often find a more even distribution of paternity among multiple mates than when just two males compete (Simmons 2001). For example, in field crickets, *Teleogryllus oceanicus*, paternity becomes increasingly evenly distributed across multiple sires as the number of sires increases (Simmons and Beveridge 2010). Thus, even when the last male to mate gains an immediate paternity advantage, in some species, such as the ladybird *Adalia bipunctata* (Haddrill et al. 2008), previous males can gain paternity in later clutches following the mixing of ejaculates within the female's sperm stores (Simmons 2001). Among social insects, those species in which queens exercise extreme polyandry tend to distribute paternity evenly across potential sires, compared with species in which queens mate with very few males where paternity is highly skewed toward a single male (Jaffe et al. 2012) (sexual selection in social insects is discussed in greater depth in Chapter 14).

Although common, multiple mating is not ubiquitous. Females are monandrous in many species of insect. Many social insect females mate only once (Boomsma 2009), as do many solitary bees (Beveridge et al. 2006), dipteran flies (Craig 1967; Riemann et al. 1967), and butterflies and moths (Ehrlich and Ehrlich 1978) (reviewed in Thornhill and Alcock 1983). That is not to say that such species are not subject to the selective agent of sperm competition. Indeed, permanent female monogamy is often triggered via seminal fluid proteins transferred with sperm in the ejaculate (Smith et al. 1989; Baer et al. 2001). Monogamy-inducing seminal proteins may be subject to strong selection (Andrés and Arnqvist 2001) and need not be in the best interest of females (Hosken et al. 2009). This serves a timely reminder that the term 'sperm competition' was coined by Parker (1970a) to describe a selective force, rather than a proximate battle between two or more motile cells. The imposition of monandry on females can be an evolved response to the selective

Box 10.1 ADAPTATIONS ARISING FOR THE AVOIDANCE OF, AND ENGAGEMENT IN, SPERM COMPETITION.[a]

Avoidance of sperm competition (defensive)

1. Pre- and post-copulatory mate guarding
2. Biochemical barriers to female remating
3. Physical barriers to female remating
4. Genital morphology
5. Copula duration
6. Copulation frequency
7. Testis size and sperm production
8. Sperm form and function

Engagement in sperm competition (offensive)

[a] Adaptations fall along a continuum between defensive and offensive means by which males battle to monopolize paternity (modified from Simmons 2001).

force of sperm competition, even though in extant taxa sperm no longer compete. Adaptations arising from sperm competition should thus be viewed as falling along a continuum, with defensive adaptations for the avoidance of sperm competition at one extreme and offensive adaptations for the engagement in sperm competition at the other (Box 10.1).

10.4 Avoidance of sperm competition

10.4.1 *Behavioural mate guarding*

We have seen how male yellow dung flies will remain with females after copulation is completed, guarding them from males that are searching for females at the site of oviposition. If a mate-searching male encounters an ovipositing pair, the guarding male will raise his middle legs and his body above that of the female so as to maximize the distance between female and rival male (Figure 10.1). If the searching male continues in his take-over attempt, the guarding male will re-engage the female in genital contact, which appears to make take-overs considerably more difficult (Parker 1970d). Theoretical modelling predicts that sperm competition should favour the evolution of mate guarding whenever (i) females show continued sexual receptivity following mating, (ii) the intensity of competition among males for access to females and male mate searching efficiency are both high, (iii) the period between insemination and oviposition is short, (iv) the probability of retaining a female when challenged is high, and (v) the last male to mate gains paternity in the brood of offspring produced by a female (Parker 1974; Yamamura and Tsuji 1989; Fryer et al. 1999). These conditions seem to be met in many insect species, and mate guarding is widespread (Table 10.2; Alcock 1994; Simmons 2001) (Figure 10.3).

In damselflies and dragonflies, the last male to mate gains almost all of the fertilizations, and females retain sexual receptivity until they have deposited their clutch. Males thus suffer a significant risk of lost paternity should a rival male discover their mate before the clutch is laid. Forms of mate guarding in these taxa vary greatly. Some species, such as *Coenagrion puella*, remain in direct contact with females, carrying them between different oviposition locations (Banks and Thompson 1985). Others, such as *Calopteryx maculatum*,

Figure 10.3 Mate guarding is a taxonomically widespread behavioural adaptation in insects for the avoidance of sperm competition. In many species males may exhibit morphological adaptations for guarding efficiency, such as the extreme sexual size dimorphism seen in these grasshoppers (Sivinski 1984; Simmons 2001).

adopt non-contact guarding, chasing rivals that come too close to their ovipositing mates (Waage 1979a). In some species, males can adopt both contact and non-contact guarding, depending on the risk of take-overs. In *Sympetrum parvulum* for example, males adopt contact mate-guarding when male density is high, flying in tandem with their mates as they oviposit (Ueda 1979). In low-density populations, however, males perch nearby, and observe their mates as they oviposit alone. Such phenotypic plasticity in mate-guarding behaviour is likely to be adaptive, because of the costs in terms of a male's time out from searching for additional females while contact guarding, and the energetic costs of this behaviour (Saeki et al. 2005).

There is now very good evidence that male insects generally adjust their mate-guarding behaviour in response to the risk of sperm competition from rival males. For example, fire bugs, *Pyrrhochoris apterus*, often occur in dense aggregations, and pairs can remain in genital contact for as long as seven days before they separate and the female lays her clutch of eggs. The duration of copulation is strongly dependent on the operational sex ratio (OSR); copulation is significantly longer when the OSR is biased toward an excess of males than when it is biased toward an excess of females (Figure 10.4). Increased copulation may indicate an increased male expenditure on sperm transfer when the risk of sperm competition from rival males is elevated (see Section 10.5.2). However, in the case of *P. apterus* (Schöfl and Taborsky 2002), and many other insect species (Simmons 2001), insemination

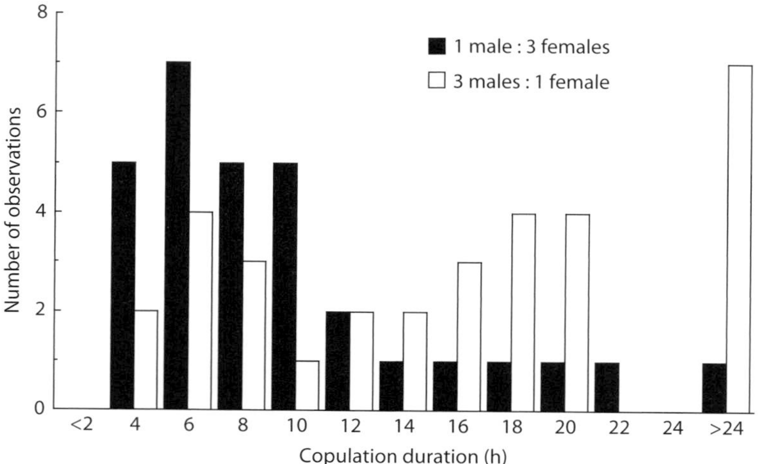

Figure 10.4 The duration of post-copulatory contact mate-guarding in firebugs, *Pyrrhochoris apterus*, depends on sex ratio. Sperm transfer is completed with just 4 h of copulation, but under a male-biased sex ratio where the risk of female remating is high, males extend the period of genital contact to a median of 15 h compared with a female-biased sex ratio where copulation lasts for a median of 7 h. (After Schöfl and Taborsky 2002.)

is completed within a few hours of pairing, and extended copulation serves to prevent rivals from gaining access to a male's mate before she lays her clutch.

Mate guarding can also occur prior to mating (Table 10.2). Such behaviour is frequently associated with female monogamy. In *Heliconius* butterflies, for example, females mate only once, and this has favoured the evolution of an unusual mating system in which males guard pupal females prior to mating, and copulate with them just before they emerge as adults (Estrada et al. 2010). The same is true of some mosquitos (Conner and Itagaki 1984) where males guard emerging females prior to copulation. Pre-copulatory mate guarding need not be associated with female monogamy, however. It occurs in a number of insects in which females remain unreceptive to further mating until they have laid their current batch of eggs. When females of the common dung fly *Sepsis cynipsea* arrive at fresh droppings to lay their eggs, they are unreceptive to mating until after oviposition. None the less, males mount females and guard them until they have laid their current batch of eggs, upon which the female may allow copulation (Parker 1972). Presumably males gain fertilizations in subsequent clutches. Similarly in locusts, *Locusta migratoria*, males will mount and ride on the backs of females for up to 10 h before the female finally lays her clutch of eggs and copulates with the guarding male. The subsequent copula duration, and paternity of the guarding male when the female produces her next clutch, is greater the longer the male has guarded (Zhu and Tanaka 2002). It may be that females also assess male quality during the guarding period, with cryptic female choice influencing her decision to accept and use sperm from the guarding male (Chapter 11). In the scarab beetle *Anomala albopilosa sakishimana*, the duration of pre-copulatory guarding depends largely on the time of day that a male finds and mounts the female, because males wait to copulate until late in the day (Arakaki et al. 2004). The probability of a female finding and mating with another male after dusk is very low, and females will oviposit early the next day. These examples illustrate how the form of mate guarding depends largely on when

Table 10.2 Examples of behavioural mate guarding in insects[a]

Order/species	Type	Duration	Cause of variation	Effect on fitness	Source
Orthoptera					
Locusta migratoria	Pre-copulatory contact	10 h		Copula duration and paternity (+)	[1]
Sphenarium purpurascens	Post-copulatory contact	17 days	Male/female size (+), female mating history (+)		[2]
Hemiptera					
Megacopta punctatisima	Copulatory	10–24 h			[3]
Pyrrhochoris apterus	Copulatory	≤7 days	Male density (+); male size (–)		[4]
Coleoptera					
Anomala albopilosa sakishimana	Pre-copulatory contact	Highly variable	Time of day paired		[5]
Papillia japonica	Post-copulatory contact	≤400 min	Male size (+) and nutritional state (+)		[6]
Tenebrio molitor	Post-copulatory contact and non-contact	Highly variable	Odour cues to male density (+); number of males (+)	Female remating (–)	[7]
Hymenoptera					
Hypoponera opacior	Pre-copulatory contact	≤800 min	Number of competing males (+); available females (–)		[8]
Lepidoptera					
Heliconius charithonia	Pre-copulatory contact				[9, 10]

Pre- and post- refer to the timing of mate guarding relative to copulation. Three forms of mate guarding are recognized: copulatory, where males remain in genital contact with the female; contact, where males remain mounted on the female but without genital contact; and non-contact, where males remain in close proximity to the female in order to ward off rival males. Causes of variation in guarding duration are noted where available. (+/–) denote sign of effect of guarding on variable in question.

[a] For a complete review see additional examples in Table 5.2 of Simmons (2001).

[1] Zhu and Tanaka (2002), [2] del Castillo (2003), [3] Hosokawa and Suzuki (2001), [4] Schöfl and Taborsky (2002), [5] Arakaki et al. (2004), [6] Saeki et al. (2005), [7] Carazo et al. (2007, 2012), [8] Kureck et al. (2011), [9] Estrada et al. (2010), and [10] Gilbert (1976).

females are receptive to remating, and when the risk of lost paternity to rival males is greatest (see also Simmons 2001).

10.4.2 *Biochemical mate guarding*

Remaining with females to prevent copulation by rival males is costly in terms of a male's ability to find and mate with additional females. As we have seen for *Drosophila*, males can avoid such costs by transferring seminal fluid proteins within their ejaculate that reduce a female's receptivity to remating. In *D. melanogaster*, the protein sex peptide directly impacts a female's sexual receptivity to the courtship of males encountered after an initial mating. Females do not regain receptivity for some days, after which they will have produced offspring sired by their previous mate (Fricke et al. 2009). However, sex peptide is costly for females because it reduces female lifespan (Wigby and Chapman 2005). More generally, a loss of sexual receptivity may not be in a female's best interests if she gains benefits from remating with additional males (Simmons and Gwynne 1991). Male adaptations to sperm competition can thereby generate significant sexual conflict (Arnqvist and Rowe 2005) (see Chapter 11). Seminal fluid proteins that influence female reproductive physiology may be subject to intense sexual selection.

Evidence for selection acting on seminal fluid proteins can be found by comparing the rates of molecular evolution of the genes encoding seminal proteins across species (for instance by comparing the rates of non-synonymous to synonymous (dN/dS) nucleotide substitutions in the DNA that encodes a given protein, Goldman and Yang 1994). Molecular studies of male reproductive accessory gland proteins—those proteins destined to be incorporated into the seminal fluid of the ejaculate—are revealing extremely high rates of protein evolution among *Drosophila* (Haerty et al. 2007), *Heliconius* butterflies (Walters and Harrison 2010) and gryllid crickets (Andrés et al. 2006). Among *Drosophila,* accessory gland protein divergence is greater in lineages with greater levels of polyandry, and thus greater selection from sperm competition, cryptic female choice, and sexual conflict (Wagstaff and Begun 2007). There is also evidence to suggest that the quantity of seminal fluid proteins produced by males may be driven by selection from sperm competition (Linklater et al. 2007; Crudgington et al. 2009).

In experimental evolution studies of *Drosophila*, the productivity of male accessory glands has been shown to evolve in direct response to increases in the strength of selection from sperm competition, imposed either by increases in the ratio of males to females in mating populations (Linklater et al. 2007) or in the degree of multiple mating by females (Crudgington et al. 2009). Thus, populations in which males suffer a greater strength of sperm competition from female remating behaviour evolve accessory glands with greater productivity, in part to more effectively prevent females from capitalizing on the greater availability of males. Comparative studies likewise provide evidence for an evolutionary link between accessory gland productivity and biochemical mate guarding. Thus, among fungus-growing ant species, there is a negative evolutionary association between accessory gland size and queen mating frequency—species in which males have large accessory glands have lower queen-mating frequency than species in which males have small accessory glands (Baer and Boomsma 2004).

Rather than influencing the receptivity of females directly, males can guard their mates by making them unattractive to rivals (Happ 1969; Yew et al. 2009; Brent and Byers 2011). In *Heliconius* butterflies, males guard their mates prior to copulation, following which females will not mate again. Female monogamy is achieved by the transfer to females of an anti-aphrodisiac pheromone, which makes them unattractive to subsequent males (Gilbert 1976). Female *Heliconius* may benefit from being rendered unattractive, as they possess specialized abdominal glands or 'stink clubs' within which they store male-derived antiaphrodisiacs, implying some control over their post-mating attractiveness or otherwise (Thornhill and Alcock 1983). Indeed, research on the pierid butterfly, *Pieris napi*, has shown how females actively release male-derived methyl-salicylate when courted by searching males, leading to those males abandoning their courtship attempts (Andersson et al. 2000, 2004). As with seminal fluid proteins, there is good evidence that sexual selection has driven the evolutionary divergence of these antiaphrodisiac pheromones. Thus male-contributed biochemical mixtures are complex and highly variable across species of *Heliconius*, and their rates of evolutionary divergence are higher in clades were females are more likely to mate multiply and thereby generate sperm competition (Estrada et al. 2011).

While biochemical mate guarding leaves males free to search for additional females, their expenditure on seminal fluid and/or anti-aphrodisiac pheromones is also likely to be costly, limiting their ability to invest in additional females. Males have been reported to become depleted of seminal fluid proteins in *D. melanogaster* (Sirot et al. 2009), and ejaculate depletion limits the ability of male *Lucilia cuprina* to suppress the receptivity of multiple females (Smith et al. 1990). As with behavioural mate guarding, we might expect males to allocate their limited resources strategically, in relation to the risk that females will encounter rival males. Indeed, there is some evidence that males might invest in biochemical mate guarding in a strategic manner. When *D. melanogaster* males were exposed to potential rivals during copulation they copulated for longer, and transferred greater quantities of sex peptide and ovulin, the two seminal fluid proteins that suppress female sexual receptivity and stimulate oviposition respectively (Wigby et al. 2009). Strategic adjustments in seminal fluid components appear to be remarkably sophisticated. When males mate with previously unmated females they transfer more ovulin than when they mate with previously mated females (Sirot et al. 2011). Once oviposition is stimulated, additional doses of ovulin do not appear to accelerate female reproduction any further, so that males benefit from withholding their limited supplies of ovulin for use with unmated females. In contrast, males always benefit from transferring sex peptide and suppressing female sexual receptivity. Accordingly, males transfer equal doses of sex peptide to both previously mated and unmated females. This level of physiological control requires an extraordinarily well co-ordinated system for obtaining and processing information about the female whom a male is courting and mating, and then using that information to fine-tune specific ejaculate components over very short time-scales.

10.4.3 *Physical barriers to female remating*

In a variety of insects the male transfers to the female substances in the ejaculate that coagulate in the female reproductive tract to form a mating plug (Simmons 2001). In other species, males transfer their ejaculate encased in a spermatophore which can similarly block the female's reproductive tract (Parker and Smith 1975). Indeed, among Lepidoptera, the incorporation of chitin into spermatophore walls aids in their persistence in the female reproductive tract (Drummond 1984). Mating plugs make it difficult for rival males to copulate with recently mated females (Parker and Smith 1975; Dickinson and Rutowski 1989; Polak et al. 2001). In some cases mating plugs can also serve as a vehicle for the biochemicals that suppress the female's long-term sexual receptivity (Baer et al. 2001).

Mating plugs can vary considerably in size, and appear to have reached an extraordinary degree of elaboration in the butterflies (Figure 10.5). In some species of papilionid for example, some 90% of a male's spermatophore can extrude outside of the female's reproductive tract, forming a 'sphragis' that serves both as a physical barrier to remating, and as a visual deterrent to rival males (Orr and Rutowski 1991). The manufacture of such mating plugs can represent a considerable cost for males (Giglioli and Mason 1966; Matsumoto and Suzuki 1992), draining resources that could otherwise be used for alternative ejaculate components such as sperm and seminal fluids. Moreover, we would expect selection to favour adaptations in males that allow them to remove or otherwise circumvent the mating plugs deposited by previous males (Parker 1984). This raises the question of how expensive mating plugs can be maintained. Theoretical models predict that plug efficacy

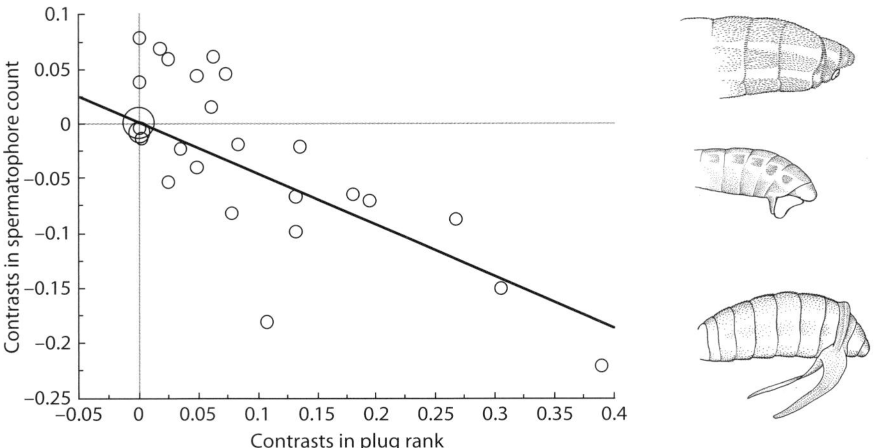

Figure 10.5 Correlated evolution of mating plugs and female remating frequency among butterflies. Species with small plugs such as *Graphium sarpedon* (Top right) have high rates of female remating, while those with increasingly enlarged plugs, such as *Atrophaneura alcinous* (middle right), or elaborated 'sphragids' such as *Euryades corethrus* (bottom right), have moderate-to-low female remating rates respectively. Thus, evolutionary increases in plug size and elaboration are associated with evolutionary decreases in female remating. (After Simmons 2001.)

should increase as the number of males relative to females increases, because males will have fewer mating opportunities, allowing them to divert resources away from additional mating capacity (Fromhage 2012). However, as the number of mating attempts per female increases, plugging efficacy should decrease. This is because at high female mating frequencies increased mating opportunities for males and increased intensity of sperm competition from rival ejaculates should favour increased investment in other components of the ejaculate—diverting resources away from plug production (Fromhage 2012). Consistent with this prediction, across species of butterflies there is a negative relationship between female mating frequency and the magnitude of male investment in mating plugs (Figure 10.5).

10.4.4 *Traumatic insemination*

Traumatic insemination must be one of the most extraordinary means of copulation found in the animal kingdom, and is perhaps best known in bed bugs and their allies (Cimicidae): rather than deliver sperm into the female's genital tract, males pierce the female's abdomen with a lanceolate adeagus and ejaculate directly into the haemocoel (Hinton 1964; Carayon 1966). Thornhill and Alcock (1983) suggested that males who bypassed the usual route for insemination might gain a selective advantage in placing their sperm closer to the site of fertilization, favouring this bizarre mode of copulation. Certainly in *Cimex lectularius* the last male to mate does have the advantage in fertilization when the female lays her clutch (Stutt and Siva-Jothy 2001), although sperm precedence is only in the region of 70% and not the complete paternity advantage afforded to male odonates who use their genitalia to remove or displace rival sperm from the site of fertilization (Simmons 2001; Waage 1979b). Traumatic insemination can be costly to female *C. lectularius* insofar as extensive wounding compromises a female's ability to maintain water balance and places a burden on the immune system during wound repair, reducing female lifespan (Stutt and Siva-Jothy 2001; Benoit et al. 2012). However, the resulting sexual conflict appears to have been resolved in this group via the evolution of the paragenital system, a subcuticular area of tissue in the abdomen—the spermalege—into which males inject their ejaculates. The spermalege greatly reduces the costs of traumatic insemination for females (Morrow and Arnqvist 2003; Reinhardt et al. 2003). Moreover, the ejaculate contains antimicrobial compounds which may protect sperm and females from microbes transmitted during copulation (Otti et al. 2009; Otti et al. 2012). Ejaculate components can delay female reproductive senescence and promote reproductive rate, thereby resolving any evolutionary conflict over traumatic insemination (Reinhardt et al. 2009). Traumatic insemination has also been discovered in the plant bug genus *Coridromius* (Tatarnic et al. 2006), and as might be expected from the sexually antagonistic co-evolution that traumatic insemination generates, there has be a co-evolutionary divergence in the complexity of both the lanceolate aedeagus and the female paragenital system (Tatarnic and Cassis 2010).

Traumatic insemination need not bypass the female's genital tract. In several insects females have been found to incur extensive wounding of the internal genital tract walls during copulation as a direct consequence of aedeagal spines (Crudgington and Siva-Jothy 2000; Blanckenhorn et al. 2002). In some of these species males transfer sperm or seminal fluids through these wounds. In the *Drosophila bipectinata* species complex, for example, the aedeagus is endowed with claw-like basal processes that penetrate the female's body wall near the genital orifice and sperm are transferred to the genital tract through these wounds. It is not known whether wounding in these species promotes male competitive fertilization success. However, in seed beetles *Callosobruchus maculatus* the genital spines of the aedeagus penetrate the internal walls of the female genital tract, allowing seminal fluid compounds to enter the female's haemolymph where they do act in promoting the paternity of the copulating male (Hotzy et al. 2012). As discussed in Chapter 11, sexual conflict generated via genital damage in these beetles has resulted in the co-evolution of male genital spines and the thickness of protective connective tissue in the female copulatory duct (Rönn et al. 2007).

10.5 Engagement in sperm competition

Chapter 9 discussed how mating with multiple males can offer significant benefits to females, for example in the form of material gifts, access to resources, or assistance in raising offspring. Thus, there may be considerable conflicts of interest between males and females over remating, and in many cases males are unsuccessful in preventing their mates from accepting additional matings from rival males. In these cases, selection should favour adaptations in rival males that allow them to pre-empt the sperm stored by females from their previous mates (Parker 1970a). In our example of the yellow dung fly we saw how a male's fertilization success could be improved by increasing copula duration and so increasing the numbers of sperm transferred to the female. In this case, greater numbers of sperm more effectively displace rival sperm from the female's sperm stores. But sperm displacement is not always possible. In many insects, females store at least some, if not all, of the sperm they receive from each of their many mating partners, so that following a series of matings sperm mix in the female's sperm storage organs and all males potentially may fertilize a proportion of the female's clutch (Simmons 1986b; Simmons et al. 2004; Haddrill et al. 2008; Simmons and Beveridge 2010; Jaffe et al. 2012). Under these conditions, selection should favour any adaptation in males, and especially in their ejaculates, that provide an advantage in the competition to fertilize the female's ova. Perhaps the most obvious way in which a male can bias paternity toward his own interests is to transfer greater quantities of sperm in an attempt to outnumber those stored from their rivals, and a rich body of theoretical work has been developed with which to predict how selection from sperm competition might be expected to shape the evolution of male ejaculate expenditure.

10.5.1 *Testes size and sperm production*

Game theory has proved particularly useful in modelling the evolution of ejaculate expenditure by males (Parker and Pizzari 2010). In sperm competition games, males are assumed to have a fixed pool of resources that can be allocated to reproduction. Resources allocated to finding, courting, and competing for females are unavailable for allocation to the production of ejaculates so that males face a fundamental life-history trade-off. They must choose between the number of females with whom they can mate and their ability to compete for fertilizations with rival males after mating has been achieved. There is now very good evidence for this assumption (Parker et al. 2013). For example, in the dung beetle *Onthophagus nigriventris*, it is possible to cauterize the imaginal cells that would otherwise be destined to become the adult thoracic horn at the prepupal stage of development. Individuals who are thus prevented from expending resources on horn development allocate more resources to testes growth than do individuals allowed to develop their horns (Simmons and Emlen 2006). Such trade-offs can have a genetic basis. In the flour beetle *Gnatocerus cornutus*, populations artificially selected for an increase in the length of the male mandibles—used in male–male contest competition—show a correlated reduction in testes mass (Yamane et al. 2010).

Sperm competition games also assume that sperm utilization conforms to a 'raffle' in which the relative numbers of sperm in the so-called 'fertilization set' that are derived from a group of males will determine directly those males' relative share of paternity. Again there is good evidence for this assumption. When two males copulate and they share paternity, the relative ejaculate sizes or copula durations often provide good predictors of relative fertilization success (Table 10.1; Simmons 2001). A general prediction of sperm competition models is that, as the risk of sperm competition—the probability that a female will mate with two males—increases across species, male expenditure on the ejaculate should increase (Figure 10.6). The maximum effect of sperm competition on ejaculate expenditure will occur when there is a 'fair raffle'—when a given sperm from each competing male is equally likely to be utilized at the time of fertilization. As the raffle becomes increasingly loaded toward one or other of a female's mating partners, the fertilization success of that male becomes independent of the numbers of sperm transferred. When the raffle is heavily loaded, the risk of sperm competition is predicted to have little or no effect on ejaculate expenditure (Figure 10.6). Loading of the raffle can arise in a number of ways—for example, when males possess adaptations for the removal of rival sperm prior to insemination of their own sperm. We would therefore not predict a relationship between the probability of a double-mating and ejaculate expenditure among species of damselflies and dragonflies where males remove rival sperm prior to insemination. Equally, when the mechanisms of cryptic female choice discussed in Chapter 11 determine which males are allowed to fertilize a female's ova, we would not expect the risk of double mating by females to drive the evolution of male ejaculate expenditure.

Experimental evolution studies offer empirical support for the predictions of sperm competition games. In *D. melanogaster*, populations evolving under greater levels of female

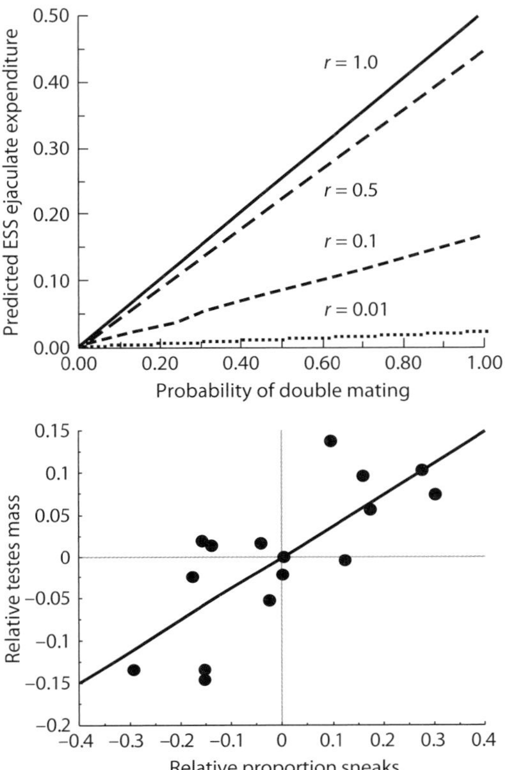

Figure 10.6 (Upper) Between-species predictions from sperm competition game theory. The predicted evolutionarily stable strategy (ESS) is for males of a species to increase ejaculate expenditure as the probability of a double mating—sperm competition risk—increases. However, whether they do so depends critically on the loading of the raffle (r). When the raffle is fair ($r = 1$) sperm competition risk will have its greatest effect, but when the raffle is loaded toward one male so that $r \approx$ zero, risk should have no impact on the ESS and all males should expend minimally on their ejaculates (after Parker et al. 1997). (Lower) Supporting these theoretical predictions, across 16 species of Onthophagine dung beetles, relative testes mass—controlling for body mass—increases as the proportion of males sneaking copulations with guarded females increases. The observed relationship is robust to control for phylogeny. (After Simmons et al. 2007b.)

multiple mating showed divergence in accessory gland productivity, with no changes in testes size (Linklater et al. 2007). In this species the raffle appears to be loaded heavily toward the second male to mate, with seminal fluid proteins playing the principle role in the displacement of rival sperm and the delaying of female remating. However, in dung beetles, *Onthophagus taurus*, sperm utilization conforms to a fair raffle (Simmons et al. 2004). In experimentally evolving populations of beetles where females were allowed to mate multiply, males evolved larger testes and greater competitive fertilization success compared to populations in which monogamy was enforced (Simmons and García-González 2008). Likewise in the moth, *Plodia interpunctella*, populations evolving under a male-biased sex ratio evolved increased numbers of sperm per ejaculate compared with populations evolving under a female-biased sex ratio (Ingleby et al. 2010). Importantly, these patterns of micro-evolutionary change are reflected in macro-evolutionary patterns of covariation between testes size and level of sperm competition among both dung beetles (Simmons et al. 2007b) and butterflies (Gage 1994).

In dung beetles of the genus *Onthophagus*, some males adopt alternative mating tactics whereby they sneak copulations with females guarded by large-horned males (Chapter 7). The strength of selection from sperm competition among species of *Onthophagus* is thus reflected by the proportion of the male population that adopts the sneaking tactic. Consistent with sperm competition games, evolutionary increases in the proportion of males adopting the sneaking tactic are associated with evolutionary increases in testes mass among species of this genus (Figure 10.6). Moreover, within species there is an asymmetry in the strength of selection from sperm competition acting on the different male tactics. By the very nature of their mating tactic, sneaks will always be subject to sperm competition whereas guards will be subject to sperm competition less frequently, dependent on the frequency of sneaks. This asymmetry in selection pressures predicts that sneaks should invest more in sperm production than guards (Parker 1990); indeed, within species of *Onthophagus*, sneaking males do have larger testes than their mate-guarding conspecifics (Simmons et al. 2007b).

While increased numbers of sperm can certainly improve a male's prospects of gaining fertilizations simply by outnumbering sperm from rival males, we might also expect selection to act strongly on the quality, and thus fertilization efficacy, of a male's sperm. Indeed, across insects the proportion of sperm in the ejaculate that is fertilization competent—measured as the proportion of sperm alive or viable—has been shown to be positively associated with the degree of multiple mating by females (Hunter and Birkhead 2002). We will consider how sexual selection acts on sperm form and function in more detail in Section 10.5.3. But first let us consider how sperm competition influences phenotypic plasticity in ejaculate expenditure within species.

10.5.2 *Strategic ejaculation*

We have seen how costs of avoiding sperm competition can generate phenotypic plasticity in male expenditure on mate guarding. Thus, males will guard their mates for less time or transfer smaller quantities of receptivity-inhibiting seminal fluid proteins when the risk of double mating is reduced. By so doing they can save resources and time that can be spent on gaining additional matings. Ejaculates can also be extremely costly for males to produce. Male insects can become depleted of sperm and seminal fluid and require considerable time out from mating to replenish spent resources (Simmons 2001; Vahed 2007b; Radhakrishnan et al. 2009; Sirot et al. 2009). Given these costs, we might also expect males to allocate resources to their ejaculates strategically, in order to maximize the numbers of matings they can achieve.

Sperm competition games have been developed to predict how males should invest in a given mating when the levels of sperm competition vary (Parker and Pizzari 2010) (Figure 10.7). At low levels of sperm competition, when females may or may not mate with more than one male, males are predicted to increase their expenditure on the ejaculate with increasing *risk* of sperm competition—the probability of competition with one other male. However, as the levels of sperm competition increase, so that females always mate with two or more males, male expenditure on their ejaculates should decrease with increasing *intensity* of sperm competition—the number of males in competition (Figure 10.7). This is because a male's fitness return per unit investment in his ejaculate should decline as the number of males sharing paternity rises. Under these conditions it should pay males to save their costly ejaculates for females with which they will suffer lower intensity of sperm competition (Parker et al. 1996). Males are thus predicted to evolve sophisticated patterns of sperm allocation in response to the immediate levels of sperm competition they face. It is worth stressing here that a male's response to risk and intensity of sperm competition are qualitatively different; males should increase expenditure with increasing risk and decrease expenditure with increasing intensity (Parker and Pizzari 2010) (Figure 10.7).

There is now considerable evidence that male insects do adjust their ejaculate expenditure in a strategic manner (Table 10.3). Quantitative meta-analysis of 15 insect species showed that male insects generally transfer more sperm when exposed to rival males while

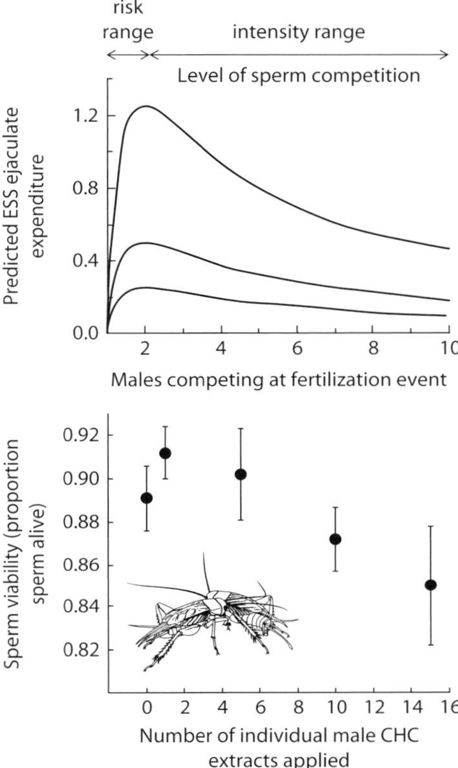

Figure 10.7 (Upper) Within-species predictions from sperm competition game theory. At low levels of sperm competition, where females mate with just two males with increasing probability, males should increase ejaculate expenditure with increasing risk. However, at high levels of sperm competition, where all females mate with two or more males, those males should reduce expenditure with increasing sperm competition intensity. (After Parker et al. 1996.) (Lower) Male field crickets, *Teleogryllus oceanicus*, adjust the quality of their ejaculates as predicted by sperm competition game theory. They increase the proportion of viable sperm in their ejaculate when they perceive the odour of just one rival male, but reduce the proportion of sperm viable when mating with females to which increasing numbers of male odours have been applied. (After Thomas and Simmons 2008.) ESS, evolutionarily stable strategy; CHCs, cuticular hydrocarbons.

they are copulating (Kelly and Jennions 2011). However, the predicted effect of sperm competition intensity on ejaculate expenditure is less clear. For 29 species of insect, there was a general effect of female mating status on ejaculate expenditure, with males generally transferring less sperm to already mated females (Kelly and Jennions 2011). Unfortunately we are unable to read the 'minds' of male insects, so it is not possible to say whether males 'interpret' a mated female as representing a high risk or low intensity of sperm competition. Likewise those studies that have tried to assess the effect of sperm competition intensity often manipulate the number of rivals present, which may signal high risk rather than intensity of sperm competition (Engqvist and Reinhold 2005). As yet, few studies have manipulated the number of males physically involved in sperm competition in order to test the intensity prediction. Male field crickets, *Teleogryllus oceanicus*, adjust their ejaculate expenditure when exposed to rival males, increasing the quality of sperm transferred rather than the quantity. Thus, when males are given experience of rival males in the mating environment, the proportion of viable sperm in the ejaculate is elevated (Simmons et al. 2007a; Gray and Simmons 2013). In these crickets, seminal fluid imparts viability to sperm, and sperm viability is a key determinant of a male's competitive fertilization success (García-González and Simmons 2005b), so it seems likely that males adjust seminal fluid composition in response to the risk of sperm competition (Simmons and Beveridge 2011).

Like most insects, these crickets secrete lipids, known as cuticular hydrocarbons or CHCs, on to their cuticles. These CHCs are individually distinct and serve a communication function. CHCs are easily rubbed off onto substrates that males and females come

Table 10.3 Ejaculation responses to correlates of risk or intensity of sperm competition[1]

Species	Correlate	Response	Source
Orthoptera			
Gryllus bimaculatus	Learned predictors of risk	Spermatophore size (+)	[1]
Gryllus veletis	Exposure to rival males	Sperm numbers (+ with one, – with multiple males)	[2]
Gryllus texensis	Exposure to rival males	No significant effect on sperm numbers	[2]
Teleogryllus oceanicus	Exposure to rival males, their odour or song	Sperm viability (+ with one male,— with multiple males); testes and accessory gland mass (+)	[3–5]
Chorthippus parallelus	Male–male encounters	Ejaculate size (+)	[6]
Locusta migratoria	OSR during mating	Sperm numbers (+)	[7]
Mantodea			
Pseudomantis albofimbriata	OSR	Male development rate (–); sperm numbers (+)	[8]
Hemiptera			
Phyllomorpha laciniata	OSR	Copula duration (+); sperm numbers (+)	[9]
Cimex lectularius	Chemical cues to rivals	Copula duration (—); ejaculate size (—)	[10]
Nysius huttoni	OSR, male density	Copula duration (+); sperm numbers (+)	[11]
Coleoptera			
Callosobruchus chinensis	Rearing density	Sperm numbers (+)	[12]
Diptera			
Drosophila bifurca	Exposure to rival males	Sperm production (+)	[13]
Drosophila melanogaster	Exposure to rival males, their odour, or song	Copula duration (+); seminal fluid proteins (+); sperm (+)	[14–17]
Drosophila subobscura	Exposure to rival males	Copula duration (+)	[18]
Drosophila acanthoptera		Copula duration (+)	[18]
Merosargus cingulatus	Male density, female size	Copula duration (+); fertilization success (+)	[19]
Sepsis cynipsea	Female age and mating history	Copula duration (+); ejaculate transfer (+)	[20]
Scatophaga stercoraria	Rearing density	Testes size (+)	[21]
Hymenoptera			
Trichogramma turkestanica	Increasing number of rival males	Sperm numbers (–)	[22]
Lepidoptera			
Pieris napi	Male density, male pheromone	Ejaculate weight (+)	[23]
Danaus plexippus	Female mating history	Eupyrene sperm numbers (+)	[24]

[a] For a complete review see additional examples in Table 7.2 of Simmons (2001).

[1] Lyons and Barnard (2006), [2] Schaus and Sakaluk (2001), [3] Bailey et al. (2010), [4] Simmons et al. (2007a), [5] Thomas and Simmons (2007, 2008), [6] Reinhardt (2001), [7] Reinhardt and Arlt (2003), [8] Allen et al. (2011), [9] García-González and Gomendio (2004), [10] Siva-Jothy and Stutt (2003), [11] Wang et al. (2008), [12] Yamane and Miyatake (2005), [13] Bjork et al. (2007), [14] Bretman et al. (2010, 2011, 2012), [15] Sirot et al. (2011), [16] Wigby et al. (2009), [17] Garbaczewska et al. (2013), [18] Lizé et al. (2012), [19] Barbosa (2011, 2012), [20] Martin and Hosken (2002), [21] Stockley and Seal (2001), [22] Martel et al. (2008), [23] Larsdotter Mellström and Wiklund (2009), and [24] Solensky and Oberhauser (2009).

into contact with, including each other. When male CHCs are applied to the cuticle of virgin females, subsequent males treat these females as if they were mated, increasing the viability of sperm in their ejaculate when they mate with them (Thomas and Simmons 2007). However, increasing the number of individual male CHCs applied to a virgin female results in a continuous decline in the quality of sperm that focal males ejaculate when allowed to mate with these females (Thomas and Simmons 2008). Thus, as predicted by sperm competition games, males of this cricket seem to respond to perceptions of increasing risk by increasing ejaculate expenditure, but reduce their ejaculate expenditure in response to increasing intensity of sperm competition (Figure 10.7).

10.5.3 *Sperm form and function*

We noted earlier how sperm competition might act not just on the numbers of sperm ejaculated, but also on the quality of those sperm, or more specifically on sperm traits that contribute to their ability to win fertilizations. Sperm length is one trait that has been the focus of considerable research attention in terms of sperm competitive ability. Across species of butterflies, for example, evolutionary increases in the degree of female multiple mating—and therefore the level of sperm competition—are associated with evolutionary increases in sperm length (Gage 1994). Likewise, among species of moth, evolutionary increases in testes size—now a widely used proxy for the level of sperm competition—are associated with evolutionary increases in sperm length (Morrow and Gage 2000). Among eusocial Hymenoptera too, sperm competition appears to exert a significant influence on sperm evolution. Thus, within- and between-male variability in sperm length decreases with increasing queen mating frequency, though sperm length itself appears unaffected (Fitzpatrick and Baer 2011).

These patterns of macro-evolutionary change suggest that sperm form and function may be subject to selection through sperm competition. Returning to fruit flies, we can see why this might be the case. During copulation, males ejaculate into the female's bursa copulatrix and sperm must then migrate from the bursa to the sperm storage organ, the seminal receptacle. Accessory gland proteins in the seminal fluid play a large role in the mobilization of sperm and its storage in the seminal receptacle (Neubaum and Wolfner 1999). However, sperm form and function also play a role here. In a remarkable series of experiments it has been possible to use flies that have been genetically modified to express either Green Fluorescent Protein or Red Fluorescent Protein, and thereby record the behaviour of sperm within their competitive environment, the female's reproductive tract (Manier et al. 2010) (Figure 10.8). Thus, immediately after copulation, sperm in the bursa are those from the last male to mate whereas sperm in the seminal receptacle are stored from a previous mating (Figure 10.8). Following copulation sperm swim between the bursa and the seminal receptacle, so that with time sperm in both sites become a mixture of those from the most recent male and a female's previous mates. The process is terminated when the female dumps sperm from her bursa, at which time the majority of sperm in the seminal receptacle are derived from the female's most recent mate. Following the displacement process, sperm are utilized for fertilization in direct proportion to their numerical representation in the seminal receptacle, so that the last male to mate gains the majority of fertilizations. Importantly, longer sperm and sperm with slower swimming speeds are better able to enter and remain within the seminal receptacle than are short sperm or sperm with fast swimming speeds (Lüpold et al. 2012). In consequence, the second male advantage at fertilization is greater for males with long and slow-swimming sperm, generating selection for increased sperm length.

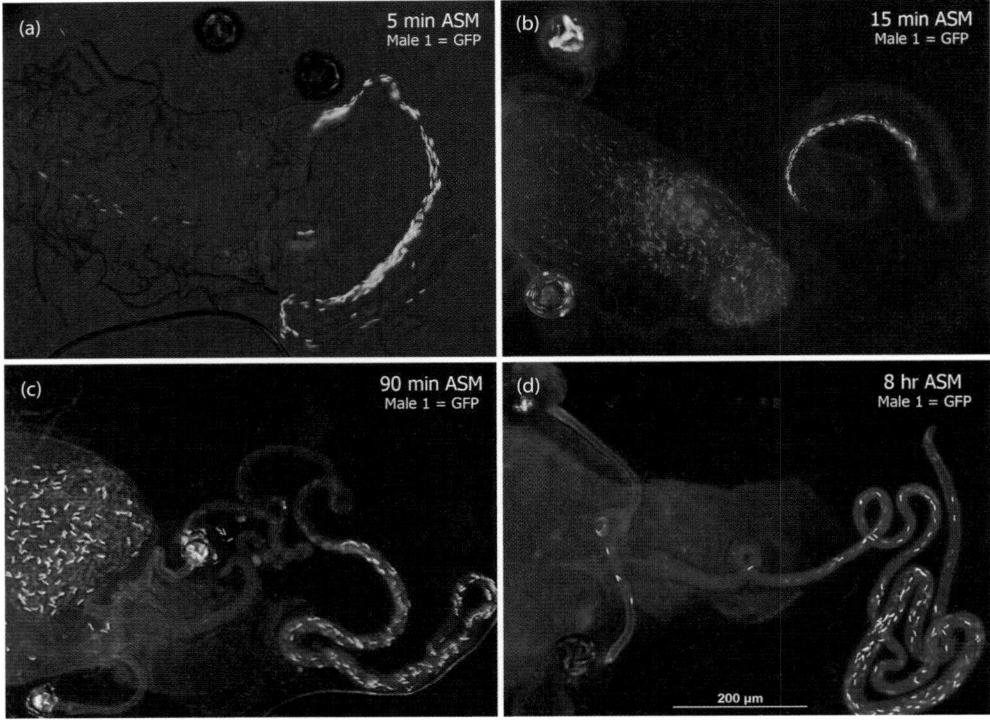

Figure 10.8 Visualization of sperm displacement in *Drosophila melanogaster*. (a) Sperm of the first male to mate express Green Fluorescent Protein (GFP) and can be seen in the seminal receptacle of the female. At copulation the second male delivers his sperm, which express Red Fluorescent Protein (RFP), into the bursa (b). (c) RFP-expressing sperm have entered the seminal receptacle and swim up and down its length, influencing the swimming speed of rival GFP-expressing sperm that leave the seminal receptacle and enter the bursa. Sperm are eventually dumped by the female from the bursa (d), stopping the displacement process, and leaving predominantly second-male RFP-expressing sperm in the seminal receptacle that is then utilized for fertilization. (From Manier et al. 2010.)

Drosophila are also of considerable interest because they produce giant sperm—the sperm of *D. melanogaster* are 1.9 mm in length, whereas those of *D. bifurca* can reach 58 mm, considerably longer than the fly itself. The production of giant sperm is costly, however, because species producing longer sperm are limited to producing fewer of them (Pitnick 1996). Among species where sperm numbers are more important for competitive fertilization success, we might expect such a trade-off to favour the production of many shorter sperm. Indeed, this seems to be the case in both the field cricket *Gryllus bimaculatus* and the dung beetle *O. taurus*, where males that produce relatively short sperm fertilize more eggs when in competition with males that produce relatively longer sperm (Gage and Morrow 2003; García-González and Simmons 2007b). These examples show us that the form of selection acting on insect sperm can be specific to the mechanisms of sperm storage and utilization of a given species or species group. Perhaps not surprisingly then, among the insects, sperm exhibit a remarkable phenotypic diversity.

In some insects, males produce more than one kind of sperm. For instance, butterflies and moths produce two forms of sperm, fertilization-competent eupyrene sperm and anucleate (non-fertilizing) apyrene sperm. Although they are not fertilization competent, apyrene sperm are highly motile and may represent up to 90% of the male ejaculate in some species (Simmons 2001; Higginson and Pitnick 2010). While relatively little is known of their functional significance, apyrene sperm have been shown to suppress female remating activity in some species (Cook and Wedell 1999; Wedell et al. 2009), and so may reduce sperm competition from future rivals in much the same way as seminal fluid proteins. In *D. pseudoobscura*, non-fertile sperm morphs protect their fertilizing sibs from toxic challenges met in the female reproductive tract (Holman and Snook 2008) whereas in the moth, *Bombyx mori*, apyrene sperm are essential for the transport of sperm to the site of fertilization (Sahara and Takemura 2003). Conjugation of sperm, as pairs or considerably larger groups, is widespread in some insect groups (Higginson and Pitnick 2010; Higginson et al. 2012a). Sperm conjugates have been found to swim faster than single sperm (Hayashi 1998), raising the intriguing idea that sperm may co-operate in competing for fertilizations (Pizzari and Foster 2008; Higginson and Pitnick 2010). However, such ideas are only recently coming to the attention of sperm competition researchers, and there is much work to be done in this area.

10.6 Concluding remarks

Multiple mating by females can generate intense selection on male reproductive behaviour, physiology, and morphology. However, females should not be seen as passive vehicles within which males fight out their battles for fertilizations. In reality, the types of adaptations to sperm competition in males depend very much on female influences over insemination, sperm storage, and sperm utilization. For example, although male genitalia serve the dual function of sperm removal and insemination in damselflies (Waage 1979b), we now know that in some species the effective removal of sperm depends on the ability of males to stimulate the appropriate mechanoreceptors in the female's reproductive tract during copulation (Córdoba-Aguilar 1999). The seminal fluid proteins ejaculated by male *Drosophila* interact with, and their action may be modified by, female reproductive tract proteins (Wolfner 2009). Finally, evolutionary changes in sperm form and function are increasingly found to correlate with evolutionary changes in the reproductive tracts of females (Morrow and Gage 2000; Miller and Pitnick 2002; Higginson et al. 2012b). In Chapter 11 we consider more fully the role of cryptic female choice in the evolution of male and female reproductive biology, and the conflicts of interest that may arise between females and their mates as males fight for genetic representation in future generations.

CHAPTER 11

Cryptic female choice

Göran Arnqvist

11.1 Introduction

In most scorpionflies (Panorpidae) and hangingflies (Bittacidae), mating is no easy feat for males. A male first needs to capture a suitable prey item, to offer as a nuptial gift to the highly polyandrous females, and then emit a sex pheromone from glands on the back of the abdomen. If a female is attracted to the pheromone, she appears to inspect the nuptial gift and may accept the copulation at which point she is offered the gift from the male. The female then feeds on the prey item throughout mating (Figure 11.1). Females of several species have been shown to benefit directly from consuming the gift and they use the size of the prey offered by males as a key criterion for female mate choice, outright rejecting males who offer small prey items (e.g. Thornhill 1976, 1983; Gwynne 1984). Thus, as we saw in Chapter 8, males that are able to offer direct benefits to females in the form of larger gifts achieve a higher mating success. However, research conducted by Randy Thornhill in the 1970s on the hangingfly *Hylobittacus apicalis* (see Thornhill and Alcock 1983 for references) revealed a much less obvious layer of variation in male reproductive success. Average copulation duration in this species is about 20 min and, starting from about 5 min into copulation, sperm are gradually transferred to the female. Importantly, females that receive a relatively small nuptial gift tend to shorten the duration of copulation to less than the 20 min required for complete sperm transfer. Females that mate for less than 20 min not only receive less sperm but also do not show refractory behaviour after copulation: they continue to be receptive and search out other males. The size of the nuptial gift not only affects male mating success, through the female tendency to accept males with large prey items as gifts, but also affects male per mating competitive fertilization success, through the female tendency to terminate copulations prematurely if the gift is small. Thus, males with small nuptial gifts will not fare well in sperm competition: they will achieve poor offensive and defensive competitive fertilization success (Chapter 10). Following some very similar findings in *Harpobittacus nigriceps* a few years later, Thornhill (1983) realized that female choice for males with large prey items may occur at different times: either prior to mating or after copulation is initiated. He termed the latter 'cryptic' female choice and noted that it operates not on variance in male mating success, but on variance in the number of fertilizations males achieve per mating. This chapter is dedicated to this form of sexual selection.

The Evolution of Insect Mating Systems. Edited by David M. Shuker and Leigh W. Simmons.
© The Royal Entomological Society 2014. Published 2014 by Oxford University Press.

Figure 11.1 Hangingfly males (*Harpobittacus similis*) capture an arthropod prey item that is offered to the female (to the right) as a nuptial gift, which she consumes during the 20 min copulation required for complete sperm transfer. If the prey item is not sufficiently large, the female terminates the copulation prematurely which reduces male competitive fertilization success both through a reduced number of sperm transferred and because the female then immediately seeks additional mates. This female behaviour constitutes a cryptic female choice (CFC) trait, which is under direct selection as females gain considerable nutritional benefits from nuptial gifts (Thornhill and Alcock 1983). Incidentally, this system also illustrates a sexual conflict over copulation duration (Thornhill and Alcock 1983; Arnqvist and Rowe 2005): if the prey item is small, the optimal copulation duration is longer for males than females, and females terminate copulations. If they prey item is large, the optimal copulation duration is longer for females than males, and males terminate copulations and take the gift back for re-use in subsequent mating. (Photo by Darryl Gwynne.)

Chapter 10 describes how sperm competition can be seen as the post-copulatory equivalent of overt pre-copulatory male–male competition, discussed in Chapter 6. By the same token, cryptic female choice (CFC) forms the post-copulatory equivalent of pre-copulatory female mate choice, discussed in Chapter 8. It seems particularly fitting to discuss CFC in the current volume, because Thornhill and Alcock (1983) was actually the first text that recognized and explicitly discussed these different layers of mate choice. However, Thornhill's (1983) insight was apparently much ahead of its time and it was not until the publication of the landmark book on CFC by Eberhard (1996) more than a decade later that CFC became more widely recognized as an important form of sexual selection (Figure 11.2). In this chapter, we introduce and define the concept of CFC, give a few empirical examples of work involving insects, discuss some more conceptual issues, and point to areas in need of more research. This chapter is not meant to be an exhaustive review of the topic. For those interested in a richer empirical review, Eberhard (1996, 1997) is still excellent reading (see also Pitnick et al. 2009; Pitnick and Hosken 2010).

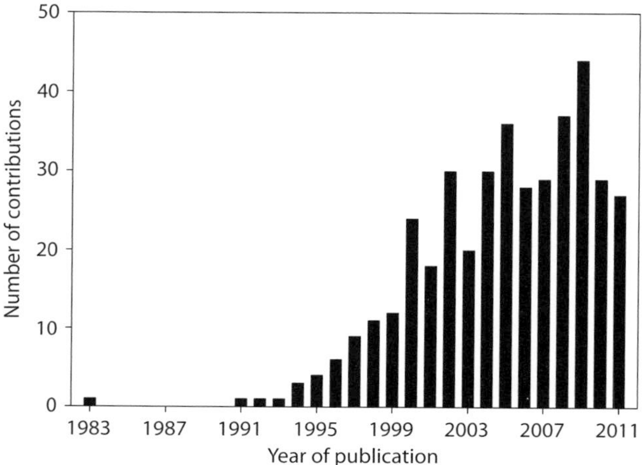

Figure 11.2 The number of published scientific articles using the term 'cryptic female choice' between 1983 and 2011 (Web of Science). Although Thornhill and Alcock (1983) and Thornhill (1983) introduced this novel concept in the early 1980s, research in this domain did not take off until a decade later.

11.2 Definition

Here, we define CFC as a pattern of non-random post-copulatory fertilization success among male phenotypes that is caused by a trait in females (Thornhill 1983; Eberhard 1996). As such, it is one of several forms of sexual selection. This seemingly simple definition actually has no less than six noteworthy features. First, CFC involves only those processes that occur after copulation is initiated. Second, variation in male fertilization success can be brought about by several routes, such as differential female sperm 'use', differential female reproductive rate or differential female remating behaviour (Eberhard 1996). Third, it involves variation in reproductive success across males that is related to, or associated with, traits in males: some male phenotypes do better than others. This is important simply because variation in male reproductive success does not necessarily translate into phenotypic selection on reproductive traits (Arnold and Wade 1984a). Unless it does, there will be no sexual selection (Sutherland 1987).

Fourth, and perhaps most importantly, this definition emphasizes the fact that CFC must be generated by a female trait (henceforth, CFC trait). This may seem like a rather trivial point, but placing a female CFC trait in the centre of this form of sexual selection greatly facilitates our ability to think clearly about the evolution of CFC and its ramifications. CFC traits come in many forms. Eberhard (1996) lists no less than 20 different ways in which CFC can be brought about and many of these can be thought of as being caused by one or several CFC traits. We suggest that CFC traits can be broadly categorized into morphological, behavioural, and physiological CFC traits.

If, for example, a long and winding internal reproductive duct in females prejudices paternity in favour of males with long or fast-swimming sperm, we would consider this duct a morphological CFC trait. If females terminate mating earlier with males that provide a smaller nuptial gift, thus biasing fertilization success in favour of males that provide large gifts (Thornhill 1983), this would exemplify a behavioural CFC trait. Similarly, female machinery for the uptake and usage of sperm that is triggered by particular seminal neuropeptides, thus favouring the uptake of sperm from males that transfer the most efficient neuropeptides in their seminal fluid, may serve as an example of a physiological CFC trait.

Fifth, the definition of CFC does not specify why CFC traits have evolved or why they are maintained in a given species. In other words, CFC is not restricted to include only CFC traits with a particular evolutionary history or only those CFC traits that are maintained by a particular form of selection (see below). Finally, CFC does not require that females somehow assess and actively or directly 'treat' males with different phenotypes differently, as is sometimes believed. Any female trait that biases fertilization success among males will constitute a CFC trait (Wiley and Poston 1996; Arnqvist and Rowe 2005).

11.3 The relationship between sperm competition and cryptic female choice

One could hold that all sperm competition (Chapter 10), at least in internally fertilizing taxa, is actually CFC. By the above definition, this is true in a strict sense. Because male gametes compete within a female 'environment', any non-random sperm competition among males will be contingent upon the CFC traits that collectively make up the female 'environment' in which sperm are competing, even if this 'environment' is uniform across females. In this basic sense, all post-copulatory sexual selection is CFC (Pitnick and Hosken 2010). In fact, the classic problem of clearly delineating pre-copulatory male–male competition and pre-copulatory female mate choice (Andersson 1994) is aggravated in the post-copulatory sexual selection domain, simply because the typical arena of male–male competition is the female herself, and several of the issues discussed in Chapter 10 could have been placed in the current chapter. However, we suggest that empirical research efforts aimed at distinguishing between sperm competition and CFC are unlikely to advance our understanding of post-copulatory sexual selection. This is in part because it is extremely difficult (Pitnick and Hosken 2010) and in part because it is conceptually somewhat misguided. All post-copulatory sexual selection by necessity involves both male and female traits and the challenge lies in understanding how these interact to determine male competitive fertilization success and, even more importantly, why female CFC traits evolve. Some would argue that we need not devote much attention to CFC, because the female arena is omnipresent and therefore should not be our focal interest. By contrast, in this chapter we suggest that a greater focus on CFC is likely to improve our understanding of post-copulatory sexual selection. In particular, a better understanding of the forces of selection that are responsible for the maintenance and evolution of CFC traits (Maklakov and Arnqvist 2009) is necessary for an improved understanding of the processes of diversification of various male adaptations that result from post-copulatory sexual selection (Chapter 10).

11.4 Cryptic female choice traits

According to the definition given above, CFC represents a non-random pattern of fertilization that is generated by a specific process (Kokko et al. 2003). The evolution of a non-random pattern of fertilization is somewhat elusive. For example, it might be difficult to conceptualize the notion that selection, which is a process in itself, acts on a pattern of fertilization and it is even more difficult to try to measure selection on CFC. In contrast, traits are tangible. Theoretical models of the evolution of female choice, whether overt or cryptic, can be seen as modelling female choice as a trait, either directly (e.g. Gavrilets et al. 2001) or indirectly as an emergent property of an underlying trait (e.g. Iwasa and Pomiankowski 1991). We suggest that embracing the view that CFC is an emergent property of CFC traits will both enable informative empirical studies and help to clarify conceptually the evolution of CFC. It is sobering to focus on CFC traits: traits can be measured and manipulated, and phenotypic selection acts on traits rather than on patterns of fertilization.

There has been much discussion, and some confusion, over the various ways in which female choice can evolve. For example, whether selection resulting in female choice is dominated by direct or indirect selection (Kirkpatrick and Barton 1997; Arnqvist and Rowe 2005) or the importance of 'good-genes' or 'Fisherian run-away' selection (Eberhard 1993, 1996) has been much debated. We can use quantitative genetics to provide a schematic but formal version of this issue. Consider three traits, z_{1-3}, that represent a focal CFC trait (z_1), a male trait favoured by CFC (z_2), and general viability (z_3). The predicted short-term evolution of the mean value of any three traits z_{1-3} is given by the general multivariate representation

$$\Delta\begin{bmatrix}\bar{z}_1\\ \bar{z}_2\\ \bar{z}_3\end{bmatrix}=\begin{bmatrix}V_1 & C_{12} & C_{13}\\ C_{21} & V_2 & C_{23}\\ C_{31} & C_{32} & V_3\end{bmatrix}\bullet\begin{bmatrix}\beta_1\\ \beta_2\\ \beta_3\end{bmatrix} \qquad [1]$$

where V represents the genetic variance of the traits (diagonal), C the genetic covariance between the traits (off-diagonal), and β the selection gradient on each trait. With regards to the evolution of a female choice trait, however, β_1 and β_2 should be discounted by a factor of 0.5 if we assume that these traits are sex-limited in expression and thus only under selection in one of the two sexes. The predicted evolution of a CFC trait (change in mean CFC trait phenotype) is then given by

$$\Delta\bar{z}_1 = V_1\beta_1\frac{1}{2}+C_{12}\beta_2\frac{1}{2}+C_{13}\beta_3 \qquad [2]$$

The first term in this equation describes the effects of direct selection on the evolution of the CFC trait. The second term describes the component of indirect selection on the CFC trait that is due to selection on the male trait, through genetic covariance between the CFC trait and the male trait. This is what we normally think of when referring to 'Fisherian run-away' selection. The third term describes the component of indirect selection on the CFC trait that is due to selection on viability, through genetic covariance between the CFC trait and viability. This is what we normally think of when referring to 'good-genes' selection. Although this representation is highly simplified, for example because it omits additional factors such as mutational effects (Kirkpatrick and Barton 1997; Fuller et al. 2005) and assumes that genetic variance in viability is not sexually antagonistic, it has important heuristic value. For example, it illustrates that these forms of selection are in no way mutually exclusive alternatives: we expect them all to act simultaneously. The issue at hand lies in determining their relative importance. Because cross-trait genetic covariances are invariably lower than trait-specific genetic variances, we generally expect direct selection (the first term) to be a stronger force than indirect selection (Kirkpatrick and Barton 1997). This expectation is strengthened by the fact that many of the known CFC traits in insects are likely to experience quite strong direct selection (see below). Further, because z_1 exerts selection on z_2, genetic variation in z_1 will tend to become statistically associated with genetic variation in z_2 (i.e. assortative mating will generate linkage disequilibrium). Because we therefore expect C_{12} to be positive, some degree of 'Fisherian run-away' selection should almost always be present. Moreover, whereas there will always be strong selection on viability (β'_3), we expect the genetic covariance between the CFC trait and viability (C_{13}) to be very low because of a low additive genetic variance in viability (Ellegren and Sheldon 2008). Thus, 'good-genes' selection on CFC traits should be omnipresent but is likely to be a weak force of selection. Sexually antagonistic genetic variation in viability (where different alleles encode for high viability in the two sexes) will, if present, further reduce the strength of 'good-genes' selection on CFC traits and may even entirely nullify its effect (Pischedda and Chippindale 2006; Bilde et al. 2009). Finally, because all evolution of a CFC trait is adaptive under this conceptual representation, if we define adaptive as evolution caused by phenotypic selection, it illustrates that the term 'adaptive [cryptic] female choice' is ambiguous.

One might ask where the roots of direct selection on CFC traits (β'_1) lie. It has been suggested elsewhere (Arnqvist 2006) that most known CFC traits originally represent pre-existing sensory biases that have been shaped by natural selection (*sensu* Fuller et al. 2005) and that CFC traits are thus likely, at least periodically, to be under relatively strong natural selection. For example, female spermatophore or gift consumption involves direct nutritional benefits to females (e.g. Vahed 1998); female receptor molecules that respond to male seminal fluid proteins mediate vital physiological reproductive functions within females (e.g. Chapman 2001); and female morphological and/or immunological 'hostility' towards the male ejaculate likely plays important roles in fighting pathogens (e.g. McGraw et al. 2004) and possibly also in reducing the risk of polyspermy (e.g. Arnqvist and Rowe 2005). Seen this way, the female 'environment' is a multidimensional trait space in which we predict that any male adaptations that increase competitive fertilization success will evolve through sensory exploitation. Such male adaptations can be costly to females, because they interfere with a female adaptation (see below), and we then expect direct selection on CFC traits to be strengthened by selection for resistance in females (Arnqvist and Rowe 2005; Arnqvist 2006). The resulting sexually antagonistic co-evolution will be driven primarily by direct selection, but may also involve indirect selection.

This simple model also illustrates why a focus on CFC traits can help provide conceptual clarity to this field. It highlights a somewhat neglected irony: despite the fact that our incomplete understanding of the evolution of CFC almost entirely reflects a lack of understanding of the forces of selection acting upon CFC traits (Kirkpatrick 1987b; Kokko et al. 2006; Maklakov and Arnqvist 2009) the vast majority of research in this field is focused on, or even entirely limited to, male traits. We suggest that continued research along this path is unlikely to help improve our understanding of the evolution of CFC: explaining the evolution of male post-copulatory adaptations (Chapter 10) is often not difficult. Given certain limitations set by costs and constraints, males are predicted to evolve whatever adaptations increase their net reproductive success. The main problem lies in understanding the relative importance of the various forces of selection that contribute to the evolution of CFC traits, and we suggest that here should be the focus of our efforts. In cases and model systems where we can identify and quantify CFC trait phenotypes, we should strive to provide empirical estimates of the key parameters of the heuristic model above (Charmantier and Sheldon 2006; Maklakov and Arnqvist 2009; Schäfer et al. 2013). Unfortunately, estimating all of the components of selection (Eq. 2) is required for a full understanding of the evolution of CFC traits. For example, in the spotted cucumber beetle *Diabrotica undecimpunctata*, males that deliver high-frequency copulatory courtship behaviour are favoured by CFC (Tallamy et al. 2002). Moreover, the frequency of copulatory courtship behaviour is heritable across males (Tallamy et al. 2003). However, the fact that the male trait shows significant genetic variance (i.e. a positive V_2 in Eq. 1) is clearly insufficient to conclude that indirect selection is important in maintaining CFC in females. Focusing on variation in male traits alone will simply not allow us to fully understand the evolution and maintenance of CFC traits. Similarly, a few studies have demonstrated a significant genetic covariance between the dimensionality of the female reproductive tract and sperm traits (i.e. C_{12} in Eq. 2) in insects, such as fruit flies (Miller and Pitnick 2002) and dung beetles (Simmons and Kotiaho 2007). Although this is a necessary condition for indirect selection to occur at all, in this case through 'Fisherian run-away' selection, it does not allow us to disentangle the relative importance of direct and indirect selection.

11.5 How can we detect cryptic female choice?

The last few decades have seen the growth of an empirical body of research in sexual selection clearly demonstrating that overt mate choice is very widespread indeed (Andersson 1994; Andersson and Simmons 2006). Yet, it has also taught us that it is often difficult to unambiguously separate mate choice from mate competition as alternative sources of non-random mating. With this in mind, it should come as no surprise that the empirical challenges with the study of CFC are significant: unveiling and documenting processes that are cryptic to our eyes is very challenging and can sometimes seem virtually impossible (Bussiere et al. 2006). Despite these difficulties, however, there are many very fine demonstrations of CFC in insects (Eberhard 1996, 1997; Pitnick et al. 2009; Pitnick and Hosken 2010), often involving clever experimental designs applied in amenable model systems. The empirical strategies so far used to demonstrate CFC have largely followed either of a few distinct routes.

One method by which CFC has been successfully revealed employs a crossed experimental design, where the inferential focus lies on an interaction between males and females. The logic is straightforward (see Pitnick and Brown 2000). For example, imagine a double mating experiment where a set of males, or male genotypes, are mated with a set of females, or female genotypes, and the extent of last male paternity is measured. In a linear model of last male sperm precedence from such data, the male term parameterizes the extent to which certain males have a higher competitive fertilization success than others (such variation could be due either to sperm competition or to CFC). The female term measures the extent to which females vary in the overall degree of last male sperm precedence. The interaction term tests whether relative competitive fertilization success among males varies across females. If it does, it strongly suggests that variation in female traits must 'affect' relative fertilization success among males and a significant interaction thus provides evidence for the existence of CFC (Pitnick and Brown 2000). This basic inferential rationale has been successfully used to demonstrate CFC for specific male genotypes in, for example, the seed beetle *Callosobruchus maculatus*, both when CFC involves last male sperm precedence (Wilson et al. 1997; Bilde et al. 2009) and female reproductive rate (Fricke et al. 2006). Similar results have been documented in several other insects, such as *Drosophila melanogaster* (Clark et al. 1999), *Tribolium castaneum* (Nilsson et al. 2003), *Callosobruchus chinensis* (Harano and Miyatake 2007b), and *Musca domestica* (Andrés and Arnqvist 2001).

Although studies based on assessing male × female interactions have been valuable in documenting the existence of CFC, they suffer from three limitations. First, male traits favoured by CFC and female CFC traits are typically a 'black box' in these studies: males and females are simply classified as belonging to a discrete genotype or population. However, crossed classification designs can be constructed to allow the incorporation of quantitative measures of male and female traits. For example, Arnqvis and Danielsson (1999) used a hierarchically structured and crossed mating design to show that the strength of post-copulatory sexual selection on male genital morphology was contingent upon female body size in water striders (Figure 11.3) and García-González and Simmons (2007b) used a similar mating design to demonstrate that competitive fertilization success was biased toward males with relatively short sperm but that the extent of this bias depended on the size of the female sperm storage organ. Further, Yamane and Miyatake (2012) showed that male seminal fluid substances are involved in CFC by documenting male × female interactions for female reproductive responses to injections of male accessory gland extracts.

A second, and perhaps the most important, limitation is that this strategy will reveal CFC only if females vary in their CFC 'preferences' or CFC 'criteria' (Pitnick and Brown 2000). If all females bias fertilizations towards the same male phenotypes, such that CFC preference functions are unanimous, there will be no male × female interaction and this method will fail to detect CFC even if it is strong. Any effect of CFC will then appear as a 'male' effect in statistical models and be statistically and conceptually conglomerated with variance across males in overall sperm competitiveness. Yet CFC should not

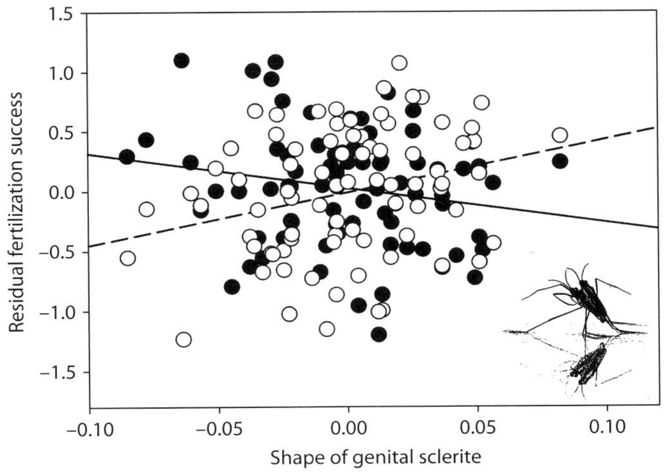

Figure 11.3 Male genital morphology is related to male competitive fertilization success in water striders (*Gerris lateralis*). The effect of the shape of the dorsal genital sclerite on the fertilization success of the second male to mate with a given female (i.e. P_2) depends upon female body size. Males with a more curved dorsal sclerite have highest fertilization success in relatively large females (filled symbols and solid line) whereas males with a straighter dorsal sclerite have highest fertilization success in relatively small females (open symbols and dashed line) (two-way interaction; $P = 0.021$; Arnqvis and Danielsson 1999). This interaction reveals cryptic female choice (CFC) on male genital morphology and implies that some size-related trait in females functions as a CFC trait. One possibility is that the overall size of the female reproductive tract, or some aspect that scales allometrically with size, selects for different optimal genital configurations in males. If this is true, natural selection on overall size in females may constitute a significant source of selection on the CFC trait. (Data from Arnqvist and Danielsson 1999; drawing by Görel Marklund.)

require variance in cryptic female preference functions any more than female mate choice requires variance in female mate preference functions. If we regard a female mate preference for large males which is shared among all females as female mate choice (Chapter 8), it would seem that we should also regard a cryptic female preference for males that transfer large amounts of sperm which is shared among all females as CFC. For these reasons, methods focused on male × female interactions are most useful when CFC is qualitative rather than quantitative and favours male genotypes that are in some sense 'compatible' with particular female genotypes (Pitnick and Brown 2000; Puurtinen et al. 2005), such that the underlying female responses to male trait variation are complex and idiosyncratic (Andrés and Arnqvist 2001; Bjork et al. 2007b). A final caveat is that, in theory, male × female interactions can also be due to strategic sperm allocation by males (Chapter 10). If some male genotypes, for example, transfer more sperm when mating with some female genotypes than with others, this may generate male × female interactions for male competitive fertilization success (Pitnick and Brown 2000; Reinhardt 2006).

Another set of methods used to document CFC involves experimental manipulations, and these methods can provide some of the clearest evidence for CFC when putative male traits favoured by CFC have been identified. Pitnick and Brown (2000) first suggested that manipulation of either male traits or females' perception of male traits could allow effects of CFC to be separated from those due to sperm competition. Edvardsson and Arnqvist (2000) tried to do just this. They manipulated female perception of male copulatory courtship behaviour (Eberhard 1994) in a study of *Tribolium castaneum* by tarsal ablation in males, which prevented males from reaching the edge of the female elytra with their manipulated legs during mating (males rub the lateral edges of the females' elytra with their tarsi during copulation). They found a positive relationship between the intensity of the copulatory courtship behaviour and relative fertilization success among unmanipulated males. In contrast, this pattern was absent among manipulated males where female perception of male behaviour differed from that actually performed. Their experimental manipulation thus showed that female perception of male copulatory courtship behaviour, rather than the male behaviour itself, apparently affected competitive male fertilization success (Figure 11.4). As an example of a manipulation of the male trait, Barbosa (2009) showed that female soldier flies (*Merosargus cingulatus*) delay oviposition after having mated with males that were experimentally rendered unable to perform copulatory courtship behaviour. The CFC trait itself can also be experimentally manipulated when identified. For example, in many crickets, females remove and consume externally attached spermatophores before insemination is complete. This CFC behaviour is non-random with respect to male phenotypes (e.g. Sakaluk 1997; Bussiere et al. 2006) and experimental manipulations in which females are prevented from removing the spermatophore have shown that spermatophore removal indeed reduces male fertilization success (e.g. Sakaluk 1984; Simmons 1987; Sakaluk and Eggert 1996).

Another type of experimental 'manipulation' involves staged double matings that vary mate relatedness (Tregenza and Wedell 2002) or mate pre-copulatory attractiveness (Bussiere et al. 2006; Fedina and Lewis 2007). Tregenza and Wedell (2002) originally showed that female sperm use in the field cricket *Gryllus bimaculatus* is non-random with regards to the relatedness of their mate: unrelated males (non-siblings) achieve a higher competitive fertilization success than do related (sibling) males. Similar experiments have since been conducted in a few other insects. These suggest that CFC in favour of unrelated males, presumably representing adaptive inbreeding avoidance, may be a widespread (e.g. Mack et al. 2002; Simmons et al. 2006) but by no means universal (e.g. Jennions et al. 2004; Ala-Honkola et al. 2011) phenomenon in insects. Subsequent research has used molecular

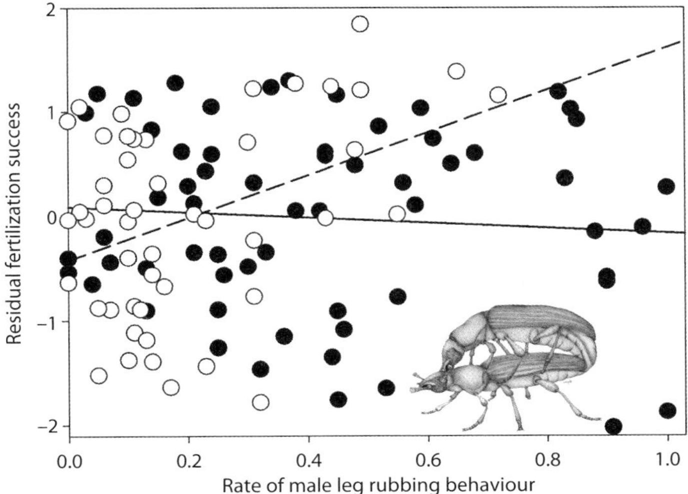

Figure 11.4 In the flour beetle, *Tribolium castaneum*, males rub their tarsi along the edges of the female elytra during copulation. This form of male copulatory courtship behaviour is apparently under cryptic female choice (CFC). Among normal males (open symbols and dashed line), the rate at which males rub the females was related to male competitive fertilization success in a series of double mating experiments. However, if one or two legs were ablated, such that males did not fully reach the females' elytra, males transferred the same number of sperm to females but male competitive fertilization success was reduced and, more importantly, the rate at which males performed 'rubbing' (in the air) was no longer related to male competitive fertilization success (filled symbols and solid line) (two-way interaction; $P = 0.019$; Edvardsson and Arnqvist 2000). Thus it seems that female perception of the male behaviour somehow affects male competitive fertilization success. Although the precise mechanism is unknown (Fedina and Lewis 2008), differential sperm utilization is one possibility. (Data from Edvardsson and Arnqvist (2000), showing the residual fertilization success after removal of the main effect of leg ablation and the effects of copulation duration of both the first and second males. Drawing by April Hobart.)

markers to show that this form of CFC results from a failure to store sperm of more closely related (Bretman et al. 2009; Tuni et al. 2013b) or less attractive (Hall et al. 2010a) males after copulation.

A final empirical strategy, advocated by Eberhard (1996), is a mechanistic 'bottom-up' approach. Such a reductionistic research strategy would aim to detect and understand CFC by first dissecting and documenting all of the underlying events and interactions that ultimately and potentially lead to CFC. Although this is no doubt possible in theory—a few attempts have been partly successful (e.g. Tallamy et al. 2002)—we are unaware of any complete and successful application of such a research programme, and the sheer complexity of reproductive morphology and physiology can make the task seem somewhat daunting. In *Drosphila melanogaster*, for example, the reproductive morphology (e.g. Pitnick et al. 1999) and the reproductive genome and transcriptome are arguably better studied and understood than in any other taxa (e.g. Chapman 2001; McGraw et al. 2004) and the fate of individual sperm can be tracked by means of transgenic techniques (Manier et al. 2010). Yet, a mechanistic, detailed and complete understanding of CFC in *Drosophila melanogaster* is currently quite distant (e.g. Bjork et al. 2007b; Ala-Honkola et al. 2011; Lupold et al. 2012; Chow et al. 2013). While a certain degree of functional and mechanistic research is no doubt often needed in order to identify CFC traits, we suggest that only studies of the fitness effects of phenotypic variation in CFC traits can provide direct information about the forms of selection responsible for the maintenance of CFC (Schäfer et al. 2013).

11.6 What do we need?

During the last three decades, a diverse set of fascinating and beautiful demonstrations of CFC in insects have been presented. We have argued here that a deeper understanding of the evolution of CFC in insects, or indeed any other taxa, requires an understanding of past and present selection on CFC traits. In theory, we can achieve this by measuring direct and indirect selection on CFC traits (Eq. 2) empirically. Empirical measures of phenotypic selection would need to focus either on natural variation in CFC trait phenotypes (e.g. Lande and Arnold 1984a) or on experimentally induced variation in CFC trait phenotypes, through phenotypic engineering (Sinervo and Basolo 1996; Travis and Reznick 1998). In practice, this endeavour is very challenging. Nevertheless, we suggest that this is the empirical route we should embark upon. An initial problem is identifying the key CFC trait in any given system: what female trait is mainly responsible for biasing post-copulatory fertilization success among male phenotypes? This problem may be aggravated by the fact that there may be several CFC traits acting at different levels or stages of CFC. In taxa where a key CFC trait can be identified, quantifying natural variation in CFC trait phenotypes represents the next challenge. This has proven difficult and has rarely been achieved even for overt female choice traits (e.g. Ritchie 1996; Charmantier and Sheldon 2006; Qvarnström et al. 2006). An alternative strategy, that may be more feasible in some cases, is to experimentally extend the range of CFC trait phenotypes by some form of experimental manipulation and study the direct and indirect effects of this manipulation (e.g. Maklakov and Arnqvist 2009). In both cases, measures of key female fitness components and some understanding of the genetic architecture of the traits involved (Eq. 2) are also required for a complete understanding of the maintenance of CFC traits.

Perhaps the study systems that comes closest to achieving this comprise those orthopterans whose females remove and consume externally attached spermatophores before insemination is complete (Gwynne 1997), which reduces male fertilization success. Experimental manipulations of this CFC behaviour suggest that it is primarily under direct selection (e.g. Wagner et al. 2001; Fleischman and Sakaluk 2004), presumably because females gain direct nutritional benefits from ejaculate consumption (for reviews see Vahed 1998, 2007a; Arnqvist and Nilsson 2000; Gwynne 2008). In these crickets therefore it may primarily be direct natural selection in females that, as a by-product, generates sexual selection by CFC among males (Sakaluk 2000; Sakaluk et al. 2006).

Another interesting example involves one of the archetypal model organisms in the study of post-copulatory sexual selection—the yellow dung fly (*Scathophaga stercoraria*) (Chapter 10). In this species, the number of female sperm storage organs (i.e. spermathecae) varies across females and this affects competitive male fertilization success (Ward 2000, 2007; Ward et al. 2008). Schäfer et al. (2013) were recently able to manipulate the number of spermathecae, through both artificial selection and environmental alteration, but found no covariation between CFC trait phenotype and subsequent offspring viability in a series of experimental multiple mating assays. Although spermathecal number shows significant genetic variance in the yellow dung fly, it also shows striking phenotypic plasticity and is developmentally integrated with key life-history traits (Berger et al. 2011; Schäfer et al. 2013). These facts suggest that variation in spermathecal number in this species is primarily the indirect result of natural selection on growth rate and other life-history traits (Berger et al. 2011).

11.7 Sperm competition and cryptic female choice: a useful distinction?

The delineation between sperm competition and CFC is a difficult one, both conceptually and empirically (Eberhard 1996; Bussiere et al. 2006; Pitnick and Hosken 2010). One pragmatic way out of this dilemma is to regard the issue as semantic and simply refer to both as post-copulatory sexual selection. However, we believe that this might be misguided and that the distinction is useful and important for at least two related reasons. First, a research focus on CFC in its own right should help increase our understanding of the evolution of CFC traits. This is important because it is arguably the part of post-copulatory sexual selection that we are furthest away from fully understanding: what forces of selection are generally responsible for the origin and maintenance of CFC traits? Second, if we wish to understand why some male traits rather than others are favoured by CFC and why these traits show certain patterns of diversification, then we need better to understand the origin and the maintenance of CFC traits. The chief reason is that different forms of selection on CFC traits should result in different types of male–female co-evolutionary dynamics or indeed in no co-evolution at all (Fuller et al. 2005; Arnqvist 2006). This is important because male–female co-evolution is potentially a powerful generator of both trait diversity and speciation. Two examples illustrate this latter point.

The first example concerns male reproductive accessory gland proteins and peptides. These substances are incorporated into the male ejaculate and, remarkably, genes encoding these proteins are some of the most rapidly evolving genes (see Chapter 10). Many of these substances act as hormones that function within females: they bind to female receptors and elicit various reproductive responses in females that elevate competitive fertilization success of the donor male (see Chapman 2001, 2008; Arnqvist and Rowe 2005; Avila et al. 2011). There is no doubt that CFC plays a major role in the evolution of male accessory gland proteins (Eberhard and Cordero 1995; Pitnick and Hosken 2010). Although little is known about the female molecules that interact with male accessory gland proteins, comparative genetic data from *Drosophila* suggest that such putative CFC trait genes also evolve rapidly (e.g. Prokupek et al. 2008; Yapici et al. 2008; Wong et al. 2012). However, our understanding of selection on those CFC traits that generate sexual selection on seminal fluid proteins in males is limited. There are good reasons to believe that sexually antagonistic co-evolution may often be responsible for this rapid evolution (e.g. Chapman et al. 1995; Wolfner 2002; Arnqvist and Rowe 2005). Consider, for example, those seminal proteins that elevate immediate female reproductive effort (i.e. gonadotrophins). Females regulate their reproductive rate by endogenously produced gonadotrophins and there is typically a positive dose-dependent reproductive rate response to the concentration of such substances. Because of life-history trade-offs, however, an intermediate amount of gonadotrophins is expected to maximize female fitness. If males capitalize on female sensory responses by evolving an ability to provide an additional dose of gonadotrophins to females in their seminal fluid, which will benefit males in many polyandrous species given their limited genetic interest in the future offspring of their mate, this will cause a depression of the fitness of their mates (Arnqvist 2006). This in turn generates direct selection for resistance in females, and diversifying sexually antagonistic co-evolution will ensue. However, the rarity of cases in which a physiological CFC trait has actually been identified

(Swanson and Vacquier 2002; Kelleher and Markow 2009; Kelleher et al. 2011) and the absence of studies of phenotypic selection on these CFC traits (Fricke et al. 2009a) render this interpretation somewhat preliminary.

The second example is genital evolution. The morphology of male intromittent genitalia shows an almost explosive diversification in many insect groups, and genital shape is often the only trait that allows closely related species to be distinguished (Eberhard 1985). There is now much evidence pointing to CFC as the main process responsible for this very rapid evolution of male genitalia in insects. Evidence comes from comparative studies across taxa (Eberhard 1985; Arnqvist 1998; Rowe and Arnqvist 2012) and studies correlating genital morphology with male fertilization success within species in beetles (House and Simmons 2003; Wenninger and Averill 2006; Hotzy and Arnqvist 2009), water striders (Arnqvist and Thornhill 1998; Arnqvis and Danielsson 1999), and praying mantids (Holwell et al. 2010), for example. In addition, studies of flies (Briceno and Eberhard 2009) and beetles (Simmons et al. 2009; Hotzy et al. 2012) have experimentally manipulated male genital morphology and have documented effects on male competitive fertilization success. In a general way, understanding the evolution of even complex male genitalia poses no fundamental conceptual problem: since post-copulatory sexual selection is relatively strong in many insects (Chapter 10; Simmons 2001; Fritzsche and Arnqvist 2013), any conceivable modification of male genital morphology that increases male competitive fertilization success will be favoured by sexual selection. However, if the female 'environment' (i.e. all aspects of the female reproductive tract) were static, as might be expected if CFC traits are under stabilizing natural selection, we would expect to see little divergent evolution of male genitalia. Genital morphology would be fine-tuned by sexual selection into an 'optimal' genital configuration and there would be little selection for morphological innovation under such a scenario. This does not seem to be what we see: the shape and complexity of intromittent male genitalia in insects generally seem to evolve more rapidly and divergently than other morphological traits (Eberhard 1985; Arnqvist 1998; Rowe and Arnqvist 2012). The most reasonable explanation for this pattern is that key aspects of the female reproductive tract (i.e. the CFC traits) are evolving, such that male genitalia are adapting not to a static but to an evolving 'environment' (Figure 11.3). Data on insects support such an interpretation: the female reproductive tract is often highly variable even within single genera, and aspects of the female reproductive tract show correlated evolution with primary sexual traits in males (e.g. Pitnick et al. 1999; Presgraves et al. 1999; Arnqvist and Rowe 2002; Rönn et al. 2007, 2011; McPeek et al. 2009; Joly and Schiffer 2010; Tatarnic and Cassis 2010; Simmons and García-González 2011; Higginson et al. 2012b). For example, Higginson et al. (2012b) showed that the size and shape of several organs and structures of the female reproductive tract show correlated evolution with sperm morphology in diving beetles, and Rönn et al. (2007) showed that female copulatory duct morphology shows correlated evolution with the shape of male genitalia in seed beetles. These types of data strongly suggest that male genital morphology and the female reproductive tract are co-evolving. Yet this does not help us understand why CFC on male genital morphology has evolved or why it is maintained. In contrast, an understanding of selection on those CFC traits that generate sexual selection on male genital morphology would help greatly (Rowe and Arnqvist 2012). This, we believe, is the greatest future challenge for research on genital evolution.

One interesting example is male intromittent genitalia that are equipped with spines or other structures that physically injure females internally during copulation (Figure 11.5). This phenomenon is widespread in insects (Eberhard 1985) and females may be left with substantial scarring in the reproductive tract in, for example, bushcrickets (von Helversen

CHAPTER 12

Parental care

Per T. Smiseth

12.1 Introduction

Parents of most insects provide no care for their offspring beyond supplying the eggs with a small amount of yolk and placing them in a location that is relatively safe from environmental hazards. The rare occurrence of post-oviposition parental care is thought to reflect the early evolution in insects of a complex suite of egg and oviposition traits that effectively protected the eggs from predators, pathogens, desiccation, and other environmental hazards without the need for costly post-oviposition parental care (Hinton 1981; Zeh et al. 1989). Yet, parents of some insects do provide post-oviposition care that enhances the offspring's survival and/or growth, often at a substantial cost to the parents' future reproductive potential. Most such insects, including the treehopper *Publilia concava*, provide relatively simple forms of care whereby one parent, typically the female, attends the eggs until hatching (Zink 2003). However, a few insect species exhibit elaborate forms of care similar to those characterizing birds and mammals. For example, in *Nicrophorus* burying beetles, male and female parents co-operate by applying antimicrobials to help preserve the vertebrate carcass that serves as a food source for the developing larvae (Rozen et al. 2008; Cotter and Kilner 2010; Arce et al. 2012), defending the brood from predators and conspecific competitors (Trumbo 2007), and provisioning the larvae with pre-digested carrion (Eggert et al. 1998; Smiseth and Moore 2002). Insects show a great deal of diversity with respect to the forms of care that parents provide, and the extent to which males and females are involved in care (Tallamy and Wood 1986; Choe and Crespi 1997; Costa 2006). This diversity in the forms and patterns of care makes insects a fascinating and important taxonomic group for studying the evolutionary causes and consequences of parental care.

The evolution of parental care in insects has attracted interest partly to identify the evolutionary causes underlying the observed diversity in parental care, and partly to identify the extent to which parental care is provided by the male, the female, or both parents (e.g. Tallamy 1984; Tallamy and Wood 1986; Trumbo 1996, 2012; Choe and Crespi 1997; Costa 2006; Wong et al. 2013). In addition, parental care in insects has attracted attention because its evolution is a crucial step toward the evolution of more advanced forms of sociality, such as eusociality in termites and many hymenopterans (Wilson 1971, 1975). Indeed, insects where parents care for offspring after hatching are often described as 'subsocial' (e.g. Wilson 1971), a term that reflects the assumption that parental care serves an intermediate step in the evolution of eusociality (Costa 2006). Finally, the evolution of parental care has attracted attention because variation in the extent to which males and

The Evolution of Insect Mating Systems. Edited by David M. Shuker and Leigh W. Simmons.
© The Royal Entomological Society 2014. Published 2014 by Oxford University Press.

females are involved in parental care is closely associated with variation in mating systems, sexual selection and sperm competition (Trivers 1972; Smith 1980; Zeh and Smith 1985; Clutton-Brock 1991; Tallamy 2000, 2001). In insects, interest in the association between parental care and mating systems has focused primarily on the evolution of male involvement in parental care (Smith 1980; Zeh and Smith 1985; Tallamy 2000, 2001; Manica and Johnstone 2004).

This chapter has five main objectives: first, to provide a brief overview of the diversity of ways by which insect parents enhance the fitness of their offspring, often at a significant cost to the parent's own fitness (Section 12.2); second, to illustrate the diversity among insects in the extent to which care is provided by the male parent, the female parent, or both (Section 12.3); third, to draw links between male and female involvement in parental care and insect mating systems; fourth, to provide a more detailed description of four case studies that illustrate these links; and fifth, to provide suggestions for future studies on insects that might help advance our understanding of the co-evolution of parental care and insect mating systems.

12.2 Forms of parental care

To provide a brief overview of the diversity in forms of parental care in insects, examples of parental care are highlighted in Table 12.1. The different forms of care in insects are arranged in chronological order in relation to the offspring's development. Examples are included

Table 12.1 Overview of some of the diversity in forms and patterns of parental care among insects[a]

Order and species	Forms of care	Pattern of care	References
Orthoptera			
Kawanaphila nartee	PG	UF/BP[b]	[1]
Dermaptera			
Forficula auricularia	NB, EA, OA, FP	UF	[2–4]
Anechura harmandi	FP	UF	[5]
Embiidina			
Antipularia urichi	NB, EA	UF	[6]
Blattodea			
Diplotera punctata	VO	UF	[7]
Phlebonetus pallen	OB	UF	[8]
Hemiptera			
Abedus herberti	EA	UM	[9]
Publilia concava	EA	UF	[10]
Rhinocoris venustus	OS	UF	[11]
Rhinocoris kumarii	OS	UF	[11]
Rhinocoris carmelita	EA	UF	[12, 11]
Rhinocoris tristis	EA	UM/UF[c]	[11–13]
Rhinocoris albopilosus	EA	UM	[11, 14]
Rhinocoris albopuctatus	EA	UM	[11, 14]
Phyllomorpha lacinata	EA[d]	UM[d]	[15, 16]
Parastriata japonensis	FP	UF	[17]
Thysanoptera			
Hoplothrips karnyi	EA	UM	[18]

Table 12.1 *Continued*

Order and species	Forms of care	Pattern of care	References
Coleoptera			
Stator limbatus	PG	UF	[19]
Mimosestes amicus	PG	UF	[20]
Lytta vesicatoria	PG	UF/BP[b]	[21]
Onthophagus taurus	FP	BP	[22]
Nicrophorus vespilloides	OA, FP, NI	BP	[23–25]
Nicrophorus quadripunctatus	OA, FP	BP	[26]
Nicrophorus pustulatus	OA, FP	BP	[27]
Coccinella septempuncta	PG	UF	[28]
Lepidoptera			
Atrophaneura alcinous	PG	UF	[28]
Diptera			
Culiseta longiareolata	OS	UF	[29]
Glossina spp.	VO	UF	[30]
Hymenoptera			
Megarhyssa atrata	OS	UF	[31]
Ampulex compressa	FP	UF	[32]
Trypoxylon politum	NB	UF	[33]
Polistes spp.	NB	UF	[34]

Forms of care: PG, provisioning of gametes; OS, oviposition-site selection; NB, nest building and burrowing; EA, egg attendance; EB, egg brooding; VO, viviparity and ovoviviparity; OA, offspring attendance; OB, offspring brooding; FP, food provisioning; NI, care after nutritional independence.

Patterns of care: UF, uniparental female care; UM, uniparental male care; BP, biparental care.

[1] Simmons (1990), [2] Lamb (1976), [3] Kölliker (2007), [4] Staerkle and Kölliker (2008), [5] Suzuki et al. (2005), [6] Edgerly (1997), [7] Stay and Coop (1974), [8] Bell et al. (2007), [9] Smith (1974, 1976, 1979, 1997), [10] Zink (2003), [11] Gilbert et al. (2010), [12] Thomas and Manica (2005), [13] Beal and Tallamy (2006), [14] Tallamy (2001), [15] Kaitala et al. (2001), [16] Gomendio and Reguera (2001), [17] Hironaka et al. (2005), [18] Crespi (1988a), [19] Fox et al. (1997), [20] Deas and Hunter (2012), [21] Eisner et al. (2002), [22] Hunt and Simmons (1997, 1998, 2000, 2002b), [23] Eggert et al. (1998), [24] Smiseth et al. (2003), [25] Rozen et al. (2008), [26] Suzuki and Nagano (2009), [27] Trumbo (2007), [28] Blum and Hilker (2002), [29] Spencer et al. (2002), [30] Glasgow (1961), [31] Le Lannic and Nénon (1999), [32] Keasar et al. (2006), [33] Brockman (1980), [34] Hansell (1987).

[a] A more comprehensive overview can be found in Wong et al. (2013). Readers should note that examples of uniparental male care and biparental care are overrepresented in this table.

[b] Male contribution to gametes may have evolved to increase mating success rather than offspring fitness.

[c] Uniparental female care only occurs following the removal of the male.

[d] This controversial example of egg brooding might be a form of intraspecific parasitism.

based on a broad-sense definition of parental care as 'any parental trait that enhances the fitness of a parent's offspring, and that is likely to have originated and/or is currently maintained for this function' (Smiseth et al. 2012). This broad-sense definition focuses on the adaptive value of parental care rather than on particular attributes of the parental traits. For example, Clutton-Brock's (1991) narrow-sense definition of parental care as 'any parental behaviour that appears likely to enhance the fitness of a parent's offspring' restricts the use of the term to strictly behavioural traits. Thus, this latter definition would include traits that are behavioural (e.g. egg attendance), while excluding traits that are morphological (e.g. viviparity), despite the fact that they might serve similar functions in terms of protecting offspring from environmental hazards. Clutton-Brock departed from his own narrow-sense definition when describing the diversity in forms of parental care by including gamete production (Clutton-Brock 1991, p. 14). To account for these different definitions of parental care being used in different contexts, Smiseth et al. (2012) suggested that the term 'parental behaviour' be used when focusing on strictly behavioural traits, and that the term 'post-oviposition parental care' be used when focusing on forms of care that occur after egg laying.

12.2.1 *Provisioning of gametes*

In most insects, attributes of the eggs, including the amount of resources allocated to each egg, provide the primary and ancestral mechanism by which parents protect their offspring against environmental hazards such as desiccation, drowning, predators, and pathogens (Zeh et al. 1989). Life-history theory predicts that an increase in resource allocation to each egg is beneficial to offspring as it enhances their survival, but costly to females as they are left with fewer resources to invest in other eggs (Smith and Fretwell 1974). Empirical studies on insects have played an important role in testing theoretical predictions concerning the evolution of gamete provisioning (Fox and Czesak 2000). For example, in the seed beetle *Stator limbatus*, females show plasticity in egg size, laying different-sized eggs depending on the host plant. Larvae have very low survival while attempting to penetrate the seed coat of the host plant *Cercidium fioridum*, whereas they have much higher survival while penetrating the seed coat of the alternative host plant *Acacia greggii*. Consequently, females produce larger but fewer eggs when they oviposit on the tougher seeds of *C. fioridum* as compared to when they oviposit on the seeds of *A. greggii* (Fox et al. 1997). Furthermore, when females are experimentally swapped between host plants, they produce progressively larger eggs on *C. fioridum* and progressively smaller eggs on *A. greggii* (Figure 12.1). Variation in egg size has important consequences for offspring fitness; when females are forced to oviposit on the tougher seeds of *C. fioridum*, larval survival to adulthood is lower when larvae hatch from the smaller eggs produced by females entrained to oviposit on *A. greggii* than when hatching from the larger eggs produced by females entrained to oviposit on *C. fioridum* (Fox et al. 1997).

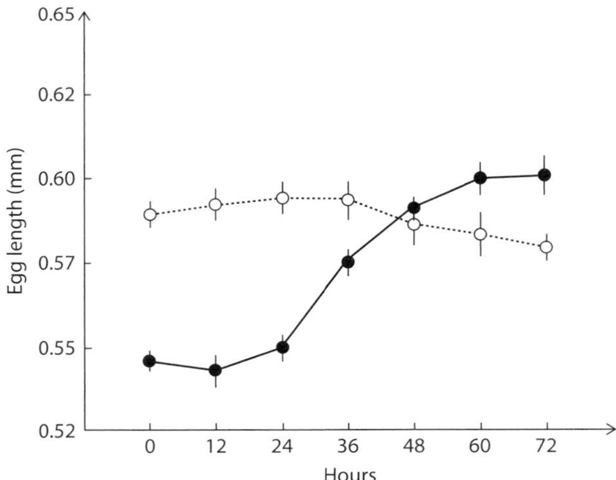

Figure 12.1 Provisioning of gametes in the seed beetle *Stator limbatus*. Egg size is plastic in this species, and adjusted to the host plant species. In this experiment, females were confined on one of two alternative host plant species—*Cercidium fioridum* or *Acacia greggii*—until they laid their first egg, after which they were switched to the other species. Females laid a relatively constant egg size for the first 24–36 h after the host switch, but then they started readjusting the egg size by laying progressively larger eggs when switched from *A. greggii* to *C. fioridum* (black symbols) and by laying progressively smaller eggs when switched from *C. fioridum* to *A. greggii* (open symbols). (Redrawn with permission from University of Chicago Press.)

Females of many insects enhance their offspring's fitness by supplying the eggs with defensive chemicals or structures that protect them from predators, parasitoids, or pathogens (Blum and Hilker 2002). Such defensive chemicals are either produced de novo or obtained through the diet (Blum and Hilker 2002; Eisner et al. 2002). For example, in the seven-spotted ladybird, *Coccinella septempuncta*, females supply their eggs with alkaloids that are produced de novo by adults, whereas in the swallowtail butterfly *Atrophaneura alcinous* females supply their eggs with bitter aristolochic acids obtained through the larval diet of birthworts (*Aristolochia* spp.) (Blum and Hilker 2002). In the seed beetle *Mimosestes amicus*, females respond to the presence of the parasitic wasp, *Uscana semifumipennis*, by producing additional inviable eggs that serve as a shield protecting viable eggs from parasitism (Deas and Hunter 2012). These inviable eggs are smaller than viable eggs, and wasp larvae developing within them suffer higher mortality than those developing within viable eggs.

In some insects, males contribute to gamete provisioning by defending resources used by females to produce eggs, providing females with nuptial gifts, transmitting nutrients or defensive chemicals through the ejaculate, or by being eaten by females (Simmons and Parker 1989; Hilker and Meiners 2002). There is evidence from some insects that such male contributions can enhance offspring fitness. For example, in the bushcricket *Kawanaphila nartee*, females that are allowed to consume the spermatophylax—a product of the male accessory glands that is attached to the sperm-containing ampulla of the spermatophore—produce larger eggs 48 h after mating than females that are experimentally prevented from consuming the spermatophylax (Simmons 1990). In the blister beetle *Lytta vesicatoria*, eggs are coated with the toxin cantharidin, which protects them from ants and carabid beetles. The production of the toxin is restricted to the male, who transmits it to the female via the ejaculate (Eisner et al. 2002). However, whether male contributions that enhance offspring fitness qualify as a form of parental care is the subject of debate (Vahed 1998). First, there is evidence to suggest that male contributions to gamete provisioning evolved as strategies for increasing male mating success (Vahed 1998), and that any positive effect on offspring fitness is an incidental by-product of selection on male mating behaviour. Second, given that males transfer their resources to the female, any effect of male contributions on offspring fitness will depend on female decisions as to whether to allocate these resources to the production of larger eggs (thereby increasing offspring fitness) or more eggs (thereby increasing the female's own fitness). Male nutrient donations may have their evolutionary origin in selection on male mating effort, but may be currently maintained because of their effects on offspring fitness (Parker and Simmons 1989; Simmons and Parker 1989). There is a need for further empirical and theoretical work to elucidate the extent to which male contributions enhance offspring fitness, and the circumstances under which selection may favour male contributions due to their effects on offspring fitness alone.

12.2.2 *Oviposition-site selection*

Female insects may also enhance the fitness of their offspring by laying their eggs nonrandomly in the environment, a phenomenon known as oviposition-site selection (Refsnider and Janzen 2010). Oviposition-site selection may increase offspring fitness by ensuring that offspring develop in sites that provide protection from a wide range of environmental hazards, including predators, parasitoids, and desiccation (Hinton 1981), and that offspring have an adequate supply of resources after hatching (Refsnider and Janzen 2010). Females of some insects have evolved an ovipositor, a specialized egg-laying tube, which allows them to oviposit their eggs in sites that are protected from predators and parasitoids, or that provide the larvae with access to resources after hatching. For example, females of the parasitic wasp *Megarhyssa atrata* have evolved a long ovipositor that can bore through 14 cm of wood to oviposit on larvae of its host, the xylophagous sawfly *Tremex columba*. The ovipositor bores through wood using a combination of rotating segments and lytic secretions that break down wood fibres (Le Lannic and Nénon 1999). Studies of insects provide good evidence that oviposition-site selection increases offspring fitness (Refsnider and Janzen 2010). For example, in the mosquito *Culiseta longiareolata*, females avoid ovipositing in pools inhabited by larval predators (Spencer et al. 2002). However, studies of some insects, such as the grass miner moth *Chromatomyia nigra*, suggest that females may also prefer ovipositing in sites that are safer for themselves rather than for their offspring (Scheirs et al. 2000).

12.2.3 *Nest building and burrowing*

Females of some insects, such as many earwigs and hymenopterans, oviposit their eggs in nests or burrows. In some such insects, parents build nests from materials they obtain from the environment, such as mud used by the mud-dauber wasp *Trypoxylon politum* (Brockman 1980). In other insects, nests are built from processed plant materials, such as paper used by wasps of the genus *Polistes* (Figure 12.2a; Hansell 1987), or from materials produced by the parents themselves, such as silk used by the webspinner *Antipularia urichi* (Edgerly 1997). Finally, females of some insects construct nesting burrows, including the European earwig, *Forficula auricularia* (Lamb 1976). Nest building and burrowing may increase offspring fitness by concealing eggs and juveniles from predators and parasitoids, or by buffering eggs and juveniles against environmental hazards, such as extreme temperatures, flooding, or desiccation.

12.2.4 *Egg attendance*

Egg attendance occurs when parents remain with their eggs at a fixed location after oviposition, usually the oviposition site. Egg attendance is the most common form of post-oviposition care in insects, and is found in many earwigs, some hemipterans and beetles, and a small number of dipterans, thrips, lepidopterans, and psocids (Crespi 1988a; Zeh and Smith 1985; Costa 2006). This form of care may increase offspring fitness by protecting eggs against a wide range of environmental hazards, including predators, oophagic conspecifics (i.e. egg cannibalism), parasitoids and pathogens, desiccation, flooding, and hypoxia. Studies on insects with egg attendance have played an important role as sources of information on the benefits and costs of parental care, as these species are tractable to parental removal experiments. For example, in the treehopper *Publilia concava*, hatching success is twice as high when the female parent is left to attend the eggs as compared to when she is experimentally removed (Zink 2003; Figure 12.3). In many insects, egg attendance is associated with parental behaviours that are directed toward specific biological or environmental hazards. For example, parents of some insects actively defend their eggs against predators or parasites, a behaviour that is often termed egg guarding. In membracid bugs, females guard their eggs when approached by a predator, and may even approach and attack the predator. In contrast, females will typically attempt to escape the same predator when not attending a clutch of eggs (Hinton 1977). Other behaviours associated with egg attendance include active grooming of eggs to remove fungi, for example in the European earwig, *Forficula auriculata* (Lamb 1976).

Figure 12.2 Examples of different forms of parental care in insects. (a) Nest building in the wasp *Polistes dominula*. In this species, foundresses and female workers build paper nests from processed plant material. (b) Egg brooding in the golden egg bug *Phyllomorpha laciniata*. The eggs might be carried either by the male that sired them or by individuals that are unrelated to the eggs. (c) Egg brooding in the ferocious water bug (*Abedus herberti*). The male is carrying a clutch of eggs that has been glued to his back by the female. (d) Viviparity in tsetse flies of the genus *Glossina*. The female has just given birth to a fully mature larva that has a similar body mass to its mother. (e) Offspring attendance in the European earwig (*Forficula auricularia*). The female stands guard over a brood of young nymphs. (f) Food provisioning in the burying beetle *Nicrophorus vespilloides*. The female parent is provisioning pre-digested carrion to a begging larva. (Photographs by Jon Carruthers, Hannes Günther, Chris Goforth, Ray Wilson, Joël Meunier and Per Smiseth.)

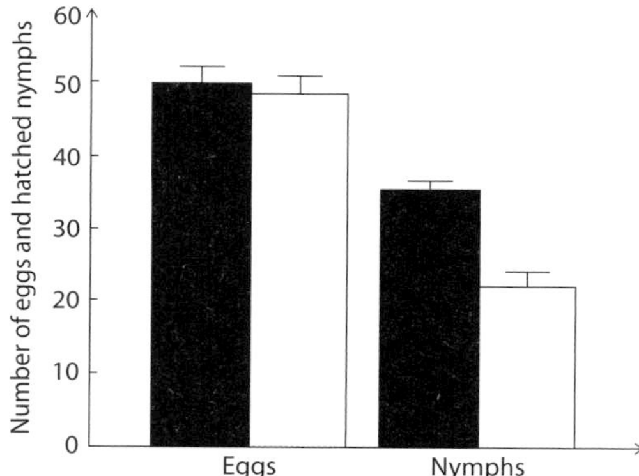

Figure 12.3 Egg attendance in the treehopper *Publilia concava*. The experimental removal of the female parent reduces the number of eggs that hatch successfully. There was no difference in the number of eggs laid between control females, which were left with the eggs until hatching (black bars), and experimental females, which were removed after laying (open bars). However, significantly fewer nymphs hatched from eggs laid by experimental females than from eggs laid by control females. (Redrawn with permission from Oxford University Press.)

12.2.5 *Egg brooding*

Egg brooding occurs in insects where parents carry their eggs after laying. Perhaps most famously, males of the giant water bug *Abedus herberti* carry eggs on their backs (Figure 12.2c; Smith 1976). Egg brooding may protect eggs against a similar range of hazards as egg attendance, and indeed this form of care might offer some advantages over egg attendance, as it allows parents to move about without abandoning the eggs. Thus, egg brooding may allow parents to move their clutches away from approaching predators, and/or track suitable conditions should they change over time. Furthermore, brooding may lower the costs of care by allowing parents to forage while caring for the eggs. A controversial case of egg brooding occurs in golden egg bugs, *Phyllomorpha laciniata*, where females deposit their eggs on the backs of conspecifics, including the male that fertilized the eggs but also conspecifics that are unrelated to the eggs (Figure 12.2b; Gomendio and Reguera 2001; Kaitala et al. 2001). This unusual example suggests that some cases of apparent egg brooding might have evolved as a form of intraspecific brood parasitism rather than as a form of parental care (Kaitala et al. 2001).

12.2.6 *Viviparity and ovoviviparity*

Viviparity and ovoviviparity are found in a wide range of insect orders, including mayflies, earwigs, psocids, thrips, cockroaches, hemipterans, neuropterans, beetles, lepidopterans, and dipterans, where females retain the fertilized eggs within their reproductive tract (Meier et al. 1999). In strictly viviparous insects, females give birth to live offspring that have hatched within the female, while in ovoviviparous insects, females oviposit eggs at an advanced stage of development that hatch during or soon after oviposition (Meier et al. 1999). Viviparity and ovoviviparity may enhance offspring fitness by providing protection against predators, parasitoids and harsh environmental conditions. All ovoviviparous insects are lecithotrophic; that is, they nourish the developing embryos solely by yolk deposited in the egg. However, viviparous species show diverse forms of embryonic provisioning, ranging from strict lecithotrophy to extreme matrotrophy, where the embryo is nourished primarily by sources other than yolk. Examples of matrotrophic insects include the Pacific beetle cockroach, *Diploptera punctata*, which nourishes broods of about 10 embryos with a milk-like secretion produced by the female's uterine lining (Stay and Coop 1974), and tsetse flies of the genus *Glossina*, where females nourish a single embryo *in utero* until giving birth to a larva that may weigh more than the female (Figure 12.2d; Glasgow 1961). Tsetse flies represent an extreme form of matrotrophy where the larva obtains all its nutrients from the female, such that its adult body size is determined completely by the amount of nutrients it obtained from the female during the gestation period.

12.2.7 *Offspring attendance*

Offspring attendance is found in a number of species from several insect orders, including orthopterans, earwigs, hemipterans, neuropterans and beetles, and occurs when parents remain with their offspring after hatching, either at a fixed location or by escorting the offspring as they move around. Offspring attendance may increase offspring fitness in much the same way as egg attendance (Section 12.2.4). Parental removal experiments with the lace bug *Gargaphia solani* show that, when there are no predators present, the vast majority of offspring survive to maturity regardless of whether the female parent is attending the offspring or not. In contrast, the presence of a guarding female improves offspring survivorship sevenfold when predators are present (Tallamy and Denno 1981). Offspring attendance is often associated with specific parental behaviours directed towards particular environmental hazards, such as predators and pathogens. For example, in the burying beetle *Nicrophorus vespilloides*, parents produce antimicrobial anal secretions that limit the growth of microbial competitors on the carrion that is used for breeding (Rozen et al. 2008; Cotter and Kilner 2010). A recent study identified the antimicrobial agent of these secretions as an insect lysozyme, which through its bactericidal effects enhances the survival of the larvae (Arce et al. 2012).

12.2.8 *Offspring brooding*

Offspring brooding is a rare form of parental care found in some cockroaches where parents carry the offspring on their bodies after hatching or birth (Bell et al. 2007). For example, females of the aquatic cockroach *Phlebonotus pallens* carry the nymphs in a brood chamber formed under the wing covers (Bell et al. 2007). Offspring brooding may increase offspring fitness in much the same way as egg brooding (Section 12.2.5).

12.2.9 *Food provisioning*

Food provisioning is an advanced form of parental care that occurs in a small number of earwigs, cockroaches, hemipterans, beetles and hymenopterans, where parents (or kin) provide the offspring with a source of food after hatching or birth. This form of care may be based on resources that are obtained directly from the environment, or specialized sources of food that are prepared by the parents. Provisioning of food directly from the environment may take the form of mass provisioning, where parents provision their offspring with food that is obtained before hatching, or progressive provisioning, where parents repeatedly feed their offspring after hatching or birth (Field 2005). One example of mass provisioning is the dung beetle *Onthophagus taurus*, where parents remove portions of dung, pack it into a ball, and bury it in an underground brood chamber that is sealed off after oviposition (Hunt and House 2011; see Figure 7.1b). In this species, the dung ball represents the entire food source for the developing larva. There is evidence that parental food provisioning enhances offspring fitness, as an offspring's adult body size (an important determinant of reproductive success; Hunt and Simmons 2001, 2002a) is largely determined by the amount of dung provided by the parents (Hunt and Simmons 1997). Another striking example of mass provisioning is the sphecid wasp *Ampulex compressa*, which is a parasitoid of the cockroach *Periplaneta americana* (Keasar et al. 2006). Once the wasp has located a host, it applies two consecutive stings directly to the host's nervous system, which serve to immobilize the prey. The wasp then cuts the host's antennae, and leads the host by one of the antennae to a burrow suitable for oviposition before laying a single egg on the cuticle (Keasar et al. 2006).

Examples of insects with progressive provisioning include the European earwig, where females regularly leave the nesting burrow after offspring have hatched to search for food for the nymphs (Lamb 1976; Section 12.5.1). Burying beetles of the genus *Nicrophorus* represents an intermediate form of provisioning where the parents raise their offspring on a vertebrate carcass that has been obtained prior to hatching, but where the parents repeatedly feed their larvae regurgitated carrion throughout offspring development (Figure 12.2f; Section 12.5.4). Provisioning of specialized food sources produced by the female parent includes the production of trophic eggs, as in the burrower bug *Parastrachia japonensis* (Hironaka et al. 2005). Perhaps the most extreme form of food provisioning is matriphagy, where the hatched offspring consume their mother after birth, which is reported from a small number of insects such as in the hump earwig *Anechura harmandi* (Suzuki et al. 2005).

12.2.10 *Care after nutritional independence*

Care for offspring after they have reached nutritional independence is a very unusual form of parental care among insects (Clutton-Brock 1991). For example, in the burying beetle *Nicrophorus vespilloides*, larvae become nutritionally independent at the age of 72 h, but female parents remain with the larvae and defend them from conspecific intruders and predators for a further 48 h (Smiseth et al. 2003). In the dung beetle *Copris lunaris* the female parent will remain with her brood balls in the nesting chamber, repairing any damage to the brood ball walls and uprighting them should they move, until her adult offspring emerge (Klemperer 1982).

12.3 Which sex cares?

Vertebrates show striking diversity with respect to whether or not males and females are involved in parental care; uniparental female care predominates in mammals, biparental care predominates in birds, and uniparental male care is common in fishes (Maynard Smith 1977; Clutton-Brock 1991; Kokko and Jennions 2012). Among the insects, uniparental female care is the predominant pattern of care, and both uniparental male care and biparental care are very rare (Tallamy 2000, 2001; Table 12.1). Uniparental male care has evolved independently on seven occasions: once among belostomatid water bugs, twice among coreid bugs, once among reduviid bugs, and three times among phlaeothripid thrips (Tallamy 2001; Tallamy et al. 2004). Biparental care is limited to termites and a small number of cockroaches (Bell et al. 2007), one species of thrip, and a small number of beetles, including ambrosia beetles, burying beetles, dung beetles, and passalid beetles (Choe and Crespi 1997; Tallamy 2001). Here we shall examine the conditions that are thought to favour the evolution of different patterns of parental care in insects. We discuss the principal benefits and costs of parental care and the reasons why they may differ for the two sexes, before briefly considering other factors that may predispose males or females to provide care for eggs and offspring.

The principal benefit of parental care is the enhanced fitness of the parent's offspring. Assuming that, ancestrally, males and females had a similar ability to provide care, males and females should be expected to have very similar benefits of parental care. However, two conditions might induce sex differences in the benefits of care: (i) males will have lower benefits than females if their paternity certainty is reduced as a consequence of sperm competition (Trivers 1972; Alonzo 2009; Alonzo and Klug 2012; Chapter 10), and (ii) males will have higher benefits than females if care enables males to attract additional mates (Tallamy 2000, 2001). Thus, sex differences in the benefits of care derive from attributes of the mating systems emerging because of sperm competition and effects of care on male mating opportunities. The principal costs of parental care are (i) increased mortality due to predation and parasites, (ii) increased energy expenditure, and (iii) loss of additional breeding opportunities while caring for current offspring (Clutton-Brock 1991). The first two costs are likely to be similar for males and females, whereas the third might differ between them due to anisogamy. As we saw in Chapter 3, males produce the smaller and thus cheaper gametes (sperm), and consequently their reproductive rate tends to be limited by access to females (Bateman 1948; Trivers 1972). In contrast, females produce the larger and thus more expensive gametes (eggs), and their reproductive rate therefore tends to be limited by access to resources that females can use to produce additional eggs (Bateman 1948; Trivers 1972).

Evolutionary game theory provides the appropriate theoretical tool for exploring the conditions favouring different patterns of parental care, since the best strategy with respect to whether or not to provide care for members of one sex depends on the strategy adopted by the other sex (Maynard Smith 1977). Based on assumptions about the benefits and costs of parental care to males and females, such models predict that uniparental female care is favoured when (i) uniparental care substantially improves offspring fitness compared to no care, and (ii) females have higher net benefits of care than males (Maynard Smith 1977). Removal experiments using insects have provided good evidence for the first prediction that uniparental female care evolves when parental care by one parent enhances offspring fitness. For example, in the treehopper *Publilia concava*, hatching success of eggs is doubled when the female is left with the eggs as compared to when the female is removed (Zink 2003; Figure 12.3). In contrast, we have limited empirical evidence concerning the second prediction, reflecting that this prediction is extremely difficult to test, as it requires estimates of the benefits and costs of care for both the caring and the non-caring sex. Nevertheless, two non-mutually exclusive hypotheses might explain why female insects often will have higher benefits of care than males. First, the common occurrence of sperm competition in insects (Simmons 2001; Chapter 10) would select against male involvement in care because it reduces the benefits of care to males. Second, male insects might have higher opportunity costs of care than females because parental care is expected to have a stronger effect on a male's ability to mate with additional females than it will have on a female's ability to acquire additional resources (Maynard Smith 1977; Chapter 3).

Uniparental male care is expected to evolve when (i) uniparental care substantially improves offspring fitness compared to no care, and (ii) males have higher net benefits of care than females (Maynard Smith 1977). Again, removal experiments using insects have provided support for the first prediction. For example, in the assassin bug *Rhinocoris tristis*, egg survival is considerably higher when males are left with the eggs than when males are removed (Gilbert et al. 2010; Section 12.5.3). There is also evidence in support of the second prediction. For example, work on *R. tristis* shows that caring males have enhanced future mating opportunities due to their increased attractiveness to females (Thomas and Manica 2005; Section 12.5.3). Furthermore, there is evidence that the evolution of male parental care is associated with paternity guards, which effectively ensure that the benefits to care are as high for males as for females (Trivers 1972). For example, in the ferocious water bug, *Abedus herberti*, a male achieves a very high paternity of the eggs oviposited on his back by demanding repeated copulations throughout oviposition (Smith 1979; Section 12.5.2).

Theory predicts that biparental care evolves when (i) care by two parents improves offspring fitness beyond uniparental care (Maynard Smith 1977), and (ii) each parent responds to a reduction in its partner's care by incomplete compensation (Houston et al. 2005). The latter prediction derives from models of sexual conflict over parental care, which predict that complete compensation by one parent would render biparental care evolutionarily unstable as the other parent would be free to reduce its contributions to care without suffering fitness costs from doing so (Houston et al. 2005). Although biparental care is extremely rare among insects, empirical work on some insects has played an important role in testing theoretical predictions concerning the evolution of biparental care. Two such species are the dung beetle *Onthophagus taurus* and the burying beetle *Nicrophorus vespilloides* (see Section 12.5.4). In *O. taurus*, there is support for the first prediction as biparental care increases the weight of dung balls, and thereby the offspring's fitness, compared to uniparental female care (Hunt and Simmons 1998, 2000). Hunt and Simmons (2002b) found support for the second prediction as females breeding on their own increased their effort by an average of 15% compared to females assisted by a male. As predicted, this compensatory response was incomplete, resulting in lighter brood masses when females breed on their own (Hunt and Simmons 2002b).

Finally, it is important to note that the evolution of sex differences in parental care may depend on factors that predispose males and females towards providing care, such as (i) the extent to which male and female parents associate with eggs at the time of oviposition (Tallamy 2001), and (ii) the existence of pre-existing traits in one sex that selection can modify into forms of parental care (Tallamy 1984). For example, in species with internal fertilization—including all sexually reproducing insects—females necessarily associate with the eggs at the time of oviposition, whereas males have the opportunity to desert earlier. Thus, uniparental female care might predominate among insects due to the asymmetry in the timing of desertion caused by internal fertilization. In contrast, male parental care might be expected to evolve in insects where males associate with the female until the time of oviposition, for example by guarding the female after copulation in order to avoid sperm competition (Tallamy 2001; Chapter 10). Interestingly, this example also illustrates how selection might modify a pre-existing trait in one sex (i.e. post-copulatory mate guarding) into a form of parental care (i.e. egg attendance) simply by extending the period of male attendance beyond oviposition.

12.4 Parental care and mating systems

The evolution of sex differences in the involvement in parental care is generally accepted to be closely associated with certain attributes of the mating system, such as intense sexual selection and sperm competition (Trivers 1972; Thornhill and Alcock 1983; Clutton-Brock 1991). Understanding the ecological and evolutionary drivers of this association is a key challenge in the study of both parental care and mating systems. This topic is important to the understanding of the evolution of parental care because, as discussed in the previous section, sex differences in the benefits and costs of parental care derive from attributes of the mating system, such as sperm competition and the potential for additional mating opportunities. Furthermore, attributes of the mating systems may induce secondary sex differences in the costs of care. For example, sexual selection may increase the mortality cost of care to males if they evolve sexual ornaments that make males more easily detected by predators. Thus, to understand the evolution of different patterns of parental care, we need to understand how aspects of the mating system drive sex differences in the benefits and costs of parental care (discussed in Section 12.3).

The evolution of parental care is also fundamental to our understanding of mating systems because theoretical arguments suggest that sex differences in the involvement in parental care are important determinants of the intensity of sexual selection (Trivers 1972; Thornhill and Alcock 1983; Clutton-Brock 1991; Chapter 3). Trivers (1972) was the first to point out that there is a close association between variation in the extent to which males and females are involved in parental care and the intensity of sexual selection (Trivers 1972). For example, in the majority of animal species, including insects such as the European earwig (*Forficula auricularia*), females alone provide post-oviposition parental care for the offspring whereas males are subject to strong sexual selection as indicated by the presence of male secondary sexual characters (Section 12.5.1). Such species are considered as having conventional sex roles; that is, females are caring while males are competing (Kokko and Jennions 2008). Meanwhile, a small number of animal species, including insects such as the ferocious water bug (*Abedus herberti*), have so-called reversed sex roles in which males have the primary responsibility for providing care for the offspring, and females compete over access to males (Section 12.5.2).

Trivers (1972) suggested that the evolution of sex roles was driven by a two-stage process whereby initial sex differences in gamete investment (anisogamy) caused sex differences in post-oviposition parental care, which in turn led to stronger sexual selection in the non-caring sex. Trivers argued that anisogamy favours uniparental female care because females are under selection to protect their greater initial investment in the zygote. He then argued that the greater female involvement in parental care leads to intense sexual selection on males because females become a limited resource over which males compete. Trivers's (1972) theoretical argument was instrumental in stimulating the growing interest in the evolution of parental care from the 1970s onwards (Clutton-Brock 1991). However, Kokko and Jennions (2008) have questioned Trivers's argument in a theoretical model, suggesting that the causal relationship between patterns of parental care and sexual selection may be the reverse. Kokko and Jennions (2008) argue that anisogamy leads to strong pre- and post-mating sexual selection on males, both of which decrease the net benefits of care to males, thereby causing selection for reduced male and increased female parental care (Kokko and Jennions 2008). Comparative studies on shorebirds and fishes support Kokko and Jennions (2008) by providing evidence that changes in mating systems drive the evolution of male and female involvement in care rather than the other way round (Reynolds and Székely 1997; Gonzalez-Voyer et al. 2008).

This section has provided a simplified account of the complex co-evolution of parental care and mating systems by suggesting that the evolution of sex differences in parental care and sex roles are caused by pre-existing variation in mating systems. As detailed above, there is evidence that mating systems exert a strong selective pressure on male and female involvement in parental care. However, there is also good reason to expect patterns of parental care to exert strong selective pressures on mating systems: decisions made by individual males and females about whether or not to provide care determine the number of mates of each sex that becomes available in the population, which in turn determines the pay-offs to alternative male and female mating strategies (Kokko and Jennions 2008). Thus, parental care and the mating system are likely to co-evolve; for example, an initial change in the level of sperm competition may lead to an evolutionary change in the amount of care provided by males, which in turn may lead to further evolutionary changes in the mating system (Alonzo 2009). Alternatively, the same ecological conditions that drive the evolution of parental care, such as food sources and predation, may drive changes in the mating system either directly or indirectly through changes in population densities or life histories. Currently, little is known about how ecological conditions contribute to changes in parental care and mating systems, and there is a need for more empirical and theoretical work to understand these complex processes. Such empirical work might be based on comparative analyses similar to those conducted on shorebirds and fishes (Reynolds and Székely 1997; Gonzalez-Voyer et al. 2008), or experiments targeting specific questions in suitable model species (Section 12.5). For example, experiments could be used to investigate how variation in ecological factors, such as resource distribution or predation, mediates changes in both parental care and attributes of the mating system.

12.5 Model systems for the study of parental care

12.5.1 *European earwig*

The European earwig, *Forficula auricularia*, is an example of an insect with conventional sex roles with caring females and competing males. This species is a nocturnal omnivore that is found throughout Eurasia, North America, and Australia. Mating normally takes place in late summer and early autumn (Lamb 1976). Male earwigs are subject to intense sexual selection as indicated by males having longer and more curved forceps than females (Radesäter and Halldórsdóttir 1993), and males using their forceps as weapons during competition over females (Radesäter and Halldórsdóttir 1993; Forslund 2003). There are two alternative male morphs in this species: macrolabic males are larger and have longer forceps relative to their size, whereas brachylabic males are smaller and have relatively shorter forceps (Chapter 7). Macrolabic males are more successful in competition over females (Radesäter and Halldórsdóttir 1993). Although males display their forceps to females during courtship, evidence is mixed as to whether females exhibit mating preferences based on male forceps length (Tomkins and Simmons 1998; Forslund 2000). There is also evidence that males risk paternity losses due to sperm competition as females routinely mate with multiple males (Lamb 1976).

In the European earwig, as in the vast majority of insects, parental care for eggs and offspring is provided by the female only (Figure 12.2e). After mating, the female ceases feeding and starts constructing a deeper nesting burrow (Lamb 1976). Once the burrow is completed, she lays a clutch of about 40–80 eggs, which are closely groomed and defended during the winter until hatching. Some females produce a single clutch only, whereas others proceed to produce a second one (Lamb 1976). After hatching, the female leaves the burrow on regular foraging trips in search of food that is brought back for the nymphs (Lamb 1976). Removal experiments show that parental care enhances the survival of eggs and nymphs (Lamb 1976; Kölliker 2007; Figure 12.4). However, parental care incurs costs to females, as caring females delay their next reproductive attempt compared to non-caring

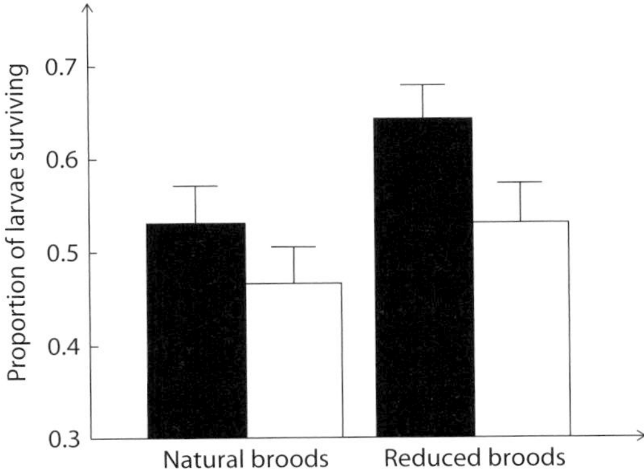

Figure 12.4 Offspring attendance in the European earwig (*Forficula auricularia*). The experimental removal of the female reduces the number of offspring that survive regardless of the number of offspring in the brood. The number of surviving nymphs in control broods, where the female was left with the brood after hatching (black bars), were significantly greater than the number of surviving nymphs in experimental broods, where the female was removed at hatching (open bars) in both natural brood and experimentally reduced broods. (Redrawn with permission from Springer.)

females (Kölliker 2007). The European earwig is among a small number of insects where parents provision food directly to individual nymphs (Staerkle and Kölliker 2008). Female food provisioning is associated with parent–offspring communication, whereby females adjust their allocation of food in response to chemical cues from nymphs that provide information about the nymph's nutritional status (Mas et al. 2009). Females prefer feeding well-fed nymphs, presumably because these nymphs have a higher reproductive value than poorly fed nymphs.

Uniparental female care is thought to be an ancient trait that is shared by virtually all earwigs (Lamb 1976; Vancassel 1984). However, it is unclear why earwigs evolved costly post-oviposition parental care, since their ancestors had evolved egg and oviposition traits that effectively neutralized the threats posed by predators and parasites (Zeh et al. 1989). The evolution of post-oviposition parental care in earwigs is associated with the loss of the ovipositor (Zeh et al. 1989), but it is unclear whether the loss of the ovipositor promoted the evolution of post-oviposition parental care or vice versa. It is also unclear why only females are involved in parental care. Presumably, the net benefits of care are lower to males than to females, either because sperm competition leads to reduced paternity or because male care interferes with male attempts to acquire additional females. However, it may also reflect that males desert females prior to oviposition. Interestingly, females become increasingly aggressive towards males after mating, driving them away from the burrow prior to oviposition (Lamb 1976). Female aggression towards males would interfere with the evolution of male assistance in care because male care could evolve only following the evolution of female tolerance of the male's presence.

12.5.2 *The ferocious water bug*

The ferocious water bug, *Abedus herberti*, is a rare example of an insect with reversed sex roles where only males provide care for the eggs (Tallamy 2000, 2001). This species is an aquatic sit-and-wait predator found in streams in south-western USA and Mexico. Males attract females through distinct pumping and push-up displays that create wave motion through the water (Smith 1997). The approach by a gravid female is followed by repeated attempts by the female to oviposit on the male's back (Smith 1997). Prior to oviposition, the female secretes a sticky mucus that covers the male's back. The male interrupts oviposition to mate after approximately every second egg laid by the female (Smith 1979). These frequent copulations serve to reduce the risk of sperm competition as the male achieves almost complete paternity certainty for the eggs oviposited on his back (Smith 1979). Oviposition lasts for about 12–48 h until the whole surface of the male's hemelytra is covered (Smith 1974).

Male ferocious water bugs have two distinct brooding behaviours (Smith 1974, 1976): (i) surface brooding, which occurs when the male sits or floats at the air–water interface such that the eggs are partially submerged in water and partially exposed to air (Figure 12.2c), and (ii) sub-surface brooding, which occurs under water and involves active push-up and pumping behaviours similar to the displays used in courtship (Smith 1974). Male brooding is essential to offspring survival and eggs perish if left unattended in open air or permanently submerged under water (Smith 1974). However, brooding also incurs significant costs to the male (Smith 1974). First, the male's energy intake is temporarily suspended because he ceases to feed while brooding (presumably to ensure that eggs and nymphs are not accidentally eaten). Second, brooding may incur energetic costs due to the significant amount of time the male spends performing active push-ups. Surface brooding may also expose the male to a substantial risk of predation, for example from fishing spiders (Smith 1997). Finally, because the space on the male's back is limited, brooding severely restricts his opportunities for siring additional eggs laid by other females, as demonstrated in the closely related *A. indentatus* (Kraus 1989). Thus, in contrast to some other insects with male parental care (Section 12.5.3), there is no evidence that male care in ferocious water bugs is maintained through sexual selection.

In the family Belastomatidae, parental care is thought to have evolved as a consequence of selection for large body size, allowing adults to feed on nutrient-rich vertebrate prey (Smith 1979). The evolution of a large adult body size may necessitate the production of large eggs that cannot obtain a sufficient oxygen supply through diffusion alone, and parental care may provide a means for ensuring an adequate oxygen supply to large eggs developing in an aquatic environment (Smith 1979). It is unclear why only males are involved in parental care, but it is thought that the net benefits of care are lower to females than to males (Smith 1997). Potentially, females may suffer greater opportunity costs of care through the loss of foraging opportunities, but this suggestion is difficult to investigate empirically as females never provide care. Ferocious water bugs illustrate two conditions that are thought to favour the evolution of uniparental male care: (i) males have a high paternity certainty for the eggs under their care (Trivers 1972), and (ii) males associate closely with the female until the time of oviposition, presumably as a means to protect their paternity (Tallamy 2001). Finally, the striking similarity between the push-up displays used by males to attract females and the sub-surface brooding behaviour used to oxygenate eggs suggests that parental behaviours may have been co-opted as sexual displays or vice versa. For example, when mating and parental behaviours are based on similar mechanisms, as in this case, females might benefit by monitoring mating displays if they provide information about a male's parental ability. Thus, one avenue for further work on this species is to establish whether push-up displays indicate male parental ability and whether females discriminate between males based on these displays.

12.5.3 *Assassin bugs*

Assassin bugs of the genus *Rhinocoris* are exceptional among insects in that this genus includes species with no post-oviposition parental care, such as *R. venustus* and *R. kumarii*;

species with uniparental female care, such as *R. carmelita*; and species with uniparental male care, such as *R. tristis*, *R. albopunctatus*, and *R. albopilosus* (Gilbert et al. 2010). Unlike the examples discussed above, assassin bugs do not conform to either conventional or reversed sex roles. For example, although *R. tristis* is a species with uniparental male parental care, field observations on marked individuals show that there is intense sexual competition among males but not among females (Thomas and Manica 2005). Males often attack other males engaged in post-copulation riding with a female, and there is no evidence of aggression between females ovipositing simultaneously in the same egg mass (Thomas and Manica 2005). There is also evidence for male mate choice, as copulations occur more frequently when females are relatively heavy and presumably more fecund (Thomas and Manica 2005).

After mating, assassin bugs provide parental care by attending and guarding eggs against parasitic wasps and insect predators. In *R. tristis*, males care for eggs for an average of 26 days, whereas in *R. carmelita*, females care for eggs for an average of 14 days (Gilbert et al. 2010). This difference reflects the fact that *R. tristis* males care for eggs laid asynchronously by several females, whereas *R. carmelita* females care for their own eggs that are laid over a shorter period of time. In both species, parental care enhances egg survival when parasitic wasps and insect predators pose a threat to the eggs (Gilbert et al. 2010; Figure 12.5a). There is evidence that parental care is costly as it reduces male survival in *R. tristis* (Gilbert et al. 2010; Figure 12.5b). In *R. tristis*, caring males continue to receive eggs from additional females, and males may simultaneously guard eggs that are laid by up to seven different

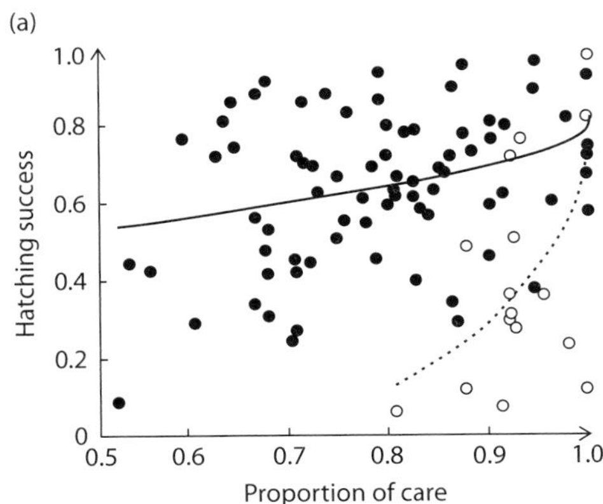

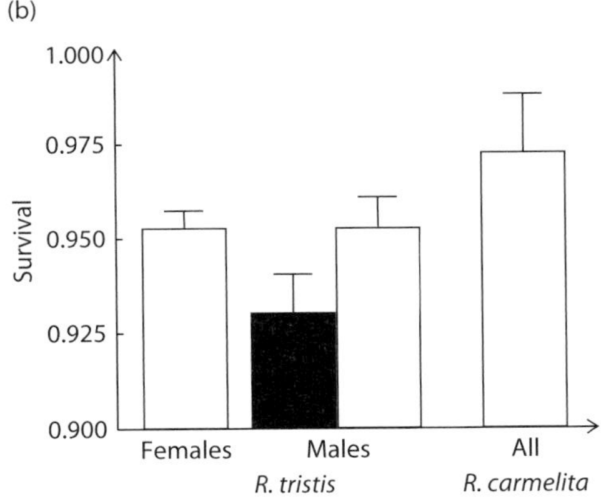

Figure 12.5 Benefits and costs of parental care in *Rhinocoris* assassin bugs. (a) Observational data from the field show that parental attendance of eggs is beneficial to the offspring, as there is a positive association between the proportion of time that parents spend attending the eggs (days spent attending the eggs relative to the duration of the incubation period) and the hatching success of eggs in *R. tristis* (black symbols) and *R. carmelita* (open symbols). The solid and dashed lines represent the estimated best-fit curves for the association in *R. tristis* and *R. carmelita*, respectively. (b) Data based on mark–release–recapture in the wild show that male parental care is costly, as the survival of caring *R. tristis* males (black bar) is lower than that of non-caring males and females, and of *R. carmelita* males and females (open bars) (Redrawn with permission from John Wiley & Sons.)

females (Gilbert et al. 2010). Parental care may even enhance the future mating opportunities of males due to their increased attractiveness to females. Indeed, some non-caring males are observed attacking caring males in a bid to take over their egg masses (Thomas and Manica 2005). These considerations suggest that there might be a sex difference in the net benefits of care between these species due to parental care and acquisition of mates not being mutually exclusive activities (Manica and Johnstone 2004).

The ultimate cause driving the divergent patterns of care in *R. tristis* and *R. carmelita* is thought to be a difference in population densities. Because *R. tristis* occurs in high densities on legumes of the genus *Stylosanthes*, males have a high probability of encountering additional females while caring for eggs at a fixed location. In contrast, *R. carmelita* occurs at much lower densities on various locally abundant plants, and males would therefore have a much lower probability of encountering additional females had they been providing care. Finally, although *R. tristis* is described as having uniparental male care, females can provide care should the male desert the eggs (Beal and Tallamy 2006). The presence of facultative care by the non-caring sex is intriguing as it greatly facilitates direct evolutionary transitions between uniparental male and uniparental female care without the need for other intermediate states such as biparental care or loss of parental care. One avenue for further research is to take advantage of the presence of facultative female parental care in *R. tristis* to obtain estimates of the relative costs and benefits of parental care for both males and females for a single species.

12.5.4 *The burying beetle Nicrophorus vespilloides*

This is a rare example of an insect with egalitarian sex roles where both sexes normally partake in parental care (there are occasional cases of uniparental female and uniparental male care). This species is an obligate scavenger of small vertebrate carcasses, and is found across Eurasia and North America. Mating typically occurs when males and females locate the carcass of a small vertebrate, but can also take place in the absence of a carcass when males release pheromones to attract females (Eggert 1992). The predominant mating system is one of social monogamy, which occurs when a dominant male and dominant female establish ownership of a carcass by driving away smaller competitors (Pukowski 1933; Otronen 1988). However, other mating systems are also found, including polygyny when several females breed communally (Eggert and Müller 1992). On larger carcasses, there may be sexual conflict as males benefit from attracting a second breeding female, whereas the first female suffers a fitness cost from sharing the carcass with another female (Eggert and Sakaluk 1995). In the closely related *N. defodiens*, females often enforce monogamy by attacking the male should he attempt to attract a second female (Eggert and Sakaluk 1995). The dominant male risks a reduction in paternity because females often store sperm from previous matings and mate with satellite males that remain near the carcass (Müller and Eggert 1989; House et al. 2008, 2009). To ensure a high certainty of paternity, the dominant male mates extremely frequently with the female and guards her against other males (Müller and Eggert 1989).

Nicrophorus vespilloides is known for its highly elaborate parental care, which involves suites of pre- and post-hatching forms of care performed by one or both parents, and behavioural interactions between parents and larvae (Figure 12.2f). Once a suitable carcass has been located, both parents co-operate to bury the carcass and remove hair or feathers, while the female lays eggs in the soil around the carcass (Pukowski 1933; Eggert and Müller 1997). Both before and after hatching, parents defend the carcass against conspecifics and other scavengers (Eggert and Müller 1997), and deposit antimicrobial secretions that help

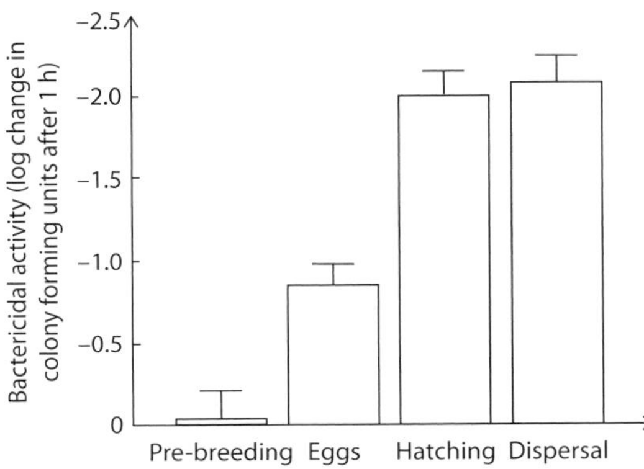

Figure 12.6 Parental antimicrobial defence in the burying beetle *Nicrophorus vespilloides*. In this species, parents deposit anal secretions on to the vertebrate carcass used for breeding. The bactericidal activity of these secretions, measured as a change in colony-forming units of *Bacillus subtilis* after 1 h of incubation, increases sharply from baseline levels before breeding (Pre-breeding) to high levels during the breeding cycle (Eggs, Hatching and Dispersal) (Redrawn with permission from John Wiley & Sons.)

preserve the carcass as a food source for the larvae (Rozen et al. 2008; Cotter and Kilner 2010; Arce et al. 2012; Figure 12.6). After hatching, parents open the carcass to allow the larvae to self-feed from it, and they provision the larvae with pre-digested carrion (Eggert et al. 1998; Smiseth et al. 2003). Parental care in *Nicrophorus vespilloides* is associated with parent–offspring communication where larvae signal their nutritional needs using tactile begging displays and parents adjust their allocation of care in response to these begging displays (Smiseth and Moore 2002, 2004a). Parental removal experiments show that parental care enhances the offspring's survival and growth (Eggert et al. 1998; Smiseth et al. 2003). There is little evidence that parental care is costly in terms of increasing the mortality or energy expenditure of parents, which might reflect the fact that parents breed underground in relative safety from predators and supplement their energy budget by feeding from the carcass (Eggert and Müller 1997). However, caring parents may suffer injuries while defending their brood against attacks from conspecific intruders, and there is evidence that parental care reduces the parents' future reproductive success (Jenkins et al. 2000; Ward et al. 2009.

Nicrophorus vespilloides is among a small group of insects in which both parents provide care (Eggert and Müller 1997). Theory predicts that biparental care should evolve when two parents are more efficient at enhancing offspring fitness than a single parent (Maynard Smith 1977; Section 12.4). Removal experiments provide little evidence for this prediction, as male removal has no detectable effect on offspring survival and growth in *N. vespilloides* (Jenkins et al. 2000; Smiseth et al. 2005). However, these studies were conducted under laboratory conditions, which exclude the risk of take-overs by conspecific competitors. Indeed, a study on the closely related *N. pustulatus* found that pairs were better than single parents at defending their brood from take-overs by conspecifics (Trumbo 2007). Finally, for biparental care to be evolutionarily stable, theory predicts that each parent should respond to a reduction in its partner's care by incomplete compensation (Houston et al. 2005; Section 12.4). In *N. vespilloides*, there is some evidence for incomplete compensation as there is a negative correlation between the amount of care provided by males and females (Smiseth and Moore 2004b). However, a removal experiment found that males compensated completely following the removal of the female whereas females did not compensate at all in response to male removal (Smiseth et al. 2005; Figure 12.7). Thus, females could potentially reduce the amount of care they provide, since males appear willing to fully compensate for the loss of female care, although for some reason they do not take advantage of this opportunity. A recent study on *N. quadripunctatus* suggests that, although males compensate fully in response to female removal, they do not compensate

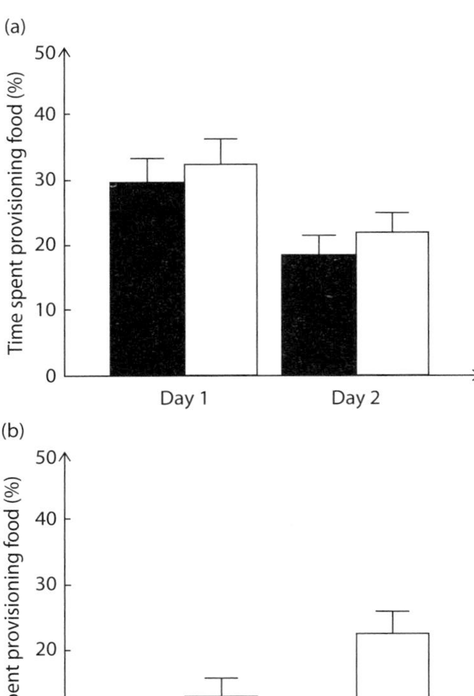

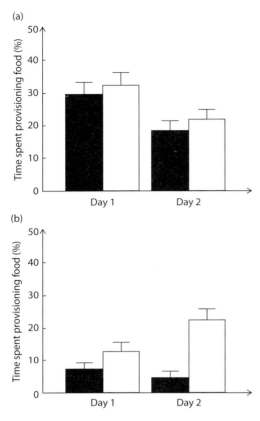

Figure 12.7 Parental food provisioning in the burying beetle *Nicrophorus vespilloides*. In this species, both parents normally provide care jointly, although females (a) spend more time on this form of care than males (b). In this experiment, the time spent provisioning food for the larvae by females and males was recorded on days 1 and 2 after hatching. In the control treatment, both parents were present and engaged in care on both days (black bars). In the experimental treatments, both parents were present and engaged in care on day 1. The partner was then removed experimentally after the observation on day 1, leaving the focal parent to provide care on its own on day 2 (open bars). There was no evidence that females compensated for the removal of the male (a), whereas males compensated for the removal of the female by increasing their level of care to match that of females (b) (Redrawn with permission from Elsevier).

in response to a reduction in the amount of care provided by the female (Suzuki and Nagano 2009). One avenue for further work on burying beetles is to obtain estimates of the future mating opportunities of males and females, the efficiency of biparental care under conditions similar to those faced in the field, and to improve our understanding of the way in which the two parents negotiate how much care each of them should provide.

12.6 Future directions

Insects provide tractable models for studies aimed at enhancing our understanding of the evolutionary and ecological causes driving the co-evolution of parental care and mating systems. There is a need to improve our understanding of how trade-offs shape the association between parental and mating behaviours. Such trade-offs are expected when individuals must allocate essential resources between mutually exclusive activities, such as parental care and mating behaviour (Low 1978). To this end, future studies should test for negative phenotypic correlations between parental and mating behaviours. Studies on life-history trade-offs often fail to find evidence for negative correlations, presumably due to confounding effects caused by individual variation in resource acquisition (van Noordwijk and de Jong 1986). Thus, it would be essential to conduct such studies on species where variation in resource acquisition can be manipulated. Another way to test for trade-offs would be to obtain estimates of genetic correlations between parental care and mating behaviours based on artificial selection experiments. Such information would be particularly valuable, as it would provide insights into how the evolution of these traits might be shaped by their genetic architecture. Insects represent uniquely tractable study systems for such experiments because they have relatively short life-cycles that allow them to be bred over subsequent generations under laboratory conditions.

There is also a need to investigate the theoretically important issue of whether sex differences in parental care determine the strength of sexual selection, as proposed originally by Trivers (1972), or whether sexual selection and sperm competition drive the evolution of sex differences in parental care as argued by Kokko and Jennions (2008). This issue could be addressed by conducting a formal comparative analysis. Comparative work on insects has been hampered by a lack of detailed phylogenies for many insect groups, and missing information on parental care and mating systems for many species. However, improvements in our knowledge of insect phylogenies and greater knowledge about parental care and mating systems in insects should help overcome this issue.

Our knowledge of the role that specific ecological factors, such as food distribution and predation risk, play in driving the co-evolution between parental care and mating systems is generally poor. One scenario is that a change in some ecological factor may cause an initial change in one trait, such as some component of the mating system, which then triggers subsequent changes in another trait, such as parental care. For example, in species breeding on rich but ephemeral resources, sexual selection might be driven by resource competition such that its strength is relatively similar in males and females (Section 12.5.4), which in turn favours the evolution of more egalitarian sex roles and biparental care (Kokko and Jennions 2008). Alternatively, resource competition might independently balance the strength of sexual selection in males and females, and select for biparental care due to the greater need to defend limited resources against conspecifics. This issue could be addressed either by conducting comparative analyses exploring the association between ecological conditions, parental care, and mating systems, or by designing targeted experiments on suitable model species. Ideal species should have highly variable mating systems and patterns of parental care, and should breed under variable ecological conditions. For example, it would be possible to conduct experiments on burying beetles of the genus *Nicrophorus* (Section 12.5.4) to explore how variation in resource availability influences both parental care and the mating system.

12.7 Conclusions

Although most insects produce eggs that are effectively protected against a wide range of environmental hazards (Hinton 1981; Zeh et al. 1989), some insects have evolved various forms of post-oviposition parental care, ranging from simple attendance of eggs to the provisioning of offspring with food after they have hatched. In most insects with parental care, such as earwigs, the female alone cares for eggs and offspring. However, in a small number of insects, males provide care for eggs and offspring, either alone as in the ferocious water bug, or jointly with the female as in burying beetles. The evolution of male and female involvement in parental care is driven by sex differences in the net benefits of care, which often derive from attributes of the mating system, such as sperm competition and the potential of additional mating opportunities. Nevertheless, many empirical and theoretical challenges remain in order to advance our understanding of how parental care and the mating system may co-evolve.

CHAPTER 13

Parasites and pathogens in sexual selection

Marlene Zuk and Nina Wedell

13.1 Introduction

So nat'ralists observe, a flea
Hath smaller fleas that on him prey;
And these have smaller fleas to bite 'em.
And so proceeds *Ad infinitum*.

As this passage from Jonathan Swift's *On Poetry: a Rhapsody* (1733) attests, the potential not only for insects to be parasites, but also for parasites to play a role in insect life, has been acknowledged for centuries. Parasites and pathogens, here defined as organisms that spend most or all of their lives on or within another living being, are ubiquitous. No species is exempt from infection, whether from viruses that enter host cells and reprogram cell machinery or parasitoid larvae that consume large amounts of tissue and come to occupy nearly the entire body cavity of their host. It is sometimes suggested that individual variation in response to pathogens, or the effect of parasites generally, is less important in insects and other invertebrates than it is in vertebrates because of the short lifespans and high reproductive rates of the latter. Insects, however, suffer from virtually all of the types of parasites that attack other animals, and they have evolved surprisingly complex immune systems to deal with the invaders (Adamo 2012).

Given the importance of pathogens in insect life, it is no surprise that infection and the response to it have influenced sexual selection and reproductive behaviour. At the time of Thornhill and Alcock (1983), research into the role of parasites and pathogens in the evolution of mating systems was in its infancy, and much has been discovered over the last 30 years. Here we consider the ways in which parasites influence insect mating systems, both directly and through selection pressure on the immune system. In addition, we discuss the role of the ubiquitous parasitic endosymbiotic bacteria *Wolbachia* and other selfish genetic elements such as sex ratio distorters in the sexual differentiation and reproduction of insect hosts.

The importance of parasites in regulating animal populations, including insects, was recognized by early ecologists such as Charles Elton, but the difficulty of observing or performing experiments on parasitism in the field hindered progress in this area for much of the twentieth century. In 1980, Peter Price published an influential book, *Evolutionary*

The Evolution of Insect Mating Systems. Edited by David M. Shuker and Leigh W. Simmons.
© The Royal Entomological Society 2014. Published 2014 by Oxford University Press.

Biology of Parasites, which helped rekindle interest in parasites both as worthwhile subjects of study in their own right and as drivers of evolution in their hosts. Parasites were also recognized as potentially playing an important part in such fundamental questions as the evolution of sexual reproduction, in which the co-evolutionary arms race between host and pathogen could provide a rationale for the continued generation of novel genotypes and ameliorate the cost of producing males. This idea is the foundation of the Red Queen hypothesis, which has found at least some support in both theoretical and empirical investigations (Jokela et al. 2009; Mostowy and Engelstaedter 2012).

13.2 Parasites and the lek paradox

In addition to their role in the evolution of sex itself (Chapter 1), parasites were also proposed as a solution to an enduring problem in sexual selection: the lek paradox (Chapter 2). Imagine a swarm of male mayflies, or a group of singing crickets or katydids. Females visit the displaying group to choose a mate, but then depart to lay their eggs, receiving no benefits from the male other than the sperm to fertilize their eggs. Male mating success may be highly skewed under such circumstances. If females choose males of high genetic quality, evolutionary theory would predict that after successive generations of such selection, heritable variation in fitness would disappear, eliminating the benefit of the preference for particular males. Yet such lek mating systems, and other less extreme forms of mate choice for certain types of individuals in species with no paternal investment, persist, leading to the paradox (Chapter 8).

A way out of this difficulty was proposed by Hamilton and Zuk (1982), who suggested that if females used male signals of parasite resistance as the basis of their mate preference, genetic variation in fitness would be continually replenished. The co-evolution between hosts and pathogens means that new genes that counter ever-evolving host defences or parasite modes of infection provide continued sources of genetic variation. The Hamilton–Zuk hypothesis makes both intra- and inter-specific predictions about the connection between parasite levels and sexual signals. Within a species, one expects preferred mates to have flashier sexual signals and fewer parasites. When comparing multiple species, one would expect selection pressure on sexual ornaments to be strongest in species infected with more parasites, as females would benefit more from choosing resistant mates in such species. Thus, one would predict that the species with the flashiest sexual signals would be infected by the greatest number of parasites.

Hamilton and Zuk (1982) noted that only certain types of parasites would be suitable for testing their idea; parasites that killed hosts too quickly would leave only resistant individuals in the population, thus ensuring that every individual encountered would be resistant and removing the need for choosiness. Relatively benign parasites, however, would not cause enough harm for choosy females to benefit by selecting mates with genes for parasite resistance.

Although most of the tests of the Hamilton–Zuk hypothesis have used vertebrates, particularly birds, as study systems, several researchers have examined the role of parasites on sexual selection and mate choice in insects. Indeed, in a number of insects, males harbouring fewer parasites have been shown to have higher mating success. For example, male *Drosophila testacea*, a fungus-eating species, were found to copulate more frequently if they were not infected with the parasitic nematode *Howardula aoronymphi* (Jaenike 1988). Male broad-horned flour beetles, *Gnatocerus cornutus*, that were experimentally infected as larvae with the rat tapeworm *Hymenolepis diminuta* developed into adults with smaller horns than uninfected males; the horns are important in male competition in this species (Demuth et al. 2012). When infected with protozoan gregarine parasites, male field crickets (*Gryllus veletis* and *G. pennsylvanicus*) replaced spermatophores more slowly than did uninfected individuals, thus limiting their mating rate and presumably subsequent mating success (Zuk 1987).

It is, of course, possible for parasites to influence host mating success without the operation of the hypothesis proposed by Hamilton and Zuk (1982). Infected individuals may simply be too depleted by the parasite to engage in sexual competition, regardless of the female's ability to use a condition-dependent secondary sexual ornament as a means of rejecting or accepting a mate. Alternatively, females may avoid infected males to decrease the possibility of direct transmission of the parasite. Multiple reasons for examining parasites in the context of mate choice and sexual selection therefore exist.

In a few cases, parasites have actually increased the mating success of infected individuals, at least in the short term. Milkweed beetles (*Labidomera clivicollis*, Chrysomelidae) are parasitized by a sub-elytral mite (*Chrysomelobia labidomerae*) which decreases host fitness when the beetles are nutritionally stressed and which is transmitted via copulation (Abbot and Dill 2001). Under laboratory conditions, parasitized male beetles came into contact with unparasitized males more often and for longer than controls, and were more likely to displace rival males in copula, thus enhancing their mating success (Abbot and Dill 2001). Abbot and Dill (2001) suggest that this increased activity is an expression of reproductive compensation, whereby organisms with reduced life expectancy invest more in short-term mating effort.

Similarly, McLachlan (1999) found that in swarms of the chironomid midge, *Paratrichocladius rufiventris*, infested by the hydracharinid mite *Unionicola ypsilophora*, mated males were more likely to be parasitized than were unmated males. This counterintuitive finding may be explained by parasite manipulation of host behaviour, with the parasite benefiting by increasing the number of individuals the host contacts and thereby enhancing its own transmission (Moore 2002). Alternatively, the midge is also subject to parasitism by a mermithid nematode that castrates its host, but the mite and worm rarely co-occur in the same individual host. McLachlan (2006) speculates that by mating with mite-bearing individuals, the midges can avoid sterile individuals.

13.3 Sexual selection and parasitoids

Parasitoids, as opposed to parasites *sensu stricto*, may play a unique role in shaping the mating system of their host. Tachinid flies from the tribe Ormiini use orthopteran mating calls to locate their hosts, and have evolved a unique hearing apparatus convergent on that of the Orthoptera (Cade 1981, Lehmann 2003). Once a gravid female fly has found a cricket or katydid, she deposits mobile larvae on and around it. The larvae then burrow inside the host, gradually consuming living host tissue for approximately 7–10 days, after which the mature larva(e) burst from the host, killing it in the process. This exploitation of the sexual signal places male crickets in a dilemma, and means that the signal itself is subject to conflicting selection pressures: sexual selection should act to increase its conspicuousness and hence attract more mates, while natural selection should act to diminish it and decrease the likelihood of detection by the parasitoid.

The presence of ormiine parasitoids has influenced the evolution of sexual behaviour in several species of crickets and katydids. Some *Gryllus integer* (now *G. texensis*) males call less per night than others; this difference is heritable, and Cade (1981) suggested that selection by the flies may have favoured either the caller or satellite strategies among male crickets that we discussed in Chapter 7. Similarly, katydids (bush crickets) subject to attack by ormiines may have shifted their calls to lower frequencies in an evolutionary attempt to escape the parasitoid (Lehmann 2003).

In one cricket species, *Teleogryllus oceanicus*, flies co-occur with the crickets only in a small part of the host's range, the Hawaiian Islands, where both parasitoid and host have been introduced. Crickets in Hawaii exhibit several differences from unparasitized populations. They are more likely to call only during darkness, when the flies are less active (Zuk et al. 1993); the call structure is altered in a manner that reflects the prevalence of the parasitoid (Rotenberry et al. 1996); and they are more sensitive to disturbance (Lewkiewicz and Zuk 2004).

More recently, a novel obligately silent morph of male *T. oceanicus* has appeared and spread on at least two of the islands. These silent crickets have modified wings that prevent calling, and now comprise approximately 90% of the males on Kauai and 50% on Oahu (Zuk et al. 2006; M. Zuk, unpublished data), an evolutionary shift that took place in fewer than twenty generations (Figure 13.1). Although the new morph is protected from the parasitoid, it faces challenges in mate attraction and courtship. Ongoing work suggests that the silent morph is more prone to act as a satellite, and that females maturing under relatively song-free conditions are more apt to accept males that do not sing (Tinghitella et al. 2009; Bailey et al. 2010). The parasitoids thus may have spurred the evolution of a novel trait in the crickets that became established because of pre-existing behavioural plasticity.

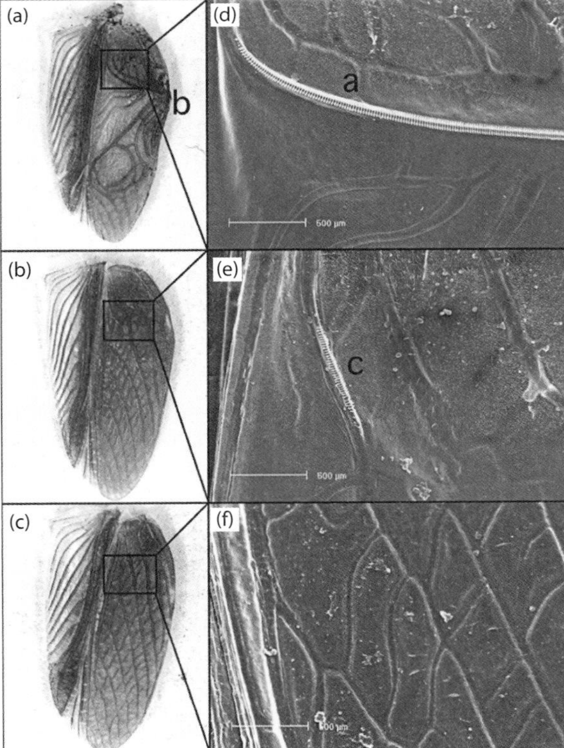

Figure 13.1 Wings from (a) male, (b) flatwing, and (c) female *Teleogryllus oceanicus*. The new morph has a misplaced file shown at (point c in panel e) and lacks the modified veins present in the normal-winged male that form the resonating structures used to amplify the sound. (From Zuk et al. 2006.)

13.4 Ecological immunology

Examining the effects of a single parasite on sexual signals can provide useful information, but it may not reveal how an insect may respond to the vast number of parasites and pathogens it may encounter over the course of its lifetime. Furthermore, how can we evaluate the importance of a particular pathogen over any other? Are females responding mainly to males capable of resisting that one species, or do they prefer males that can respond to parasites in general and avoid or resist infection by a large number of pathogens? Answering the latter question when examining parasites alone is difficult, as it is often impossible, or at least unfeasible, to measure the total parasite burden of an animal under natural conditions.

Many scientists interested in how parasites influence mate choice and mating systems have therefore begun to examine immune response rather than pathogens per se. The immune system has evolved to allow organisms to distinguish self from non-self, and in so doing respond to the parasites that an animal encounters. Assessing immunity should therefore provide an acceptable means of evaluating an animal's parasite resistance (Møller et al. 1999; Adamo 2012). Scientists interested in the effects of parasites on sexual selection were brought into the field of ecological immunology by this new direction (Sheldon and Verhulst 1996; Møller et al. 1999; Siva-Jothy et al. 2005; Lawniczak et al. 2007). Although the invertebrate immune system is simpler than that of vertebrates, insects and other arthropods still possess specialized organs and cells that defend their bearer against invasion by a variety of pathogens (Gillespie et al. 1997).

Ecological immunology is based on the principle that immune response is a necessary but costly trait in terms of energy and resources, and that these costs make immune defence a candidate for trade-offs with other life-history traits, such as reproductive effort (Zuk 1990; Sheldon and Verhulst 1996; Zuk and Stoehr 2002; French et al. 2009). The issue of costs of immunity is an important one from the perspective of sexual signalling because it helps explain how signal honesty is maintained (Kotiaho 2001). In order for signals to remain honest indicators of quality, both the signal and the immune response must be costly. These costs impose a handicap that can maintain signal reliability, as low-quality males will not be able to pay the costs associated with high signal quality (Zahavi 1975; Nur and Hasson 1984; Getty 2002). Among insects, melanin is commonly used both in the immune response, where it plays an essential role in encapsulation of foreign objects, and in ornamental traits such as wing spots in odonates and Lepidoptera (Jiang et al. 2011). The ability to produce melanin might therefore be limiting for insects, suggesting a link between ornamentation and the immune system.

Whereas the idea that immune response may be linked to sexual signals started as a development of the Hamilton–Zuk hypothesis, the hypothesis itself is not quite the same as many of the ideas put forward in ecological immunology. Both seek to put disease resistance, or immune function, in a more evolutionary context by pointing out that parasites and pathogens represent an important selection pressure to their hosts, and that defences to fight off these pathogens should be under strong selection (Zuk 1994; Sheldon and Verhulst 1996; Zuk and Stoehr 2002). However, the Hamilton–Zuk hypothesis emphasizes genetic-based parasite resistance as the driver of the process. Only a resistant male would have sufficiently low parasite burdens to produce a high-quality sexual signal. Further, the Hamilton–Zuk hypothesis relies on co-evolutionary cycles between hosts and their pathogens to generate a constant source of genetic variability in quality (Hamilton and Zuk 1982).

In contrast, ecological immunologists tend to focus more on the ways in which energy allocation to immunity versus signalling can influence sexual signal quality. This explanation does not include parasites directly, although parasites are certainly important in that they impose a cost on individuals who allocate too many resources away from the immune system. Rather than focusing on parasites, and how these organisms interact with their hosts directly, ecological immunology relies more on the costs of immunity as the all-important explanation behind signal honesty. This causes ecological immunologists to emphasize the importance of an animal's immune response, rather than resistance to a given pathogen (Westneat and Birkhead 1998). Whereas there are fundamental differences between the Hamilton–Zuk hypothesis and ecological immunology's emphasis on energetic trade-offs, both predict that sexual signalling and immunity are linked.

Much of the work examining sexual selection and immunology has attempted to map variation in immune response on to variation in sexual signals and/or mating success. Among insects, dragonflies and damselflies have been favourite subjects for such studies. For example, male American rubyspot damselflies (*Hetaerina americana*) exhibit red wing spots. Bigger males with larger spots are more successful both in territorial and mating competition than are non-territorial males (Contreras-Garduño et al. 2008). When the damselflies were given an immune challenge, the area of the spot decreased (Contreras-Garduño et al. 2008), suggesting that males must balance the cost of the ornament against immune defence (Figure 13.2). Along similar lines, more pigmented males of the damselfly *Calopteryx splendens* are preferred by females, and these males are less likely to be infected with protozoan parasites—a relationship thought to be mediated by melanin (Siva-Jothy 1999, 2000). In other *Hetaerina* species, however, this relationship between immunity and spot size differed, with some showing a positive correlation and others a negative one

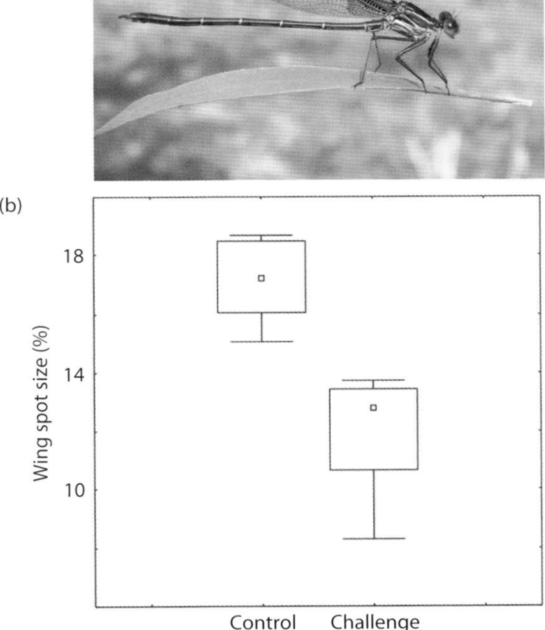

Figure 13.2 Interaction between immunity and sexual coloration. (a) American rubyspot damselfly (*Hetaerina americana*). (b) The size and number of wing spots in this species are reduced when individuals are subjected to an experimental immune challenge, reflecting the balance between investment in immunity and sexual signalling. (Courtesy of Jorge Contreras-Garduño.)

(Gonzalez-Santoyo et al. 2010), underscoring the complexity of the interaction between immune response and morphology.

Part of this complexity may stem from the need to marshal different amounts of resources at different life stages. In the cricket *Gryllus campestris*, immune challenge during the juvenile stage had long-lasting effects on the melanization and size of the harp, a sexually selected wing character, but not on other adult traits (Jacot et al. 2005). Mated female house crickets (*Acheta domesticus*) exhibited a reduced immune response compared with virgin females, suggesting that investment in reproduction trades off against investment in immunity during the egg maturation stage (Bascunan-Garcia et al. 2010). A similar trade-off between immune response and a sexually selected trait was observed in another species of field cricket, *Teleogryllus oceanicus*; males with higher sperm viability had lower levels of lysozyme activity, a component of the defence against bacteria (Simmons and Roberts 2005). In contrast, in African horned dung beetles (*Euoniticellus intermedius*), the activity of phenoloxidase, a melanin precursor, was higher in males with longer horns (Pomfret and Knell 2006b). Pomfret and Knell (2006b) suggest that high-quality males are simply able to maintain both a robust immune system and a large ornament.

It is important to appreciate that trade-offs between ornament expression and immunity may not always be apparent because of the so-called 'car–house' paradox (van Noordwijk and de Jong 1986). The paradox is named from the superficially puzzling observation that while one might expect individuals who invest heavily in one area (a house) to have fewer resources left for investment in another (a car), in real life those who live in expensive houses also drive fancy cars. Of course, the paradox arises because of unequal starting points; individuals with a larger pool of resources to begin with will have more to invest in any one area, leading to an apparent absence of a trade-off across the population (Roff and Fairbairn 2007).

A particularly intriguing example of the interaction between immune defence and sexual selection can be seen in bedbugs and related insects. These groups exhibit 'traumatic insemination', whereby the male pierces the female's exoskeleton with his intromittent organ and deposits sperm directly into the body cavity (Chapter 10). Frequent wounding from multiple mating reduces female survival (Reinhardt and Siva-Jothy 2007), and females appear to have evolved counter-defences that protect their interests (Reinhardt and Siva-Jothy 2007). In addition to providing scope for sexual conflict in general, this mating mechanism makes the female vulnerable to infection with bacteria and other microorganisms that may enter once the exoskeleton is breached. Intromission is most likely at a groove in the female's abdomen, and females possess a structure, the spermalege, containing immunologically active cells (Reinhardt et al. 2003). If the spermalege is bypassed, female bedbugs are more likely to develop infections, suggesting that the organ helps ameliorate some of the costs of traumatic insemination (Figure 13.3).

In addition, an optimal immune response may not be one that is maximally reactive, and choosing mates with the strongest immunity—or the fewest parasites—might not always be beneficial. If immune responses are very costly, then allocating too many resources into defence against parasites, as opposed to other avenues such as food acquisition or signal development, can be harmful, and intermediate levels of immunity might be best (Viney et al. 2005). Blue tits infected with moderate levels of blood parasites survived longer than those with extremely high or low parasite burdens (Stjernman et al. 2008), and Råberg and Stjernman (2003) found that selection on the antibody response to a novel antigen was stabilizing, rather than directional, supporting the idea that an intermediate immune response was optimal. Stjernman et al. (2008) note that failing to recognize that less-infected individuals are not always more fit could lead to the potentially erroneous

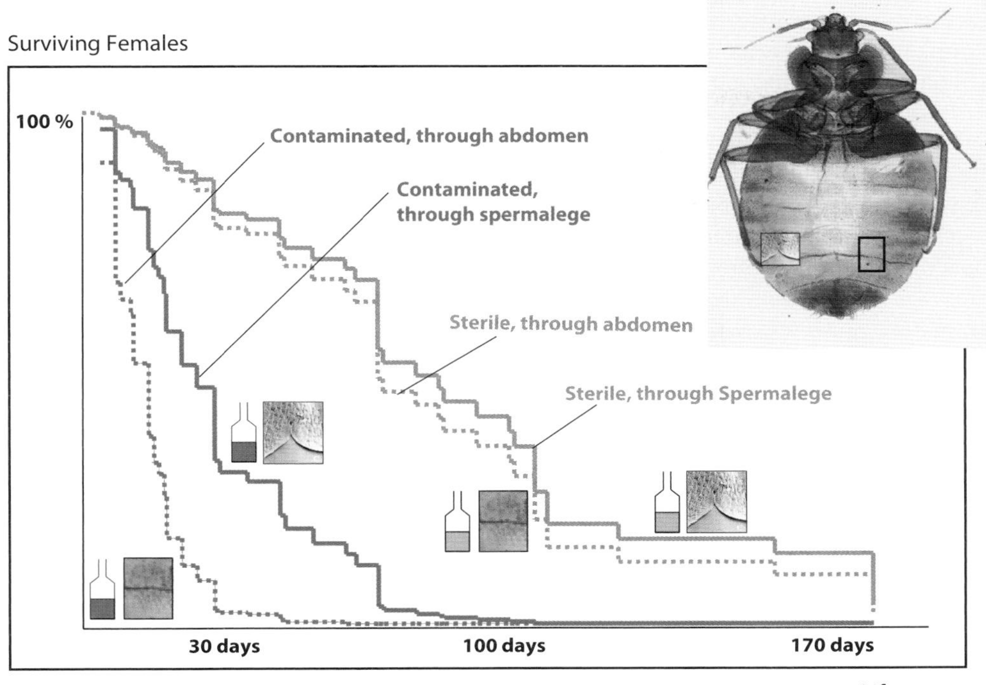

Figure 13.3 Traumatic insemination in bedbugs (*Cimex lectularius*). Survival curves for females wounded with a sterile or bacteria-contaminated needle in either their spermalege (solid lines) or their abdomen (broken lines). Introduction of bacteria via the spermalege is associated with greater survival than when bacteria are introduced directly into the abdomen. (From Reinhardt et al. 2003.)

conclusion that the parasite is harmless to its host. Most work on this notion has centred on vertebrates, with their complex multi-level immune systems. Whether such fitness benefits of intermediate immunity also apply to insects is worth further study.

The level of immune response may also be context dependent, as illustrated in the field cricket *Gryllus campestris*; males that were challenged with bacteria and then placed either in predator-free plots or in plots with predatory shrews exhibited lower survival than controls only in the plots with predators (Otti et al. 2013). This concept, that the cost of immune activation depends on the other competing demands on an organism's resources, suggests that studies of hosts under a variety of circumstances in both the field and laboratory are warranted. Insects are particularly well suited for such laboratory–field comparisons because of their small size and ease of manipulation.

Finally, hosts may deal with parasites, not by mounting an immune response in an effort to rid themselves of infection, but by tolerating the infection and thus minimizing the damage done by the pathogen (Råberg et al. 2007; Boots 2008). Such a strategy permits an animal to stay in better condition than its similarly infected conspecifics. This idea of parasite tolerance rather than resistance has long been recognized by plant biologists, but its applications to animals have often been neglected (Råberg et al. 2009). If high-quality mates are not less likely to become infected, but simply better able to tolerate the presence of parasites, preferred mates might actually show a lower immune response relative to the rest of the population because they are better able to tolerate parasites but less able to mount an effective response against them.

Such tolerance may be difficult to study, particularly in the field, as it requires demonstration of differential effects of parasite intensity on different host genotypes. If one genotype shows less decline in condition as parasite intensity increases (i.e. if there is a significant interaction between genotype and parasite intensity), then one may conclude that that genotype is more tolerant than the others (Råberg et al. 2009). Such experiments may be easier to perform in insects than in the vertebrates that are the more usual subjects for ecological immunology studies.

13.5 Genomic parasites: the impact of selfish genetic elements

Just as parasites and pathogens may have a profound impact on mating decisions, mating rate, and resources invested in reproduction, we have recently realized that the genome of an organism also may contain parasitic 'selfish' genetic elements that can directly influence host reproduction and mating system. Selfish genetic elements (SGEs) are genes present in the genome or cell of an organism that are passed on to subsequent generations at a higher frequency than the rest of the genome. This 'selfish' nature of SGEs ensures that they will accumulate in frequency within the genome (Burt and Trivers 2006). SGEs are ubiquitous in living organisms and can make up a large proportion of the genome. They often give rise to conflict with the rest of the genome by favouring their own transmission to the detriment of other genes.

There are many forms of SGEs, often classified according to their mode of action: transposons, homing endonucleases, segregation distorters, and post-segregation distorters. Transposons are transposable elements that move around by inserting themselves throughout the genome, so-called 'jumping genes'. These are DNA sequences encoding enzymes that catalyse their own movement, and they can jump between chromosomes, frequently causing mutations and changes in the amount of DNA in the genome. Homing endonucleases also use enzymes that cut DNA at particular sites in the genome and insert copies of themselves. They multiply by moving from one chromosome location to another. Segregation distorters, on the other hand, act to ensure that they are present in the majority of offspring after meiosis. Meiotic drive chromosomes are well-studied segregation distorters that alter the meiotic process so that the driving chromosome is present in >50% of the gametes. If the meiotic drive chromosome is sex-linked, this gives rise to sex ratio distortion. Post-segregation distorters, as the name implies, reduce the frequency of non-carriers after fertilization and commencement of development. One widespread group of post-segregation distorters comprises various types of cellular endosymbionts. The best studied of these is the maternally transmitted bacterium *Wolbachia pipientis*, which is frequent in arthropods. SGEs manipulate the genome in a variety of ways to ensure their own enhanced transmission success to subsequent generations.

13.5.1 *SGEs, immunity and host protection*

Interestingly, endosymbionts such as *Wolbachia* do not appear to activate the host immune response. This may be the result of *Wolbachia* going undetected within the cell, as there is no evidence that it actively suppresses the host immune system (Bourtzis et al. 2000). It is possible that the intracellular localization of such endosymbionts make them invisible to the host immune detection system. There is also evidence of endosymbiont-mediated host protection (Haine 2008). Endosymbiotic bacteria such as *Wolbachia* and *Spiroplasma* can protect their insect hosts against RNA viruses and nematode infections. For example, *Spiroplasma* protects *Drosophila neotestacea* against the sterilizing effects of a parasitic nematode (Jaenike et al. 2010). Similarly, *Wolbachia* was shown to inhibit the replication of several RNA viruses, including dengue virus, and several other vector-borne pathogens (e.g. *Plasmodium* and filarial nematodes) in mosquitoes and flies resulting in increased survival, but this protection is by no means universal in insects (Graham et al. 2012). It is unclear whether endosymbiont-mediated defences that increase host survival are traded off against investment in reproduction. It is also unclear whether carrying protective endosymbionts can affect mate choice in a similar fashion to the predictions of the Hamilton–Zuk hypothesis, with individuals choosing partners carrying resistance factors (e.g. endosymbiont-mediated 'resistance'), although it is worth noting that this should only affect male mate choice as endosymbionts are largely maternally inherited. The few studies specifically examining the influence of *Wolbachia* on mate preferences have found little support, with either no influence, as is the case in *Drosophila melanogaster* and *D. simulans* (Champion de Crespigny and Wedell 2007), or assortative mating based on the specific *Wolbachia* strains involved in the *D. paulistorum* species complex (Miller et al. 2010) (see also Section 13.6.4).

Similarly, many other SGEs do not induce an immune response, despite causing marked reductions in fitness. This may in part be explained by the existence of effective silencing mechanisms such as DNA methylation, heterochromatic formation and the action of piRNAs—small RNA-based inherited genome immune systems present in eukaryotes. These small RNAs protect the genome against the deleterious impact of SGEs such as transposons by silencing their action (Bao and Yan 2012). Silencing piRNA pathways are conserved across invertebrates and mammals and are active within the gonads of animals including insects, highlighting their importance as general protectors against SGEs. In *D. melanogaster*, for example, the action of a single transposon is sufficient to cause severe sterility, but they are kept in check by piRNAs. Transposons are very common—most wild fly populations carry a range of transposons, the majority of which are silenced, and the same is true for most other eukaryotes. In general, SGEs generate strong counter-selection favouring suppressor genes and pathways such as RNA interference mechanisms that suppress the selfish action of SGEs. However, such suppressors may often themselves have a detrimental effect on fertility, and it is conceivable that the presence and absence of specific SGEs and their suppressors will influence both pre- and post-copulatory sexual selection. Thus far, this possibility has not been thoroughly examined.

13.5.2 *SGEs compromise male fertility and competitive fertilization success*

SGEs frequently target the sex cells, as this is an effective way to increase their own transmission to subsequent generations at the expense of the rest of the genes within the genome. This manipulation can be achieved by modifying sperm during development, as is the case for many endosymbionts in flies and moths (Snook et al. 2000; Lewis et al. 2011). Other SGEs such as meiotic drivers and some transposons destroy sperm that do not carry the selfish gene during spermatogenesis (Policansky and Ellison 1970), resulting in a markedly reduced number of sperm in SGE carriers compared with non-carrying males (Price and Wedell 2008). In addition, the method of sperm killing may compromise the performance of the surviving sperm that carry the SGE (Price et al. 2008a). One consequence of such manipulations is that SGE-carrying males frequently suffer reduced fertility compared with non-carrying males (Price and Wedell 2008; Price et al. 2008a). This effect ranges from a slight decrease in fertility to a reduction of 50% or more in some species (Price and Wedell 2008). The effect size is important since a reduction of a minimum of 50% is required for fertility differences alone to stabilize an SGE's transmission advantage (Haig and Bergtrom 1995). In *D. melanogaster*, for example, the reduction in fertility of a segregation distorter is greater than expected due to the elimination of non-carrying sperm (Hartl et al. 1967), and several studies have shown that segregation distortion in males is associated with a greater fertility reduction than that expected based purely on sperm numbers (Fry and Wilkinson 2004; Price et al. 2008a).

Further evidence that segregation distorters are associated with compromised sperm production comes from studies showing that reduced hybrid male fertility is associated with the genomic region responsible for drive. For example, *Overdrive* (*Ovd*) in *D. pseudoobscura* is an X-linked locus that causes both male sterility and sex ratio distortion in crosses between USA and Bogota fly populations (Phadnis and Orr 2009). Sterility may arise for three reasons (Meiklejohn and Tao 2009). First, it may be solely a consequence of the mismatch between the drivers and the lack of corresponding suppressors. Second, it may arise due to co-evolution between the sex ratio distorters and meiotic sex chromosome activation, generating divergence between species. Third, sterility could arise due to compensatory evolution of other genes involved in gametogenesis, resulting in divergence and reproductive incompatibilities between species and populations. The sterility and reproductive incompatibility caused by such cryptic segregation distorters has even been suggested to contribute to speciation (Werren 2011). It is likely that segregation distorters are far more widespread than previously realized, as biased sex ratios will result in rapid evolution of suppressors of such distorters, resulting in 'cryptic' segregation distortion. However, it remains to be demonstrated whether reproductive incompatibilities directly affect mate choice and/or mating patterns, although they may favour female multiple mating (Zeh and Zeh 1997; Wedell 2013).

13.5.3 *SGEs and polyandry*

Compromised sperm production and sperm killing by SGEs have important implications for sexual selection via sperm competition (Chapter 10), because they can reduce the sperm competitive success of SGE-carrying male insects (Champion de Crespigny and Wedell 2006; Price and Wedell 2008; Price et al. 2008a). The reduced fertility and sperm competitive ability of SGE-carrying males provides an important link between the presence of SGEs and female mating frequency. This is predicted to promote polyandry as a female strategy to bias paternity against low-fertility SGE-carrying males and avoid passing on the SGEs to offspring (Zeh and Zeh 1996; Wedell 2013; Chapter 9). The advantage to polyandrous females comes as a reduced risk of mating only with an SGE-carrying male, although this bet-hedging strategy can only work under limited circumstances (Yasui 1998). More importantly, the reduced sperm competitive ability of carrier males means that polyandrous females are able to swamp the sperm of SGE-carrying males with that of normal males and thereby bias paternity against the former (Zeh and Zeh 1996). In this sense, polyandry represents a mechanism of cryptic female choice (Chapter 11).

Several lines of evidence support the hypothesis that SGEs may promote polyandry. Some fly populations of *D. pseudoobscura* harbour an X-linked meiotic drive element ('sex ratio', SR) resulting in female-only broods due to the elimination of all Y-chromosome sperm during spermatogenesis. The loss of sperm by SR-carrying males makes them poor sperm competitors (Price et al. 2008a). Female multiple mating should therefore be an effective strategy to undermine the transmission of SR by biasing paternity towards non-carrying males. Females that are at risk of SR are therefore predicted to evolve shorter intervals between copulations. Using experimental evolution, Price et al. (2008b) found support for this prediction (Figure 13.4). The rapid evolution of higher female remating in SR populations may in part be due to sperm and/or ejaculate depletion, which will be exacerbated in SR males under a female-biased sex ratio.

Similarly, in two species of stalk-eyed flies, females mated more frequently in populations that had a higher frequency of meiotic drive. In addition, female mating rate increased with the frequency of drive across stalk-eyed fly species (Wilkinson et al. 2003). In at least one stalk-eyed fly species, males carrying the drive allele suffer reduced sperm competitive success (Fry and Wilkinson 2004), further corroborating the link between SGEs and poor sperm competitive ability. In the fire ant *Solenopsis invicta*, one SGE called *Gp-9* determines the social organization of the ant colony by directly influencing the level of polyandry in queens (Lawson et al. 2012). Workers carrying the *Gp-9*^{b} allele only tolerate sexuals that also carry the *Gp-9*^{b} allele and kill *Gp-9*^{BB} queens. However, males carrying the *Gp-9*^{b} allele have low sperm counts, and therefore suffer reduced paternity in mixed broods compared with *Gp-9*^{B} males. In addition, the low sperm numbers of *Gp-9*^{b} males result in increased female remating rates, likely as a direct consequence of sperm limitation. Reduced paternity in sperm competition together with an inability to suppress female receptivity sharply reduced the reproductive success of *Gp-9*^{b} males (Lawson et al. 2012). Polyandry in this ant species is therefore largely determined by the reduced sperm production of *Gp-9*^{b} males, favouring female multiple mating. In general, polyandry is likely to be favoured whenever SGE-carrying males suffer reduced fertility and sperm competitive ability.

However, the link between SGEs and polyandry is not always clear. For example, in the fly *D. simulans* carrying the Riverside strain of *Wolbachia* (*wRi*), crosses between uninfected females and infected males result in >90% hatching failure due to cytoplasmic incompatibility (CI). This enhances the transmission of *Wolbachia* as all other crosses

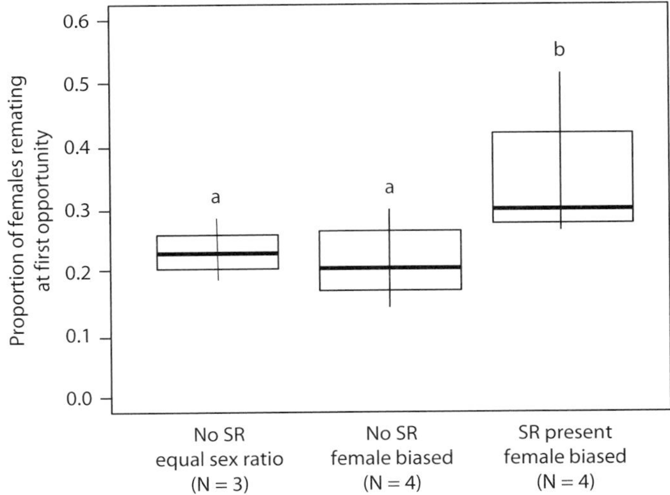

Figure 13.4 The proportion of experimental evolution population *Drosophila pseudoobscura* females remating at their first opportunity, for each selection regime ($n = 11$ lines), when offered standardized stock males, showing median, interquartile range, and range (distributions with different letters were significantly different). (Redrawn from Price et al. 2008b).

involving infected individuals are fertile and result in *Wolbachia*-infected offspring. However, infected males are poor sperm competitors relative to uninfected males (Champion de Crespigny and Wedell 2006). Uninfected females mating with at least one uninfected male are able to recuperate the fitness loss caused by CI due to the poor sperm performance of infected males, and have the same reproductive output as infected females (Champion de Crespigny et al. 2008). Uninfected females should therefore favour polyandry, and indeed, they remate sooner than infected females. However, despite the poor sperm competitive ability of infected males, this benefit cannot promote polyandry by itself, because of the relatively low probability of uninfected females mating with both an infected and uninfected male, which becomes less likely as the frequency of *Wolbachia* increases in the population and most individuals are infected (Champion de Crespigny et al. 2008). The probability of mating with an incompatible mate in conjunction with the potential costs associated with polyandry imposes severe restrictions on the conditions in which CI-inducing *Wolbachia* generate a selective advantage for polyandry. Although reproductive incompatibility avoidance has been suggested as a putative benefit of polyandry, models based on laboratory data do not support the hypothesis that CI caused by *Wolbachia* promotes the evolution of polyandry (Champion de Crespigny et al. 2008). However, these theoretical models did show that the disadvantage in sperm competition may inhibit or prevent the invasion of *Wolbachia*, which may prevent the successful establishment of CI-inducing *Wolbachia* in polyandrous populations (Champion de Crespigny et al. 2008).

SGEs may also influence male mating rates. For example, *Wolbachia*-infected *D. melanogaster* and *D. simulans* display a higher mating rate than uninfected males (Champion de Crespigny et al. 2006). Although this could be due to *Wolbachia* directly enhancing ejaculate production, such an effect seems unlikely, since *Wolbachia*-infected *D. simulans* males have reduced sperm production (Snook et al. 2000). Alternatively, this may be a strategy adopted by males to increase their reproductive success. When infected males mate with uninfected females few or no offspring are produced due to CI: infected males are only reproductively compatible with females infected with the same strain of *Wolbachia*. However, it has been shown that high male mating rate reduces the level of CI and thus enhances reproductive compatibility with uninfected females (Karr et al. 1998). It is therefore plausible that the elevated mating frequencies observed among infected males represent a male strategy to restore reproductive compatibility. Further support for this idea comes from the observation that the difference in male mating rate is related to the severity of CI induced in crosses between infected males and uninfected females, which in *D. simulans* is >90% compared to ~30% in *D. melanogaster*, with a corresponding difference in mating rate of 50% and 16% in *D. simulans* and *D. melanogaster*, respectively (Champion de Crespigny et al. 2006). It is possible then, that CI-inducing *Wolbachia* increase male mating frequency to a level at which reproductive compatibility is restored with all females in the population, thereby increasing male reproductive success. Sustained high male mating rate is likely to result in sperm depletion, which may explain why uninfected males do not show such high mating rates, whereas the benefit to infected males by reducing the level of CI might outweigh these costs.

In addition, polyandry is likely to reduce the transmission of any SGEs and therefore directly affect their frequency in insect populations. In laboratory populations of *D. pseudoobscura*, polyandry is effective at suppressing the frequency of SR. Whereas the frequency of SR was high in populations evolving under monandry, just one remating opportunity was sufficient to suppress substantially the frequency of SR (Price et al. 2010a). This finding corroborates theoretical predictions showing that more than two sires per brood are necessary to prevent invasion and persistence of sex ratio drive (Taylor and Jaenike 2002). Laboratory studies in *D. pseudoobscura* indicate that polyandry may directly enhance survival of populations harbouring sex ratio distorters. Populations evolving under monandry had a significantly higher probability of going extinct (Price et al. 2010a), indicating that monandry is deleterious in populations harbouring sex ratio distorters, further promoting polyandry. If, in nature, monandrous populations harbouring sex ratio distorters are generally at a higher risk of extinction, this would generate an association between polyandry and the presence of SGEs. This possibility may be difficult to quantify in the wild, since documenting extinction events of entire populations is difficult. Nevertheless, differences in the degree of polyandry among populations may be important in determining the frequency of many SGEs in natural populations, highlighting the dynamic link between polyandry and SGEs (Wedell 2013).

13.5.4 *SGEs and mate preferences*

Many SGEs reduce the fitness of the individuals that carry them. An obvious prediction therefore is that individuals should avoid mating with SGE carriers. However, evidence corroborating this prediction is surprisingly rare (Price and Wedell 2008). The best example comes from sexually dimorphic stalk-eyed fly species, some populations of which harbour a sex-ratio-distorting meiotic driver. In the two well-studied species, *Teleopsis dalmanni* and *T. whitei*, there are several strains of X-linked sex ratio distorters that can reach high frequencies (Presgraves et al. 1997). This results in female-biased populations, where males on average will mate with many females. Females that avoid mating with meiotic drive-carrying males will therefore increase their fitness by producing sons and their offspring that do not inherit the driving X. Some Y-chromosomes are able to resist the driving effect and importantly these are associated with wider eye-spans in males (Wilkinson et al. 1998). By preferring males with wider eye-spans, females are able to avoid mating with meiotic-drive-carrying males (Wilkinson et al. 1998).

There is also evidence that some endosymbionts can promote mate choice (Goodacre and Martin 2012). The best evidence comes from spider mites (*Tetranychus urticae*), where uninfected females (that suffer CI when mating to an infected male) prefer to mate with uninfected males, whereas infected females show no preference (Vala et al. 2004). As mentioned above, in the fly *D. paulistorum* species complex, there is evidence of assortative mating based on the specific *Wolbachia* strain they carry (Miller et al. 2010). The mechanism driving such *Wolbachia*-assortative mate preferences is not yet clear, but cuticular hydrocarbon sex pheromones may be involved (Ringo et al. 2011). In *D. melanogaster* curing of *Wolbachia* also appears to alter mate preferences, but in unpredictable ways (Markov et al. 2009), and this is not a universal finding (Champion de Crespigny and Wedell 2007).

Despite these intriguing cases, most studies examining the potential for SGE-based mate preferences have not found any conclusive evidence for it, despite the severe fitness consequences of not discriminating against SGE-carrying individuals (e.g. Champion de Crespigy and Wedell 2007; Price et al. 2012). The lack of avoidance of SGE carriers may stem from recombination breaking up any association between an ornament, the SGE, and a potential preference allele, allowing for SGE-based mate choice to evolve (Nichols and Butlin 1989; Pomiankowski and Hurst 1999). Support for this suggestion is that the only known examples that are not associated with either behavioural or pheromonal differences of SGE-carriers do indeed show reduced recombination—for example, there is little recombination between the driving and non-driving X-chromosome in stalk-eyed flies (Johns et al. 2005).

13.5.5 *Sex ratio bias*

The biased population sex ratios caused by some SGEs can directly influence the mating system and mating behaviour of insects. Many populations of the butterfly *Hypolimnas bolina* harbour a male-killing strain of *Wolbachia* (Dyson and Hurst 2004). South Pacific islands show a large variation in the frequency of male killers that is associated with differences in the degree of female sex ratio bias, with a higher frequency of male killers associated with a more severe female-biased population sex ratio (Charlat et al. 2007). The mating system therefore differs in relation to the frequency of male killers. In high-prevalence populations, males provide smaller sperm packets than in low-prevalence populations, probably due to the greater availability of females. This in turn influences female multiple mating, with females mating more frequently in such populations due to the increased severity of sperm limitation. A correspondingly higher male mating rate will then exacerbate the cycle of male fatigue, with an associated female sperm shortage that will further favour female multiple mating. At extremely high prevalence, the levels of male killing can result in severe sex ratio bias of 100 females to one male (Dyson and Hurst 2004).

A skewed population sex ratio also has direct consequences for sexual selection (Chapter 3). Increased levels of female multiple mating will increase the risk of sperm competition to males, but this risk is likely to be lower in high-prevalence populations since the number of different males mated by individual females will be reduced. In such a situation, sperm competition is reduced, and males should discriminate against already-inseminated females. Male mate choice is expected to be more prevalent under such situations, and females might change their mating behaviour to increase the likelihood of successful insemination. In populations of the butterfly *Acrea encedon* harbouring male-killing *Wolbachia*, females show sex-role reversal and engage in lekking behaviour, presumably to advertise their presence to the rare males and hence increase mating success (Jiggins et al. 2000). In populations without male killers, males and females adopt more conventional sex roles and only males engage in lekking behaviour.

13.5.6 *SGEs and sexual conflict*

Since SGEs can favour polyandry, there is clearly scope for sexual conflict over aspects of female mating patterns, as polyandry will increase the risk of sperm competition for males. This can in turn generate selection on males, as well as co-evolution between male and female mating rates. For example, in populations of *D. pseudoobscura* where females evolved increased mating rate in the presence of a sex ratio distorter (SR), males also evolved to transfer bigger ejaculates containing more sperm that is likely favoured in sperm competition (Chapter 10). In addition, males also evolved to be better at suppressing female receptivity, with males evolving in the presence of SR being better able to suppress female remating than males evolving in populations without SR (Figure 13.5). There was also evidence of a close association between the level of female remating and the ability of males in the same populations to suppress female receptivity—in populations where females evolved the highest level of remating, males were also best able to suppress female remating, whereas, in populations where females evolved the least increase in remating rate, males were least able to prevent the female from remating (Price et al. 2010b). This change is likely to be driven by conflict between males and females in these populations. Initially, there is conflict between females and the SR males favouring evolution of a higher degree of polyandry to promote sperm competition and bias paternity against these males (Figure 13.4). This in turn will result in increased conflict over female remating and all the

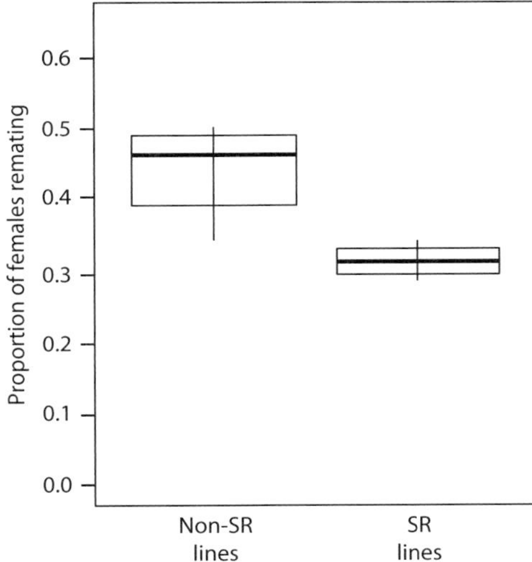

Figure 13.5 Box plot of the proportion of female *Drosophila pseudoobscura* that remated following initial mating to a male from an experimental evolution population where SR (sex ratio drive) was present or absent. (Redrawn from Price et al. 2010bb.) Female remating tendency was suppressed by mating to a male from a population evolving in the presence of SR (Mann–Whitney *U*-test: $n = 8$, $U = 0.5$, $P = 0.029$), likely due to males in these populations being faced with higher risk of sperm competition due to increased female remating rates. Median, interquartile range, and range.

males, not just SR males, in the population due to the increased risk of sperm competition. Due to the increased risk of sperm competition, males evolved ejaculates that were more successful at manipulating female reproductive physiology and behaviour (Figure 13.5). Remarkably, this occurred despite the fact that the frequency of the sex ratio distorter was <5%. This indicates that even at a low frequency, SR had a notable effect on the mating system of this fly species: it promoted evolution of increased female remating rate, in turn favouring the evolution of bigger and more 'potent' male ejaculates. This example provides a powerful illustration of SGE potency in generating selection, even when present at low frequency, by causing a chain of adaptations and counter-adaptations in females and males.

There is also evidence that SGEs can cause sexual conflict by acting as sexually antagonistic alleles through having the opposite fitness effect when expressed in males and females. In *D. melanogaster*, dichlorodiphenyltrichloroethane resistance (DDT-R) is caused by upregulation of resistance genes via a retrotransposon inserted into the promoter region of a detoxification gene (*Cyp6g1*). Remarkably, DDT-R females are more fecund and have offspring of higher fitness than susceptible females (McCart et al. 2005). Despite this fitness advantage, DDT-R did not spread until the use of pesticides, indicating a cost to males in order to balance the benefits to females. Indeed, DDT-R males are less likely to obtain matings when competing with susceptible males (Smith et al. 2011). This is the hallmark of a sexually antagonistic allele: a gene with opposite fitness effects when expressed in the two sexes. Remarkably, the relative fitness reduction in DDT-R males almost perfectly balances the fitness benefit to females when DDT-R is expressed in the same genetic background (Figure 13.6). It is possible that other SGEs also function as sexually antagonistic alleles with opposite fitness effects when expressed in males and females. This indicates that SGEs not only generate sexual selection by favouring polyandry, but may also generate conflict by acting as a sexually antagonistic allele with dramatic sex-specific fitness effects.

These situations may be more widespread than previously recognized. For example, endosymbionts are exclusively maternally inherited, as for them males represent an evolutionary dead end. Thus, any trait that favours female reproduction over that of males will also benefit the endosymbiont and hence is under positive selection in females, despite

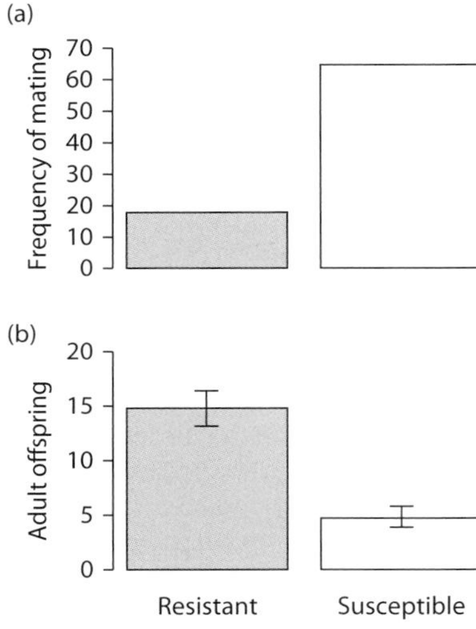

Figure 13.6 Male competitive mating success and female fecundity of dichlorodiphenyltrichloroethane (DDT)-resistant (R) and DDT-susceptible (S) *Drosophila melanogaster* males of the *Canton-S* genetic background showing that the DDT-R allele has opposing fitness effects when expressed in the two sexes. DDT-R males have reduced mating success in competition with DDT-S males (a), whereas DDT-R females enjoy higher fecundity than DDT-S females (b). The relative fitness benefit to DDT-R females almost perfectly matched the fitness cost to DDT-R males. (After McCart et al. 2005 and Smith et al. 2011)

being detrimental to males. It is therefore possible that endosymbionts favour sexually antagonistic alleles that are beneficial to females, but detrimental when expressed in males due to sex-specific selection. Hitherto there have been no examples conclusively demonstrating this possibility, but the prevalence of maternally inherited endosymbionts means that this is a real possibility. This suggestion is further corroborated by the observation that some insects have become entirely dependent on their endosymbionts for successful reproduction (Chapter 1). The wasp *Asobara tabida*, for example, is entirely reliant on its *Wolbachia* for successful oogenesis (Dedeine et al. 2001). Theoretical models have suggested that such sexually antagonistic processes could even lead to loss of CI-inducing *Wolbachia* from populations (Koehncke et al. 2009). Given the prevalence of endosymbionts such as *Wolbachia* in insects, they may also fuel sexual antagonism. The extent to which SGEs may have such sex-specific fitness effects is currently unknown, but likely to be an overlooked possibility.

Finally, the selfish nature of SGEs generates intragenomic conflict with the rest of the genome that will select for suppression and silencing to restore equality. SGEs and their corresponding suppressors are therefore locked in an antagonistic arms race. Counter-selection to restore sex ratio to unity in response to sex ratio distorters can promote evolution of novel sex determination pathways, resulting in new and different ways to determine sex. For example, feminization or male killing by endosymbionts in insects can favour evolution of nuclear resistance, resulting in new sex determination genes to restore equal sex ratio (Cordaux et al. 2011). This process can be rapid, and there is even evidence that multiple sex-determining mechanisms can coexist, as exemplified by field populations of the house fly *Musca domestica* (Denholm et al. 1985). However, changes to sex determination such as going from male heterogamety to female heterogamety (XY to ZW, or vice versa), in response to selfish distorters to restore sex ratio, will simultaneously alter the opportunity for sexual selection and sexual conflict. Selection frequently favours sex linkage of sexually selected and sexually antagonistic traits, and the sex chromosomes are known to contribute disproportionately to sexually antagonistic variation (Rice 1984). This is due to sex-linked genes residing and being exposed to selection for different lengths of time in males and females. For example, heterogamety exposes recessive alleles to selection, and thus generates differential selection on sex-linked genes when expressed in males and females. Sex chromosome linkage can therefore favour the accumulation of sexually antagonistic alleles (Rice 1984). The intragenomic conflict generated by segregation-distorting SGEs promoting a rapid turnover of sex chromosomes will thus simultaneously affect any sex-linked allele by altering their exposure to selection. Therefore, there may be a direct link between (i) the recurrent conflicts generated by SGEs and their suppressors and (ii) the exposure of sexually antagonistic alleles that are played out involving the sex chromosomes. SGEs potentially have more far-reaching impacts on insect mating systems than previously recognized.

13.6 Conclusions

We have discussed the importance of parasites and pathogens in shaping the mating systems of insects and have highlighted some of the major findings in the last 30 years since the publication of Thornhill and Alcock's seminal book. At the time, pathogens were not considered, largely due to the lack of studies examining their impact on the reproduction of their hosts. There are now numerous elegant studies demonstrating the potency of parasites and pathogens to affect host immune responses and influencing their reproductive strategies. It is now recognized that parasite resistance represents an important source maintaining genetic variation in traits involved in mate choice (Hamilton and Zuk 1982), with several experimental and comparative corroborations of this hypothesis in insects. Selective pressures by a variety of parasitoids are known to be responsible for notable shifts in the mating signals of insects, ranging from complete obliteration of calls used in mate attraction (e.g. silent crickets on Hawaii; Zuk et al. 2006) to altering the timing of reproductive activity (e.g. Zuk et al. 1993), and even fine-tuning of the signal used (Rotenberry et al. 1996). The field of ecological immunology has emerged as a vibrant new area of research in the last few decades that has revolutionized the way we view insect immunity and how it is traded off against reproductive effort. However, further studies are needed to quantify the link between host condition, immune responses, and sexual signals. It is not clear to what extent the cost of mounting an immune response affects the evolution of resistance, and whether variation in host immunity is under direct selection or whether the immune response itself may generate selection on the host by affecting sexual signals. It is likely that both processes operate simultaneously. Future studies unravelling the pathways involved in insect immunity and how they relate to reproductive effort and mating may provide exciting opportunities to evaluate this possibility (Simmons and Roberts 2005; Reinhard and Siva-Jothy 2007), and it is worth noting that these processes are likely to be sex specific (McNamara et al. 2013).

At the time of Thornhill and Alcock's book we were even less aware of the impact of genomic parasites such as *Wolbachia*, first described by Hertig and Wolbach back in 1924 and named *Wolbachia pipientis* by Hertig in 1936 after his studies on *Culex pipiens* mosquitoes in which he demonstrated their presence in the gonads of both sexes. It was suggested already in the late 1960s that cytoplasmic incompatibility could be utilized to control mosquitoes (Laven 1967), and this approach is now adopted to control dengue and malaria mosquito vectors (Bian et al. 2013). However, their impact on host reproduction was not documented until the 1970s (Yen and Barr 1971). Today *Wolbachia* are estimated to infect up to two-thirds of all insects (Hilgenboecker et al. 2008). Taking into account the prevalence of *Wolbachia* and other genomic parasites such as transposable elements and sex ratio distorters, and their impact on fertility and reproductive compatibility, it is not surprising that SGEs may have substantial impacts on insect mating patterns. However, it is worth noting that, unlike parasites, SGEs do not appear to play an important role in mate choice. We anticipate that future work will provide further evidence of the widespread impact of SGEs on insect reproductive biology. For example, the maternal inheritance of endosymbionts such as *Wolbachia*, *Spiroplasma*, and *Cardinium* indicate that they may be an overlooked source of reported nuclear–mitochondrial interactions affecting male fertility that may stem from linkage disequilibrium between certain maternal mitochondrial haplotypes and the nuclear genome. There is evidence of extensive mitochondrial allelic variation for male fertility in insects (e.g. flies Yee et al. 2013). An exciting possibility is that endosymbiots may partly be responsible for generating this ‘mitochondrial load’, reducing male fertility more generally. The challenge for the future is to identify potential SGEs and to quantify their impact as selective agents more generally to unravel how they shape insect mating systems.

CHAPTER 14

Sexual selection in social insects

Boris Baer

14.1 Introduction

Social living has evolved repeatedly and in a remarkable number of insect species. Detailed study over the last three decades has provided a solid understanding of the evolution of, and the selective forces maintaining, sociality (Bourke 2011). This is especially true for insects that are eusocial, which is the most advanced form of social living characterized by a reproductive division of labour, overlapping generations, and a co-operative care of offspring (Wilson 1971). Two phylogenetic groups of insects are characterized by the presence of a large number of eusocial species, the Isoptera (termites) (Abe et al. 2000) and the Hymenoptera (the social bees, ants and wasps) (Wilson 1971; Hölldobler and Wilson 1990). Eusociality is present in all of the ~3000 termite species known to date, and in at least 19,000 hymenopterans, where it has evolved independently at least nine times (Hughes et al. 2008). Eusociality has allowed some of these species to become highly successful ecologically and to dominate entire ecosystems (Hölldobler and Wilson 1990).

The study of social insects was driven mainly by sociobiologists using them as model systems to develop and test ideas on the evolution and maintenance of altruistic helping behaviour. The initial contributions by Hamilton (1964a,b) and Trivers and Hare (1976) and the formulation of Hamilton's rule (see Box 14.1) marked the development of kin selection theory as the basis for explaining how helping behaviour may be favoured by natural selection. The empirical evidence for kin selection that has been generated since is overwhelming (Boomsma et al. 2011) despite some criticisms and misunderstandings (Nowak et al. 2010), and sociobiology has established itself as a scientific field within the natural sciences (Wilson 1975) to study conflicts and collaborations within societies, and their evolutionary history (Ratnieks et al. 2006). There is now ample of evidence that individuals benefit from helping kin, thereby propagating their genes through relatives. Because such inclusive fitness gains increase with increasing relatedness of helpers, their genetic make-up within a colony became a central interest for sociobiologists (Boomsma and Ratnieks 1996). Despite the fact that offspring relatedness is ultimately determined by the mating system of a species, remarkably little attention was given to studying the mating systems of social insects, other than Boomsma and Ratnieks' (1996) analysis of paternity distributions among helpers.

Patterns of multiple mating and the resulting paternity distributions among offspring were also of interest to another field within evolutionary ecology: sexual selection. As we saw in Chapters 10 and 11, biases in paternity contributions were studied to explain

The Evolution of Insect Mating Systems. Edited by David M. Shuker and Leigh W. Simmons.
© The Royal Entomological Society 2014. Published 2014 by Oxford University Press.

Box 14.1 DEFINITIONS OF TECHNICAL TERMS

Cryptic female choice
Female traits that bias paternity post copulation towards preferred males. The choice is cryptic in the sense that males are unable to assess or influence female manipulations of paternity.

Eusociality
A special form of social living, characterized by a reproductive division of labour, overlapping generations, and co-operative care of offspring.

Inclusive fitness
The ability of an individual to pass on genes to the next generation, including shared genes that are passed on through an organism's close relatives.

Kin selection
A form of natural selection in which genes encoding altruistic behaviours increase in the population due to increased survival of individuals that are genetically related to altruistic helpers.

Monandry
A female copulating or remating with only a single mate.

Multiple paternity
A female uses sperm from more than one male to sire offspring. This may occur simultaneously, resulting from the presence of sperm of more than one male in the female's sexual tract, or serially, where females remate and sire individual batches of eggs with sperm of different males.

Paternity skew
A non-random bias of paternity towards one of few males. In the most extreme cases paternity skew results in the precedence of a single male, for example the first or last male that copulated with a female.

Polyandry
A female copulating with more than one male.

Sexual selection
A selective force of natural selection that provides some individuals with an advantage over others during sexual reproduction. Sexual selection results in deviations from randomly expected paternity.

Single paternity
A female uses sperm from only a single male to sire all her offspring.

Sperm competition
The competition between the ejaculates from two or more males for the fertilization of a given set of ova.

Social insects
Insects that spend at least part of their life cycle in conspecific social aggregations. Insect societies can be subdivided into different groups such as eusocial, para- or presocial, sub- or semisocial, or co-operative breeding systems. Sexual selection likely operates rather differently in these various social forms.

the fitness consequences of male–male competition and/or cryptic female choice (Parker 1970; Eberhard 1996a; Simmons 2001). Despite this common interest in the genetic make-up among offspring, kin selection research developed largely independently from research conducted on sexual selection (Boomsma 2007) because studies on kin selection were more concerned with the consequences of paternity distributions. Sexual selection research, on the other hand, was focused on explaining how paternity distributions are generated through pre- and post-copulatory sexual selection, and their effects on the evolution of individual life-history traits. Students of sexual selection did not use social insects as their model systems for a number of reasons, such as the fleetingness of reproductive behaviour in many social insects or the lack of obvious extravagant secondary structures that readily identify the presence of sexual selection. Combined with the initial research focus to generate empirical confirmation for kin selection theory, this led to the eclipse of the study of sexual selection in social insects. As a consequence, kin selection and sexual selection research developed in parallel and researchers in their respective fields started to use reproductive terminology in different contexts and based on different definitions. We therefore provide definitions of technical terms used throughout this chapter in Box 14.1. As both fields developed into areas of major research strength, a few empirical and theoretical papers discussing the importance of sexual selection for eusocial species eventually emerged (Hölldobler and Bartz 1985; Simmons 2001; Baer 2003, 2005, 2011; Boomsma et al. 2005; Hartke and Baer 2011). The importance of sexual selection for the evolution of social insect mating systems is a subject that is slowly developing, however, and one with considerable future potential.

Rather than further elaborating on the theoretical aspects of the effect of mating systems on social evolution, here we take a different approach and summarize what we currently know about sexual selection in eusocial insects. We start by introducing social insect mating systems and point out that they share some common characteristics that can be assumed to have evolved in response to their eusocial lifestyle. We then provide an overview of sexually selected traits in social insects that have been reported in the literature to illustrate that they have evolved and operate across a wide range of species. These traits are subject to sexual selection acting both before and after copulation, and through both male–male competition and female choice. Finally we discuss the implications of sexual selection on social living and possible consequences of sexual selection for kin selection.

14.2 The mating biology of social insects

Understanding sexual selection acting on any species requires detailed information about that species' mating biology. The general characteristics of the reproductive biology of social insects have been covered by a number of reviews (Starr 1984; Hölldobler and Bartz 1985; Page 1986; Boomsma and Ratnieks 1996; Baer 2003, 2005, 2011; Hartke and Baer 2011). In brief, the vast majority of social insects reproduce sexually, but clonal or semi-clonal systems have been described for a small number of species as well (Pearcy et al. 2004; Fournier et al. 2005; Himler et al. 2009). Furthermore, only one or very few males monopolize the paternity of broods in the majority of social insects. This can be expected from inclusive fitness theory and continuous selection to maintain high relatedness among helpers (Boomsma and Ratnieks 1996); eusociality evolved from solitary ancestors with close to single paternity (Hughes et al. 2008; see Figure 14.1).

Typically, pair formation and/or copulations occur during a relatively short period of time, early in the life of the reproductive castes (Boomsma et al. 2005). As soon as

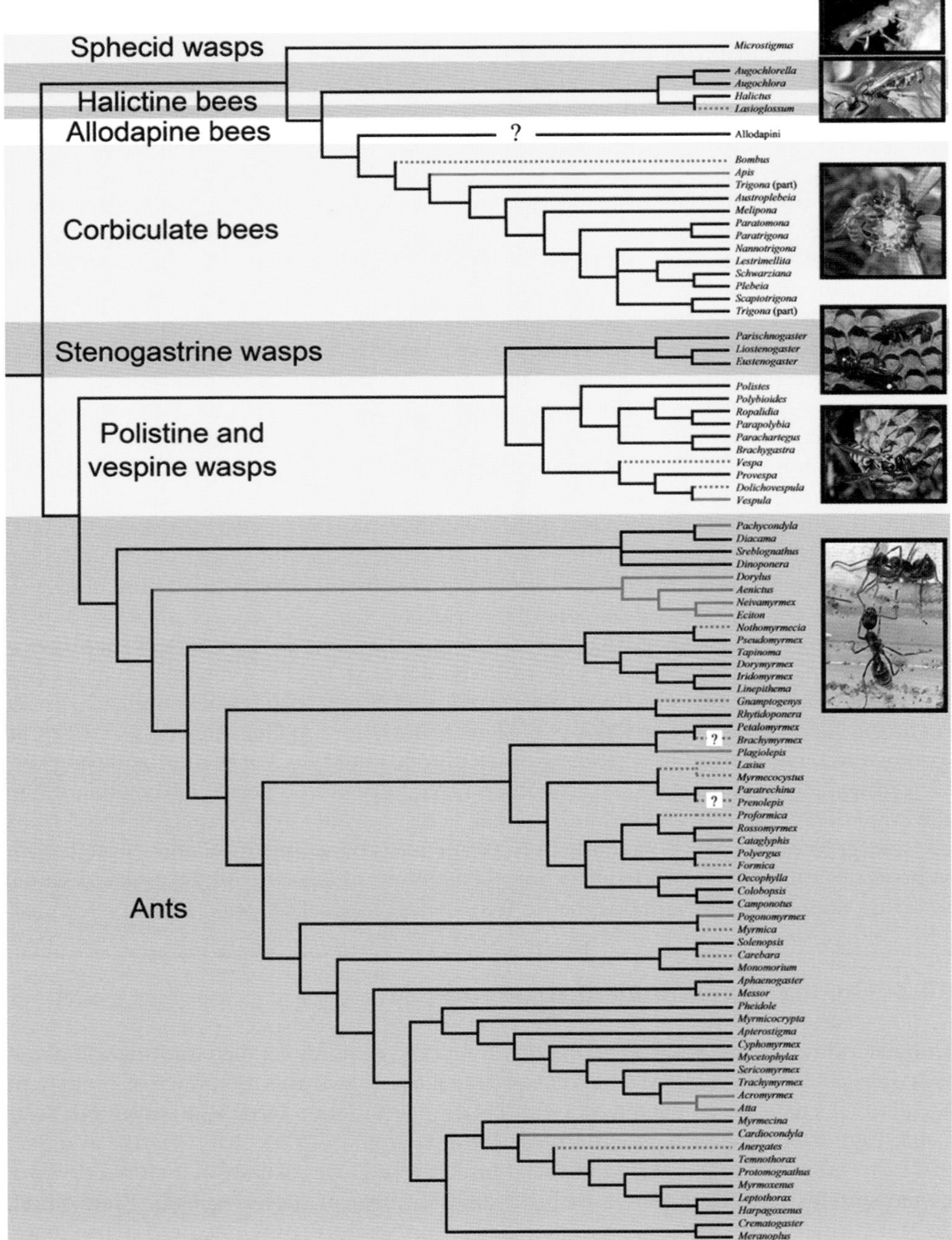

Figure 14.1 The phylogeny of eusocial ants, bees, and wasps for which paternity-frequency data are available. Each independent origin of eusociality is indicated by alternately coloured clades. Clades exhibiting high polyandry (>2 effective mates) have solid red branches, those exhibiting facultative low polyandry (>1 but <2 effective mates) have dotted red branches, and genera with single paternity have solid black branches. Mating-frequency data are not available for the allodapine bees. (Reproduced with permission from William Hughes and the American Association for the Advancement of Science as published in Hughes et al. 2008.)

females (queens) start to lay an initial set of eggs, eusocial insect societies tend to become genetically closed systems, where no additional genetic contributions are accepted. This is applicable to the majority of social insects known so far (Chapter 1), although termites and hymenopterans differ in the way the genetic closure is realized during the eusocial life stage. In termites, a king and a queen establish themselves as a pair of primary reproductives, and kings continuously produce and contribute sperm to females (Hartke and Baer 2011). In the social hymenopterans, fathers are not physically present during the colony phase because they die during or shortly after copulation, and so only survive as stored sperm inside the queen's body. These characteristics resulted in the evolution of specific adaptations to accommodate long-term fecundity and partner commitment, especially in species with large and long-lived colonies where queens need to be capable of initially storing several hundred million sperm and using them economically for several decades to produce tens of millions of offspring (Weber 1972; Pamilo 1991; den Boer et al. 2009).

The presence of only one or very few patrilines in worker offspring resulted in the notion that social insects have mostly monogamous mating systems where queens only commit and mate to a single partner (Strassmann 2001). However, paternity distributions among offspring are not accurate predictors for queen mating frequencies because paternities may be biased towards the one or few males that are most competitive or most preferred by females (Boomsma and Ratnieks 1996; Baer 2011; Jaffe et al. 2012). A comparative study on paternity skew in polyandrous social Hymenoptera indeed revealed that in highly polyandrous species, which produce large and long-lived colonies, paternities are equalized among sires, perhaps because of the well-documented fitness benefits associated with genetic diversity among helpers (Baer and Schmid Hempel 1999; Tarpy 2003; Hughes and Boomsma 2005). However, in species where queens mate with few males, paternities are strongly biased towards one or very few males (Jaffe et al. 2012). This finding is interesting for several reasons. First, it implies that inclusive fitness of helpers is maximized in species with low queen mating frequencies. In this situation, sexual selection favours the necessary traits that result in high paternity skew and relatedness, which is also necessary for social evolution. Sexually selected reproductive traits could therefore have been more important for social evolution than previously acknowledged. Second, the presence of strong sexual selection resulting in the complete monopolization of paternity by the most competitive or most attractive male implies that polyandry may also be present in species that are currently thought to be monandrous based on paternity analyses alone. Observational data on queen mating frequencies indeed indicate that female multiple mating is more widespread than suggested from molecular studies on paternity (Boomsma and Ratnieks 1996; Baer 2011). For example, in a comparative study of ants for which observational data on the copulation behaviour of queens are available, 78% of species were reported to mate polyandrously (Baer 2011). As we saw in Chapter 10, among non-social insects strong or complete last-male paternity is widespread. Post-copulatory sexual selection could therefore explain this mismatch between observed mating frequencies and effective paternity frequency in social insects, and thereby provide a more complete picture of the mating systems of social insects and the role of sexual selection within a kin selection framework. Furthermore, whereas polyandry was originally assumed to be restricted to only a few, highly derived, clades within the bees, ants, wasps, and termites (Strassmann 2001), it is phylogenetically widespread (Figure 14.1) as indicated by an increasing number of eusocial Hymenoptera being confirmed to be polyandrous (Hughes et al. 2008). In summary, although there is general agreement that sexually selected traits evolved within social insects, further work is required to quantify the general importance of sexual selection in social insects and to understand how their social lifestyle influenced the way sexually selected traits evolved.

14.3 Sexual selection in social insects

Identifying sexually selected traits and understanding their consequences for the evolution of social insect mating systems has received significant scientific scrutiny only recently (Baer 2003, 2005, 2011; Boomsma 2007). Apart from the fact that sociobiologists have not been interested in the precise mechanisms that create the genetic make-up of colony members, the characteristics of social insect mating biology as outlined in the last paragraph has made it difficult to identify the presence and effects of sexual selection in social species. Because reproductive events such as mating flights and copulations occur only rarely, and typically cannot be induced under laboratory conditions, studying the mating biology of social insects has proved rather challenging (Baer 2003). Furthermore, as copulations often occur on the wing, any manipulations of pair formation or copulation are limited. Finally, a eusocial lifestyle prolongs generation times, because queens have to go through several initial rounds of producing worker offspring after colony foundation

Figure 14.2 Pair of copulating bumblebees (*Bombus terrestris*). Our ability to observe copulations in bumblebees and to manipulate copulations using artificial insemination methods has provided the opportunity to unravel the mating biology of these species, and has revealed a number of their reproductive traits to be under sexual selection. (Photo by B. Baer.)

before a new generation of sexual offspring can be produced. This restricts the opportunities to study cross-generational evolutionary dynamics of reproductive traits. As a consequence, students of sexual selection generally choose more tractable species as their model systems.

The ability to observe and manipulate matings in some social insects such as bumblebees has stimulated studies that have gained a more detailed understanding of their reproductive biology. Honeybee reproduction has also been studied in great detail (see Baer 2005 and references therein), due to their economic importance and the development of an artificial insemination technique for controlled breeding of managed stocks (Woyke 1962; Ruttner 1976). Honeybees illustrate nicely how methodological limitations to study social insect reproduction can be successfully overcome. Encouraged by the progress with bees, researchers eventually expanded the range of species studied, for example to *Cardiocondyla* (Heinze et al. 1998; Mercier et al. 2007), *Atta* leaf-cutting ants (Baer and Boomsma 2006) and more recently paper wasps (Tibbetts and Sheehan 2011; Izzo and Tibbetts 2012). An improved understanding of the mating biology of social insects eventually also revealed a number of reproductive traits under sexual selection. An overview of known sexually selected traits in social insects is provided in Table 14.1, illustrating that sexual selection is present in a number of species, although there is a strong bias towards the Hymenoptera. Nonetheless, the absence of termites in Table 14.1 is merely a result of a lack of studies on termite reproduction (Hartke and Baer 2011). More research on their mating biology seems warranted, since the little information available indicates that sexual selection might also be present in these species.

As Table 14.1 also illustrates, the processes and patterns of sexual selection in social insects are comparable with other insects and occur both before and after copulation, in terms of male–male competition and/or female choice. Although we tend to think of anonymous mass swarmings as the typical mode of reproduction for many social insects, a wide range of pre-copulatory behaviours of social insect males are known, such as guarding and patrolling of conspecific nest entrances, as well as males waiting at strategic points to increase their chances of encountering females. Because operational sex ratios in social insects can be highly male-biased (Beekman and Van Stratum 1998), male–male competition is expected to be strong and has been taken to absolute extremes in some species such as honeybees where around 25,000 drones (Page 1986) from several hundred different colonies may compete for a single virgin queen at mating sites (Baudry et al. 1998); the numerical sex ratio in the honeybee *A. mellifera* may therefore exceed 20,000:1 (Page and Metcalf 1982). Strong male–male competition in mating arenas is also found in ants (Hölldobler and Bartz 1985) where sexual selection for larger body size and shape increases male reproductive success (Abell et al. 1999). Some social insects such as bumblebees, wasps, and ants produce sex pheromones which seem not only to be involved in species recognition, but are also important for female choice (see reviews by Ayasse et al. 2001; Baer 2005). Some social insect males actively fight with each other or even kill each other for access to females, as in *Cardiocondyla* ants (Anderson et al. 2003). Male ornamentation has not been studied in great detail in social insects but has evolved in the paper wasp *Polistes dominulus* (Izzo and Tibbetts 2012), where males possess a black dorsal spot on their abdomen. The spot acts as a signal of male quality used both for female choice and male–male competition. Males with larger spots win dominance fights and thereby increase mating success, especially since queens prefer males with larger, more irregular, spots (Figure 14.3). Furthermore, facial patterns of queens determine the outcomes of queen contests and consequently the ownership of the nest, where queens with more facial spots have a selective advantage (Tibbetts and Shorter 2009; Tibbetts and Sheehan 2011). The work on wasps by

Table 14.1 Overview of different sexually selected traits in social bees, wasps, ants, and termites[a]

Level	Actor and recipient	Trait	Effect	Groups	Examples from the literature
Pre-copulatory	Male–male	Territory establishment	Males fight for and establish territories and defend them, excluding competitors. Sex pheromones are used to mark territories and to attract queens.	Bees, wasps	[1–3]
Pre-copulatory	Male–male	Enforced copulation	Males copulate with queen pupae before they emerge and disperse	Ants, bees, wasps	[4]
Pre-copulatory	Male–male	Copulation duration	Males increase time spent for mating in the presence of competing males; longer copulations reduce queen remating rates	Ants, bees	[5, 6]
Pre-copulatory	Male–male	Weaponry	Males possess morphological structures for fighting with rivals, such as sabre-shaped mandibles	Ants	[7]
Pre-copulatory	Male–male	Male size	Larger males are more likely to win copulations	Ants, bees	[8, 9]
Pre-copulatory	Male–male	Male fighting	Males monopolize access to queens by killing rival males	Ants	[10]
Pre-copulatory	Male–male	Fluctuating asymmetry	Developmental stability in males is advantageous for reaching mating sites	Bees	[11]
Pre-copulatory	Female–male	Male quality	Queens prefer unrelated, non-infected males or males with larger fat-bodies	Termites	[12]
Pre-copulatory	Male–male and female–male	Male ornamentation	Male ornamentation determines outcomes of male–male contests and female choice	Wasps	[1]
Pre-copulatory	Male–male	Male genital morphology	Male genitalia attached to queens prevent female remating	Bees	[13]
Pre-copulatory	Male–male	Territoriality	Males exclude other males from mating arenas	Ants, bees	[14]
Pre-copulatory	Male–male	Male dimorphism	Males of different morphs compete for females in different environments	Ants	[15]
Pre-copulatory	Male–female	Tandem runs	Mate choice by queens (?)	Termites	[16]
Pre-copulatory	Female–male	Female social status	Queens of specific age, weight or hormonal status or high-ranking mothers have higher mating success	Wasps	
Pre-copulatory	Female–male	Male killing	Queens kill males that are trying to copulate with her	Bees	[14]
Pre-copulatory	Female–female	Queen fighting	Queens monopolize access to males and reproduction by killing rival virgin (sister) queens	Bees, ants	[17]
Pre-copulatory	Female–female	Reproductive skew	Queens of high rank have higher mating success	Wasps	

Post-copulatory	Male–male and/or male–female	Strategic ejaculation	Male transfer ejaculates directly into sperm storage organs to avoid female choice	Ants, bees	[2, 18]
Post-copulatory	Male–male and/or female–male	Paternity skew	Differential sperm storage success of males results in biased paternity in offspring	Ants, bees, wasps	[19]
Post-copulatory	Male–male	Sperm incapacitation	Killing of rival sperm	Ants, bees	[20]
Post-copulatory	Male–male	Sperm length	Sperm length variation is lower in polyandrous social insects compared to species with single paternity	Ants	[21]
Post-copulatory	Male–male	Sperm viability	Males facing sperm competition produce more viable sperm	Bees	[22]
Post-copulatory	Male–male	Sperm length	Sperm length influences storage success	Bees	[14]
Post-copulatory	Male–female	Mating plugs/signs	Reduction of queen willingness to remate or blocking of her sexual tract	Ants (?), bees	[23–26]
Post-copulatory	Male–female	Mate guarding	Males mate guard females, in some species by sacrificing their own lives—'suicidal mate guarders'	Ants, bees	[2, 27, 28]
Post-copulatory	Male–female	Seminal fluid investment	Males with larger accessory glands monopolize paternity	Ants	[29]
Post-copulatory	Male–female	Sperm number	Males transfer substantially more sperm to females than required for a complete colony cycle	Bees	[30]
Post-copulatory	Female–male	Cryptic female choice	Females neutralize sperm-damaging effects of seminal fluid on rival sperm	Ants	[20]
Post-copulatory	Female–male	Female choice	Queens preferentially store sperm of a specific length	Bees	[14]
Post-copulatory	Female–male	Sperm selection	Queens preferentially store sperm of some males and discard unwanted ejaculates	Bees (?)	[2]
Post-copulatory	Female–male	Sperm dumping	Queens do not store sperm but expel it	Bees	[31]

[a] Sexual selection was originally defined by Darwin (1871) as 'advantages (i.e. traits), which certain individuals have over others of the same sex and species solely in respect to reproduction'. I here distinguish between traits that evolved in response to conflict over reproduction between males, between males and females, and between females. Furthermore, traits under sexual selection may influence pre-copulatory mating success or post-copulatory paternity success.

[1] Izzo and Tibbetts (2012), [2] Baer (2005), [3] Alcock (2000), [4] Foitzik et al. (2002), [5] Yamauchi et al. (2001), [6] Brown et al. (2002), [7] Heinze et al. (1998), [8] Berg et al. (1997), [9] Abell et al. (1999), [10] Anderson et al. (2003), [11] Jaffe and Moritz (2010), [12] Shellman-Reeve (2001), [13] Da Silva et al. (1972), [14] Baer (2003), [15] Heinze and Hoelldobler (1993), [16] Hartke and Baer (2011), [17] Graham (2005), [18] Baer and Boomsma (2006), [19] Jaffe et al. (2012), [20] den Boer et al. (2010), [21] Fitzpatrick and Baer (2011), [22] Hunter and Birkhead (2002), [23] Baer et al. (2000), [24] Baer et al. (2001), [25] Mikheyev (2003), [26] Winston (1991), [27] Foitzik et al. (2002), [28] Monnin and Peeters (1998, [29] Baer and Boomsma (2004), [30] Röseler (1973), and [31] Ruttner (1956).

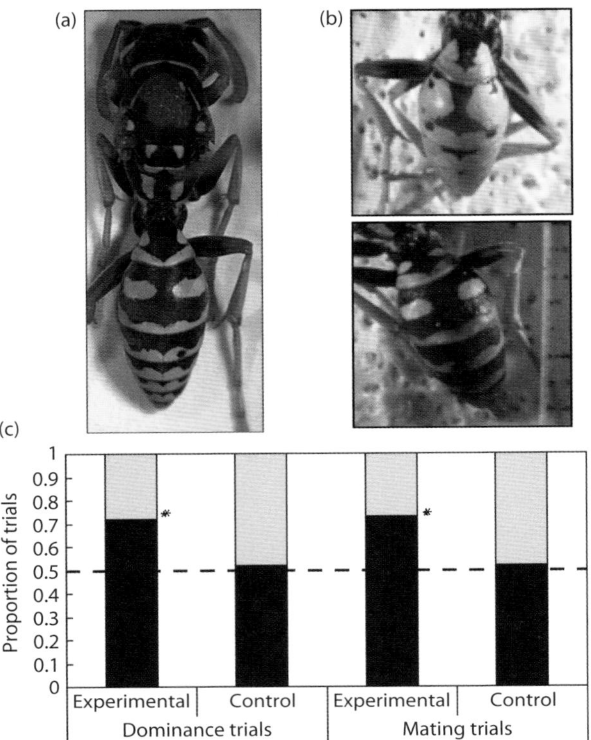

Figure 14.3 In the paper wasp *Polistes dominulus*, males possess black dorsal spots (a). The spots are extremely variable between males (b) and act as a signal of male quality that determines the outcomes of male–male contests and female choice. Males with artificially enlarged spots are more competitive as they are more likely to win male dominance trials and are also preferred by females during mating trials (c). (Photos kindly provided by Elisabeth Tibbetts; (c) reprinted with permission from Izzo and Tibbetts 2012).

Tibbetts and co-workers is remarkable, not only because it provides the first empirical evidence for sexually selected ornamentation in social insects, but it also combines the study of male–male competition with female choice.

In some bumblebee species, males occupy and defend territories where they leave scent marks to attract queens, but queens are choosy and often reject males that attempt to copulate (Duvoisin et al. 1999). Queens reject males by bending their abdomen and refusing males access to their sting apparatus, which is the initial step in establishing copulation. In *Bombus hypnorum*, queens also sting males to death if they try to copulate too vigorously. The reasons why queens accept specific mates has not been studied in great detail but female choice seems crucial in these species because male contributions such as sperm and seminal fluids influence female hibernation success and fitness (Baer and Schmid-Hempel 2001; Gerloff and Schmid-Hempel 2005). The time spent in copula varies between species from a few minutes up to several hours (Baer 2003; Brown and Baer 2005). The ejaculate is typically transferred shortly after the onset of copulation (Duvoisin et al. 1999), indicating that males prolong time spent in copula to mate guard queens and avoid queen remating (see also Chapter 10). Such behaviours are also known from ants where, in the most extreme cases, males become suicidal maters that are decapitated and their abdomens act as a specific form of mate guarding (Monnin and Peeters 1998; Baer 2011). Finally, males transfer mating plugs, mating 'signs' (a form of plug produced by honeybee drones for instance), or spermatophores into the reproductive tract of queens (Figure 14.4). In at least one species, *B. terrestris*, a single component within the mating plug changes the female's willingness to (re)mate and therefore acts as an anti-aphrodisiac (Baer et al. 2001).

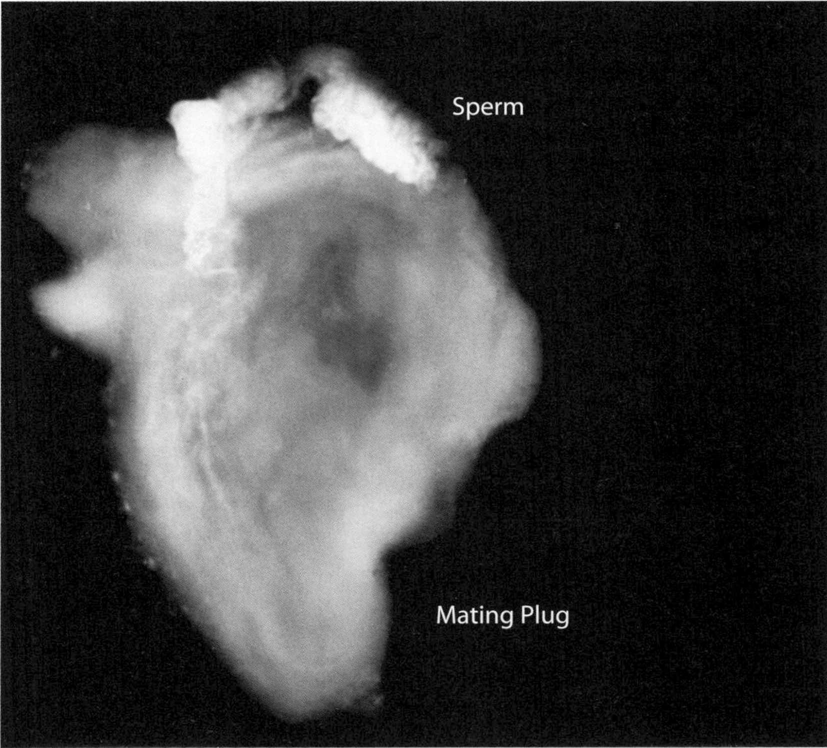

Figure 14.4 Bumblebee ejaculate dissected from the sexual tract of a queen shortly after copulation had been terminated. The ejaculate consists of two clearly distinguished components, a smaller fraction of sperm and seminal fluid (white mass at top of image) and a larger fraction making up the 'mating plug'. Similar to seminal fluid, the plug also originates from the accessory glands but is transferred to the female after sperm and seminal fluid. The plug is slowly dissolved within the female's sexual tract and reduces the queen's willingness to mate. (Photo by B. Baer.)

As a consequence, although bumblebee queens have been shown to benefit from multiple paternity due to increased worker heterogeneity resulting in lower parasitism, most bumblebee queens produce offspring sired by only a single male (Schmid-Hempel and Schmid-Hempel 2000). Overall, bumblebees illustrate rather nicely that single paternity is achieved through a number of traits under sexual selection, demonstrating again that paternity distributions among offspring are unreliable markers for predicting the presence or absence of sexual selection in a species.

Similar observations are also available from ants such as the Argentine ant *Linepithema humile*, where queens mate with up to 20 males, although only a single male contributes to offspring (Keller and Passera 1992). Likewise in the ant, *Leptothorax gredleri*, queens mate with up to four males but only a single father is detected in worker offspring (Oberstadt and Heinze 2003). These examples illustrate that although queens ultimately commit to only a single or very few fathers, sexual selection could be important in these species to determine these few or single winners. This is also in the interest of queens, because they only have a single opportunity for mate choice early in life, and any trait that will increase their chance of storing sperm from a superior male will be favoured. All such traits are by definition under sexual selection.

Ejaculates consist of sperm as well as glandular secretions that make up the seminal fluid, which can be of central importance in determining male fertility and reproductive success in many species (Poiani 2006; Avila et al. 2011). Seminal fluid can thus be subject to sexual selection (Chapman et al. 1995; Simmons 2001; Kubli 2003), and has received considerable research interest in non-social insects (Wolfner 2002; Ram and Wolfner 2007) (Chapter 10). In insects generally, major components of seminal fluid are produced by the male accessory glands (Gillott 1996; Gillott 2003). Accessory glands can be relatively large in some social insect species as seen in Figure 14.5, and have been studied in a number of species (Colonello and Hartfelder 2005).

The male accessory glands of social insects produce mating plugs, mating signs or spermatophores, but apart from bumblebees their functioning has not been investigated in great detail. In honeybees, males are suicidal maters and produce and leave a 'mating sign' in the queen, which consists of a large amount of accessory gland secretion and parts of the male's endophallus, but the functional significance of the 'mating sign' and its effect on paternity remains unclear (Baer 2005; Wilhelm et al. 2011). Similarly male accessory glands also produce spermatophores, but whether and how they bias paternities has not yet been studied.

A comparative study of male accessory glands across attine ants confirmed their presence in species with single and multiple paternity (Baer and Boomsma 2004). However,

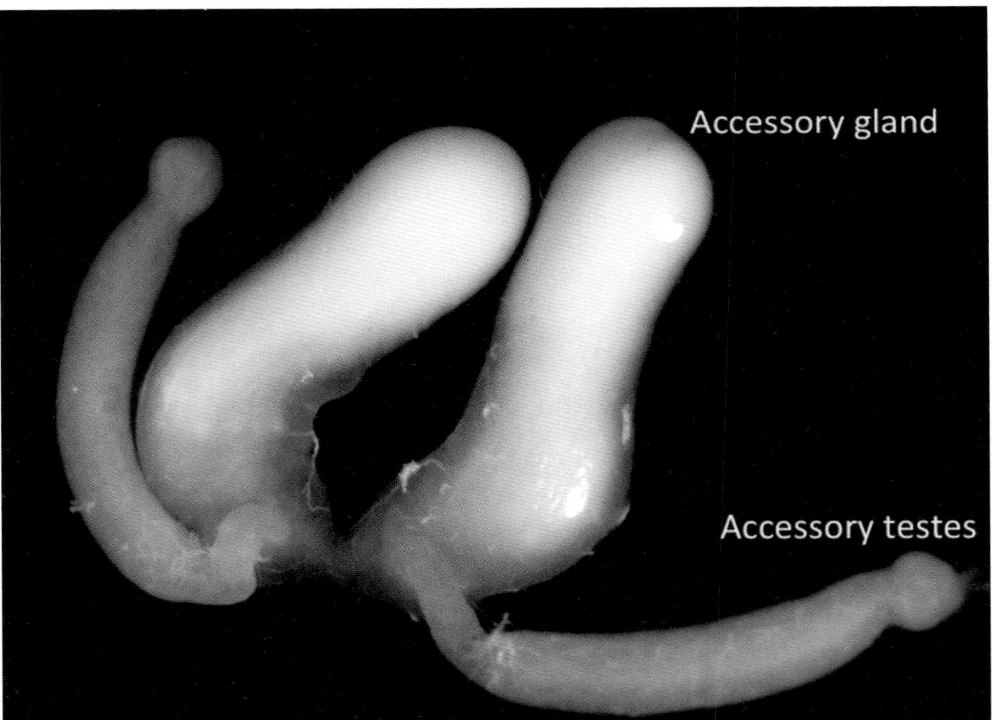

Figure 14.5 Internal organs of a mature honeybee male. In the majority of bees, ants, and wasps, spermatogenesis is terminated before males eclose, and they are born with a fixed amount of sperm that cannot be replenished later in life. During maturation early in the life of males, sperm are transferred from the testes to the accessory testes (also referred to as seminal vesicles) and the testes degenerate. Social insect males also have accessory glands that produce seminal fluid, mating signs, mating plugs, and spermatophores. In some species they are of substantial size, indicating substantial investments by males to bias paternity.

male investments into these glands differ among species. Males of polyandrous fungus-growing ants have smaller glands relative to body size compared to monandrous species. Similar observations have been reported for honeybees, where male accessory glands are small in species where males transfer sperm directly to the spermatheca (Baer 2005). These observations from bees and ants imply that males reduce investment into male accessory glands if they have lost the ability to monopolize or influence paternity via seminal fluid or mating plugs. However, sexual selection maintains large gland sizes in monandrous ant species in order to monopolize paternity of a single male (suggesting that female monandry is strongly influenced by male seminal components in these species), providing further support for the idea that sexual selection is present and could even be strong in species with single paternity (Chapter 10).

Apart from manipulating female mating behaviour, seminal fluid also contains components that influence ejaculates of rival males. In the polyandrous ants *Atta colombica* and *Acromyrmex echinatior* and the honeybee *A. mellifera*, seminal fluid incapacitates sperm of rival males (den Boer et al. 2010). In species where queens require large numbers of viable sperm, prolonged warfare between ejaculates may not be in the interest of queens, especially if queens do not receive excess amounts of sperm. This is the case in *A. colombica* where males transfer their ejaculates directly to the spermatheca. Queens have therefore lost the ability to select sperm prior to admission to the storage organ but require large amounts of viable sperm to build and maintain colonies of several million workers. Any hostility among ejaculates resulting in storage of dead sperm is against the interest of queens and the spermathecal fluid of queens indeed terminates hostile interactions among competing ejaculates (den Boer et al. 2010). In species with single paternity such as bumblebees and *Trachymyrmex zeteki*, where males have sexually selected large accessory glands to monopolize females, seminal fluid does not incapacitate rival sperm, which is expected if competing ejaculates do not typically co-occur within the sexual tract of a female.

Studies identifying the molecular basis of seminal fluid have been conducted in bumblebees (Baer et al. 2000), honeybees (Baer et al. 2009), fire ants (Mikheyev 2003) and leaf-cutting ants (E. Paynter, unpublished observations). With the availability of an increasing number of social insect genomes, modern proteomic approaches became feasible and have provided large-scale insights into the make-up of seminal fluid proteins. In the honeybee *A. mellifera* more than 70 seminal fluid proteins have been identified and a subset of them have been hypothesized to be the products of sexual selection, either by affecting sperm of competing males and being responsible for the observed incapacitation of rival sperm (den Boer et al. 2010) or by influencing female physiology and behaviour (Baer et al. 2001). More experimental work is required to fully understand the molecular interplay between proteins or proteomic networks of seminal fluid and/or sperm with competing ejaculates or the female. Apart from proteomic approaches, genomic methods such as next-generation sequencing can provide spectacular insights into the functioning of sexual conflicts on the molecular scale (Chapter 4).

14.4 Conclusions

Research over the last three decades has revealed that sexual selection is present in a number of social insects, in species that have single and multiple paternity, and is expressed through a variety of traits both before and after copulation. More recent work has used the knowledge gained to develop a theoretical framework that links sexual selection and kin selection but this process remains ongoing and requires more research, especially on the empirical side. There is increasing support for the idea that sexual selection has been an important selective force contributing to the evolution of eusociality, if it provided the necessary mechanisms to bias paternity towards a most competitive or a most preferred male, and thereby increased offspring relatedness and inclusive fitness gains. It is remarkable that to this point, our knowledge of sexual selection in eusocial insects is fully compatible with the expectations from kin selection theory and even provides intriguing examples and justifications for the importance of high degrees of relatedness among helpers in insect societies.

CHAPTER 15

The Evolution of Insect Mating Systems

John Alcock and Randy Thornhill

More than 30 years ago we wrote *The Evolution of Insect Mating Systems*. To be invited to comment on the current collection of chapters that update our book is both flattering and sobering—flattering because it is nice to think that we may have stimulated others to study insect mating systems, and sobering because the invitation reminds us of how much time has elapsed since our book originally appeared. In the interim between then and now, one of us (R.T.) has shifted his research from insects to human beings while the second author (J.A.) has retired.

A great deal else has changed since 1983. One of the virtues of the present volume is to point out the major advances in the field of insect behavioural ecology. No longer is it necessary to proclaim the virtues of the adaptationist approach as a foundation for behavioural research. This approach was still considered relatively novel in 1983 even though George C. Williams (1966) had written *Adaptation and Natural Selection* 17 years previously while E. O. Wilson's *Sociobiology* (1975) and Richard Dawkins's *The Selfish Gene* (1976) had been available for the better part of a decade. Despite these now widely acclaimed books, not many entomologists of the era were ready to use natural selection theory to explore adaptation in insect behaviour and physiology. In addition, most academic biologists investigating behavioural questions in the early 1980s ignored insects in favour of birds and other vertebrates. Insect behavioural ecology was therefore underexplored, a deficiency that has been remedied as is clear from this new book.

Because an adaptationist approach to insect behaviour has been used so widely over the past three decades, we now know much that was either uncertain or untouched at the time our book was published. Consider the issue of polyandry, a topic that concerns many authors in this volume. In the early 1980s, females were believed to mate rarely with several males (in pair-bonding vertebrates particularly) and for good reason as explained by Emlen and Oring (1977). As these authors and others writing at about the same time pointed out, males would seem to lose fitness by sharing a female with other males and so one would expect that instead males would seek to monopolize the reproductive output of as many females as possible. In many species, males do indeed follow this rule. However, this argument fails to consider the important role females play in mating systems. We now know that even in apparently monogamous birds, matings by females outside the pair bond are commonplace, a discovery based on innovative use of genetic technology in the study of paternity that became widely used in the 1980s and beyond (Westneat et al. 1990). Snook (Chapter 9) provides an update on the development of genetic techniques that enable researchers to discover who fathered whom. Without this technological

The Evolution of Insect Mating Systems. Edited by David M. Shuker and Leigh W. Simmons.
© The Royal Entomological Society 2014. Published 2014 by Oxford University Press.

advance, we would be forced to rely on indirect and potentially misleading clues about the paternity of a brood, which could generate misleading conclusions about the female's mating system.

However, even in 1983 we knew that multiple mating by females occurred in a wide range of insects, and so in our book we considered four possible advantages that females might gain by mating with different partners: sperm replenishment, acquiring of material benefits, securing genetic benefits, and avoiding the costs of repelling sexually driven males. That polyandrous females may secure material benefits (paternal care of young or resource donations) was relatively quickly accepted in part because of an early and convincing demonstration that female hangingflies make acceptance of a male's sperm contingent upon the quality and quantity of his nuptial gift (Thornhill 1976). Far more controversial has been the possibility that females might secure better genes by mating with several males. Nevertheless, as discussed by Snook (Chapter 9), there is now abundant evidence that females accept more than one sexual partner even though they possess sufficient sperm, do not receive any apparent material benefit from their 'secondary' mates, and are not harassed into mating more than once. To take one example, Zuk and Wedell (Chapter 13) argue that the presence of various selfish genetic elements (SGEs) that attempt to propagate themselves at the expense of the rest of the genome (and the individual in which they reside) can explain some cases of polyandry in insects. By mating with several males, females can potentially swamp any sperm that they may have received from males carrying SGEs.

Polyandry is merely one aspect of female reproductive behaviour that has been subject to much greater attention over the past 30 years as behavioural ecologists have increasingly come to realize that females are not merely passive vessels for the receipt of male sperm but are active, adaptive strategists when it comes to sexual reproduction. Polyandry is one possible consequence of female mate choice—which is a topic worthy of investigation in its own right, as indicated above and demonstrated in this volume. One of the major findings from research on female mate choice is the existence of *cryptic* mate choice, one example of which involves the ability of females to choose among the sperm received from several males when fertilizing their eggs, an event that takes place out of sight *after* a female has mated with multiple partners. Hangingflies yielded early evidence that females can make choices of this sort (Thornhill 1983) and now cryptic mate choice is believed to be widespread, although as Arnqvist (Chapter 11) explains, clear demonstrations of this kind of mate choice are challenging to document.

Nonetheless, it seems likely that female insects often have the means to choose among the sperm provided to them by several partners, an ability that has led some researchers to conclude incorrectly that the queens of social insects are monogamous. As Baer (Chapter 14) points out, it may be misleading to claim that a female was monogamous if her brood had a single sire. But if the female, a queen ant for example, had mated with several males while only using one of her partner's sperm to fertilize her eggs she would appear to be monogamous when in fact she was not. Instead such a polyandrous queen exhibits cryptic mate choice by selectively using one male's sperm to fertilize all her eggs rather than also using the gametes of her other sexual partners. By restricting the genetic make-up of her daughters through cryptic sperm utilization, a queen increases the degree of relatedness of her workers, which in turn makes kin-directed altruism more likely to evolve and to be maintained by kin selection (Hamilton 1964a, b). Here is an important subject that was not discussed in our book at all.

Consider also the lek paradox, an issue first identified by Borgia (1979) that does not appear by name in our book although we did discuss the nature of the problem: how can females in lek mating systems be selected to maintain a strong preference for males with certain extreme traits when strong female choice in the past should have eliminated any genetic variation that underlies the preferred male traits, thereby eliminating the fitness benefits for choosy females? The lek paradox is discussed in several chapters of this book in which different hypotheses for its solution are presented. Shuker (Chapter 2), Ritchie and Butlin (Chapter 3) and Hunt and Sakaluk (Chapter 8) discuss the genic capture hypothesis, which is based on the notion that if traits reflecting the condition of the individual exist and if the development of these traits depends on the action of many loci, then female preferences for certain male characteristics will not necessarily eliminate the genetic variation underlying the ability of males to produce these condition-dependent attributes.

Zuk and Wedell (Chapter 13) provide an alternative proposal, which is also discussed by Shuker (Chapter 2) and Hunt and Sakaluk (Chapter 8). Imagine that parasites are co-evolving with their hosts, in which case males in the parasite-infested populations that happen to have the ability to combat the currently common parasitic genotype will have higher fitness (and so will their mating partners that produce parasite-resistant offspring). If relatively parasite-free males can invest in extravagant ornaments or other signals of their resistance, then females 'should' evolve a continuing preference for males whose appearance or behaviour is associated with a reduced parasite load. Genetic variation affecting male attributes will be maintained to the extent that parasite pressure changes over time through the co-evolution of parasite and host with selection for more effective parasites constantly selecting for more highly effective parasite-resistant hosts.

Hunt and Sakaluk (Chapter 8) review yet another solution to the lek paradox based on the possibility that the environments occupied by males of a given species will vary such that the best male genotype in one environment may not be the best in another. Gene flow between environments will tend to maintain genetic variation and thereby maintain the adaptive value of female mate choice that matches the particular environment in which females are choosing among the competing males.

The main point that we wish to make here is simply that thinking about the lek paradox from an adaptationist perspective has resulted in several novel explanations for the phenomenon that were not considered in *The Evolution of Insect Mating Systems*. Moreover, the lek paradox arises because of a concern about the adaptive basis for female mate choice, which was once considered unlikely but now is the focus of so much research that male–male competition is no longer the primary focus of behavioural ecologists interested in sexual selection (Hunt and Sakaluk, Chapter 8).

However, despite the strong interest exhibited by many contributors in female adaptations for successful reproduction, male reproductive attributes are not ignored in this book. Although prior to 1983, the traits of male insects received more attention than female attributes, much remained to be learned about such things as the evolution of weapons employed by males as they compete for mates. Emlen (Chapter 6) illuminates the nature of resource competition in relation to male weaponry in insects by making a distinction between scrambles (in which males try to sequester valuable resources before others can get them) and duels (in which males defend key resources violently in order to keep others away from these resources).

Although sperm competition by males was one of the first phenomena to be identified and studied from an adaptationist perspective (Parker 1970e), there are still many issues allied with sperm competition that are worth exploring. Simmons (Chapter 10) examines some of these problems by reviewing, for example, how game theory can help make sense of why males might regulate the quantity of materials transferred to a mate, or why males might include two structurally different forms of sperm in their ejaculates.

Sperm competition is also involved in the evolution of the alternative mating behaviours that are so widespread among male insects, a conclusion that becomes more apparent with every passing year. Buzzato and co-authors (Chapter 7) concern themselves with the ways in which alternative phenotypes can be reversed—a new area of research—as well as considering the likelihood that in many species, females as well as males possess the flexibility to adopt one or another reproductive tactic in ways that elevate their fitness. Here we have another instance of the expanded interest in female reproductive behaviour that has occurred post 1983.

We are far from exhausting the list of novel issues that are examined in these pages, topics that we did not consider largely because the critical research had yet to be done during the time we were writing our book.

Should mating system diversity be approached from the perspective of Emlen and Oring (1977) or should we rely on models that incorporate Bateman's gradient in conjunction with the concept of the operational sex ratio (Kokko and co-authors, Chapter 3)? Or should we employ the statistical techniques of Lande and Arnold (1983) in order to account for the evolution of different mating systems (Hunt and Sakaluk, Chapter 8)? Resolution of these questions about how mating systems of insects can be best understood, or whether insect mating systems should be studied with a combination of these approaches, may eventually contribute to an encompassing general theory of mating systems across all animal taxa.

And what about the links between the proximate (genetic–developmental) and ultimate (evolutionary) components of insect mating systems? As Ritchie and Butlin explain (Chapter 4), there have been major new genetic advances from quantitative trait locus analysis to expression analyses to complete genome descriptions for insects, all of potential importance to evolutionary biologists. Indeed, if we wish to determine how selection acts on specific proximate physiological mechanisms, we would do well to understand the operation of these mechanisms, including those that underlie the production of gametes, an area untouched by Thornhill and Alcock, as Moore (Chapter 5) notes.

Then there is the problem of sexual selection itself (Shuker, Chapter 2) with a host of issues surrounding the degree to which it can be differentiated from natural selection and the extent to which Fisherian processes can account for mate choice and the evolution of extreme ornaments. This is to say nothing of sexual reproduction as well, which may be an adaptation to thwart co-evolving parasites; indeed parasites surely have had multiple evolutionary effects on insect reproduction (Zuk and Wedell, Chapter 13). The word 'parasite' does not appear in the index to our book, nor does 'ecological immunity', which is yet another adaptationist sub-discipline that developed in the post-1983 era.

Although we did not or could not explore many important topics in *The Evolution of Insect Mating Systems*, we are confident that further advances in the understanding of insect reproductive biology will arise from the application of an adaptationist approach to the unresolved issues in this field, as illustrated by the chapters in this book. Because the authors of these chapters have identified significant remaining questions about insect mating systems, the book seems certain to stimulate additional productive research that we look forward to learning about in the years ahead.

References

Abbot P, Dill LM (2001) Sexually transmitted parasites and sexual selection in the milkweed leaf beetle, *Labidomera clivicollis*. *Oikos* 92:91–100.

Abe T, Bignell DE, Higashi M (2000) *Termites: Evolution, Sociality, Symbioses, Ecology*. Kluwer Academic Publishers, Dordrecht.

Abell AJ, Cole BJ, Reyes R, Wiernasz DC (1999) Sexual selection on body size and shape in the western harvester ant, *Pogonomyrmex occidentalis cresson*. *Evolution* 53:535–545.

Adachi-Hagimori T, Miura K, Abe Y (2011) Gene flow between sexual and asexual strains of parasitic wasps: a possible case of sympatric speciation caused by a parthenogenesis-inducing bacterium. *Journal of Evolutionary Biology* 24:1254–1262.

Adamo S (2012) The importance of physiology for ecoimmunology: lessons from the insects. In GE Demas, RJ Nelson (eds), *Ecoimmunology*, pp. 413–439. Oxford University Press, Oxford.

Ala-Honkola O, Manier MK, Lüpold S, Pitnick S (2011) No evidence for postcopulatory inbreeding avoidance in *Drosophila melanogaster*. *Evolution* 65:2699–2705.

Alcock J (1987) Leks and hilltopping in insects. *Journal of Natural History* 21:319–328.

Alcock J (1994) Postinsemination associations between males and females in insects: the mate-guarding hypothesis. *Annual Review of Entomology* 39:1–21.

Alcock J (1995) Persistent size variation in the anthophorine bee *Centris pallida* (Apidae) despite a large male mating advantage. *Ecological Entomology* 20:1–4.

Alcock J (1996) Provisional rejection of three alternative hypotheses on the maintenance of a size dichotomy in Dawson's burrowing bee, *Amegilla dawsoni* (Apidae, Apinae, Anthophorini). *Behavioral Ecology and Sociobiology* 39:181–188.

Alcock J (1997) Competition from large males and the alternative mating tactics of small males of Dawson's burrowing bee (*Amegilla dawsoni*) (Apidae, Apinae, Anthophorini). *Journal of Insect Behavior* 10:99–113.

Alcock J (2000) Possible causes of variation in territory tenure in a lekking pompilid wasp (*Hemipepsis ustulata*) (Hymenoptera). *Journal of Insect Behavior* 13:439–453.

Alcock J, Jones CE, Buchmann SL (1977) Male mating strategies in bee *Centris pallida* Fox (Anthophoridae–Hymenoptera). *American Naturalist* 111:145–155.

Alexander RD, Bigelow RS (1960) Allochronic speciation in field crickets, and a new species, *Acheta veletis*. *Evolution* 14:334–346.

Ali JG, Tallamy DW (2010) Female spotted cucumber beetles use own cuticular hydrocarbon signature to choose immunocompatible mates. *Animal Behaviour* 80:9–12.

Allen LE, Barry KL, Holwell GI, Herberstein ME (2011) Perceived risk of sperm competition affects juvenile development and ejaculate expenditure in male praying mantids. *Animal Behaviour* 82:1201–1206.

Almbro M, Kullberg C (2009) The downfall of mating: the effect of mate-carrying and flight muscle ratio on the escape ability of a pierid butterfly. *Behavioral Ecology and Sociobiology* 63:413–420.

Alonzo SH (2009) Social and coevolutionary feedbacks between mating and parental investment. *Trends in Ecology and Evolution* 25:99–108.

Alonzo SH, Klug H (2012) Paternity, maternity, and parental care. In NJ Royle, PT Smiseth, M Kölliker (eds), *The Evolution of Parental Care*, pp. 189–205. Oxford University Press, Oxford.

Amin MR, Bussière LF, Goulson D (2012) Effect of male age and size on mating success in the bumblebee *Bombus terrestris*. *Journal of Insect Behaviour* 25:362–374.

Andersen JC, Gruwell, ME, Morse GE, Normark BB (2010) Cryptic diversity of the *Aspidiotus nerii* complex in Australia. *Annals of the Entomological Society of America* 103:844–854.

Anderson C, Cremer S, Heinze J (2003) Live and let die: Why fighter males of the ant *Cardiocondyla* kill each other but tolerate their winged rivals. *Behavioral Ecology* 14:54–62.

Andersson J, Borg-Karlson A-K, Wiklund C (2000) Sexual cooperation and conflict in butterflies: a male-transferred anti-aphrodisiac reduces harassment of recently mated females. *Proceedings of the Royal Society of London B* 267:1271–1275.

Andersson J, Borg-Karlson A-K, Wiklund C (2004) Sexual conflict and anti-aphrodisiac titre in a polyandrous butterfly: male ejaculate tailoring and absence of female control. *Proceedings of the Royal Society of London B* 271:1765–1770.

Andersson M (1994) *Sexual Selection*. Princeton University Press, Princeton, NJ.

Andersson M, Iwasa Y (1996) Sexual selection. *Trends in Ecology and Evolution* 11:53–58.

Andersson M, Simmons LW (2006) Sexual selection and mate choice. *Trends in Ecology and Evolution* 21:296–302.

Andrés JA, Arnqvist G (2001) Genetic divergence of the seminal signal-receptor system in houseflies: the footprints of sexually antagonistic coevolution? *Proceedings of the Royal Society of London B* 268:399–405.

Andrés JA, Maroja LS, Bogdanowicz SM, et al. (2006) Molecular evolution of seminal proteins in field crickets. *Molecular Biology and Evolution* 23:1574–1584.

Angelo G, Gilst MRV (2009) Starvation protects germline stem cells and extends reproductive longevity in *C. elegans*. *Science* 326:954–958.

Arakaki N, Kishita M, Nagayama A, et al. (2004) Precopulatory mate guarding by the male green chafer, *Anomala albopilosa sakishimana* Nomura (Coleoptera: Searabaeidae). *Applied Entomology and Zoology* 39:455–462.

Arbuthnott D (2009) The genetic architecture of insect courtship behavior and premating isolation. *Heredity* 103:15–22.

Arce AN, Johnston PR, Smiseth PT, Rozen DE (2012) Mechanisms and fitness effects of antibacterial defenses in a carrion beetle. *Journal of Evolutionary Biology* 25:930–937.

Archetti M (2005) Accumulation of deleterious mutations in hybridogenetic organisms. *Journal of Theoretical Biology* 234:151–152.

Arnold SJ (1994) Bateman's principles and the measurement of sexual selection in plants and animals. *American Naturalist* 144:S126–S149.

Arnold SJ, Duvall D (1994) Animal mating systems: a synthesis based on selection theory. *American Naturalist* 143:317–348.

Arnold SJ, Wade MJ (1984a) On the measurement of natural and sexual selection: theory. *Evolution* 38:709–719.

Arnold SJ, Wade MJ (1984b) On the measurement of natural and sexual selection: applications. *Evolution* 38:720–734.

Arnott G, Elwood RW (2009) Assessment of fighting ability in animal contests. *Animal Behaviour* 77:991–1004.

Arnqvist G (1989) Multiple mating in a water strider: mutual benefits or intersexual conflict. *Animal Behaviour* 38:749–756.

Arnqvist G (1992) Pre-copulatory fighting in a water strider: inter-sexual conflict or mate assessment? *Animal Behaviour* 43:559–567.

Arnqvist G (1998) Comparative evidence for the evolution of genitalia by sexual selection. *Nature* 393:784–786.

Arnqvist G (2006) Sensory exploitation and sexual conflict. *Philosophical Transactions of the Royal Society B* 361:375–386.

Arnqvis G, Danielsson I (1999) Copulatory behavior, genital morphology, and male fertilization success in water striders. *Evolution* 53:147–156.

Arnqvist G, Nilsson T (2000) The evolution of polyandry: multiple mating and female fitness in insects. *Animal Behaviour* 60:145–164.

Arnqvist G, Rowe L (2002) Antagonistic coevolution between the sexes in a group of insects. *Nature* 415:787–789.

Arnqvist G, Rowe L (2005) *Sexual Conflict*. Princeton University Press, Princeton, NJ.

Arnqvist G, Thornhill R (1998) Evolution of animal genitalia: patterns of phenotypic and genotypic variation and condition dependence of genital and non-genital morphology in water strider (Heteroptera: Gerridae: Insecta). *Genetical Research* 71:193–212.

Aspi J (1992) Female mate choice and mating system among boreal *Drosophila virilis* group species. *Acta University of Ouluensis* A:241.

Attisano A, Moore AJ, Moore PJ (2012) Reproduction–longevity trade-offs reflect diet, not adaptation. *Journal of Evolutionary Biology* 25:873–880.

Austad SN (1984) A classification of alternative reproductive behaviors and methods for field-testing ESS models. *American Zoologist* 24:309–319.

Avila FW, Sirot LK, Laflamme BA, et al. (2011) Insect seminal fluid proteins: identification and function. *Annual Review of Entomology* 56:21–40.

Ayasse M, Paxton RJ, Tengo J (2001) Mating behavior and chemical communication in the order Hymenoptera. *Annual Review of Entomology* 46:31–78.

Ayroles JF, Carbone MA, Stone EA, et al. (2009) Systems genetics of complex traits in *Drosophila melanogaster*. *Nature Genetics* 41:299–307.

Bacigalupe LD, Crudgington HS, Hunter F, et al. (2007) Sexual conflict does not drive reproductive isolation in experimental populations of *Drosophila pseudoobscura*. *Journal of Evolutionary Biology* 20:1763–1771.

Baer B (2003) Bumblebees as model organisms to study male sexual selection in social insects. *Behavioral Ecology and Sociobiology* 54:521–533.

Baer B (2005) Sexual selection in *Apis* bees. *Apidologie* 36:187–200.

Baer B (2011) The copulation biology of ants (Hymenoptera: Formicidae). *Myrmecological News* 14:55–68.

Baer B, Boomsma JJ (2004) Male reproductive investment and queen mating frequency in fungus growing ants. *Behavioral Ecology* 15:426–432.

Baer B, Boomsma JJ (2006) Mating biology of the leaf-cutting ants *Atta colombica* and *A. cephalotes*. *Journal of Morphology* 267:1165–1171.

Baer B, Schmid-Hempel P (1999) Experimental variation in polyandry affects parasite loads and fitness in a bumblebee. *Nature* 397:151–154.

Baer B, Schmid-Hempel P (2001) Unexpected consequences of polyandry for parasitism and fitness in the bumblebee, *Bombus terrestris*. *Evolution* 55:1639–1643.

Baer B, Maile R, Schmid-Hempel P, et al. (2000) Chemistry of a mating plug in bumblebees. *Journal of Chemical Ecology* 26:1869–1875.

Baer B, Morgan ED, Schmid-Hempel P (2001) A non-specific fatty acid within the bumblebee mating plug prevents females from remating. *Proceedings of the National Academy of Sciences of the USA* 98:3926–3928.

Baer B, Schmid-Hempel P, Hoeg JT, Boomsma JJ (2003) Sperm length, sperm storage and mating system characteristics in bumblebees. *Insectes Sociaux* 50:101–108.

Baer B, Heazlewood JL, Taylor NL, et al. (2009) The seminal fluid proteome of the honeybee *Apis mellifera*. *Proteomics* 9:2085–2097.

Bailey WJ, Field G (2000) Acoustic satellite behaviour in the Australian bushcricket *Elephantodeta nobilis* (Phaneropterinae, Tettigoniidae, Orthoptera). *Animal Behaviour* 59:361–369.

Bailey RI, Innocenti P, Morrow EH, et al. (2011) Female *Drosophila melanogaster* gene expression and mate choice: the X chromosome harbours candidate genes underlying sexual isolation. *PLoS One* 6:e17358.

Bailey NW, Moore AJ (2012) Runaway sexual selection without genetic correlations: social environments and flexible mate choice initiate and enhance the Fisher process. *Evolution* 66:2674–2684.

Bailey NW, Gray B, Zuk M (2010) Acoustic experience shapes alternative mating tactics and reproductive investment in male field crickets. *Current Biology* 20:845–849

Baker RH, Wilkinson GS (2001) Phylogenetic analysis of sexual dimorphism and eye-span allometry in stalk-eyed flies (Diopsidae). *Evolution* 55:1373–1385.

Bakker TCM (1993) Positive genetic correlation between female preference and preferred male ornament in sticklebacks. *Nature* 363:255–257.

Banks MJ, Thompson DJ (1985) Lifetime mating success in the damselfly *Coenagrion puella*. *Animal Behaviour* 33:1175–1183.

Bao J, Yan W (2012) Male germline control of transposable elements. *Biology of Reproduction* 86:1–14.

Barbosa F (2009) Cryptic female choice by female control of oviposition timing in a soldier fly. *Behavioral Ecology* 20:957–960.

Barbosa F (2011) Copulation duration in the soldier fly: the roles of cryptic male choice and sperm competition risk. *Behavioral Ecology* 22:1332–1336.

Barbosa F (2012) Males responding to sperm competition cues have higher fertilization success in a soldier fly. *Behavioral Ecology* 23:815–819.

Barbosa P, Krischik V, Lance D (1989) Life-history traits of forest-inhabiting flightless Lepidoptera. *American Midland Naturalist* 122:262–274.

Bargmann CI (2006) Comparative chemosensation from receptors to ecology. *Nature* 444:295–301.

Barry KL, Holwell GI, Herberstein ME (2011) A paternity advantage for speedy males? Sperm precedence patterns and female re-mating frequencies in a sexually cannibalistic praying mantid. *Evolutionary Ecology* 25:107–119.

Barry KL, Kokko H (2010) Male mate choice: why sequential choice can make its evolution difficult. *Animal Behaviour* 80:163–169.

Bascunan-Garcia AP, Lara C, Cordoba-Aguilar, A (2010) Immune investment impairs growth, female reproduction and survival in the house cricket, *Acheta domesticus*. *Journal of Insect Physiology* 56:204–211.

Bastock M, Manning A (1955) The courtship of *Drosophila melanogaster*. *Behaviour* 8:85–111.

Bateman AJ (1948) Intra-sexual selection in *Drosophila*. *Heredity* 2:349–368.

Bateson P (1983) *Mate Choice*. Cambridge University Press, Cambridge.

Baudry E, Solignac M, Garnery L, et al. (1998). Relatedness among honeybees (*Apis mellifera*) of a drone congregation. *Proceedings of the Royal Society of London B* 26:2009–2014.

Beal CA, Tallamy DW (2006) A new record of amphisexual care in an insect with exclusive paternal care: *Rhinocoris tristis* (Heteroptera: Reduviidae). *Journal of Ethology* 24:305–307.

Beani L, Turillazzi S (1988) Alternative mating tactics in males of *Polistes dominulus* (Hymenoptera, Vespidae). *Behavioral Ecology and Sociobiology* 22:257–264.

Beekman M, Van Stratum P (1998) Bumblebee sex ratios: why do bumblebees produce so many males? *Proceedings of the Royal Society of London B* 265:1535–1543.

Bell G (1982) *The Masterpiece of Nature*. University of California Press, Berkeley.

Bell WJ, Roth LM, Nalepa CA (2007) *Cockroaches: Ecology, Behavior, and Natural History*. John Hopkins University Press, Baltimore.

Bengtsson BO, Lofstedt C (2007) Direct and indirect selection in moth pheromone evolution: population genetical simulations of asymmetric sexual interactions. *Biological Journal of the Linnean Society* 90:117–123.

Benoit JB, Jajack AJ, Yoder JA (2012) Multiple traumatic insemination events reduce the ability of bed bug females to maintain water balance. *Journal of Comparative Physiology B* 182:189–198.

Bentsen CL, Hunt J, Jennions MD, Brooks R (2006) Complex multivariate sexual selection on male acoustic signaling in a wild population of *Teleogryllus commodus*. *American Naturalist* 167:E102–E116.

Berg S, Koeniger N, Koeniger G, Fuchs S (1997) Body size and reproductive success of drones (*Apis mellifera* L). *Apidologie* 28:449–460.

Berger D, Bauerfeind SS, Blanckenhorn WU, Schafer MA (2011) High temperatures reveal cryptic genetic variation in a polymorphic female sperm storage organ. *Evolution* 65:2830–2842.

Bergland AO (2011) Mechanisms of nutrient-dependent reproduction in dipteran insects. In T Flatt, A Heyland (eds), *Mechanisms of Life History Evolution*, pp. 127–136. Oxford University Press, Oxford.

Bergland AO, Genissel A, Nuzhdin SV, Tatar M (2008) Quantitative trait loci affecting phenotypic plasticity and the allometric relationship of ovariole number and thorax length in *Drosophila melanogaster*. *Genetics* 180:567–582.

Bergsten J, Miller KB (2007) Phylogeny of diving beetles reveals a coevolutionary arms race between the sexes. *PLoS One* 2:e522.

Beukeboom LW, Vrijenhoek RC (1998) Evolutionary genetics and ecology of sperm-dependent parthenogenesis. *Journal of Evolutionary Biology* 11:755–782.

Beveridge M, Simmons LW, Alcock J (2006) Genetic breeding system and investment patterns within nests of Dawson's burrowing bee (*Amegilla dawsoni*) (Hymenoptera: Anthophorini). *Molecular Ecology* 15:3459–3467.

Bian G, Joshi D, Dong Y, et al. (2013) Establishment of a stable *Wolbachia* infection in *Anopheles stephensi* induces refractoriness to *Plasmodium* infections. *Science* 340:748–751.

Bilde T, Friberg U, Maklakov AA, et al. (2008) The genetic architecture of fitness in a seed beetle: assessing the potential for indirect genetic benefits of female choice. *BMC Evolutionary Biology* 8:295.

Bilde T, Foged A, Schilling N, Arnqvist G (2009) Postmating sexual selection favors males that sire offspring with low fitness. *Science* 324:1705–1706.

Billeter J-C, Levine JD (2013) Who is he and what is he to you? Recognition in *Drosophila melanogaster*. *Current Opinion in Neurobiology* 23:17–23.

Billeter J-C, Rideout EJ, Dornan AJ, Goodwin SF (2006) Control of male sexual behavior in *Drosophila* by the sex determination pathway. *Current Biology* 16:766–776.

Billeter J-C, Jagadeesh S, Stepek N, et al. (2012) *Drosophila melanogaster* females change mating behaviour and offspring production based on social context. *Proceedings of the Royal Society of London B* 279:2417–2425.

Bilto, DT, Thompson A, Foster GN (2008) Inter- and intrasexual dimorphism in the diving beetle *Hydroporus memnonius* Nicolai (Coleoptera: Dytiscidae). *Biological Journal of the Linnean Society* 94:685–697.

Birkhead TR (2010) How stupid not to have thought of that: post-copulatory sexual selection. *Journal of Zoology* 281:78–93.

Birkhead TR, Moller AP (1998) *Sperm Competition and Sexual Selection*. Academic Press, London.

Bjork A, Dallai R, Pitnick S (2007a) Adaptive modulation of sperm production rate in *Drosophila bifurca*, a species with giant sperm. *Biology Letters* 3:517–519.

Bjork A, Starmer WT, Higginson DM, et al. (2007b) Complex interactions with females and rival males limit the evolution of sperm offence and defence. *Proceedings of the Royal Society of London* B 274:1779–1788.

Blanckenhorn WU, Hosken DJ, Martin OY, et al. (2002) The costs of copulating in the dung fly *Sepsis cynipsea*. *Behavioral Ecology* 13:353–358.

Blows MW (2007) A tale of two matrices: multivariate approaches to evolutionary biology. *Journal of Evolutionary Biology* 20:1–8.

Blows MW, Brooks R (2003) Measuring nonlinear selection. *American Naturalist* 162:815–820.

Blows MW, Chenoweth SF, Hine E (2004) Orientation of the genetic variance–covariance matrix and the fitness surface for multiple male sexually selected traits. *American Naturalist* 163:329–346.

Blum MS, Hilker M (2002) Chemical protection of insect eggs. In M Hilker, T Meiners (eds), *Chemoecology of Insect Eggs and Egg Deposition*, pp. 61–90. Blackwell, Oxford.

Blyth JE, Gilburn AS (2006) Extreme promiscuity in a mating system dominated by sexual conflict. *Journal of Insect Behavior* 19:447–455.

Boake CRB (1989) Repeatability: its role in evolutionary studies of mating behaviour. *Evolutionary Ecology* 3:173–182.

Boake CRB (1991) Coevolution of senders and receivers of sexual signals: genetic coupling and genetic coevolution. *Trends in Ecology and Evolution* 6:225–227.

Boake CRB, Arnold SJ, Breden F, et al. (2002) Genetic tools for studying adaptation and the evolution of behavior. *American Naturalist* 160:S143–S159.

Boerjan B, Tobback J, De Loof A, et al. (2011) Fruitless RNAi knockdown in males interferes with copulation success in *Schistocerca gregaria*. *Insect Biochemistry and Molecular Biology* 41:340–347.

Boivin G (2013) Sperm as a limiting factor in mating success in Hymenoptera parasitoids. *Entomologia Experimentalis et Applicata* 146:149–155.

Boivin G, Jacob S, Damiens D (2005) Spermatogeny as a life-history index in parasitoid wasps. *Oecologia* 143:198–202.

Bonduriansky R (2001) The evolution of male mate choice in insects: a synthesis of ideas and evidence. *Biological Reviews* 76:305–339.

Boomsma JJ (2007) Kin selection versus sexual selection: why the ends do not meet. *Current Biology* 17:R673–R683.

Boomsma JJ (2009) Lifetime monogamy and the evolution of eusociality. *Philosophical Transactions of the Royal Society of London B* 364:3191–3207.

Boomsma JJ, Ratnieks FLW (1996) Paternity in eusocial Hymenoptera. *Philosophical Transactions of the Royal Society of London B* 351:947–975.

Boomsma JJ, Baer B, Heinze J (2005) The evolution of male traits in social insects. *Annual Review of Entomology* 50:395–420.

Boomsma JJ, Beekman M, Cornwallis CK, et al. (2011) Only full-sibling families evolved eusociality. *Nature* 471:E4–E5.

Boots M (2008)Fight or learn to live with the consequences? *Trends Ecol Evol* 23:248–250.

Borgia G (1979) Sexual selection and the evolution of mating systems. In MS Blum, NA Blum (eds), *Sexual Selection and Reproductive Competition*, pp. 19–80. Academic Press, New York.

Bourke AFG (2011) *Principles of Social Evolution*. Oxford University Press, Oxford.

Bourtzis K, Pettigrew MM, O'Neill SL (2000) *Wolbachia* neither induces nor suppresses transcripts encoding antimicrobial peptides. *Insect Molecular Biology* 9:635–639.

Bousquet F, Nojima T, Houot B, et al. (2012) Expression of a desaturase gene, *desat1*, in neural and nonneural tissues separately affects perception and emission of sex pheromones in *Drosophila*. *Proceedings of the National Academy of Sciences of the USA* 109:249–254.

Boyle M, Wong C, Rocha M, Jones DL (2007) Decline in self-renewal factors contributes to aging of the stem cell niche in the *Drosophila* testis. *Cell Stem Cell* 1:470–478.

Bradbury JW, Andersson MB (1987) *Sexual Selection: Testing the Alternatives*. John Wiley & Sons, Chichester.

Bradbury JW, Vehrencamp SL (2011) *Principles of Animal Communication*. Sinauer Associates, Massachusetts.

Bradshaw WE, Holzapfel CM (2010) Circadian clock genes, ovarian development and diapause. *BMC Biology* 8:115.

Brakefield PM, El Filali E, van der Laan R, et al. (2001) Effective population size, reproductive success and sperm precedence in the butterfly, *Bicyclus anyana*, in captivity. *Journal of Evolutionary Biology* 14:148–156.

Brandt Y, Swallow JG (2009) Do the elongated eye stalks of Diopsid flies facilitate rival assessment? *Behavioral Ecology and Sociobiology* 63:1243–1246.

Bray S, Amrein H (2003) A putative *Drosophila* pheromone receptor expressed in male-specific taste neurons is required for efficient courtship. *Neuron* 39:1019–1029.

Brent CS, Byers JA (2011) Female attractiveness modulated by a male-derived antiaphrodisiac pheromone in a plant bug. *Animal Behaviour* 82:937–943.

Bretman A, Tregenza T (2005) Measuring polyandry in wild populations: a case study using promiscuous crickets. *Molecular Ecology* 14:2169–2179.

Bretman A, Newcombe D, Tregenza T (2009) Promiscuous females avoid inbreeding by controlling sperm storage. *Molecular Ecology* 18:3340–3345.

Bretman A, Fricke C, Hetherington P, et al. (2010) Exposure to rivals and plastic responses to sperm competition in *Drosophila melanogaster*. *Behavioral Ecology* 21:317–321.

Bretman A, Rodriguez-Munoz R, Walling C, et al. (2011a) Fine-scale population structure, inbreeding risk and avoidance in a wild insect population. *Molecular Ecology* 20:3045–3055.

Bretman A, Westmancoat JD, Gage MJG, Chapman T (2011b) Males use multiple, redundant cues to detect mating rivals. *Current Biology* 21:617–622.

Bretman A, Westmancoat JD, Gage MJG, Chapman T (2012) Individual plastic responses by males to rivals reveal mismatches between behaviour and fitness outcomes. *Proceedings of the Royal Society of London B* 279:2868–2876.

Briceno RD, Eberhard WG (2009) Experimental modifications imply a stimulatory function for male tsetse fly genitalia, supporting cryptic female choice theory. *Journal of Evolutionary Biology* 22:1516–1525.

Brockman HJ (1980) Diversity in the nesting behavior of mud-daubers (*Trypoxylon politum* Say; Sphecidae). *Florida Entomologist* 63:53–64.

Brockmann HJ (2008) Alternative reproductive tactics in insects. In RF Oliveira, M Taborsky, HJ Brockmann (eds), *Alternative Reproductive Tactics: An Integrative Approach*, pp. 177–223. Cambridge University Press, Cambridge.

Brodt J, Tallamy DW, Ali JG (2005) Female choice by scent recognition in the spotted cucumber beetle. *Ethology* 112:300–306.

Brommer JE, Fricke C, Edward DA, Chapman T (2012) Interactions between genotype and sexual conflict environment influence transgenerational fitness in *Drosophila melanogaster*. *Evolution* 66:517–531.

Brooks R, Hunt J, Blows MW, et al. (2005) Experimental evidence for multivariate stabilizing sexual selection. *Evolution* 59:871–880.

Brown JL (1964) The evolution of diversity in avian territorial systems. *The Wilson Bulletin* 76:160–169.

Brown SW (1964) Automatic frequency response in the evolution of male haploidy and other coccid chromosome systems. *Genetics* 49:797–817.

Brown MJF, Baer B (2005) The evolutionary significance of long copulation duration in bumblebees. *Apidologie* 36:157–167.

Brown MJF, Baer B, Schmid-Hempel R, Schmid-Hempel P (2002) Dynamics of multiple-mating in the bumblebee *Bombus hypnorum*. *Insectes Sociaux* 49:315–319.

Brown L, Macdonell J, Fitzgerald VJ (1985) Courtship and female choice in the horned beetle *Bolitotherus cornutus* (Coleoptera: Tenebrionidae). *Annals of the Entomological Society of America* 78:423–427.

Brun LO, Stuart J, Gaudichon V, et al. (1995) Functional haplodiploidy: a mechanism for the spread of insecticide resistance in an important international insect pest. *Proceedings of the National Academy of Sciences of the USA* 92:9861–9865.

Buchner P (1965) *Endosymbiosis of Animals with Plant Microorganisms*. Wiley Interscience, New York.

Bull JJ (1979) An advantage for the evolution of male haploidy and systems with similar genetic transmission. *Heredity* 43:361–381.

Bull JJ, Charnov EL (1985) On irreversible evolution. *Evolution* 39:1149–1155.

Bulmer MG (1983) Models for the evolution of protandry in insects. *Theoretical Population Biology* 23:314–322.

Bundgaard J, Barker JSF, Frydenberg J, Clark AG (2004) Remating and sperm displacement in a natural population of *Drosophila buzzatii* inferred from mother–offspring analysis of microsatellite loci. *Journal of Evolutionary Biology* 17:376–381.

Büning J (1994) *The Insect Ovary: Ultrastructure, Previtellogenic Growth and Evolution*. Springer, Berlin.

Burk T (1983) Male aggression and female choice in a field cricket (*Teleogryllus oceanicus*): the importance of courtship song. In DT Gwynne, GK Morris (eds), *Orthopteran Mating Systems: Sexual Selection in a Diverse Group of Insects*, pp. 97–119. Westview Press, Boulder, CO.

Burkhardt D, de la Motte I (1988) Big 'antlers' are favoured: female choice in stalk-eyed flies (Diptera, Insecta), field collected harems and laboratory experiments. *Journal of Comparative Physiology A* 162:649–652.

Burkhardt D, de la Motte I, Lunau K (1994) Signalling fitness: larger males sire more offspring. Studies of the stalk-eyed fly *Cyrtodiopsis whitei* (Diopsidae, Diptera). *Journal of Comparative Physiology A* 174:61–64.

Burt A (1995) The evolution of fitness. *Evolution* 49:1–8.

Burt A, Trivers A (2006) *Genes in Conflict: the Biology of Selfish Genetic Elements*. Harvard University Press, Harvard.

Burton-Chellew MN, Beukeboom LW, West SA, Shuker DM (2007) Laboratory evolution of polyandry in the parasitoid wasp *Nasonia vitripennis*. *Animal Behaviour* 74:1147–1154.

Bussiere LF, Hunt J, Jennions MD, Brooks R (2006) Sexual conflict and cryptic female choice in the black field cricket, *Teleogryllus commodus*. *Evolution* 60:792–800.

Bussière LF, Gwynne DT, Brooks R (2008) Contrasting sexual selection on males and females in a role-reversed swarming dance fly, *Rhamphomyia longicauda* (Diptera: Empididae). *Journal of Evolutionary Biology* 21:1683–1691.

Butlin RK (1993) The variability of mating signals and preferences in the brown planthopper, *Nilaparvata lugens* (Homoptera: Delphacidae). *Journal of Insect Behaviour* 6:125–140.

Butlin RK, Ritchie MG (1989) Genetic coupling in mate recognition systems: what is the evidence? *Biological Journal of the Linnean Society* 37:237–246.

Buzatto BA, Simmons LW, Tomkins JL (2012a) Genetic variation underlying the expression of a polyphenism. *Journal of Evolutionary Biology* 25:748–758.

Buzatto BA, Tomkins JL, Simmons LW (2012b) Maternal effects on male weaponry: female dung beetles produce major sons with longer horns when they perceive higher population density. *BMC Evolutionary Biology* 12:118.

Cade WH (1980) Alternative male reproductive behaviors. *Florida Entomologist* 63:30–45.

Cade WH (1981) Alternative male strategies: genetic differences in crickets. *Science* 212:563–564.

Cade WH (1984) Genetic variation underlying sexual behavior and reproduction. *American Zoologist* 24:355–366.

Cade WH, Wyatt DR (1984) Factors affecting calling behavior in field crickets, *Teleogryllus* and *Gryllus* (age, weight, density, and parasites). *Behaviour* 88:61–75.

Cahan SH (2003) Reproductive division of labor between hybrid and nonhybrid offspring in a fire ant hybrid zone. *Evolution* 57:1562–1570.

Calabrese JM, Fagan WF (2004) Lost in time, lonely, and single: reproductive asynchrony and the Allee effect. *American Naturalist* 164:25–37.

Calabrese JM, Ries L, Matter SF, et al. (2008) Reproductive asynchrony in natural butterfly populations and its consequences for female matelessness. *Journal of Animal Ecology* 77:746–756.

Cameron E, Day T, Rowe L (2003) Sexual conflict and indirect benefits. *Journal of Evolutionary Biology* 16:1055–1060.

Cameron E, Day T, Rowe L (2007) Sperm competition and the evolution of ejaculate composition. *American Naturalist* 169:E158–E172.

Carayon J (1966) Traumatic insemination and the paragenital system. In RL Usinger (ed.), *Monograph of Cimicidae*, pp. 81–166. Entomological Society of America, Baltimore.

Carazo P, Font E, Alfthan B (2007) Chemosensory assessment of sperm competition levels and the evolution of internal spermatophore guarding. *Proceedings of the Royal Society of London B* 274:261–267.

Carazo P, Fernández-Perea R, Font E (2012) Quantity estimation based on numerical cues in the mealworm beetle (*Tenebrio molitor*). *Frontiers in Psychology* 3:502.

Carbone SS, Rivera AC (1998) Sperm competition, cryptic female choice and prolonged mating in the Eucalyptus snout-beetle, *Gonipterus scutellatus* (Coleoptera, Curculionidae). *Etología* 6:33–40.

Carbone SS, Rivera AC (2003) Fertility and paternity in the Eucalyptus snout-beetle Gonipterus scutellatus: females might benefit from sperm mixing. *Ethology, Ecology and Evolution* 15:283–294.

Cardé RT, Baker TC (1984) Sexual communication with pheromones. In WJ Bell, RT Cardé (eds), *Chemical Ecology of Insects*, pp. 355–383. Chapman & Hall, London.

Cardé RT, Roelofs WL, Harrison RG, et al. (1978) European corn-borer–pheromone polymorphism or sibling species. *Science* 199:555–556.

Carney GE (2007) A rapid genome-wide response to *Drosophila melanogaster* social interactions. *BMC Genomics* 8:288.

Carranza J (2010) Sexual selection and the evolution of evolutionary theories. *Animal Behaviour* 79:E5–E6.

Champion de Crespigny FE, Wedell N (2006) *Wolbachia* infection reduces sperm competitive ability in an insect. *Proceedings of the Royal Society of London B* 273:1455–1458.

Champion de Crespigny FE, Wedell N (2007) Mate preferences in *Drosophila* infected with *Wolbachia*? *Behavioral Ecology and Sociobiology* 61:1229–1235.

Champion de Crespigny FE, Pitt T, Wedell N (2006) Increased male mating rate in *Drosophila* is associated with *Wolbachia* infection. *Journal of Evolutionary Biology* 19:1964–1972.

Champion de Crespigny FE, Hurst LD, Wedell N (2008) Do *Wolbachia* associated incompatibilities promote polyandry? *Evolution* 62:107–122.

Chang AS (2004) Conspecific sperm precedence in sister species of *Drosophila* with overlapping ranges. *Evolution* 58:781–789.

Chapman AD (2009) *Numbers of Living Species in Australia and the World*. Australian Biodiversity Information Services, Toowoomba.

Chapman RF (2013) *The Insects: Structure and Function*, SJ Simpson, AE Douglas (eds). Cambridge University Press, Cambridge.

Chapman T (2001) Seminal fluid-mediated fitness traits in *Drosophila*. *Heredity* 87:511–521.

Chapman T (2008) The soup in my fly: evolution, form and function of seminal fluid proteins. *PLoS Biology* 6:1379–1382.

Chapman T, Liddle LF, Kalb JM, et al. (1995) Cost of mating in *Drosophila melanogaster* females is mediated by male accessory gland products. *Nature* 373:241–244.

Chapman T, Arnqvist G, Bangham J, Rowe L (2003a) Sexual conflict. *Trends in Ecology and Evolution* 18:41–47.

Chapman T, Bangham J, Vinti G, et al. (2003b) The sex peptide of *Drosophila melanogaster*: female post-mating responses analyzed by using RNA interference. *Proceedings of the National Academy of Sciences of the USA* 100:9923–9928.

Charalambous M, Butlin RK, Hewitt GM (1994) Genetic variation in male song and female song preference in the grasshopper *Chrothippus brunneus* (Orthoptera: Acrididae). *Animal Behaviour* 47:399–411.

Charlat S, Reuter M, Dyson EA, et al. (2007) Male killing bacteria trigger a cycle of increasing male fatigue and female promiscuity. *Current Biology* 17:273–277.

Charlesworth B, Charlesworth D (2010) *Elements of Evolutionary Genetics*. Roberts & Co., Greeenwood Village, CO.

Charmantier A, Sheldon BC (2006) Testing genetic models of mate choice evolution in the wild. *Trends in Ecology and Evolution* 21:417–419.

Chenoweth SF, Blows MW (2003) Signal trait sexual dimorphism and mutual sexual selection in *Drosophila serrata*. *Evolution* 57:2326–2334.

Chenoweth SF, Blows, MW (2005) Contrasting mutual sexual selection on homologous signal traits in *Drosophila serrata*. *American Naturalist* 165:281–289.

Chenoweth SF, Blows MW (2006) Dissecting the complex genetic basis of mate choice. *Nature Reviews* 7:681–692.

Chenoweth SF, McGuigan K (2010) The genetic basis of sexually selected variation. *Annual Review of Ecology, Evolution, and Systematics* 41:81–101.

Chenoweth SF, Rundle HD, Blows MW (2008) Genetic constraints and the evolution of display trait sexual dimorphism by natural and sexual selection. *American Naturalist* 171:22–34.

Chenoweth SF, Rundle HD, Blows MW (2010) The contribution of selection and genetic constraints to phenotypic divergence. *American Naturalist* 175:186–196.

Chenoweth SF, Hunt J, Rundle HD (2012) Analyzing and comparing the geometry of individual fitness surfaces. In EI Svensson, R Calsbeek (eds), *The Adaptive Landscape in Evolutionary Biology*, pp. 126–150. Oxford University Press, Oxford.

Cheron B, Monnin T, Federici P, Doums C (2011) Variation in patriline reproductive success during queen production in orphaned colonies of the thelytokous ant *Cataglyphis cursor*. *Molecular Ecology* 20:2011–2022.

Chippindale AK, Gibson JR, Rice WR (2001) Negative genetic correlation for adult fitness between sexes reveals ontogenetic conflict in *Drosophila*. *Proceedings of the National Academy of Sciences of the USA* 98:1671–1675.

Choe JC, Crespi BJ (1997) *The Evolution of Social Behavior in Insects and Arachnids*. Cambridge University Press, Cambridge.

Chow CY, Wolfner MF, Clark AG (2013) Large neurological component to genetic differences underlying biased sperm use in *Drosophila*. *Genetics* 193:177–185.

Christy JH (1995) Mimicry, mate choice, and the sensory trap hypothesis. *American Naturalist* 146:171–181.

Christy JH, Salmon M (1984) Ecology and evolution of mating systems of fiddler crabs (Genus *Uca*). *Biological Reviews* 59:483–509.

Clark AG, Begun DJ, Prout T (1999) Female × male interactions in *Drosophila* sperm competition. *Science* 283:217–220.

Clutton-Brock TH (1991) *The Evolution of Parental Care*. Princeton University Press, Princeton, NJ.

Clutton-Brock T (2007) Sexual selection in males and females. *Science* 318:1882–1885.

Clutton-Brock, T (2009) Sexual selection in females. *Animal Behaviour* 77:3–11.

Clutton-Brock TH, Parker GA (1992) Potential reproductive rates and the operation of sexual selection. *Quarterly Reviews of Biology* 67:437–456.

Clyne PJ, Warr CG, Freeman MR, et al. (1999) A novel family of divergent seven-transmembrane proteins: candidate odorant receptors in *Drosophila*. *Neuron* 22:327–338.

Colegrave N, Kotiaho JS, Tomkins JL (2002) Mate choice or polyandry: reconciling genetic compatibility and good genes sexual selection. *Evolutionary Ecology Research* 4:911–917.

Collins RD, Cardé RT (1989a) Heritable variation in pheromone response of the pink bollworm, *Pectinophora gossypiella* (Lepidoptera: Gelechiidae). *Journal of Chemical Ecology* 15:2647–2659.

Collins RD, Cardé RT (1989b) Selection for altered pheromone-component ratios in the pink bollworm moth, *Pectinophora gossypiella* (Lepidoptera: Gelechiidae). *Journal of Insect Behaviour* 2:609–621.

Collins RD, Cardé RT (1990) Selection for increased pheromone response in the male pink bollworm, *Pectinophora gossypiella* (Leptidoptera: Gelechiidae). *Behavior Genetics* 20:325–331.

Collins RD, Rosenblum SL, Cardé RT (1990) Selection for increased pheromone titre in the pink bollworm moth, *Pectinophora gossypiella* (Lepidoptera: Gelechiidae). *Physiological Entomology* 15:141–147.

Colonello NA, Hartfelder K (2005) She's my girl—male accessory gland products and their function in the reproductive biology of social bees. *Apidologie* 36:231–244.

Coltman DW (2008) Molecular ecological approaches to studying the evolutionary impact of selective harvesting in wildlife. *Molecular Ecology* 17:221–235.

Conner J (1988) Field measurements of natural and sexual selection in the fungus beetle *Bolitotherus cornutus*. *Evolution* 42:736–749.

Conner WE, Itagaki H (1984) Pupal attendance in the crabhole mosquito *Deinocerites cancer*: the effects of pupal sex and age. *Physiological Entomology* 9:263–267.

Conner WE, Roach B, Benedict E, et al. (1990) Courtship pheromone production and body size as correlates of larval diet in males of the arctiid moth, *Utetheisa ornatrix*. *Journal of Chemical Ecology* 16:543–552.

Contreras-Garduño J, Canales-Lazcana J, Córdoba-Aguilar A (2006) Wing pigmentation, immune ability, fat reserves and territorial status in males of the rubyspot damselfly, *Hetaerina americana*. *Journal of Ethology* 24:165–173.

Contreras-Garduño J, Buzatto BA, Serrano-Meneses MA, et al. (2008) The size of the red wing spot of the American rubyspot as a heightened condition-dependent ornament. *Behavioral Ecology* 19:724–732.

Contreras-Garduño J, Córdoba-Aguilar A, Lanz-Mendoza H, Cordero Rivera A (2009) Territorial behaviour and immunity are mediated by juvenile hormone: the physiological basis of honest signalling? *Functional Ecology* 23:157–163.

Contreras-Garduño J, Córdoba-Aguilar A, Azpilicueta-Amorín M, Cordero-Rivera, A (2011) Juvenile hormone favors sexually-selected traits but impairs fat reserves and abdomen mass in males and females. *Evolutionary Ecology* 25:845–856.

Cook RM (1973) Courtship processing in *Drosophila melanogaster*. I. Selection for receptivity to wingless males. *Animal Behaviour* 21:338–348.

Cook JM (1993) Experimental tests of sex determination in *Goniozus nephantidis* (Hymenoptera: Bethylidae). *Heredity* 71:130–137.

Cook JM, Bean D (2006) Cryptic male dimorphism and fighting in a fig wasp. *Animal Behaviour* 71:1095–1101.

Cook PA, Wedell N (1999) Non-fertile sperm delay female remating. *Nature* 397:486.

Cordaux R, Bouchon D, Grève P (2011) The impact of endosymbionts on the evolution of host sex-determination mechanisms. *Trends in Genetics* 27:332–341.

Córdoba-Aguilar A (1999) Male copulatory sensory stimulation induces female ejection of rival sperm in a damselfly. *Proceedings of the Royal Society of London B* 266:779–784.

Córdoba-Aguilar A, Cordero-Rivera A (2005) Evolution and ecology of Calopterygidae (Zygoptera: Odonata): status of knowledge and research perspectives. *Neotropical Entomology* 34:861–879.

Corley LS, Cotton S, McConnell E, et al. (2006) Highly variable sperm precedence in the stalk-eyed fly, *Teleopsis dalmanni*. *BMC Evolutionary Biology* 6:53.

Cornell SJ, Tregenza T (2007) A new theory for the evolution of polyandry as a means of inbreeding avoidance. *Proceedings of the Royal Society of London B* 274:2873–2879.

Costa JT (2006) *The Other Insect Societies*. Belknap/Harvard University Press, Cambridge, MA.

Cothran RD, Chapman K, Stiff AR, Relyea RA (2012) "Cryptic" direct benefits of mate choice: choosy females experience reduced predation risk while in precopula. *Behavioral Ecology and Sociobiology* 66:905–913.

Cotter SC, Kilner RM (2010) Sexual division of antibacterial resource defence in breeding burying beetles, *Nicrophorus vespilloides*. *Journal of Animal Ecology* 79:35–43.

Craig GB (1967) Mosquitoes: female monogamy induced by male accessory gland substance. *Science* 156:1499–1501.

Crespi BJ (1988a) Risks and benefits of lethal male fighting in the colonial, polygynous thrips *Hoplothrips karnyl* (Insecta: Thysanoptera). *Behavioral Ecology and Sociobiology* 22:293–301.

Crespi BJ (1988b) Adaptation, compromise, and constraint—the development, morphometrics, and behavioral basis of a fighter-flier polymorphism in male *Hoplothrips karnyi* (Insecta, Thysanoptera). *Behavioral Ecology and Sociobiology* 23:93–104.

Crespi BJ (1988c). Alternative male mating tactics in a thrips—effects of sex-ratio variation and body size. *American Midland Naturalist* 119:83–92.

Cronin H (1991) *The Ant and the Peacock: Altruism and Sexual Selection from Darwin to Today*. Cambridge University Press, Cambridge.

Crook JH, Crook SJ (1988) Tibetan polyandry: problems of adaptation and fitness. In L Betzig, M Borgerhoff Mulder, T Turkel (eds), *Human Reproductive Behaviour: A Darwinian Perspective*, pp. 97–114. Cambridge University Press, Cambridge.

Crossley SA (1974) Changes in mating behaviour produced by selection for ethological isolation between ebony and vestigial mutants of *Drosophila melanogaster*. *Evolution* 28:631–647.

Crow JF (1988) The ultraselfish gene. *Genetics* 118:389–391.

Crudgington HS, Siva-Jothy MT (2000) Genital damage, kicking and early death. *Nature* 407:855–856.

Crudgington HS, Fellows S, Badcock NS, Snook RR (2009) Experimental manipulation of sexual selection promotes greater male mating capacity but does not alter sperm investment. *Evolution* 63:926–938.

Cruickshank RH, Thomas RH (1999) Evolution of haplodiploidy in dermanyssine mites (Acari: Mesostigmata). *Evolution* 53:1796–1803.

Cuellar O (1994) Biogeography of parthenogenetic animals. *Compte Rendu des Seances de la Societe de Biogeographie* 70:1–13.

Curtsinger JW, Heisler IL (1988) A diploid "sexy son" model. *American Naturalist* 132:437–453.

Curtsinger JW, Heisler IL (1989) On the consistency of sexy-son models: a reply to Kirkpatrick. *American Naturalist* 134:978–981.

Daguerre JB (1931) Costumbres nupicales del *Diloboderus abderus* Sturm. *Revista de la Sociedad Entomologia Argentina* 3:253–256.

Dalton JE, Kacheria TS, Knott SRV, et al. (2010) Dynamic, mating-induced gene expression changes in female head and brain tissues of *Drosophila melanogaster*. *BMC Genomics* 11:541

Danforth BN (1991) The morphology and behavior of dimorphic males in *Perdita portalis* (Hymenoptera: Andrenidae). *Behavioral Ecology and Sociobiology* 229:235–247.

Danielsson I (2001) Antagonistic pre- and post-copulatory sexual selection on male body size in a water strider (*Gerris lacustris*). *Proceedings of the Royal Society of London B* 268:77–81.

Darwin C (1859) *On the Origins of Species by Means of Natural Selection*. John Murray, London.

Darwin C (1871) *The Descent of Man and Selection in Relation to Sex*. John Murray, London.

Da Silva DLN, Zucchi R, Kerr WE (1972) Biological and behavioural aspects of the reproduction in some species of *Melipona* (Hymenoptera, Apidae, Meliponinae). *Animal Behaviour* 20:123–132.

David P, Bjorksten T, Fowler K, Pomiankowski A (2000) Condition-dependent signaling of genetic variation in stalk-eyed flies. *Nature* 406:186–188.

Davies NB (1991) Mating systems. In JR Krebs, NB Davies (eds), *Behavioural Ecology: An Evolutionary Approach*, pp. 263–294. Blackwell, Oxford.

Davies NB, Krebs JR, West SA (2012) *An Introduction to Behavioural Ecology*. John Wiley & Sons, Oxford.

Dawkins MS, Guilford T (1996) Sensory bias and the adaptiveness of female choice. *American Naturalist* 148:937–942.

Dawkins R (1976) *The Selfish Gene*. Oxford University Press, Oxford.

Dawkins R, Krebs JR (1979) Arms races between and within species. *Proceedings of the Royal Society of London B* 205:489–511.

Dawson G (2007) *Darwin, Literature and Victorian Respectability*. Cambridge University Press, Cambridge.

Day T (2000) Sexual selection and the evolution of costly female preferences: spatial effects. *Evolution* 54:715–730.

Deas JB, Hunter MS (2012) Mothers modify eggs into shields to protect offspring from parasitism. *Proceedings of the Royal Society of London B* 279:847–853.

Dedeine F, Vavre F, Fleury F, et al. (2001) Removing symbiotic *Wolbachia* specifically inhibits oogenesis in a parasitic wasp. *Proceedings of the National Academy of Sciences of the USA* 98:6247–6252.

de Jong MCM, Sabelis MW (1991) Limits to runaway sexual selection: the wallflower paradox. *Journal of Evolutionary Biology* 4:637–655.

del Castillo RC (2003) Body size and multiple copulations in a neotropical grasshopper with an extraordinary mate-guarding duration. *Journal of Insect Behavior* 16:503–522.

del Castillo RC, Núnez-Farfán J (2002) Female mating success and risk of pre-reproductive death in a grasshopper. *Oikos* 96:217–224.

Delcourt M, Blows MW, Rundle HD (2010) Quantitative genetics of female mate preference in an ancestral and a novel environment. *Evolution* 64:2758–2766.

Demary KC, Lewis SM (2007) Male courtship attractiveness and paternity success in *Photinus greeni* fireflies. *Evolution* 61:431–439.

Demir E, Dickson BJ (2005) Fruitless splicing specifies male courtship behavior in *Drosophila*. *Cell* 121:785–794.

Demont M, Buser CC, Martin OY, Bussière LF (2011) Natural levels of polyandry: differential sperm storage and temporal changes in sperm competition intensity in wild yellow dung flies. *Functional Ecology* 25:1079–1090.

Demuth JP, Naidu A, Mydlarz LD (2012) Sex, war, and disease: the role of parasite infection on weapon development and mating success in a horned beetle (*Gnatocerus cornutus*). *PLoS One* 7:e28690.

den Boer SPA, Baer B, Boomsma JJ (2010) Seminal fluid mediates ejaculate competition in social insects. *Science* 327:1506–1509.

den Boer SPA, Baer B, Dreier S, et al. (2009) Prudent sperm use by leaf-cutter ant queens. *Proceedings of the Royal Society of London B* 276:3945–3953.

Denholm I, Franco MG, Rubini PG, Vecchi M (1985) Geographical variation in house-fly (*Musca domestica* L) sex determinants within the British Isles. *Genetics Research* 47:19–27.

De Winter AJ (1992) The genetic basis and evolution of acoustic mate recognition signals in a *Ribautodelphax* planthopper (Homoptera, Delphacidae) 1. The female call. *Journal of Evolutionary Biology* 5:249–264.

Dickinson JL, Rutowski RL (1989) The function of the mating plug in the chalcedon checkerspot butterfly. *Animal Behaviour* 38:154–162.

Dickson BJ (2008) Wired for sex: the neurobiology of *Drosophila* mating decisions. *Science* 322:904–909.

Dobata S, Sasaki T, Mori H, et al. (2009) Cheater genotypes in the parthenogenetic ant *Pristomyrmex punctatus*. *Proceedings of the Royal Society of London B* 276:567–574.

Dodson GN (1997) Resource defense mating system in antlered flies, *Phytalmia* spp. (Diptera: Tephritidae). *Annals of the Entomological Society of America* 90:496–504.

Donelson NC, van Staaden MJ (2005) Alternate tactics in male bladder grasshoppers *Bullacris membracioides* (Orthoptera: Pneumoridae). *Behaviour* 142:761–778.

Donnell DM, Corley LS, Chen G, Strand MR (2004) Caste determination in a polyembryonic wasp involves inheritance of germ cells. *Proceedings of the National Academy of Sciences of the USA* 101:10095–10100.

Do Thi Khanh H, Bressac C, Chevrier C (2005) Male sperm donation consequences in single and double matings in *Anisopteromalus calandrae*. *Physiological Entomology* 30:29–35.

Dow M (1977) Selection for mating success of yellow mutant *Drosophila melanogaster*: biometrical genetic analysis. *Heredity* 38:161–168.

Drosophila 12 Genomes Consortium (2007) Evolution of genes and genomes on the *Drosophila* phylogeny. *Nature* 450:203–218.

Drummond BA (1984) Multiple mating and sperm competition in the lepidotera. In RL Smith (ed.), *Sperm Competition and the Evolution of Animal Mating Systems*, pp. 547–572. Academic Press, London.

Drummond-Barbosa D, Spradling AC (2001) Stem cells and their progeny respond to nutritional changes during *Drosophila* oogenesis. *Developmental Biology* 231:265–278.

Dunn DW, Crean CS, Wilson CL, Gilburn AS (1999) Male choice, willingness to mate and body size in seaweed flies (Diptera: Coelopidae). *Animal Behaviour* 57:847–853.

Dunning Hotopp JC, Clark ME, Oliveira DC, et al. (2007) Widespread lateral gene transfer from intracellular bacteria to multicellular eukaryotes. *Science* 317:1753–1756.

Dussourd DE, Harvis CA, Meinwald J, Eisner T (1991) Pheromonal advertisement of a nuptial gift by a male moth (*Utetheisa ornatrix*). *Proceedings of the National Academy of Sciences of the USA* 88:9224–9227.

Duvoisin N, Baer B, Schmid-Hempel P (1999) Sperm transfer and male competition in a bumblebee. *Animal Behaviour* 58:743–749.

Dyson EA, Hurst GDD (2004) Persistence of an extreme sex ratio bias in natural populations. *Proceedings of the National Academy of Sciences of the USA* 101:6520–6523.

Eberhard WG (1978) Fighting behavior of male *Golofa porteri* beetles (Scarabaeidae: Dynastinae). *Psyche* 83:292–298.
Eberhard WG (1979) The function of horns in *Podischnus agenor* (Dynastinae) and other beetles. In MS Blum, NA Blum (eds), *Sexual Selection and Reproductive Competition in Insects*, pp. 231–259. Academic Press, New York.
Eberhard WG (1982) Beetle horn dimorphism: making the best of a bad lot. *American Naturalist* 119:420–426.
Eberhard WG (1985) *Sexual Selection and Animal Genitalia*. Harvard University Press, Cambridge, MA.
Eberhard WG (1987) Use of horns in fights by the dimorphic males of *Ageopsis nigricollis* Coleoptera Scarabeidae Dynastinae. *Journal of the Kansas Entomological Society* 60:504–509.
Eberhard WG (1993) Evaluating models of sexual selection—genitalia as a test-case. *American Naturalist* 142:564–571.
Eberhard WG (1994) Evidence for widespread courtship during copulation in 131 species of insects and spiders, and implications for cryptic female choice. *Evolution* 48:711–733.
Eberhard WG (1996) *Female Control: Sexual Selection by Cryptic Female Choice*. Princeton University Press, Princeton, NJ.
Eberhard WG (1997) Sexual selection by cryptic female choice in insects and arachnids. In CJ Choe, BJ Crespi (eds), *The Evolution of Mating Systems in Insects and Arachnids*, pp. 32–57. Cambridge University Press, Cambridge.
Eberhard WG (1998) Sexual behavior of *Acanthocephala declivis guatemalana* (Hemiptera: Coreidae) and the allometric scaling of their modified hind legs. *Annals of the Entomological Society of America* 91:863–871.
Eberhard WG (2009) Postcopulatory sexual selection: Darwin's omission and its consequences. *Proceedings of the National Academy of Sciences of the USA* 106:10025–10032.
Eberhard WG, Cordero C (1995) Sexual selection by cryptic female choice on male seminal products—a new bridge between sexual selection and reproductive physiology. *Trends in Ecology and Evolution* 10:493–496.
Eberhard WG, Garcia CJM, Lobo J (2000) Size-specific defensive structures in a horned weevil confirm a classic battle plan: avoid fights with larger opponents. *Proceedings of the Royal Society of London B* 267:1129–1134.
Edgerly JS (1997) Life beneath silk walls: a review of the primitively social Embiidina. In JC Choe, BJ Crespi (eds), *Social Behavior in Insects and Arachnids*, pp. 14–25. Cambridge University Press, Cambridge.
Edvardsson M, Arnqvist G (2000) Copulatory courtship and cryptic female choice in red flour beetles *Tribolium castaneum*. *Proceedings of the Royal Society of London B* 267:559–563.
Edvardsson M, Arnqvist G (2005) The effects of copulatory courtship in differential allocation in the red flour beetle *Tribolium castaneum*. *Journal of Insect Behaviour* 18:313–322.
Edward DA, Chapman T (2011) The evolution and significance of male mate choice. *Trends in Ecology and Evolution* 26:647–654.
Eggert A-K (1992) Alternative male mate-finding tactics in burying beetles. *Behavioral Ecology* 3:243–254.
Eggert A-K, Müller JK (1992) Joint breeding in female burying beetles. *Behavioral Ecology and Sociobiology* 31:237–242.
Eggert A-K, Müller JK (1997) Biparental care and social evolution in burying beetles: lessons from the larder. In JC Choe, BJ Crespi (eds), *Social Behavior in Insects and Arachnids*, pp. 216–236. Cambridge University Press, Cambridge.
Eggert A-K, Sakaluk SK (1995) Female-coerced monogamy in burying beetles. *Behavioral Ecology and Sociobiology* 37:147–153.
Eggert A-K, Reinking M, Müller JK (1998) Parental care improves offspring survival and growth in burying beetles. *Animal Behaviour* 55:97–107.
Ehrlich AH, Ehrlich PR (1978) Reproductive strategies in the butterflies: I. Mating frequency, plugging, and egg number. *Journal of the Kansas Entomological Society* 51:666–697.
Eisner T, Rossini C, González A, et al. (2002) Paternal investment in egg defence. In M Hilker, T Meiners (eds), *Chemoecology of Insects Eggs and Egg Deposition*, pp. 91–116. Blackwell, Oxford.
Ekblom R, Galindo J (2011) Applications of next generation sequencing in molecular ecology of non-model organisms. *Heredity* 107:1–15.

Ellegren H (2007) Characteristics, causes and evolutionary consequences of male-biased mutation. *Proceedings of the Royal Society of London B* 274:1–10.

Ellegren H, Parsch J (2007) The evolution of sex-biased genes and sex-biased gene expression. *Nature Reviews Genetics* 8:689–698.

Ellegren H, Sheldon BC (2008) Genetic basis of fitness differences in natural populations. *Nature* 452:169–175.

Ellis LL, Carney GE (2009) *Drosophila melanogaster* males respond differently at the behavioural and genome-wide levels to *Drosophila melanogaster* and *Drosophila simulans* females. *Journal of Evolutionary Biology* 22:2183–2191.

Ellis LL, Carney GE (2011) Socially-responsive gene expression in male *Drosophila melanogaster* is influenced by the sex of the interacting partner. *Genetics* 187:157–169.

Elwood RW, Arnott G (2012) Understanding how animals fight with Lloyd Morgan's canon. *Animal Behaviour* 84:1095–1102.

Emlen DJ (1997) Alternative reproductive tacts and male-dimorphism in the horned beetle *Onthophagus acuminatus* (Coleoptera: Scarabaeidae). *Behavioral Ecology and Sociobiology* 41:335–341.

Emlen DJ (2001) Costs and the diversification of exaggerated animal structures. *Science* 291: 1534–1536.

Emlen DJ (2008) The evolution of animal weapons. *Annual Review of Ecology, Evolution, and Systematics* 39:387–413.

Emlen DJ, Nijhout HF (2000) The development and evolution of exaggerated morphologies in insects. *Annual Review of Entomology* 45:661–708.

Emlen ST, Oring LW (1977) Ecology, sexual selection, and the evolution of mating systems. *Science* 197:215–223.

Emlen DJ, Philips TK (2006) Phylogenetic evidence for an association between tunneling behavior and the evolution of horns in dung beetles (Coleoptera: Scarabaeidae: Scarabaeinae). *Coleopterists Society Monographs* 5:47–56.

Emlen DJ, Lavine LC, Ewen-Campen B (2007) On the origin and evolutionary diversification of beetle horns. *Proceedings of the National Academy of Sciences of the USA* 104:8661–8668.

Emlen DJ, Warren IA, Johns A, et al. (2012) A mechanism of extreme growth and reliable signaling in sexually selected ornaments and weapons. *Science* 337:860–864.

Endler JA (1986) *Natural Selection in the Wild*. Princeton University Press, Princeton, NJ.

Endler JA, Basolo AL (1998) Sensory ecology, receiver biases and sexual selection. *Trends in Ecology and Evolution* 13:415–420.

Engelhard G, Foster SP, Day TH (1989) Genetic differences in mating success and female choice in seaweed flies (*Coelopa frigida*). *Heredity* 62:123–131.

Engels S, Sauer KP (2006) Love for sale and its fitness benefits: nuptial gifts in the scorpionfly *Panorpa vulgaris* represent paternal investment. *Behaviour* 143:825–837.

Engelstadter J (2008) Constraints on the evolution of asexual reproduction. *Bioessays* 30: 1138–1150.

Enquist M, Leimar O (1983) Evolution of fighting behaviour: decision rules and assessment of relative strength. *Journal of Theoretical Biology* 102:387–410.

Enquist M, Leimar O (1987) Evolution of fighting behaviour: the effect of variation in resource value. *Journal of Theoretical Biology* 127:187–205.

Engqvist L (2007) Nuptial food gifts influence female egg production in the scorpionfly *Panorpa cognata*. *Ecological Entomology* 32:327–332.

Engqvist L, Reinhold K (2005) Pitfalls in experiments testing predictions from sperm competition theory. *Journal of Evolutionary Biology* 18:116–123.

Engqvist L, Sauer KP (2001) Strategic male mating effort and cryptic male choice in a scorpionfly. *Proceedings of the Royal Society of London B* 268:729–735.

Espinedo CM, Gabor CR, Aspbury AS (2010) Males, but not females, contribute to sexual isolation between two sympatric species of *Gambusia*. *Evolutionary Ecology* 24:865–878.

Estrada C, Yildizhan S, Schulz S, Gilbert LE (2010) Sex-specific chemical cues from immatures facilitate the evolution of mate guarding in *Heliconius* butterflies. *Proceedings of the Royal Society of London B* 277:407–413.

Estrada C, Schulz S, Yildizhan S, Gilbert LE (2011) Sexual selection drives the evolution of antiaphrodisiac pheromones in butterflies. *Evolution* 65:2843–2854.

Etges WJ, De Oliveira CC, Noor MAF, Ritchie MG (2010) Genetics of incipient speciation in *Drosophila mojavensis*. III. Life-history divergence in allopatry and reporductive isolation. *Evolution* 64:3549–3569.

Ewen-Campen B, Schwager EE, Extavour CGM (2010) The molecular machinery of germ line specification. *Molecular Reproduction and Development* 77:3–18.

Fagerström T, Wiklund C (1982) Why do males emerge before females? Protandry as a mating strategy in male and female butterflies. *Oecologia* 52:164–166.

Fairbairn DJ, Blanckenhorn WU, Szekely T (2007) *Sex, Size & Gender Roles*. Oxford University Press, Oxford.

Falconer DS (1965) Inheritance of liability to certain diseases estimated from incidence among relatives. *Annals of Human Genetics* 29:51–76.

Fang S, Ting CT, Lee CR, et al. (2009) Molecular evolution and functional diversification of fatty acid desaturases after recurrent gene duplication in *Drosophila*. *Molecular Biology and Evolution* 26:1447–1456.

Fawcett TW, Kuijper B, Weissing FJ, Pen I (2011) Sex-ratio control erodes sexual selection, revealing evolutionary feedback from adaptive plasticity. *Proceedings of the National Academy of Sciences of the USA* 108:15925–15930.

Feaver M (1983) Pair formation in the katydid *Orchelimum nigripes* (Orthoptera: Tettigoniidae). In DT Gwynne, GK Morris (eds), *Orthopteran Mating Systems: Sexual Competition in a Diverse Group of Insects*, pp. 205–239. Westview Press, Boulder, CO.

Fedina TY, Lewis SM (2007) Female mate choice across mating stages and between sequential mates in flour beetles. *Journal of Evolutionary Biology* 20:2138–2143.

Fedina TY, Lewis SM (2008) An integrative view of sexual selection in *Tribolium* flour beetles. *Biological Reviews* 83:151–171.

Felsenstein J (1974) The evolutionary advantage of recombination. *Genetics* 78:737–756.

Ferveur J-F (2010) *Drosophila* female courtship and mating behaviors: sensory signals, genes, neural structures and evolution. *Current Opinion in Neurobiology* 20:764–769.

Field J (2005) The evolution of progressive provisioning. *Behavioral Ecology* 16:770–778.

Fincke OM (2004) Polymorphic signals of harassed female odonates and the males that learn them support a novel frequency-dependent model. *Animal Behaviour* 67:833–845.

Fincke OM, Jödicke R, Paulson D, Schultz TD (2005) The evolution and frequency of female color morphs in Holarctic Odonata: why are male-like morphs typically the minority? *International Journal of Odonatology* 8:183–212.

Fischer K, Fiedler K (2001) Resource-based territoriality in the butterfly *Lycaena hippothoe* and environmentally induced behavioural shifts. *Animal Behaviour* 61:723–732.

Fisher RA (1930) *The Genetical Theory of Natural Selection*. Clarendon Press, Oxford.

Fitzjohn RG, Maddison WP, Otto SP (2009) Estimating trait-dependent speciation and extinction rates from incompletely resolved phylogenies. *Systematic Biology* 58:595–611.

Fitzpatrick JL, Baer B (2011) Polyandry reduces sperm length variation in social insects. *Evolution* 65:3006–3012.

Fitzpatrick MJ, Ben-Shahar Y, Smid HM, et al. (2005) Candidate genes for behavioural ecology. *Trends in Ecology and Evolution* 20:96–104.

Flatt T, Heyland A (2011) *Mechanisms of Life History Evolution: the Genetics and Physiology of Life History Traits and Trade-Offs*. Oxford University Press, Oxford.

Flatt T, Tu M-P, Tatar M (2005) Hormonal pleiotropy and the juvenile hormone regulation of *Drosophila* development and life history. *Bioessays* 27:999–1010.

Fleischman RR, Sakaluk SK (2004) No direct or indirect benefits to cryptic female choice in house crickets (*Acheta domesticus*). *Behavioral Ecology* 15:793–798.

Foitzik S, Heinze J, Oberstadt B, Herbers JM (2002) Mate guarding and alternative reproductive tactics in the ant *Hypoponera opacior*. *Animal Behaviour* 63:597–604.

Folstad I, Karter AJ (1992) Parasites, bright males, and the immunocompetence handicap. *American Naturalist* 139:603–622.

Forslund P (2000) Male–male competition and large size mating advantage in European earwigs, *Forficula auricularia*. *Animal Behaviour* 59:753–762.

Forslund P (2003) An experimental investigation into status-dependent male dimorphism in the European earwig, *Forficula auricularia*. *Animal Behaviour* 65:309–316.

Forsyth A, Alcock J (1990) Female mimicry and resource defense polygyny by males of a tropical rove beetle, *Leistotrophus versicolor* (Coleoptera, Staphylinidae). *Behavioral Ecology and Sociobiology* 26:325–330.

Forsyth A, Montgomerie R D (1987) Alternative reproductive tactics in the territorial damselfly *Calopteryx maculata*—sneaking by older males. *Behavioral Ecology and Sociobiology* 21:73–81.

Foucaud J, Estoup A, Loiseau A, et al. (2010) Thelytokous parthenogenesis, male clonality and genetic caste determination in the little fire ant: new evidence and insights from the lab. *Heredity* 105:205–212.

Fournier D, Estoup A, Orivel J, et al. (2005) Clonal reproduction by males and females in the little fire ant. *Nature* 435:1230–1234.

Fox CW, Czesak ME (2000) Evolutionary ecology of progeny size in arthropods. *Annual Review of Entomology* 45:341–369.

Fox CW, Thakar MS, Mousseau TA (1997) Egg size plasticity in a seed beetle: an adaptive maternal effect. *American Naturalist* 149:149–163.

French BW, Cade WH (1989) Sexual selection at varying population-densities in male field crickets, *Gryllus veletis* and *Gryllus pennsylvanicus*. *Journal of Insect Behavior* 2:105–121.

French SS, Moore MC, Demas GE (2009) Ecological immunology: the organism in context. *Integrated and Comparative Biology* 49:246–253.

Frentiu FD, Chenoweth SF (2010) Clines in cuticular hydrocarbons in two *Drosophila* species with independent population histories. *Evolution* 64:1784–1794.

Friberg M, Vongvanich N, Borg-Karlson A-K, et al. (2008) Female mate choice determines reproductive isolation between sympatric butterflies. *Behavioral Ecology and Sociobiology* 62:873–886.

Fricke C, Arnqvist G, Amaro N (2006) Female modulation of reproductive rate and its role in postmating prezygotic isolation in *Callosobruchus maculatus*. *Functional Ecology* 20:60–368.

Fricke C, Perry J, Chapman T, Rowe L (2009a) The conditional economics of sexual conflict. *Biology Letters* 5:671–674.

Fricke C, Wigby S, Hobbs R, Chapman T (2009b) The benefits of male ejaculate sex peptide transfer in *Drosophila melanogaster*. *Journal of Evolutionary Biology* 22:275–286.

Fritzsche K, Arnqvist G (2013) Homage to Bateman: sex roles predict sex differences in sexual selection. *Evolution* 67:1926–1936.

Fromhage L (2012) Mating unplugged: a model for the evolution of mating plug (dis-) placement. *Evolution* 66:31–39.

Fromhage L, Schneider JM (2006) Emasculation to plug up females: the significance of pedipalp damage in *Nephila fenestrata*. *Behavioral Ecology* 17:353–357.

Fromhage L, Jacobs K, Schneider JM (2007) Monogynous mating behaviour and its ecological basis in the golden orb spider *Nephila fenestrata*. *Ethology* 113:813–820.

Fry CL (2006) Juvenile hormone mediates a trade-off between primary and secondary sexual traits in stalk-eyed flies. *Evolution and Development* 8:191–201.

Fry CL, Wilkinson GS (2004) Sperm survival in female stalkeyed flies depends on seminal fluid and meiotic drive. *Evolution* 58:1622–1626.

Fryer T, Cannings C, Vickers GT (1999) Sperm competition. II—Post-copulatory guarding. *Journal of Theoretical Biology* 197:343–360.

Fuchikawa T, Sanada S, Nishio R, et al. (2010) The clock gene cryptochrome of *Bactrocera cucurbitae* (Diptera: Tephritidae) in strains with different mating times. *Heredity* 104:387–392.

Fujisaki K (1992) A male fitness advantage to wing reduction in the oriental chinch bug, *Cavelerius saccharivorus* Okajima (Heteroptera, Lygaeidae). *Researches on Population Ecology* 34:173–183.

Fuller MT, Spradling AC (2007) Male and female *Drosophila* germline stem cells: two versions of immortality. *Science* 316:402–404.

Fuller RC, Houle D, Travis J (2005) Sensory bias as an explanation for the evolution of mate preferences. *American Naturalist* 166:437–446.

Funk DH, Tallamy DW (2000) Courtship role reversal and deceptive signals in the long-tailed dance fly, *Rhamphomyia longicuada*. *Animal Behaviour* 59:411–421.

Funk DH, Sweeney BW, Jackson JK (2010) Why stream mayflies can reproduce without males but remain bisexual: a case of lost genetic variation. *Journal of the North American Benthological Society* 29:1258–1266.

Gadgil M (1972) Male dimorphism as a consequence of sexual selection. *American Naturalist* 106:574–580.

Gage MJG (1994) Associations between body size, mating pattern, testis size and sperm lengths across butterflies. *Proceedings of the Royal Society of London B* 258:247–254.

Gage MJG (1998) Influences of sex, size, and symmetry on ejaculate expenditure in a moth. *Behavioral Ecology* 9:592–597.

Gage MJG, Morrow EH (2003) Experimental evidence for the evolution of numerous, tiny sperm via sperm competition. *Current Biology* 13:754–757.

Gage MJG, Parker GA, Nylin S, Wiklund C (2002) Sexual selection and speciation in mammals, butterflies and spiders. *Proceedings of the Royal Society of London B* 269:2309–2316.

Gagné RJ (2010) *Update for a Catalog of the Cecidomyiidae (Diptera) of the World*. Systematic Entomology Laboratory, Agricultural Research Service, US Department of Agriculture, Washington, DC.

Garbaczewska M, Billeter J-C, Levine JD (2013) *Drosophila melanogaster* males increase the number of sperm in their ejaculate when perceiving rival males. *Journal of Insect Physiology* 59:306–310.

García-González F, Gomendio M (2004) Adjustment of copula duration and ejaculate size according to the risk of sperm competition in the golden egg bug (*Phyllomorpha laciniata*). *Behavioral Ecology* 15:23–30.

García-González F, Núñez Y, et al. (2003) Sperm competition mechanisms, confidence of paternity, and the evolution of paternal care in the golden egg bug (*Phyllomorpha laciniata*). *Evolution* 57:1078–1088.

García-González F, Simmons LW (2005a) The evolution of polyandry: intrinsic sire effects contribute to embryo viability. *Journal of Evolutionary Biology* 18:1097–1103.

García-González F, Simmons LW (2005b) Sperm viability matters in insect sperm competition. *Current Biology* 15:271–275.

García-González F, Simmons LW (2007a) Paternal indirect genetic effects on offspring viability and the benefits of polyandry. *Current Biology* 17:32–36.

García-González F, Simmons LW (2007b) Shorter sperm confer higher competitive fertilization success. *Evolution* 61:816–824.

Gardner A, Ross L (2011) The evolution of hermaphroditism by an infectious male-derived cell lineage: an inclusive-fitness analysis. *American Naturalist* 178:191–201.

Gavrilets S (2000) Rapid evolution of reproductive barriers driven by sexual conflict. *Nature* 403:886–889.

Gavrilets S, Hayashi TI (2006) The dynamics of two- and three-way sexual conflicts over mating. *Philosophical Transactions of the Royal Society of London B* 361:345–354.

Gavrilets S, Arnqvist G, Friberg U (2001) The evolution of female mate choice by sexual conflict. *Proceedings of the Royal Society of London B* 268:531–539.

Gay L, Hosken DJ, Eady P, et al. (2011) The evolution of harm—effect of sexual conflicts and population size. *Evolution* 65:725–737.

Gerloff CU, Schmid-Hempel P (2005) Inbreeding depression and family variation in a social insect, *Bombus terrestris* (Hymenoptera: Apidae). *Oikos* 111:67–80.

Gershman SG, Hunt J, Sakaluk SK (2013) Food fight: sexual conflict over free amino acids in the nuptial gifts of male decorated crickets. *Journal of Evolutionary Biology* 26:693–704.

Gershman SG, Mitchell C, Sakaluk SK, Hunt J (2012) Biting off more than you can chew: sexual selection on the free amino acid composition of the spermatophylax in decorated crickets. *Proceedings of the Royal Society of London B* 279:2531–2538.

Gershman SG, Hunt J, Sakaluk SK (2013) Food fight: sexual conflict over free amino acids in the nuptial gifts of male decorated crickets. *Journal of Evolutionary Biology* 26:693–704.

Getty T (1998) Handicap signalling: when fecundity and viability do not add up. *Animal Behaviour* 56:127–130.

Getty T (2002) Signaling health versus parasites. *American Naturalist* 159:363–371.

Ghiselli F, Milani L, Scali V, Passamonti M (2007) The *Leptynia hispanica* species complex (Insecta Phasmida): polyploidy, parthenogenesis, hybridization and more. *Molecular Ecology* 16:4256–4268.

Gibert P, Capy P, Imasheva A, et al. (2004) Comparative analysis of morphological traits among *Drosophila melanogaster* and *D. simulans*: genetic variability, clines and phenotypic plasticity. *Genetica* 120:165–179.

Gibson JR, Chippindale AK, Rice WR (2002) The X chromosome is a hot spot for sexually antagonistic fitness variation. *Proceedings of the Royal Society of London B* 269:499–505.

Giglioli MEC, Mason GF (1966) The mating plug in anopheline mosquitoes. *Proceedings of the Royal Entomological Society of London A* 41:123–129.

Gilbert LE (1976) Postmating female odour in *Heliconius* butterflies: a male-contributed antiaphrodisiac? *Science* 193:419–420.

Gilbert JDJ, Thomas LK, Manica A (2010) Quantifying the benefits and costs of parental care in assassin bugs. *Ecological Entomology* 35:639–651.

Gilburn AS, Day TH (1994a) Sexual dimorphism, sexual selection and the αβ chromosomal inversion polymorphism in the seaweed fly, *Coelopa frigida*. *Proceedings of the Royal Society of London B* 257:303–309.

Gilburn AS, Day TH (1994b) The inheritance of female mating behaviour in the seaweed fly, *Coelopa frigida*. *Genetical Research* 64:19–25.

Gilburn AS, Crean CS, Day TH (1996) Sexual selection in natural populations of seaweed flies: variation in the offspring fitness of females carrying different inversion karyotypes. *Proceedings of the Royal Society of London* B 263:249–256.

Gilburn AS, Foster SP, Day TH (1992) Female mating preference for large size in *Coelopa frigida* (seaweed fly). *Heredity* 69:209–212.

Gilburn AS, Foster SP, Day TH (1993) Genetic correlation between a female mating preference and the preferred male character in seaweed flies (*Coelopa frigida*). *Evolution* 47:1788–1795.

Gilchrist AS, Partridge L (2000) Why it is difficult to model sperm displacement in *Drosophila melanogaster*: the relation between sperm transfer and copula duration. *Evolution* 54:534–542.

Gillespie JP, Kanost MR, Trenczek T (1997) Biological mediators of insect immunity. *Annual Review of Entomology* 42:611–643.

Gillott C (1996) Male insect accessory glands: functions and control of secretory activity. *Invertebrate Reproduction and Development* 30:199–205.

Gillott C (2003) Male accessory gland secretions: modulators of female reproductive physiology and behaviour. *Annual Review of Entomology* 48:163–184.

Giorgini M, Bernardo U, Monti MM, et al. (2010) Rickettsia symbionts cause parthenogenetic reproduction in the parasitoid wasp *Pnigalio soemius* (Hymenoptera: Eulophidae). *Applied Environmental Microbiology* 76:2589–2599.

Gioti A, Wigby S, Wertheim B, et al. (2012) Sex peptide of *Drosophila melanogaster* males is a global regulator of reproductive processes in females. *Proceedings of the Royal Society of London B* 279:4423–4432.

Giron D, Ross K, Strand M (2007) Presence of soldier larvae determines the outcome of competition in a polyembryonic wasp. *Journal of Evolutionary Biology* 20:165–172.

Giurfa M (2007) Behavioral and neural analysis of associative learning in the honeybee: a taste from the magic well. *Journal of Comparative Physiology* A 193:801–824.

Glasgow JP (1961) Selection for size in tsetse flies. *Journal of Animal Ecology* 30:87–94.

Gleason JM, Ritchie MG (2004) Do Quantitative Trait Loci (QTL) for a courtship song difference between *Drosophila simulans* and *D. sechellia coincide with candidate genes and intraspecific QTL? Genetics* 166:1303–1311.

Glesener RR, Tilman D (1978) Sexuality and the components of environmental uncertainty: clues from geographic parthenogenesis in terrestrial animals. *American Naturalist* 112:659–673.

Godfray HCJ, Cook JM (1997) Mating systems of parasitoid wasps. In Choe JC, Crespi BJ (eds), *The Evolution of Mating Systems in Insects and Arachnids*, pp. 211–211. Cambridge University press, Cambridge.

Goldman N, Yang Z (1994) A codon-based model of nucleotide substitution for protein-coding DNA sequences. *Molecular Biology and Evolution* 11:725–736.

Gomendio M, Reguera P (2001) Egg carrying in the golden egg bug (*Phyllomorpha laciniata*): parental care, parasitism, or both? Reply to Kaitala et al. *Behavioral Ecology* 12:369–373.

Gomez-Zurita J, Funk DJ, Vogler AP (2006) The evolution of unisexuality in *Calligrapha* leaf beetles: molecular and ecological insights on multiple origins via interspecific hybridization. *Evolution* 60:328–347.

Gonzalez JM, Matthews RW (2008) Female and male polymorphism in two species of *Melittobia* parasitoid wasps (Hymenoptera: Eulophidae). *Florida Entomologist* 91:162–169.

Gonzalez-Santoyo I, Cordoba-Aguilar A, Gonzalez-Tokman DM, Lanz-Mendoza, H (2010) Phenoloxidase activity and melanization do not always covary with sexual trait expression in *Hetaerina* damselflies (Insecta: Calopterygidae). *Behaviour* 147:1285–1307

Gonzalez-Voyer A, Fitzpatrick JL, Kolm N (2008) Sexual selection determines parental care patterns in cichlid fishes. *Evolution* 62:2015–2026.

Good JM (2012) The conflict with and the escalating war between sex chromosomes. *PLoS Genetics* 8:e1002955.

Goodacre SL, Martin OY (2012) Modification of insect and arachnid behaviours by vertically transmitted endosymbionts: infections as drivers of behavioural change and evolutionary novelty. *Insects* 3:246–261.

Gordon S, Strand M (2009) The polyembryonic wasp *Copidosoma floridanum* produces two castes by differentially parceling the germ line to daughter embryos during embryo proliferation. *Development Genes and Evolution* 219:445–454.

Goubault M, Batchelor TP, Romani R, et al. (2008) Volatile chemical release by bethylid wasps: identity, phylogeny, anatomy and behaviour. *Biological Journal of the Linnean Society* 94:837–852.

Goudeau J, Bellemin S, Toselli-Mollereau E, et al. (2011) Fatty acid desaturation links germ cell loss to longevity through NHR-80/HNF4 in *C. elegans*. *PLoS Biology* 9:e1000599.

Gowaty PA, Kim Y-K, Rawlings J, Anderson WW (2010) Polyandry increases offspring viability and mother productivity but does not decrease mother survival in *Drosophila pseudoobscura*. *Proceedings of the National Academy of Sciences of the USA* 107:13771–13776.

Gowaty PA, Kim Y-K, Anderson WW (2012) No evidence of sexual selection in a repetition of Bateman's classic study of *Drosophila melanogaster*. *Proceedings of the National Academy of Sciences of the USA* 109:11740–11745.

Grace JL, Shaw KL (2011) Coevolution of male mating signal and female preference during early lineage divergence of the Hawaiian cricket, *Laupala cerasina*. *Evolution* 65:2184–2196.

Grafen A (1984) Natural selection, kin selection and group selection. In Krebs JR, Davies NB (eds), *Behavioural Ecology: An Evolutionary Approach*, 2nd edn, pp. 62–84. Blackwell, Oxford.

Grafen A (1987) The logic of divisively asymmetric contests: respect for ownership and the desperado effect. *Animal Behaviour* 35:462–467.

Grafen A (1990) Sexual selection unhandicapped by the Fisher process. *Journal of Theoretical Biology* 144:473–516.

Graham MJ (2005) *The Hive and The Honey Bee*. Dadant & Sons, Hamilton, IL.

Graham RI, Grzywacz D, Mushobozi WL, Wilson K (2012) *Wolbachia* in a major African crop pest increases susceptibility to viral disease rather than protects it. *Ecology Letters* 15:993–1000.

Grant CA, Chapman T, Pomiankowski A, Fowler K (2005) No detectable genetic correlation between male and female mating frequency in the stalk-eyed fly *Cyrtodiopsis dalmanni*. *Heredity* 95:444–448.

Gray DA (1999) Intrinsic factors affecting female choice in house crickets: time cost, female age, nutritional condition, body size, and size-relative reproductive investment. *Journal of Insect Behavior* 12:691–700.

Gray DA, Cade WH (1999) Quantitative genetics of sexual selection in the field cricket, *Gryllus integer*. *Evolution* 53:848–854.

Gray B, Simmons LW (2013) Acoustic cues alter perceived sperm competition risk in the field cricket *Teleogryllus oceanicus*. *Behavioral Ecology* 24:982–986.

Grbic M, Ode PJ, Strand MR (1992) Sibling rivalry and brood sex ratios in polyembryonic wasps. *Nature* 360:254–256.

Green D, Sarikaya D, Extavour C (2011) Counting in oogenesis. *Cell and Tissue Research* 344:207–212.

Greenacre ML, Ritchie MG, Byrne BC, Kyriacou CP (1993) Female song preference and the period gene in *Drosophila melanogaster*. *Behaviour Genetics* 23:85–90.

Greenfield MD, Shelly TE (1985) Alternative mating strategies in a desert grasshopper—evidence of density-dependence. *Animal Behaviour* 33:1192–1210.

Greenfield MD, Danka RG, Gleason JM, et al. (2012) Genotype × environment interaction, environmental heterogeneity and the lek paradox. *Journal of Evolutionary Biology* 25:601–613.

Grether GF (1996) Intrasexual competition alone favors a sexually selected dimorphic ornament in the rubyspot damselfly *Hetaerina americana*. *Evolution* 50:1949–1957.

Grimaldi D, Engel MS (2005) *Evolution of the Insects*. Cambridge University Press, Cambridge.

Gromko MH, Newport MEA (1988) Genetic basis for remating in *Drosophila melanogaster*. III. Correlated responses to selection for female remating speed. *Behavior Genetics* 18:633–643.

Gromko MH, Gilbert DG, Richmond RC (1984) Sperm transfer and use in the multiple mating system of *Drosophila*. In RL Smith (ed.), *Sperm Competition and the Evolution of Animal Mating Systems*, pp. 371–426. Academic Press, London.

Groot AT, Classen A, Staudacher H, et al. (2010) Phenotypic plasticity in sexual communication signal of a noctuid moth. *Journal of Evolutionary Biology* 23:2731–2738.

Groot AT, Staudacher H, Barthel A, et al. (2013) One quantitative trait locus for intra- and interspecific variation in a sex pheromone. *Molecular Ecology* 22:1065–1080.

Gross MR (1996) Alternative reproductive strategies and tactics: diversity within sexes. *Trends in Ecology and Evolution* 11:92–98.

Grosse-Wilde E, Svatos A, Krieger J (2006) A pheromone-binding protein mediates the bombykol-induced activation of a pheromone receptor *in vitro*. *Chemical Senses* 31:547–555.

Gullan PJ, Cranston P S (2011) *The Insects: an Outline of Entomology*. Wiley–Blackwell, Chichester.

Gwynne DT (1984) Nuptial feeding-behavior and female choice of mates in *Harpobittacus similis* (Mecoptera, Bittacidae). *Journal of the Australian Entomological Society* 23:271–276.

Gwynne DT (1997) The evolution of edible 'sperm sacs' and other forms of courtship feeding in crickets, katydids and their kin (Orthoptera: Ensifera). In J Choe, BJ Crespi (eds), *The Evolution of Mating Systems in Insects and Arachnids*, pp. 110–139. Cambridge University Press, Cambridge.

Gwynne DT (2008) Sexual conflict over nuptial gifts in insects. *Annual Review of Entomology* 53:83–101.

Gwynne DT, Simmons LW (1990) Experimental reversal of courtship roles in an insect. *Nature* 346:172–174.

Haddrill PR, Shuker DM, Amos W, et al. (2008) Female multiple mating in wild and laboratory populations of the two-spot ladybird, *Adalia bipunctata*. *Molecular Ecology* 17:3189–3197.

Haerty W, Jagadeeshan S, Kulathinal RJ, et al. (2007) Evolution in the fast lane: Rapidly evolving sex-related genes in *Drosophila*. *Genetics* 177:1321–1335.

Haig D, Bergstrom CT (1995) Multiple mating, sperm competition and meiotic drive. *Journal of Evolutionary Biology* 8:265–282.

Haine ER (2008) Symbiont-mediated protection. *Proceedings of the Royal Society of London B* 275:353–361.

Halffter G, Edmonds WD (1982) *The Nesting Behavior of Dung Beetles (Scarabaeidae): An Ecological and Evolutive Approach*. Instituto de Ecologia, Mexico DF, Mexico City.

Hall MD, Bussière LF, Hunt J, Brooks R (2008) Experimental evidence that sexual conflict influences the opportunity, form and intensity of sexual selection. *Evolution* 62:2305–2315.

Hall MD, Bussière LF, Demont M, et al. (2010a) Competitive PCR reveals the complexity of post-copulatory sexual selection in *Teleogryllus commodus*. *Molecular Ecology* 19:610–619.

Hall MD, Lailvaux SP, Blows MW, Brooks RC (2010b) Sexual conflict and the maintenance of multivariate genetic variation. *Evolution* 64:1697–1703.

Halliday TR (1978) Sexual selection and mate choice. In Krebs JRD, Davies NB (eds), *Behavioural Ecology: An Evolutionary Approach*. pp. 180–213. Blackwell, Oxford.

Halliday TR (1983) The study of mate choice. In Bateson, P (ed.), *Mate Choice*. pp.3–32. Cambridge University Press, Cambridge.

Halliday TR, Arnold SJ (1987) Multiple mating by females: a perspective from quantitative genetics. *Animal Behaviour* 35:939–941.

Hamilton WD (1964a) The genetical evolution of social behavior I. *Journal of Theoretical Biology* 7:1–16.

Hamilton WD (1964b) The genetical evolution of social behavior II. *Journal of Theoretical Biology* 7:17–32.

Hamilton WD (1978) Evolution and diversity under bark. In LA Mound, N Waloff (eds), *Diversity of Insect Faunas*, pp. 154–175. Blackwell, Oxford.

Hamilton WD (1979) Wingless and fighting males in fig wasps and other insects. In MS Blum, NA Blum (eds), *Reproductive Competition, Mate Choice, and Sexual Selection in Insects*, pp. 167–220. Academic Press, New York.

Hamilton WD (1980) Sex versus non-sex versus parasite. *Oikos* 35:282–290.

Hamilton WD (2001) *Narrow Roads of Gene Land, Volume 2: The Evolution of Sex*. Oxford University Press, Oxford.

Hamilton WD, Zuk M (1982) Heritable true fitness and bright birds: a role for parasites? *Science* 213:384–387.

Hammer M, Menzel R (1995) Learning and memory in the honeybee. *Journal of Neuroscience* 15:1617–1630.

Hammerstein P, Parker GA (1987) Sexual selection: games between the sexes. In JW Bradbury, MB Andersson (eds), *Sexual Selection: Testing the Alternatives*, pp. 119–142. Wiley, Chichester.

Han CS, Jablonski PG (2009) Female genitalia concealment promotes intimate male courtship in a water strider. *PLoS One* 4:e5793.

Han CS, Jablonski PG (2010) Male water striders attract predators to intimidate females into copulation. *Nature Communications* 1:52.

Hanley RS (2001) Mandibular allometry and male dimorphism in a group of obligately mycophagous beetles (Insecta: Coleoptera: Staphylinidae: Oxyporinae). *Biological Journal of the Linnean Society* 72:451–459.

Hansell M (1987) Nest building as a facilitating and limiting factor in the evolution of eusociality in the Hymenoptera. *Oxford Surveys in Evolutionary Biology* 4:155–181.

Hanski I, Cambefort Y (1991) *Dung Beetle Ecology*. Princeton University Press, Princeton, NJ.

Hansson BS, Löfstedt C, Roelofs WL (1987) Inheritance of olfactory response to sex pheromone components in *Ostrinia nubialis*. *Naturwissenschaften* 74:497–499.

Hansson BS, Tòth M, Löfstedt C, et al. (1990) Pheromone variation among eastern European and western Asian populations of the turnip moth *Agrotis segetum*. *Journal of Chemical Ecology* 16:1611–1622.

Happ GM (1969) Multiple sex pheromones of the mealworm beetle *Tenebrio molitor*. *Nature* 222:180–181.

Harano T, Miyatake T (2007a) No genetic correlation between the sexes in mating frequency in the bean beetle, Callosobruchus chinensis. *Heredity* 99:295–300.

Harano T, Miyatake T (2007b) Interpopulation variation in female remating is attributable to female and male effects in *Callosobruchus chinensis*. *Journal of Ethology* 25:49–55.

Harari AR, Landolt PJ, O'Brien CW, Brockmann HJ (2003) Prolonged mate guarding and sperm competition in the weevil *Diaprepes abbreviatus* (L.). *Behavioral Ecology* 14:89–96.

Harari AR, Zahavi T, Thiéry D (2011) Fitness cost of pheromone production in signaling female moths. *Evolution* 65:1572–1582.

Hardin PE (2011) Molecular genetic analysis of circadian timekeeping in *Drosophila*. *Advances in Genetics* 74:141–173.

Härdling R, Kaitala A (2005) The evolution of repeated mating under sexual conflict. *Journal of Evolutionary Biology* 18:106–115.

Härdling R, Karlsson K (2009) The dynamics of sexually antagonistic coevolution and the complex influences of mating system and genetic correlation. *Journal of Theoretical Biology* 260:276–282.

Harris WE, Moore PJ (2005) Sperm competition and male ejaculate investment in *Nauphoeta cinerea*: effects of social environment during development. *Journal of Evolutionary Biology* 18:474–480.

Harshman LG, Prout T (1994) Sperm displacement without sperm transfer in *Drosophila melanogaster*. *Evolution* 48:758–766.

Hartfield M, Keightley PD (2012) Current hypotheses for the evolution of sex and recombination. *Integrative Zoology* 7:192–209.

Hartfield M, Otto SP, Keightley PD (2010) The role of advantageous mutations in enhancing the evolution of a recombination modifier. *Genetics* 184:1153–1164.

Hartke TR, Baer B (2011) The mating biology of termites: a comparative review. *Animal Behaviour* 82:927–936.

Hartl DL, Brown SW (1970) The origin of male haploid genetic systems and their expected sex ratio. *Theoretical Population Biology* 1:165–190.

Hartl DL, Hiraizumi Y, Crow JF (1967) Evidence for sperm dysfunction as the mechanism of segregation distortion in *Drosophila melanogaster*. *Proceedings of the National Academy of Sciences of the USA* 58:2240–2245.

Harvey JA, Corley LS, Strand MR (2000) Competition induces adaptive shifts in caste ratios of a polyembryonic wasp. *Nature* 406:183–186.

Hayashi F (1998) Sperm co-operation in the fishfly, *Parachauliodes japonicus*. *Functional Ecology* 12:347–350.

Hazel WN, Smock R (1993) Modeling selection on conditional strategies in stochastic environments. In J Yoshimura, CW Clark (eds), *Adaptation in Stochastic Environments*, pp. 147–154. Springer, Berlin.

Hazel WN, Smock R (2000) Inheritance in the conditional strategy revisited. *Journal of Theoretical Biology* 204:307–309.

Hazel WN, Smock R, Johnson MD (1990) A polygenic model for the evolution and maintenance of conditional strategies. *Proceedings of the Royal Society of London B* 242:181–187.

Hazel WN, Smock R, Lively CM (2004) The ecological genetics of conditional strategies. *American Naturalist* 163:888–900.

Head ML, Hunt J, Jennions MD, Brooks R (2005) The indirect benefits of mating with attractive males outweigh the direct costs. *PLoS Biology* 3:e33.

Head ML, Hunt J, Brooks R (2006) Genetic association between male attractiveness and female differential allocation. *Biology Letters* 2:341–344.

Heinze J, Hoelldobler B (1993) Fighting for a harem of queens: physiology of reproduction in *Cardiocondyla* male ants. *Proceedings of the National Academy of Sciences of the USA* 90:8412–8414.

Heinze J, Keller L (2000) Alternative reproductive strategies: a queen perspective in ants. *Trends in Ecology and Evolution* 15:508–512.

Heinze J, Hoelldobler B, Yamauchi K (1998) Male competition in *Cardiocondyla* ants. *Behavioral Ecology and Sociobiology* 42:239–246.

Heimpel GE, de Boer JG (2008) Sex determination in the Hymenoptera. *Annual Review of Entomology* 53:209–230.

Heisler IL (1984a) A quantitative genetic model for the origin of mating preferences. *Evolution* 38:1283–1295.

Heisler IL (1984b) Inheritance of female mating preferences for yellow locus genotypes in *Drosophila melanogaster*. *Genetical Research* 44:133–149.

Heisler IL (1985) Quantitative genetic models of female choice based on 'arbitrary' male characters. *Heredity* 55:187–198.

Henry L, Schwander T, Crespi BJ (2012) Deleterious mutation accumulation in asexual *Timema* stick insects. *Molecular Biology and Evolution* 29:401–408.

Herrick G, Seger J (1999) Imprinting and paternal genome elimination in insects. In R Ohlsson (ed.), *Genomic Imprinting: An Interdisciplinary Approach*. Springer, Berlin.

Hertig M (1936) The Rickettsia, *Wolbachia pipientis* (Gen. et sp. n.), and associated inclusions of the mosquito, *Culex pipiens*. *Parasitology* 28:453–486.

Hertig M, Wolbach SB (1924) Studies on rickettsia-like micro-organisms in insects. *Journal of Medical Research* 44:329–374.

Higginson DM, Pitnick S (2010) Evolution of intra-ejaculate sperm interactions: do sperm cooperate? *Biological Reviews* 86:249–270.

Higginson AD, Reader T (2009) Environmental heterogeneity, genotype-by-environment interactions and the reliability of sexual traits as indicators of mate quality. *Proceedings of the Royal Society of London* B 276:1153–1159.

Higginson DM, Morin S, Nyboer ME, et al. (2005) Evolutionary trade-offs of insect resistance to *Bacillus thuringiensis* crops: fitness cost affecting paternity. *Evolution* 59:915–920.

Higginson DM, Miller KB, Segraves KA, Pitnick S (2012a) Convergence, recurrence and diversification of complex sperm traits in diving beetles (Dytiscidae). *Evolution* 66:1650–1661.

Higginson DM, Miller KB, Segraves KA, Pitnick S (2012b) Female reproductive tract form drives the evolution of complex sperm morphology. *Proceedings of the National Academy of Sciences of the USA* 109:4538–4543.

Hilgenboecker K, Hammerstein P, Schlattmann P, et al. (2008) How many species are infected with *Wolbachia*? A statistical analysis of current data. *FEMS Microbiology Letters* 281:215–220.

Hilker M, Meiners T (eds) (2002) *Chemoecology of Insect Eggs and Egg Deposition*. Blackwell, Oxford.

Himler AG, Caldera EJ, Baer BC, et al. (2009) No sex in fungus-farming ants or their crops. *Proceedings of the Royal Society of London B* 276:2611–2616.

Hine E, Chenoweth SF, Blows MW (2004) Multivariate quantitative genetics and the lek paradox: genetic variance in male sexually selected traits of *Drosophila serrata* under field conditions. *Evolution* 58:2754–2762.

Hine E, Chenoweth SF, Rundle HD, Blows MW (2009) Characterizing the evolution of genetic variance using genetic covariance tensors. *Philosophical Transactions of the Royal Society of London B* 364:1567–1578.

Hine E, McGuigan K, Blows MW (2011) Natural selection stops the evolution of male attractiveness. *Proceedings of the National Academy of Sciences of the USA* 108:3659–3664.

Hinton HE (1964) Sperm transfer in insects and the evolution of haemocoelic insemination. In KC Highnam (ed.), *Insect Reproduction*, pp. 95–107. Symposia of the Royal Entomological Society of London No. 2.

Hinton HE (1977) Subsocial behaviour and biology of some Mexican membracid bugs. *Ecological Entomology* 2:61–79.

Hinton HE (1981) *The Biology of Insect Eggs*. Pergamon Press, Oxford.

Hironaka M, Nomakuchi S, Iwakuma S, Filippi L (2005) Trophic egg production in a subsocial shield bug, *Parastrachia japonensis* Scott (Heteroptera: Parastrachiidae), and its functional value. *Ethology* 111:1089–1102.

Hissmann K (1990) Strategies of mate finding in the European field cricket (*Gryllus campestris*), at different population-densities—a field-study. *Ecological Entomology* 15:281–291.

Hockham LR, Graves JA, Ritchie MG (2004) Sperm competition and the level of polyandry in a bushcricket with large nuptial gifts. *Behavioral Ecology and Sociobiology* 57:149–154.

Hodin J (2009) She shapes events as they come: plasticity in female reproduction. In DW Whitman, TN Ananthakrishnan (eds), *Phenotypic Plasticity of Insects: Mechanism and Consequences*, pp. 423–521. Science Publishers, Enfield, NH.

Hodin J, Riddiford LM (2000) Different mechanisms underlie phenotypic plasticity and interspecific variation for a reproductive character. *Evolution* 54:1638–1653.

Hoffmann AA, Reynolds KT, Nash MA, Weeks AR (2008) A high incidence of parthenogenesis in agricultural pests. *Proceedings of the Royal Society of London B* 275:2473–2481.

Höglund J, Alatalo RV (1995) *Leks*. Princeton University Press, Princeton, NJ.

Hoikkala A, Aspi J, Suvanto L (1998) Male courtship song frequency as an indicator of male genetic quality in an insect species, *Drosophila montana*. *Proceedings of the Royal Society of London B* 265:503–508.

Holland B, Rice WR (1998) Chase-away sexual selection: antagonistic seduction versus resistance. *Evolution* 52:1–7.

Hölldobler B, Bartz SH (1985) Sociobiology of reproduction in ants. In B Hölldobler, M Lindauer (eds), *Experimental Behavioural Ecology and Sociobiology*, pp. 237–257. Sinauer Associates, Sunderland, MA.

Hölldobler B, Wilson EO (1990) *The Ants*. Springer, Berlin.

Holm E (1993) On the genera of African Cetoniinae: II. Eudicella White, and the related genera with horned males (Coleoptera: Scarabaeidae). *Journal of African Zoology* 107:65–81.

Holman L, Kokko H (2013) The consequences of polyandry for population viability, extinction risk and conservation. *Philosophical Transactions of the Royal Society of London B* 368:20120053.

Holman L, Snook RR (2008) A sterile sperm caste protects brother fertile sperm from female-mediated death in *Drosophila pseudoobscura*. *Current Biology* 18:292–296.

Holman L, Freckleton RP, Snook RR (2008) What use is an infertile sperm? A comparative study of sperm-heteromorphic *Drosophila*. *Evolution* 62:374–385.

Holt RD (2009) Bringing the Hutchinsonian niche into the 21st century: ecological and evolutionary perspectives. *Proceedings of the National Academy of Sciences of the USA* 106:19659–19665.

Holwell GI, Winnick C, Tregenza T, Herberstein ME (2010) Genital shape correlates with sperm transfer success in the praying mantis *Ciulfina klassi* (Insecta: Mantodea). *Behavioral Ecology and Sociobiology* 64:617–625.

Hongo Y (2007) Evolution of male dimorphic allometry in a population of the Japanese horned beetle *Trypoxylus dichotomus septentrionalis*. *Behavioral Ecology and Sociobiology* 62:245–253.

Hood ME, Antonovics J (2004) Mating within the meiotic tetrad and the maintenance of genomic heterozygosity. *Genetics* 166:1751–1759.

Horner VL, Wolfner MF (2008) Transitioning from egg to embryo: triggers and mechanisms of egg activation. *Developmental Dynamics* 237:527–544.

Hosken DJ (2001) Sex and death: microevolutionary trade-offs between reproductive and immune investment in dung flies. *Current Biology* 11:R379–R380.

Hosken DJ, Stockley P (2003) Benefits of polyandry: a life history perspective *Evolutionary Biology* 33:173–194.

Hosken DJ, Stockley P, Tregenza T, Wedell N (2009) Monogamy and the battle of the sexes. *Annual Review of Entomology* 54:361–378.

Hosken DJ, Ward PI (2001) Experimental evidence for testis size evolution via sperm competition. *Ecology Letters* 4:10–13.

Hosokawa T, Suzuki N (2001) Significance of prolonged copulation under the restriction of daily reproductive time in the stink bug *Megacopta punctatissima* (Heteroptera: Plataspidae). *Annals of the Entomological Society of America* 94:750–754.

Hosoya T, Araya K (2005) Phylogeny of Japanese stag beetles (Coleoptera: Lucanidae) inferred from 16S mtrRNA gene sequences, with reference to the evolution of sexual dimorphism of mandibles. *Zoological Science* 22:1305–1318.

Hotzy C, Arnqvist G (2009) Sperm competition favors harmful males in seed beetles. *Current Biology* 19:404–407.

Hotzy C, Polak M, Rönn JL, Arnqvist G (2012) Phenotypic engineering unveils the function of genital morphology. *Current Biology* 22:2258–2261.

House CM, Simmons LW (2003) Genital morphology and fertilization success in the dung beetle *Onthophagus taurus*: an example of sexually selected male genitalia. *Proceedings of the Royal Society of London B* 270:447–455.

House CM, Evans GMV, Smiseth PT, et al. (2008) The evolution of repeated mating in the burying beetle, *Nicrophorus vespilloides*. *Evolution* 62:2004–2014.

House CM, Walling CA, Stamper CE, Moore AJ (2009) Females benefit from multiple mating but not multiple mates in the burying beetle *Nicrophorus vespilloides*. *Journal of Evolutionary Biology* 22:1961–1966.

Houston T (1970) Discovery of an apparent male soldier caste in a nest of a halictine bee (Hymenoptera: Halictidae), with notes on the nest. *Australian Journal of Zoology* 18:345–351.

Houston TF, Maynard GV (2012) An unusual new paracolletine bee, *Leiproctus* (*Ottocolletes*), *muelleri* suben. & sp. nov. (Hymenoptera:Colletidae): with notes on nesting biology and in-buurow nest guarding by macrocephalic males. *Australian Journal of Entomology* 51:248–257.

Houston AI, Székely T, McNamara JM (2005) Conflict between parents over care. *Trends in Ecology and Evolution* 20:33–38.

Howard RS, Lively CM (1994) Parasitism, mutation accumulation and the maintenance of sex. *Nature* 367:554–557.

Hoy RR, Hahn J, Paul RC (1977) Hybrid cricket auditory behavior—evidence for genetic coupling in animal communication. *Science* 195:82–84.

Hughes WOH, Boomsma JJ (2005) Genetic diversity and disease resistance in leafcutting ant societies. *Evolution* 58:1251–1260.

Hughes WOH, Oldroyd BP, Beekman M, Ratnieks FLW (2008) Ancestral monogamy shows kin selection is key to the evolution of eusociality. *Science* 320:1213–1216.

Huigens ME, Luck RF, Klaassen RHG, et al. (2000) Infectious parthenogenesis. *Nature* 405:178–179.

Hunt J, Blows MW, Zajitschek F, et al. (2007) Reconciling strong stabilizing selection with the maintenance of genetic variation in a natural population of black field crickets (*Teleogryllus commodus*). *Genetics* 177:875–880.

Hunt J, Breuker CJ, Sadowski, JA, Moore AJ (2009) Male–male competition, female mate choice and their interaction: determining total sexual selection. *Journal of Evolutionary Biology* 22:13–26.

Hunt J, House CM (2011) The evolution of parental care in the onthophagine dung beetles. In LW Simmons, TJ Ridsdill-Smith (eds), *Ecology and Evolution of Dung Beetles*, pp. 152–176. Wiley–Blackwell, Oxford.

Hunt J, Simmons LW (1997) Patterns of fluctuating asymmetry in beetle horns: an experimental examination of the honest signalling hypothesis. *Behavioral Ecology and Sociobiology* 41:109–114.

Hunt J, Simmons LW (1998) Patterns of parental provisioning covary with male morphology in horned beetle (*Onthophagus taurus*), (Coleoptera: Scarabaeidae). *Behavioral Ecology and Sociobiology* 42:447–451.

Hunt J, Simmons LW (2000) Maternal and paternal effects on offspring phenotype in the dung beetle *Onthophagus taurus*. *Evolution* 54:936–941.

Hunt J, Simmons LW (2001) Status-dependent selection in the dimorphic beetle *Onthophagus taurus*. *Proceedings of the Royal Society of London B* 268:2409–2414.

Hunt J, Simmons LW (2002a) Confidence of paternity and paternal care: covariation revealed through the experimental manipulation of the mating system in the beetle *Onthophagus taurus*. *Journal of Evolutionary Biology* 15:784–795.

Hunt J, Simmons LW (2002b) Behavioural dynamics of biparental care in the dung beetle *Onthophagus taurus*. *Animal Behaviour* 64:65–75.

Hunt J, Kotiaho JS, Tomkins JL (1999) Dung pad residence time covaries with male morphology in the dung beetle *Onthophagus taurus*. *Ecological Entomology* 24:174–180.

Hunt J, Simmons LW, Kotiaho JS (2002) A cost of maternal care in the dung beetle *Onthophagus taurus*? *Journal of Evolutionary Biology* 15:57–64.

Hunt J, Bussière LF, Jennions MD, Brooks R (2004) What is genetic quality? *Trends in Ecology and Evolution* 19:329–333.

Hunter FM, Birkhead TR (2002) Sperm viability and sperm competition in insects. *Current Biology* 12:121–123.

Hunter MS, Perlman SJ, Kelly SE (2003) A bacterial symbiont in the *Bacteroidetes* induces cytoplasmic incompatibility in the parasitoid wasp *Encarsia pergandiella*. *Proceedings of the Royal Society of London B* 270:2185–2190.

Ide J-Y, Kondoh M (2000) Male–female evolutionary game on mate-locating behaviour and evolution of mating systems in insects. *Ecology Letters* 3:433–440.

Ikeda H, Maruo O (1982) Directional selection for pulse repetition rate of the courtship sound and correlated responses occurring in several characters of *Drosophila mercatorum*. *Japan Journal of Genetics* 57:241–258.

Immonen E, Ritchie MG (2011) Animal communication: flies' ears are tuned in. *Current Biology* 21:R278–R280.

Immonen E, Ritchie MG (2012) The genomic response to courtship song stimulation in female *Drosophila melanogaster*. *Proceedings of the Royal Society of London B* 279:1359–1365.

Immonen E, Hoikkala A, Kazem AJN, Ritchie MG (2009) When are vomiting males attractive? Sexual selection on condition-dependent nuptial feeding in *Drosophila subobscura*. *Behavioral Ecology* 20:289–295.

Ims RA (1988) The potential for sexual selection in males: effect of sex ratio and spatiotemporal distribution of receptive females. *Evolutionary Ecology* 2:338–352.

Ingleby FC, Hunt J, Hosken DJ (2010a) The role of genotype-by-environment interactions in sexual selection. *Journal of Evolutionary Biology* 23:2031–2045.

Ingleby FC, Lewis Z, Wedell N (2010b) Level of sperm competition promotes evolution of male ejaculate allocation patterns in a moth. *Animal Behaviour* 80:37–43.

Inoda T, Hardling R, Kubota S (2012) The inheritance of intrasexual dimorphism in female diving beetles (Coleoptera: Dytiscidae). *Zoological Science* 29:505–509.

Ishiwata K, Sasaki G, Ogawa J, et al. (2011) Phylogenetic relationships among insect orders based on three nuclear protein-coding gene sequences. *Molecular Phylogenetics and Evolution* 58:169–180.

Ivy TM (2007) Good genes, genetic compatibility and the evolution of polyandry: use of the diallel cross to address competing hypotheses. *Journal of Evolutionary Biology* 20:479–487.

Ivy TM, Sakaluk SK (2005) Polyandry promotes enhanced offspring survival in decorated crickets. *Evolution* 59:152–159.

Ivy TM, Sakalaluk SK (2007) Sequential mate choice in decorated crickets: females use a fixed internal threshold in pre- and postcopulatory choice. *Animal Behaviour* 74:1065–1072.

Iwasa Y, Odendaal FJ, Murphy DD, et al. (1983) Emergence patterns in male butterflies: a hypothesis and a test. *Theoretical Population Biology* 23:363–379.

Iwasa Y, Pomiankowski A, Nee S (1991) The evolution of costly mate preferences II. The 'handicap' principle. *Evolution* 45:1431–1442.

Iwasa Y, Pomiankowski A (1994) The evolution of mate preferences for multiple sexual ornaments. *Evolution* 48:853–867.

Iwasa Y, Pomiankowski A (1999) Good parent and good genes models of handicap evolution. *Journal of Theoretical Biology* 200:97–109.

Iyengar VK, Eisner T (1999) Female choice increases offspring fitness in an arctiid moth (*Utetheisa ornatrix*). *Proceedings of the National Academy of Sciences of the USA* 96:15013–15016.

Iyengar VK, Rossini C, Eisner T (2001) Precopulatory assessment of male quality in an arctiid moth (*Utetheisa ornatrix*): hydroxydanaidal is the only criterion of choice. *Behavioral Ecology and Sociobiology* 49:283–288.

Iyengar VK, Reeve HK, Eisner T (2002) Paternal inheritance of a female moth's mating preference. *Nature* 419:830–832.

Izzo AS, Tibbetts EA (2012) Spotting the top male: sexually selected signals in male *Polistes dominulus* wasps. *Animal Behaviour* 83:839–845.

Jacobs AC, Zuk M (2012) Sexual selection and parasites: do mechanisms matter? In G Demas, R Nelson (eds,) *Ecoimmunology*, pp. 468–496. Oxford University Press, Oxford.

Jacot A, Scheuber H, Kurtz J, Brinkhof MWG (2005) Juvenile immune status affects the expression of a sexually selected trait in field crickets. *Journal of Evolutionary Biology* 18:1060–1068.

Jaenike J (1978) An hypothesis to account for the maintenance of sex within populations. *Evolutionary Theory* 3:191–194.

Jaenike J (1988) Parasitism and male mating success in *Drosophila testacea*. *American Naturalist* 131:774–780.

Jaenike J, Unckles R, Cockburn SN, et al. (2010) Adaptation via symbiosis: recent spread of a *Drosophila* defensive symbiont. *Science* 329:212–215.

Jaffe R, Moritz RFA (2010) Mating flights select for symmetry in honeybee drones (*Apis mellifera*). *Naturwissenschaften* 97:337–343.

Jaffe R, Garcia-Gonzalez F, den Boer SPA, et al. (2012) Patterns of paternity skew among polyandrous social insects: what can they tell us about the potential for sexual selection? *Evolution* 66:3778–3788.

Jamieson BGM, Dallai R, Afzelius B (1999) *Insects: their Spermatozoa and Phylogeny*. Science Publishers, Enfield, NH.

Jang Y, Greenfield MD (1998) Absolute versus relative measurements of sexual selection: assessing the contributions of ultrasonic signal characters to mate attraction in lesser wax moths, *Achroia grisella* (Lepidoptera: Pyralidae). *Evolution* 52:1383–1393.

Jang Y, Greenfield MD (2000) Quantitative genetics of female choice in an ultrasonic pyralid moth, *Achroia grisella*: variation and evolvability of preference along multiple dimensions of the male advertisement signal. *Heredity* 84:73–80.

Jarosch A, Stolle E, Crewe RM, Moritz RF (2011) Alternative splicing of a single transcription factor drives selfish reproductive behavior in honeybee workers (*Apis mellifera*). *Proceedings of the National Academy of Sciences of the USA* 108:15282–15287.

Jenkins EV, Morris C, Blackman S (2000) Delayed benefits of paternal care in the burying beetle *Nicrophorus vespilloides*. *Animal Behaviour* 60:443–451.

Jennions MD, Backwell PRY (1996) Residency and size affect fight duration and outcome in the fiddler crab *Uca annulipes*. *Biological Journal of the Linnean Society* 57:293–306.

Jennions MD, Kokko H (2010) Sexual selection. In DF Westneat, CW Fox (eds), *Evolutionary Behavioral Ecology*, pp. 343–364. Oxford University Press, Oxford.

Jennions MD, Petrie M (2000) Why do females mate multiply? A review of the genetic benefits. *Biological Reviews* 75:21–64.

Jennions MD, Hunt J, Graham R, Brooks R (2004) No evidence for inbreeding avoidance through postcopulatory mechanisms in the black field cricket, *Teleogryllus commodus*. *Evolution* 58:2472–2477.

Jennions MD, Møller AP, Petrie M (2001) Sexually selected traits and adult survival: A meta-analysis. *Quarterly Review of Biology* 76:3–36.

Jennions MD, Kokko H, Klug H (2012) The opportunity to be misled in studies of sexual selection. *Journal of Evolutionary Biology* 25:591–598.

Jervis MA, Ellers J, Harvey JA (2008) Resource acquisition, allocation, and utilization in parasitoid reproductive strategies. *Annual Review of Entomology* 53:361–385.

Jia F-Y, Greenfield MD, Collins RD (2000) Genetic variance of sexually selected traits in waxmoths: maintenance by genotype × environment interaction. *Evolution* 54:953–967.

Jiang H, Vilcinskas A, Kanost MR (2011) Immunity in lepidopteran insects. *Advances in Experimental Medicine and Biology* 708:181–204.

Jiang Y, Bolnick DI, Kirkpatrick M (2013) Assortative mating in animals. *American Naturalist* 181:E125–E138.

Jiggins FM, Hurst GDD, Majerus MEN (2000) Sex-ratio distorting *Wolbachia* causes sex-role reversal in its butterfly host. *Proceedings of the Royal Society of London B* 267:69–73.

Johns PM, Wolfenbarger LL, Wilkinson GS (2005) Genetic linkage between a sexually selected trait and X chromosome meiotic drive. *Proceedings of the Royal Society of London B* 272:2097–2103.

Johnson C (1975) Polymorphism and natural selection in ischnuran damselflies. *Evolutionary Theory* 1:81–90.

Johnson MT, Fitzjohn RG, Smith SD, et al. (2011) Loss of sexual recombination and segregation is associated with increased diversification in evening primroses. *Evolution* 65:3230–3240.

Johnstone RA (1995) Honest advertisement of multiple qualities using multiple signals. *Journal of Theoretical Biology* 177:87–94.

Johnstone RA, Hinde CA (2006) Negotiation over offspring care—how should parents respond to each other's efforts? *Behavioral Ecology* 17:818–827.

Johnstone RA, Keller L (2000) How males can gain by harming their mates: sexual conflict, seminal toxins, and the cost of mating. *American Naturalist* 156:368–377.

Jokela J, Dybdahl MF, Lively CM (2009) The maintenance of sex, clonal dynamics, and host–parasite coevolution in a mixed population of sexual and asexual snails. *American Naturalist* 174:S43–S53.

Joly D, Schiffer M (2010) Coevolution of male and female reproductive structures in *Drosophila*. *Genetica* 138:105–118.

Jones DL (2007) Aging and the germ line: where mortality and immortality meet. *Stem Cell Reviews* 3:192–200.

Jones AG (2009) On the opportunity for sexual selection, the Bateman gradient and the maximum intensity of sexual selection. *Evolution* 63:1673–1684.

Jones AG, Ratterman NL (2009) Mate choice and sexual selection: what have we learned since Darwin? *Proceedings of the National Academy of Sciences of the USA* 106:10001–10008.

Jones AG, Rosenqvist G, Berglund A, et al. (2000) The Bateman gradient and the cause of sexual selection in a sex-role-reversed pipefish. *Proceedings of the Royal Society of London B* 267:677–680.

Jones AG, Arguello JR, Arnold SJ (2002) Validation of Bateman's principles: a genetic study of sexual selection and mating patterns in the rough-skinned newt. *Proceedings of the Royal Society of London* B 269:2533–2539.

Jones RT, Salazar PA, Jiggins CD, Joron M (2012) Evolution of a mimicry supergene from a multilocus architecture. *Proceedings of the Royal Society of London B* 279:316–325.

Joron M, Frezal L, Jones RT, et al. (2011) Chromosomal rearrangements maintain a polymorphic supergene controlling butterfly mimicry. *Nature* 477:203–206.

Judge KA (2010) Female social experience affects the shape of sexual selection on males. *Evolutionary Ecology Research* 12:389–402.

Kaczmarczyk AN, Kopp A (2011) Germline stem cell maintenance as a proximate mechanism of life-history trade-offs? *BioEssays* 33:5–12.

Kaitala A, Härdling R, Katvala M, et al. (2001) Is nonparental egg carrying parental care? *Behavioral Ecology* 12:367–368.

Kajtoch L, Lachowska-Cierlik D (2009) Genetic constitution of parthenogenetic form of *Polydrusus inustus* (Coleoptera: Curculionidae)—hints of hybrid origin and recombinations. *Folia Biolica* 57:149–156.

Kambysellis M, Heed W (1971) Studies of oogenesis in natural populations of Drosophilidae. I. Relation of ovarian development and ecological habitats of the Hawaiian species. *American Naturalist* 105:31–49.

Kamimura Y (2010) Copulation anatomy of *Drosophila melanogaster* (Diptera: Drosophilidae): wound-making organs and their possible roles. *Zoomorphology* 129:163–174.

Karr TL, Yang W, Feder ME (1998) Overcoming cytoplasmic incompatibility in *Drosophila*. *Proceedings of the Royal Society of London B* 265:391–395.

Kawecki TJ, Lenski RE, Ebert D, et al. (2012) Experimental evolution. *Trends in Ecology and Evolution* 27:547–560.

Kearney M, Moussalli A (2003) Geographic parthenogenesis in the Australian arid zone: II. Climatic analysis of orthopteroid insects of the genera *Warramaba* and *Sipyloidea*. *Evolutionary Ecology Research* 5:977–997.

Kearns PWE, Tomlinson IPM, Veltman CJ, O'Donald P (1992) Non-random mating in *Adalia bipunctata* (the two-spot ladybird). II. Further tests for female mating preference. *Heredity* 68:385–389.

Keasar T, Sheffer N, Glusman G, Libersat F (2006) Host-handling behavior: an innate component of foraging behavior in the parasitoid wasp *Ampulex compressa*. *Ethology* 112:699–706.

Keays MC, Barker D, Wicker-Thomas C, Ritchie MG (2011) Signatures of selection and sex-specific expression variation of a novel duplicate during the evolution of the *Drosophila* desaturase gene family. *Molecular Ecology* 20:3617–3630.

Kehl T, Karl I, Fischer K (2013) Old-male paternity advantage is a function of accumulating sperm and last-male precedence in a butterfly. *Molecular Ecology* 22:4289–4297.

Kelleher ES, Markow TA (2009) Duplication, selection and gene conversion in a *Drosophila mojavensis* female reproductive protein family. *Genetics* 181:1451–1465.

Kelleher ES, Clark NL, Markow TA (2011) Diversity-enhancing selection acts on a female reproductive protease family in four subspecies of *Drosophila mojavensis*. *Genetics* 187:865–876.

Keller L, Passera L (1992) Mating system, optimal number of matings, and sperm transfer in the Argentine ant *Iridomyrmex humilis*. *Behavioral Ecology and Sociobiology* 31:359–366.

Keller L, Reeve H (1995) Why do females mate with multiple males? The sexually selected sperm hypothesis. *Advances in the Study of Behavior* 24:291–315.

Kelly CD (2006a) Fighting for harems: assessment strategies during male–male contests in the sexually dimorphic Wellington tree weta. *Animal Behaviour* 72:727–736.

Kelly CD (2006b) The relationship between resource control, association with females and male weapon size in a male dominance insect. *Ethology* 112:362–369.

Kelly CD (2008a) Identifying a causal agent of sexual selection on weaponry in an insect. *Behavioral Ecology* 19:184–192.

Kelly CD (2008b) Sperm investment in relation to weapon size in a male trimorphic insect? *Behavioral Ecology* 19:1018–1024.

Kelly CD, Adams DC (2010) Sexual selection, ontogenetic acceleration, and hypermorphosis generates male trimorphism in Wellington tree weta. *Evolutionary Biology* 37:200–209.

Kelly CD, Jennions MD (2011) Sexual selection and sperm quantity: meta-analyses of strategic ejaculation. *Biological Reviews* 86:863–884.

Kemp DJ (2008) Female mating biases for bright ultraviolet iridescence in the butterfly *Eurema hecabe* (Pieridae). *Behavioral Ecology* 19:1–8.

Kemp DJ, Wiklund C (2001) Fighting without weaponry: a review of male–male contest competition in butterflies. *Behavioral Ecology and Sociobiology* 49:429–442.

Kirkpatrick M (1982) Sexual selection and the evolution of female choice. *Evolution* 36:1–12.

Kirkpatrick M (1985) Evolution of female choice and male parental investment in polygynous species: the demise of the "Sexy Son". *American Naturalist* 125:788–810.

Kirkpatrick M (1986) The handicap mechanism of sexual selection does not work. *American Naturalist* 127:222–240.

Kirkpatrick M (1987a) Sexual selection by female choice in polygynous animals. *Annual Review of Ecology and Systematics* 18:43–70.

Kirkpatrick M (1987b) The evolutionary forces acting on female mating preferences in polygynous animals. In JW Bradbury, MB Andersson (eds), *Sexual Selection: Testing the Alternatives*, pp. 67–82. John Wiley & Sons, Chichester.

Kirkpatrick M (2010) How and why chromosome inversions evolve. *PLoS Biology* 8:e1000501

Kirkpatrick M, Barton NH (1997) The strength of indirect selection on female mating preferences. *Proceedings of the National Academy of Sciences of the USA* 94:1282–1286.

Kirkpatrick M, Ryan MJ (1991) The evolution of mating preferences and the paradox of the lek. *Nature* 350:33–38.

Kirkpatrick M, Servedio MR (1999) The reinforcement of mating preferences on an island. *Genetics* 151:865–884.

Kirkpatrick M, Price T, Arnold SJ (1990) The Darwin–Fisher theory of sexual selection in monogamous birds. *Evolution* 44:180–193.

Kishi S, Nishida T (2008) Optimal investment in sons and daughters when parents do not know the sex of their offspring. *Behavioral Ecology and Sociobiology* 62:607–615.

Klemperer HG (1982) Parental behaviour in *Copris lunaris* (Coleoptera, Scarabaeidae): care and defense of brood balls and nest. *Ecological Entomology* 7:155–167.

Klug H, Heuschele J, Jennions MD, Kokko H (2010) The mismeasurement of sexual selection. *Journal of Evolutionary Biology* 23:447–462.

Knell RJ, Pomfret JC, Tomkins JL (2004) The limits of elaboration: curved allometries reveal the constraints on mandible size in stag beetles. *Proceedings of the Royal Society of London B* 271:523–528.

Knowlton N, Keller BD (1982) Symmetric fights as a measure of escalation potential in a symbiotic, terratorial snapping shrimp. *Behavioral Ecology and Sociobiology* 10:289–292.

Kock D, Sauer KP (2007) High variation in sperm precedence and last male advantage in the scorpionfly *Panorpa germanica* L. (Mecoptera, Panorpidae): Possible causes and consequences. *Journal of Insect Physiology* 53:1145–1150.

Kock D, Hardt C, Epplen JT, Sauer KP (2006) Patterns of sperm use in the scorpionfly *Panorpa germanica* L. (Mecoptera: Panorpidae). *Behavioral Ecology and Sociobiology* 60:528–535.

Koehncke A, Telschow A, Werrent JH, Hammerstein P (2009). Life and death of an influential passenger: *Wolbachia* and the evolution of CI-modifiers by their hosts. *PLoS One* 4:e4425.

Koepfer HR (1987) Selection for sexual isolation between geographic forms of *Drosophila mojavensis*. II. Effects of selection on mating preferences and propensity. *Evolution* 41:1409–1413.

Kokko H, Heubel K (2008) Condition-dependence, genotype-by-environment interactions and the lek paradox. *Genetica* 132:209–216.

Kokko H, Jennions MD (2008) Parental investment, sexual selection and sex ratios. *Journal of Evolutionary Biology* 21:919–948.

Kokko H, Jennions MD (2012) Sex differences in parental care. In NJ Royle, PT Smiseth, M Kölliker (eds), *The Evolution of Parental Care*, pp. 101–116. Oxford University Press, Oxford.

Kokko H, Johnstone RA (2002) Why is mutual mate choice not the norm? Operational sex ratios, sex roles and the evolution of sexually dimorphic and monomorphic signalling. *Philosophical Transactions of the Royal Society of London B* 357:319–330.

Kokko H, Mappes J (2013) Multiple mating by females is a natural outcome of a null model of mate encounters. *Entomologia Experimentalis et Applicata* 146:26–37.

Kokko H, Monaghan P (2001) Predicting the direction of sexual selection. *Ecology Letters* 4:159–165.

Kokko H, Ots I (2006) When not to avoid inbreeding. *Evolution* 60:467–475.

Kokko H, Wong BBM (2007) What determines sex roles in mate searching? *Evolution* 61:1162–1175.

Kokko H, Brooks R, McNamara JM, Houston AI (2002) The sexual selection continuum. *Proceedings of the Royal Society of London B* 269:1331–1340.

Kokko H, Brooks R, Jennions MD, Morley J (2003) The evolution of mate choice and mating biases. *Proceedings of the Royal Society of London B* 270:653–664.

Kokko H, Jennions MD, Brooks R (2006) Unifying and testing models of sexual selection. *Annual Review of Ecology Evolution and Systematics* 37:43–66.

Kokko H, Klug H, Jennions MD (2012) Unifying cornerstones of sexual selection: operational sex ratio, Bateman gradient, and the scope for competitive investment. *Ecology Letters* 15:1340–1351.

Kolics B, Acs Z, Chobanov DP, et al. (2012) Re-visiting phylogenetic and taxonomic relationships in the genus *Saga* (Insecta: Orthoptera). *PLoS One* 7:e42229.

Kölliker M (2007) Benefits and costs of earwig (*Forficula auricularia*) family life. *Behavioral Ecology and Sociobiology* 61:1489–1497.

Kondrashov AS (1982) Selection against harmful mutations in large sexual and asexual populations. *Genetical Research* 40:325–332.

Kondrashov AS (1993) Classification of hypotheses on the advantage of amphimixis. *Journal of Heredity* 84:372–387.

Koning JW, Jamieson IG (2001) Variation in size of male weaponry in a harem-defence polygynous insect, the mountain stone weta *Hemideina maori* (Orthoptera: Anostostomatidae). *New Zealand Journal of Zoology* 28:109–117.

Kostarakos K, Hartbauer M, Römer H (2008) Matched filters, mate choice and the evolution of sexually selected traits. *PLoS One* 3:e3005.

Kotiaho JS (2001) Costs of sexual traits: a mismatch between theoretical considerations and empirical evidence. *Biological Reviews* 76:365–376.

Kozlowski MW (2004) Reduction in last male sperm precedence caused by competitive assault on a mating male in *Gastrophysa viridula*, a highly polygamous leaf beetle. *Ethology Ecology and Evolution* 16:15–23.

Kraaijeveld K, Kraaijeveld-Smit FJL, Komdeur J (2007) The evolution of mutual ornamentation. *Animal Behaviour* 74:657–677.

Kraaijeveld K, Kraaijeveld-Smit FJL, Maan ME (2011) Sexual selection and speciation: the comparative evidence revisited. *Biological Reviews* 86:367–377.

Kraaijeveld K, Zwanenbur B, Hubert B, et al. (2012) Transposon proliferation in an asexual parasitoid. *Molecular Ecology* 21:3898–3906.

Kraus WF (1989) Is male back space limiting? An investigation into the reproductive demography of the giant water bug, *Abedus indentatus* (Heteroptera: Belostomatidae). *Journal of Insect Behavior* 2:623–648.

Kronforst MR, Young LG, Kapan DD, et al. (2006) Linkage of butterfly mate preference and wing color preference cue at the genomic location of wingless. *Proceedings of the National Academy of Sciences of the USA* 103:6575–6580.

Krupp JJ, Kent C, Billeter JC, et al. (2008) Social experience modifies pheromone expression and mating behavior in male *Drosophila melanogaster*. *Current Biology* 18:1373–1383.

Kubli E (2003) Sex-peptides: seminal peptides of the *Drosophila* male. *CMLS Cell and Molecular Life Sciences* 60:1689–1704.

Kuijper B, Pen I (2010) The evolution of haplodiploidy by male-killing endosymbionts: importance of population structure and endosymbiont mutualisms. *Journal of Evolutionary Biology* 23:40–52.

Kuijper B, Pen I, Weissing FJ (2012) A guide to sexual selection theory. *Annual Review of Ecology, Evolution, and Systematics* 43:287–311.

Kukuk PF (1996) Male dimorphism in *Lasioglossum* (*Chilalictus*) *hemichalceum*: the role of larval nutrition. *Journal of the Kansas Entomological Society* 69:147–157.

Kukuk PF, Schwarz M (1987) Intranest behavior of the communal sweat bee *Lasioglossum* (*Chilalictus*) *erythrurum* (Hymenoptera, Halictidae). *Journal of the Kansas Entomological Society* 60:58–64.

Kukuk PF, Schwarz M (1988) Macrocephalic male bees as functional reproductives and probable guards. *Pan-Pacific Entomologist* 64:131–138.

Kuo T-H, Fedina TY, Hansen I, et al. (2012) Insulin signaling mediates sexual attractiveness in *Drosophila*. *PLoS Genetics* 8:640–650.

Kureck IM, Neumann A, Foitzik S (2011) Wingless ant males adjust mate-guarding behaviour to the competitive situation in the nest. *Animal Behaviour* 82:339–346.

Kvarnemo C, Ahnesjö I (1996) The dynamics of operational sex ratios and competition for mates. *Trends in Ecology and Evolution* 11:404–408.

Kvarnemo C, Simmons LW (2013) Polyandry as a mediator of sexual selection before and after mating. *Philosophical Transactions of the Royal Society B* 368:20120042.

Lamb RJ (1976) Parental behaviour in the Dermaptera with special reference to *Forficula auricularia* (Dermaptera: Forficulidae). *Canadian Entomologist* 108:609–619.

LaMunyon CW (1997) Increased fecundity, as a function of multiple mating, in an arctiid moth, *Utetheisa ornatrix*. *Ecological Entomology* 22:69–73.

Lande R (1979) Quantitative genetic analysis of multivariate evolution, applied to brain:body size allometry. *Evolution* 33:402–416.

Lande R (1981) Models of speciation by sexual selection on polygenic traits. *Proceedings of the National Academy of Sciences of the USA* 78:3721–3725.

Lande R, Arnold SJ (1983) The measurement of selection on correlated characters. *Evolution* 37:1210–1226.

Lane JE, Forrest MNK, Willis CKR (2011) Anthropogenic influences on natural animal mating systems. *Animal Behaviour* 81:909–917.

Langellotto GA, Denno RF, Ott JR (2000) A trade-off between flight capability and reproduction in males of a wing-dimorphic insect. *Ecology* 81:865–875.

Lanier GN, Birch MC, Schmitz RE, Furniss MM (1972) Pheromones of *Ips pini* (Coleoptera: Scolytidae): variation in response among three populations. *Canadian Journal of Entomology* 104:1917–1923.

Larsdotter Mellström, H, Wiklund C (2009) Males use sex pheromone assessment to tailor ejaculates to risk of sperm competition in a butterfly. *Behavioral Ecology* 20:1147–1151.

Larsdotter Mellström H, Wiklund C (2010) What affects mating rate? Polyandry is higher in the directly developing generation of the butterfly *Pieris napi*. *Animal Behaviour* 80:413–418.

Lassance J-M, Groot AT, Lienard MA, et al. (2010) Allelic variation in a fatty-acyl reductase gene causes divergence in moth sex pheromones. *Nature* 466:486–489.

Lassance J-M, Bogdanowicz SM, Wanner KW, et al. (2011) Gene genealogies reveal differentiation at sex pheromone olfactory loci in pheromone strains of the European corn borer *Osrtinia nubilalis*. *Evolution* 65:1583–1593.

Lassance J-M, Liénard MA, Antony B, et al. (2013) Functional consequences of sequence variation in the pheromone biosynthetic gene pgFAR for *Ostrinia* moths. *Proceedings of the National Academy of Sciences of the USA* 110:3967–3972.

Laven H (1967) Eradication of *Culex pipiens fatigans* through cytoplasmic incompatibility. *Nature* 216:383–384.

Lawniczak MKN, Begun DJ (2004) A genome-wide analysis of courting and mating responses in *Drosophila melanogaster* females. *Genome* 47:900–910.

Lawniczak MKN, Barnes AI, Linklater JR, et al. (2007) Mating and immunity in invertebrates. *Trends in Ecology and Evolution* 22:48–55.

Lawson LP, Vander Meer RK, Shoemaker D (2012) Male reproductive fitness and queen polyandry are linked to variation in the supergene Gp-9 in the fire ant *Solenopsis invicta*. *Proceedings of the Royal Society of London B* 279:3217–3222.

LeBas NR (2006) Female finery is not for males. *Trends in Ecology and Evolution* 21:170–173.

LeBas NR, Hockham LR, Ritchie MG (2003) Nonlinear and correlational sexual selection on 'honest' female ornamentation. *Proceedings of the Royal Society of London B* 270:2159–2165.

LeBas NR, Hockham LR, Ritchie MG (2004) Sexual selection in the gift-giving dance fly, *Rhamphomyia sulcata*, favors small males carrying small gifts. *Evolution* 58:1763–1772.

Lehmann GUC (2003) Review of biogeography, host range and evolution of acoustic hunting in Ormiini (Insecta, Diptera, Tachinidae), parasitoids of night-calling bushcrickets and crickets (Insecta, Orthoptera, Ensifera). *Zoologischer Anzeiger* 242:107–120.

Lehmann GUC (2007) Density-dependent plasticity of sequential mate choice in a bushcricket (Orthoptera: Tettigoniidae). *Australian Journal of Zoology* 55:123–130.

Lehmann GUC (2012) Weighing costs and benefits of mating in bushcrickets (Insecta: Orthoptera: Tettigoniidae), with an emphasis on nuptial gifts, protandry and mate density. *Frontiers in Zoology* 9:19.

Lehmann GUC, Siozios S, Bourtzis K, et al. (2011) Thelytokous parthenogenesis and the heterogeneous decay of mating behaviours in a bushcricket. *Journal of Zoological Systematics and Evolutionary Research* 49:102–109.

Lehtonen J, Kokko H (2012) Positive feedback and alternative stable states in inbreeding, cooperation, sex roles and other evolutionary processes. *Philosophical Transactions of the Royal Society of London B* 367:211–221.

Lehtonen J, Jennions MD, Kokko H (2012) The many costs of sex. *Trends in Ecology and Evolution* 27:172–178.

Lehtonen J, Schmidt DJ, Heubel K, Kokko H (2013) Evolutionary and ecological implications of sexual parasitism. *Trends in Ecology and Evolution* 28:297–306.

Le Lannic J, Nénon J-P (1999) Functional morphology of the ovipositor in *Megarhyssa atrata* (Hymenoptera, Ichneumonidae) and its penetration into wood. *Zoomorphology* 119:73–79.

Lemos WP, Serrão JE, Ramalho FS, et al. (2005) Effect of diet on male reproductive tract of *Podisus nigrispinus* (Dallas) (Heteroptera: Pentatomidae). *Brazilian Journal of Biology* 65:91–96.

Leniaud L, Darras H, Boulay R, Aron S (2012) Social hybridogenesis in the clonal ant *Cataglyphis hispanica*. *Current Biology* 22:1188–1193.

Lessells CK (2006) The evolutionary outcome of sexual conflict. *Philosophical Transactions of the Royal Society of London* B 361:301–307.

Lewis SM, Cratsley CK (2008) Flash signal evolution, mate choice, and predation in fireflies. *Annual Review of Entomology* 53:293–321.

Lewis Z, Champion de Crespigny FE, Sait SM, Tregenza T, Wedell N (2011) *Wolbachia* lowers fertile sperm transfer in a moth. *Biology Letters* 7:187–189.

Lewkiewicz DA, Zuk M (2004) Latency to resume calling after disturbance in the field cricket, *Teleogryllus oceanicus*, corresponds to population-level differences in parasitism risk. *Behavioral Ecology and Sociobiolology* 55:569–573.

Linklater JR, Wertheim B, Wigby S, Chapman T (2007) Ejaculate depletion patterns evolve in response to experimental manipulation of sex ratio in *Drosophila melanogaster*. *Evolution* 61:2027–2034.

Liu HF, Kubli E (2003) Sex-peptide is the molecular basis of the sperm effect in *Drosophila melanogaster*. *Proceedings of the National Academy of Sciences of the USA* 100:9929–9933.

Lively CM (1986) Canalization versus developmental conversion in a spatially-variable environment. *American Naturalist* 128:561–572.

Lively CM (2011) The cost of males in non-equilibrium populations. *Evolutionary Ecology Research* 13:105–111.

Lively CM, Johnson SG (1994) Brooding and the evolution of parthenogenesis: strategy models and evidence from aquatic invertebrates. *Proceedings of the Royal Society of London B* 256:89–95.

Lively CM, Lloyd DG (1990) The cost of biparental sex under individual selection. *American Naturalist* 135:489–500.

Lizé A, Doff RJ, Smaller EA, et al. (2012) Perception of male–male competition influences *Drosophila* copulation behaviour even in species where females rarely remate. *Biology Letters* 8:35–38.

Loewe L (2009) A framework for evolutionary systems biology. *BMC Systems Biology* 3:27

Löfstedt C, Löfqvist J, Lanne BS, et al. (1986) Pheromone dialects in European turnip moths *Agrotis segetim*. *Oikos* 46:250–257.

Loher W, Rence B (1978) Mating-behavior of *Teleogryllus commodus* (Walker) and its central and peripheral control. *Zeitschrift Fur Tierpsychologie* 46:225–259.

Longair RW (2004) Tusked males, male dimorphism and nesting behavior in a subsocial afrotropical wasp, *Synagris cornuta*, and weapons and dimorphism in the genus (Hymenoptera: Vespidae: Eumeninae). *Journal of the Kansas Entomological Society* 77:528–557.

Lorch PD, Proulx S, Rowe L, Day T (2003) Condition-dependent sexual selection can accelerate adaptation. *Evolutionary Ecology Research* 5:867–881.

Low BS (1978) Environmental uncertainty and the parental strategies of marsupials and placentals. *American Naturalist* 112:197–213.

Luck N, Dejonghe B, Fruchard S, et al. (2007) Male and female effects on sperm precedence in the giant sperm species *Drosophila bifurca*. *Genetica* 130:257–265.

Lüpold S, Manier MK, Ala-Honkola O, et al. (2011) Male *Drosophila melanogaster* adjust ejaculate size based on female mating status, fecundity, and age. *Behavioral Ecology* 22:184–191.

Lüpold S, Manier MK, Berben KS, et al. (2012) How multivariate ejaculate traits determine competitive fertilization success in *Drosophila melanogaster*. *Current Biology* 22:1667–1672.

Lynch M (1984) Destabilizing hybridization, general-purpose genotypes and geographical parthenogenesis. *Quarterly Review of Biology* 59:257–290.

Lynch M, Walsh B (1998) *Genetics and Analysis of Quantitative Traits*. Sinauer, Sunderland, MA.

Lyons C, Barnard CJ (2006) A learned response to sperm competition in *Gryllus bimaculatus* (de Geer). *Animal Behaviour* 72:673–680.

Mack PD, Hammock BA, Promislow DEL (2002) Sperm competitive ability and genetic relatedness in *Drosophila melanogaster*: similarity breeds contempt. *Evolution* 56:1789–1795.

Mack PD, Kapelnikov A, Heifetz Y, Bender M (2006) Mating-responsive genes in reproductive tissues of female *Drosophila melanogaster*. *Proceedings of the National Academy of Sciences of the USA* 103:10358–10363.

Mackay TFC (2001) Quantitative trait loci in *Drosophila*. *Nature Reviews Genetics* 2:11–20.

Mackay TFC, Heinsohn SL, Lyman RF, et al. (2005) Genetics and genomics of *Drosophila* mating behavior. *Proceedings of the National Academy of Sciences of the USA* 102:6622–6629.

Magnuson-Ford K, Otto SP (2012) Linking the investigations of character evolution and species diversification. *American Naturalist* 180:225–245.

Majerus MEN (2003) *Sex Wars: Genes, Bacteria, and Biased Sex Ratios*. Princeton University Press, Princeton, NJ.

Majerus MEN, O'Donald P, Weir J (1982) Female mating preference is genetic. *Nature* 300:521–523.

Majerus MEN, O'Donald P, Kearns PWE, Ireland H (1986) Genetics and evolution of female choice. *Nature* 321:164–167.

Maklakov AA, Arnqvist G (2009) Testing for direct and indirect effects of mate choice by manipulating female choosiness. *Current Biology* 19:1903–1906.

Maklakov AA, Bilde T, Lubin Y (2005) Sexual conflict in the wild: elevated mating rate reduces female lifetime reproductive success. *American Naturalist* 165:S38–S45.

Manica A, Johnstone RA (2004) The evolution of paternal care with overlapping broods. *American Naturalist* 164:517–530.

Manier MK, Belote JM, Berden KS, et al. (2010) Resolving mechanisms of competitive fertilization success in *Drosophila melanogaster*. *Science* 328:354–357.

Mank JE, Wedell N, Hosken DJ (2013) Polyandry and sex-specific gene expression. *Philosophical Transactions of the Royal Society of London* 368:20122047.

Marcillac F, Grosjean Y, Ferveur J-F (2005) A single mutation alters production and discrimination of *Drosophila* sex pheromones. *Proceedings of the Royal Society of London B* 272:303–309.

Marden JH, Rollins R A (1994) Assessment of energy reserves by damselflies engaged in aerial contests for mating territories. *Animal Behaviour* 48:1023–1030.

Margolis J, Spradling A (1995) Identification and behavior of epithelial stem cells in the *Drosophila* ovary. *Development* 121:3797–3807.

Markov AV, Lazebny OE, Goryacheva II, et al. (2009) Symbiotic bacteria affect mating choice in *Drosophila melanogaster*. *Animal Behaviour* 77:1011–1017.

Markow TA (1988) Reproductive behavior of *Drosophila melanogaster* and *D. nigrospiracula* in the field and in the laboratory. *Journal of Comparative Psychology* 102:169–173.

Martel V, Damiens D, Boivin G (2008) Strategic ejaculation in the egg parasitoid *Trichogramma turkestanica* (Hymenoptera: Trichogrammatidae). *Ecological Entomology* 33:357–361.

Martin OY, Hosken DJ (2002) Strategic ejaculation in the common dung fly *Sepsis cynipsea*. *Animal Behaviour* 63:541–546.

Martin OY, Hosken DJ (2003) The evolution of reproductive isolation through sexual conflict. *Nature* 423:979–982.

Mas F, Haynes KF, Kölliker M (2009) A chemical signal of offspring quality affects maternal care in a social insect. *Proceedings of the Royal Society of London B* 276:2847–2853.

Matsumoto K, Suzuki N (1992) Effectiveness of the mating plug in *Atrophaneura alcinous* (Lepidoptera: Papilionidae). *Behavioral Ecology and Sociobiology* 30:157–163.

Matsuura K (2011) Sexual and asexual reproduction in termites. In Bignell D, Roisin Y, Lo N (eds), *Biology of Termites: A Modern Synthesis*, pp. 255–277. Springer, Dordrecht.

Matsuura K, Vargo EL, Kawatsu K, et al. (2009) Queen succession through asexual reproduction in termites. *Science* 323:1687.

Matthews RW, Gonzalez JM, Matthews JR, Deyrup LD (2009) Biology of the parasitoid *Melittobia* (Hymenoptera: Eulophidae). *Annual Review of Entomology* 54:251–266.

Maynard Smith J (1974) The theory of games and the evolution of animal conflicts. *Journal of Theoretical Biology* 47:209–221.

Maynard Smith J (1977) Parental investment: a prospective analysis. *Animal Behaviour* 25:1–9.

Maynard Smith J (1978) *The Evolution of Sex*. Cambridge University Press, Cambridge.

Maynard Smith J (1982) *Evolution and the Theory of Games*. Cambridge University Press, Cambridge.

Maynard Smith J (1991) Theories of sexual selection. *Trends in Ecology and Evolution* 6:146–151.

Maynard Smith J, Brown RLW (1986) Competition and body size. *Theoretical Population Biology* 30:166–179.

Maynard Smith J, Parker GA (1976) The logic of asymmetric contests. *Animal Behaviour* 24:159–175.

Maynard Smith J, Price GR (1973) The logic of animal conflict. *Nature* 246:15–18.

McCart C, Buckling A, Ffrench-Constant RH (2005) DDT resistance in flies carries no cost. *Current Biology* 15:R587–589.

McCartney J, Kokko H, Heller K-G, Gwynne, DT (2012) The evolution of sex differences in mate searching when females benefit: new theory and a comparative test. *Proceedings of the Royal Society of London B* 279:1225–1232.

McGlothlin JW (2010) Combining selective episodes to estimate lifetime nonlinear selection. *Evolution* 64:1377–1385.

McGraw LA, Gibson G, Clark AG, Wolfner MF (2004). Genes regulated by mating, sperm, or seminal proteins in mated female *Drosophila melanogaster*. *Current Biology* 14:1509–1514.

McGraw LA, Clark AG, Wolfner MF (2008) Post-mating gene expression profiles of female *Drosophila melanogaster* in response to time and to four male accessory gland proteins. *Genetics* 179:1395–1408.

McLachlan A (1999) Parasites promote mating success: the case of a midge and a mite. *Animal Behaviour* 57:1199–1205.

McLachlan AJ (2006) You are looking mitey fine: parasites as direct indicators of fitness in the mating system of a host species. *Ethology Ecology and Evolution* 18:233–239.

McLeod CJ, Wang L, Wong C, Jones DL (2010) Stem cell dynamics in response to nutrient availability. *Current Biology* 20:2100–2105.

McNamara KB, Elgar MA, Jones TM (2009) Large spermatophores reduce female receptivity and increase male paternity success in the almond moth, *Cadra cautella*. *Animal Behaviour* 77:931–936.

McNamara KB, van Lieshout E, Jones TM, Simmons LW (2013) Age-dependent trade-offs between immunity and male, but not female, reproduction. *Journal of Animal Ecology* 82:235–244.

McPeek MA, Shen L, Farid H (2009) The correlated evolution of three-dimensional reproductive structures between male and female damselflies. *Evolution* 63:73–83.

Mead LS, Arnold SJ (2004) Quantitative genetic models of sexual selection. *Trends in Ecology and Evolution* 19:264–271.

Meier N, Käppeli SC, Hediger Niessen M, et al. (2013) Genetic control of courtship behavior in the housefly: evidence for a conserved bifurcation of the sex-determining pathway. *PLoS One* 8:e62476.

Meier R, Kotrba M, Ferrar P (1999) Ovoviviparity and viviparity in the Diptera. *Biological Reviews* 74:199–258.

Meiklejohn CD, Tao Y (2009) Genetic conflict and sex chromosome evolution. *Trends in Ecology and Evolution* 25:215–223.

Meirmans S, Meirmans PG, Kirkendall LR (2012) The costs of sex: facing real-world complexities. *Quarterly Review of Biology* 87:19–40.

Mendelson TC, Shaw KL (2005) Rapid speciation in an arthropod. *Nature* 433:375–376.

Mendelson TC, Shaw KL (2012) The (mis)concept of species recognition. *Trends in Ecology and Evolution* 27:421–427.

Menzel R (1993) Associative learning in honey-bees. *Apidologie* 24:157–168.

Menzel R, Muller U (1996) Learning and memory in honeybees: from behavior to neural substrates. *Annual Review of Neuroscience* 19:379–404.

Mercier JL, Lenoir JC, Eberhardt A, et al. (2007) Hammering, mauling, and kissing: stereotyped courtship behavior in *Cardiocondyla* ants. *Insectes Sociaux* 54:403–411.

Merrill RM, Van Schooten B, Scott JA, Jiggins CD (2011) Pervasive genetic associations between traits causing reproductive isolation in *Heliconius* butterflies. *Proceedings of the Royal Society of London B* 278:511–518.

Michalczyk Ł, Millard AL, Martin OY, et al. (2011) Inbreeding promotes female promiscuity. *Science* 333:1739–1742.

Mikheyev AS (2003) Evidence for mating plugs in the fire ant *Solenopsis invicta*. *Insectes Sociaux* 50:401–402.

Milam EL (2010). *Looking for a few Good Males: Female Choice in Evolutionary Biology*. John Hopkins University Press, Baltimore.

Milani L, Ghiselli F, Pellecchia M, et al. (2010) Reticulate evolution in stick insects: the case of *Clonopsis* (Insecta Phasmida). *BMC Evolutionary Biology* 10:258.

Millar CD, Lambert DM (1986) Laboratory-induced changes in the mate recognition system of *Drosophila pseudoobscura*. *Behaviour Genetics* 16:285–294.

Miller CW, Moore AJ (2007) A potential resolution to the lek paradox through indirect genetic effects. *Proceedings of the Royal Society of London B* 274:1279–1286.

Miller GT, Pitnick S (2002) Sperm-female coevolution in *Drosophila*. *Science* 298:1230–1233.

Miller KB (2003) The phylogeny of diving beetles (Coleoptera: Dytiscidae) and the evolution of sexual conflict. *Biological Journal of the Linnean Society* 79:359–388.

Miller WJ, Ehrman L, Schneider D (2010) Infectious speciation revisited: impact of symbiont-depletion on female fitness and mating behavior of *Drosophila paulistorum*. *PLoS Pathology* 6:12.

Misra S, Crosby M, Mungall C, et al. (2002) Annotation of the *Drosophila melanogaster* euchromatic genome: a systematic review. *Genome Biology* 3:RESEARCH0083.

Miyatake T (1995) Territorial mating aggregation in the bamboo bug, *Notobitus meleagris*, Fabricius (Heteroptera: Coreidae). *Journal of Ethology* 13:185–189.

Miyatake T (1997) Functional morphology of the hind legs as weapons for male contests in *Leptoglossus australis* (Heteroptera: Coreidae). *Journal of Insect Behavior* 10:727–735.

Miyatake T (2011) Insect quality control: synchronized sex, mating system, and biological rhythm. *Applied Entomology and Zoology* 46:3–14.

Miyatake T, Matsumoto A, Matsuyama T, et al. (2002) The period gene and allochronic reproductive isolation in *Bactrocera cucurbitae*. *Proceedings of the Royal Society of London B* 269:2467–2472.

Moczek AP (2010) Phenotypic plasticity and diversity in insects. *Philosophical Transactions of the Royal Society of London B* 365:593–603.

Moczek AP, Emlen DJ (2000) Male horn dimorphism in the scarab beetle, *Onthophagus taurus*: Do alternative reproductive tactics favour alternative phenotypes? *Animal Behaviour* 59:459–466.

Moczek AP, Nijhout HF (2004) Trade-offs during the development of primary and secondary sexual traits in a horned beetle. *American Naturalist* 163:184–191.

Moehring AJ, Llopart A, Elwyn S, et al. (2006) The genetic basis of prezygotic reproductive isolation between *Drosophila santomea* and *D. yakuba* due to mating preference. *Genetics* 173:215–223.

Monnin T, Peeters C (1998) Monogyny and regulation of worker mating in the queenless ant *Dinoponera quadriceps*. *Animal Behaviour* 55:299–306.

Moore AJ (1989) Sexual selection in *Nauphoeta cinerea*: inherited mating preference? *Behaviour Genetics* 19:717–724.

Moore AJ (1990) The evolution of sexual dimorphism by sexual selection: the separate effects of intrasexual selection and intersexual selection. *Evolution* 44:315–331.

Moore AJ, Pizzari T (2005) Quantitative genetic models of sexual conflict based on interacting phenotypes. *American Naturalist* 165:S88–S97.

Moore AJ, Wilson P (1992) The evolution of sexually dimorphic earwig forceps: Social interactions among adults of the toothed earwig, *Vostox apicedentatus*. *Behavioral Ecology* 4:40–48.

Moore AJ, Brodie ED, Wolf JB (1997) Interacting phenotypes and the evolutionary process. 1. Direct and indirect genetic effects of social interactions. *Evolution* 51:1352–1362.

Moore AJ, Haynes KF, Preziosi RF, Moore PJ (2002) The evolution of interacting phenotypes: genetics and evolution of social dominance. *American Naturalist* 160:S186–S197.

Moore J (2002) *Parasites and the Behavior of Animals*. Oxford University Press, Oxford.

Moore PJ, Moore AJ (2001) Reproductive aging and mating: the ticking of the biological clock in female cockroaches. *Proceedings of the National Academy of Sciences of the USA* 98:9171–9176.

Morbey YE, Ydenberg RC (2001) Protandrous arrival timing to breeding areas: a review. *Ecology Letters* 4:663–673.

Morehouse NI, Rutowski RL (2010) In the eyes of the beholders: female choice and avian predation risk associated with an exaggerated male butterfly color. *American Naturalist* 176:768–784.

Morgan-Richards M, Trewick SA (2005) Hybrid origin of a parthenogenetic genus? *Molecular Ecology* 14:2133–2142.

Morrow EH, Arnqvist G (2003) Costly traumatic insemination and a female counter-adaptation in bed bugs. *Proceedings of the Royal Society of London B* 270:2377–2381.

Morrow EH, Gage MJG (2000) The evolution of sperm length in moths. *Proceedings of the Royal Society of London B* 267:307–313.

Mostowy R, Engelstaedter J (2012) Host–parasite coevolution induces selection for condition-dependent sex. *Journal of Evolutionary Biology* 25:2033–2046.

Moulds MS (1978) Field observations on the behavior of a north Queensland species of *Phytalmia* (Diptera: Tephritidae). *Journal of the Australian Entomological Society* 16:347–352.

Møller AP, Christe P, Lux E (1999) Parasitism, host immune function, and sexual selection. *Quarterly Review of Biology* 74:3–20.

Muller HJ (1964) The relation of recombination to mutational advance. *Mutation Research* 1:2–9.

Müller JK, Eggert A-K (1989) Paternity assurance by 'helpful' males: adaptations to sperm competition in burying beetles. *Behavioral Ecology and Sociobiology* 24:245–249.

Müller JK, Braunisch V, Hwang W, Eggert A-K (2006) Alternative tactics and individual reproductive success in natural associations of the burying beetle, *Nicrophorus vespilloides*. *Behavioral Ecology* 18:196–203.

Munguia-Steyer R, Cordoba-Aguilar A, Romo-Beltran A (2010) Do individuals in better condition survive for longer? Field survival estimates according to male alternative reproductive tactics and sex. *Journal of Evolutionary Biology* 23:175–184.

Mustaparta H, Tømmerås BA, Lanier GN (1985) Pheromone receptor cell specificity in interpopulational hybrids of *Ips pini* (Coleoptera: Scolytidae). *Journal of Chemical Ecology* 11:999–1007.

Narbonne P, Roy R (2006) Regulation of germline stem cell proliferation downstream of nutrient sensing. *Cell Division* 1:29.

Narraway C, Hunt J, Wedell N, Hosken DJ (2010) Genotype-by-environment interactions for female preference. *Journal of Evolutionary Biology* 23:2550–2587.

Nation JL, Bowers WS (1982) Fatty acid composition of milkweed bugs *Oncopeltus fasciatus* and the influence of diet. *Insect Biochemistry* 12:455–459.

Neaves W, Baumann P (2011) Unisexual reproduction among vertebrates. *Trends in Genetics* 27:81–88.

Neff BD, Pitcher TE (2005) Genetic quality and sexual selection: an integrated framework for good genes and compatible genes. *Molecular Ecology* 14:19–38.

Neiman M, Hehman G, Miller JT, et al. (2010) Accelerated mutation accumulation in asexual lineages of a freshwater snail. *Molecular Biology and Evolution* 27:954–963.

Neubaum DM, Wolfner MF (1999) Mated *Drosophila melanogaster* females require a seminal fluid protein, Acp36DE, to store sperm efficiently. *Genetics* 153:845–857.

Neumann P, Hartel S, Kryger P, et al. (2011) Reproductive division of labour and thelytoky result in sympatric barriers to gene flow in honeybees (*Apis mellifera* L.). *Journal of Evolutionary Biology* 24:286–294.

Neville M, Goodwin SF (2012) Genome-wide approaches to understanding behaviour in *Drosophila melanogaster*. *Briefings in Functional Genomics* 11:395–404.

Nichols RA, Butlin RK (1989) Does runaway sexual selection work in finite populations. *Journal of Evolutionary Biology* 2:299–313.

Niehuis O, Buellesbach J, Gibson JD, et al. (2013) Behavioural and genetic analyses of *Nasonia* shed light on the evolution of sex pheromones. *Nature* 494:345–348.

Nilsson T, Fricke C, Arnqvist G (2003) The effects of male and female genotype on variance in male fertilization success in the red flour beetle (*Tribolium castaneum*). *Behavioral Ecology and Sociobiology* 53:227–233.

Normark BB (2003) The evolution of alternative genetic systems in insects. *Annual Review of Entomology* 48:397–423.

Normark BB (2004a) The biology of demons. *Evolution* 58:676–679.

Normark BB (2004b) Haplodiploidy as an outcome of coevolution between male-killing cytoplasmic elements and their hosts. *Evolution* 58:790–798.

Normark BB (2006) Maternal kin groups and the evolution of asymmetric genetic systems—genomic imprinting, haplodiploidy, thelytoky. *Evolution* 60:631–642.

Normark BB (2009) Unusual genetic and gametic systems. In Hosken D, Pitnick S, Birkhead T (eds), *Sperm Biology*, pp. 507–538. Academic Press, London.

Normark BB, Johnson NA (2011) Niche explosion. *Genetica* 139:551–564.

Normark BB, Lanteri AA (1998) Incongruence between morphological and mitochondrial-DNA characters suggests hybrid origins of parthenogenetic weevil lineages (genus *Aramigus*). *Systematic Biology* 47:459–478.

Normark BB, Moran NA (2000) Testing for the accumulation of deleterious mutations in asexual eukaryote genomes using molecular sequences. *Journal of Natural History* 34:1719–1729.

Nothnagle PJ, Schultz JC (1987) What is a forest pest? In Barbosa P, Schultz JC (eds), *Insect Outbreaks*, pp. 241–268. Academic Press, San Diego.

Nowak MA, Tarnita CE, Wilson EO (2010) The evolution of eusociality. *Nature* 466:1057–1062.

Nunney L (1989) The maintenance of sex by group selection. *Evolution* 43:245–257.

Nur N, Hasson O (1984) Phenotypic plasticity and the handicap principle. *Journal of Theoretical Biology* 110:275–297.

Nylin S, Gotthard K (1998) Plasticity in life-history traits. *Annual Review of Entomology* 43:63–83.

Oberstadt B, Heinze J (2003) Mating biology and population structure of the ant, *Leptothorax gredleri*. *Insectes Sociaux* 50:340–345.

O'Donald P, Majerus MEN (1992) Non-random mating in *Adalia bipunctata* (the two-spot ladybird). III. New evidence of genetic preference. *Heredity* 69:521–526.

Oliveira RF, Taborsky M, Brockmann HJ (2008) *Alternative Reproductive Tactics: an Integrative Approach*. Cambridge University Press, Cambridge.

Olsson M, Madsen T, Shine R (1997) Is sperm really so cheap? Costs of reproduction in male adders, *Vipera berus*. *Proceedings of the Royal Society of London B* 264:455–459.

Orgogozo V, Broman KW, Stern DL (2006) High-resolution quantitative trait locus mapping reveals sign epistasis controlling ovariole number between two *Drosophila* species. *Genetics* 173:197–205.

Orr HA (2001) The genetics of species differences. *Trends in Ecology and Evolution* 16:343–350.

Orr AG, Rutowski RL (1991) The function of the shpragis in *Cressida cressida* (Fab) (Lepidoptera, Papilionidae): a visual deterrent to copulation attempts. *Journal of Natural History* 25:703–710.

Östergren G (1945) Parasitic nature of extra fragment chromosomes. *Botaniska Notiser* 2:157–163.

Otronen M (1988) The effects of body size on the outcome of fights in burying beetles (*Nicrophorus*). *Annales Zoologici Fennici* 25:191–201.

Otti O, Naylor RA, Siva-Jothy MT, Reinhardt K (2009) Bacteriolytic activity in the ejaculate of an insect. *American Naturalist* 174:292–295.

Otti O, McTighe AP, Reinhardt K (2012) In vitro antimicrobial sperm protection by an ejaculate-like substance. *Functional Ecology* 27:219–226.

Otti O, Gantenbein-Ritter I, Jacot A, Brinkhof MWG (2013) Immune response increases predation risk. *Evolution* 66:732–739.

Otto SP (2009) The evolutionary enigma of sex. *American Naturalist* 174:S1–S14.

Otto SP, Gerstein AC (2006) Why have sex? The population genetics of sex and recombination. *Biochemical Society Transactions* 34:519–522.

Owens IPF, Thompson DBA (1994) Sex differences, sex ratios and sex roles. *Proceedings of the Royal Society of London B* 258:93–99.

Page RE (1986) Sperm utilization in social insects. *Annual Review of Entomolology* 31:297–320.

Page RE, Metcalf RA (1982) Multiple mating, sperm utilization and social evolution. *American Naturalist* 119:263–282.

Paland S, Lynch M (2006) Transitions to asexuality result in excess amino acid substitutions. *Science* 311:990–992.

Pamilo P (1991) Life span of queens in the ant *Formica exsecta*. *Insectes Sociaux* 38:111–120.

Panhuis TM, Wilkinson GS (1999) Exaggerated male eye span influences contest outcome in stalk-eyed flies (Diopsidae). *Behavioral Ecology and Sociobiology* 46:221–227.

Parker GA (1970a) Sperm competition and its evolutionary consequences in the insects. *Biological Reviews* 45:525–567.

Parker GA (1970b) Sperm competition and its evolutionary effect on copula duration in the fly *Scatophaga stercoraria*. *Journal of Insect Physiology* 16:1301–1328.

Parker GA (1970c) The reproductive behaviour and the nature of sexual selection in *Scatophaga stercoraria* L. (Diptera: Scatophagidae). VII. The origin and evolution of the passive phase. *Evolution* 24:774–788.

Parker GA (1970d) The reproductive behaviour and the nature of sexual selection in *Scatophaga stercoraria* L. (Diptera: Scatophagidae). IV. Epigamic recognition and competition between males for the possession of females. *Behaviour* 37:114–139.

Parker GA (1970e) Sperm competition and its evolutionary consequences in the insects. *Biological Reviews* 45:525–567.

Parker GA (1971) The reproductive behaviour and the nature of sexual selection in *Scatophaga stercoraria* L. (Diptera: Scatophagidae). VI. The adaptive significance of emigration from the oviposition site during the phase of genital contact. *Journal of Animal Ecology* 40:215–233.

Parker GA (1972) Reproductive behaviour of *Sepsis cynipsea* (L.) (Diptera: Sepsidae). II. The significance of the precopulatory passive phase and emigration. *Behaviour* 41:242–250.

Parker GA (1974) Courtship persistence and female-guarding as male time investment strategies. *Behaviour* 48:157–184.

Parker GA (1978) Evolution of competitive mate searching. *Annual Review of Entomology* 23:173–196.

Parker GA (1979) Sexual selection and sexual conflict. In MS Blum, NA Blum (eds), *Sexual Selection and Reproductive Competition in Insects*, pp. 123–166. Academic Press, London.

Parker GA (1983a) Arms races in evolution—an ESS to the opponent-independent costs game. *Journal of Theoretical Biology* 101:619–648.

Parker GA (1983b) Mate quality and mating decisions. In PPG Bateson (ed.), *Mate Choice*, pp. 141–166. Cambridge University Press, Cambridge.

Parker GA (1984) Sperm competition and the evolution of animal mating strategies. In RL Smith (ed.), *Sperm Competition and the Evolution of Animal Mating Systems*, pp. 2–60. Academic Press, London.

Parker GA (1990) Sperm competition games: sneaks and extra-pair copulations. *Proceedings of the Royal Society of London B* 242:127–133.

Parker GA (2006) Sexual conflict over mating and fertilization: an overview. *Philosophical Transactions of the Royal Society of London B* 361:235–259.

Parker GA, Partridge L (1998) Sexual conflict and speciation. *Philosophical Transactions of the Royal Society of London B* 353:261–274.

Parker GA, Pizzari T (2010) Sperm competition and ejaculate economics. *Biological Reviews* 85:897–934.

Parker GA, Simmons LW (1989) Nuptial feeding in insects: theoretical models of male and female interests. *Ethology* 82:3–26.

Parker GA, Simmons LW (1994) Evolution of phenotypic optima and copula duration in dungflies. *Nature* 370:53–56.

Parker GA, Simmons LW (1996) Parental investment and the control of sexual selection: predicting the direction of sexual competition. *Proceedings of the Royal Society of London B* 263:315–321.

Parker GA, Simmons LW (2000) Optimal copula duration in yellow dung flies: Ejaculatory duct dimensions and size-dependent sperm displacement. *Evolution* 54:924–935.

Parker GA, Smith JL (1975) Sperm competition and the evolution of the precopulatory passive phase behaviour in *Locusta migratoria migratorioides*. *Journal of Entomology A* 49:155–171.

Parker GA, Ball MA, Stockley P, Gage MJG (1996) Sperm competition games: individual assessment of sperm competition intensity by group spawners. *Proceedings of the Royal Society of London B* 263:1291–1297.

Parker GA, Ball MA, Stockley P, Gage MJG (1997) Sperm competition games: a prospective analysis of risk assessment. *Proceedings of the Royal Society of London B* 264:1793–1802.

Parker GA, Lessells CM, Simmons LW (2013) Sperm competition games: a general model for precopulatory male–male competition. *Evolution* 67:95–109.

Parmesan C (2006) Ecological and evolutionary responses to recent climate change. *Annual Review of Ecology, Evolution, and Systematics* 37:637–669.

Partridge L, Parker, GA (1999) Sexual conflict and speciation. In AE Magurran, R M May (eds), *Evolution of Biological Diversity*, pp. 130–159. Oxford University Press, Oxford.

Partridge L, Hoffmann A, Jones JS (1987) Male size and mating success in *Drosophila melanogaster* and *D. pseudoobscura* under field conditions. *Animal Behaviour* 35:468–476.

Payne RJH (1998) Gradually escalating fights and displays: the cumulative assessment model. *Animal Behaviour* 56:651–662.

Pearcy M, Aaron S, Doums C, Keller L (2004) Conditional use of sex and parthenogenesis for worker and queen production in ants. *Science* 306:1780–1783.

Perry JC, Sharpe DMT, Rowe L (2009) Condition-dependent female remating resistance generates sexual selection on male size in a ladybird beetle. *Animal Behaviour* 77:743–748.

Peterson G, Hall JC, Rosbash M (1988) The period gene of *Drosophila* carries species-specific behavioural instructions. *EMBO Journal* 7:3939–3947.

Phadnis N, Orr HA (2009) A single gene causes both male sterility and segregation distortion in *Drosophila* hybrids. *Science* 323:376–379.

Phillips PC, Arnold SJ (1989) Visualizing multivariate selection. *Evolution* 43:1209–1222.

Philips TK, Pretorius E, Scholtz CH (2004) A phylogenetic analysis of dung beetles (Scarabaeinae: Scarabaeidae): unrolling an evolutionary history. *Invertebrate Systematics* 18:53–88.

Pischedda A, Chippindale AK (2006) Intralocus sexual conflict diminishes the benefits of sexual selection. *PloS Biology* 4:e356.

Pitnick S (1996) Investment in testes and the cost of making long sperm in *Drosophila*. *American Naturalist* 148:57–80.

Pitnick S, Brown WD (2000) Criteria for demonstrating female sperm choice. *Evolution* 54: 1052–1056.

Pitnick S, Hosken DJ (2010) Postcopulatory sexual selection. In DF Westneat, CW Fox (eds), *Evolutionary Behavioural Ecology*, pp. 379–399. Oxford University Press, Oxford.

Pitnick S, Markow T, Spicer GS (1999) Evolution of multiple kinds of female sperm-storage organs in *Drosophila*. *Evolution* 53:1804–1822.

Pitnick S, Wolfner M, Suarez S (2009) Sperm–female interactions. In TR Birkhead, DJ Hosken, S Pitnick (eds), *Sperm Biology: An Evolutionary Perspective*, pp. 247–304. Academic Press, London.

Pizzari T, Foster KR (2008) Sperm sociality: cooperation, altruism, and spite. *PLoS Biology* 6:e130.

Pizzari T, Gardner A (2012) The sociobiology of sex: inclusive fitness consequences of inter-sexual interactions. *Philosophical Transactions of the Royal Society B* 367:2314–2323.

Plaistow S, Siva-Jothy MT (1996) Energetic constraints and male mate-securing tactics in the damselfly *Calopteryx splendens xanthostoma* (Charpentier). *Proceedings of the Royal Society of London B* 263:1233–1239.

Plaistow SJ, Johnstone RA, Colegrave N, Spencer M (2004) Evolution of alternative mating tactics: conditional versus mixed strategies. *Behavioral Ecology* 15:534–542.

Platt AP, Allen JF (2001) Sperm precedence and competition in doubly-mated *Limenitis arthemis-astyanax* butterflies (Rhopalocera: Nymphalidae). *Annals of the Entomological society of America* 94:654–663.

Poiani A (2006) Complexity of seminal fluid: a review. *Behavioral Ecology and Sociobiology* 60: 289–310.

Poissant J, Wilson AJ, Coltman DW (2010) Sex-specific genetic variance and the evolution of sexual dimorphism: a systematic review of cross-sex genetic correlations. *Evolution* 64:97–107.

Polak M, Wolf LL, Starmer WT, Barker JSF (2001) Function of the mating plug in *Drosophila hibisci* Bock. *Behavioral Ecology and Sociobiology* 49:196–205.

Policansky D, Ellison J (1970) "Sex ratio" in *Drosophila pseudoobscura*: spermiogenic failure. *Science* 169: 888–889.

Pomfret JC, Knell RJ (2006a) Sexual selection and horn allometry in the dung beetle *Euoniticellus intermedius*. *Animal Behaviour* 71, 567–576.

Pomfret JC, Knell RJ (2006b) Immunity and the expression of a secondary sexual trait in a horned beetle. *Behavioral Ecology* 17:466–472.

Pomfret J, Knell RJ (2008) Crowding, sex ratio and horn evolution in a South African beetle community. *Proceedings of the Royal Society of London B* 275:315–321.

Pomiankowski A (1987) Sexual selection—the handicap principle does work sometimes. *Proceedings of the Royal Society of London B* 231:123–145.

Pomiankowski A (1987) The costs of choice in sexual selection. *Journal of Theoretical Biology* 128:195–218.

Pomiankowski A, Hurst LD (1999) Driving sexual preference. *Trends in Ecology and Evolution* 14: 425–426.

Pomiankowski A, Møller AP (1995) A resolution of the lek paradox. *Proceedings of the Royal Society of London B* 260:21–29.

Pomiankowski A, Iwasa Y, Nee S (1991) The evolution of costly mate preferences I. Fisher and biased mutation. *Evolution* 45:1422–1430.

Poon A, Chao L (2004) Drift increases the advantage of sex in RNA bacteriophage φ6. *Genetics* 166:19–24.

Presgraves DC, Severence E, Wilkinson GS (1997) Sex chromosome meiotic drive in stalk-eyed flies. *Genetics* 147:1169–1180.

Presgraves DC, Baker RH, Wilkinson GS (1999) Coevolution of sperm and female reproductive tract morphology in stalk-eyed flies. *Proceedings of the Royal Society of London B* 266:1041–1047.

Price PW (1980) *Evolutionary Biology of Parasites*. Princeton University Press, Princeton, NJ.

Price T, Schluter D, Heckman NE (1993) Sexual selection when the female directly benefits. *Biological Journal of the Linnean Society* 48:187–211.

Price TAR, Wedell N (2008) Selfish genetic elements and sexual selection: their impact on male fertility. *Genetica* 132:295–307.

Price TAR, Bretman AJ, Avent TD, et al. (2008a) Sex ratio distorter reduces sperm competitive ability in an insect. *Evolution* 62:1644–1652.

Price TAR, Hodgson DJ, Lewis Z, et al. (2008b) Selfish genetic elements promote polyandry in a fly. *Science* 322:1241–1243.

Price TAR, Hurst GDD, Wedell N (2010a) Polyandry prevents extinction. *Current Biology* 20:471–475.

Price TAR, Lewis Z, Smith DT, et al. (2010b) Sex ratio drive promotes sexual conflict and sexual coevolution in the fly *Drosophila pseudoobscura*. *Evolution* 64:1504–1509.

Price TAR, Lewis Z, Smith DT, et al. (2012) No evidence of mate discrimination against males carrying a sex ratio distorter in *Drosophila pseudoobscura*. *Behavioral Ecology and Sociobiology* 66:561–568.

Priest NK, Galloway LF, Roach DA (2008) Mating frequency and inclusive fitness in *Drosophila melanogaster*. *American Naturalist* 171:10–21.

Prokop ZM, Michalczyk Ł, Drobniak SM, et al. (2012) Meta-analysis suggests choosy females get sexy sons more than 'good genes'. *Evolution* 66:2665–2673.

Prokupek A, Hoffmann F, Eyun SI, et al. (2008) An evolutionary expressed sequence tag analysis of *Drosophila* spermatheca genes. *Evolution* 62:2936–2947.

Prum RO (2010) The Lande–Kirkpatrick mechanism is the null model of evolution by intersexual selection: implications for meaning, honesty and design in intersexual signals. *Evolution* 64:3085–3100.

Pukowski E (1933) Ökologische Untersuchungen an Necrophorus F. *Zeitschrift für Morphologie und Ökologie der Tiere* 27:518–586.

Punzalan D, Rodd FH, Rowe L (2008) Contemporary sexual selection on sexually dimorphic traits in the ambush bug *Phymata americana*. *Behavioural Ecology* 19:860–870.

Punzalan D, Rodd FH, Rowe L (2010) Temporal variation in patterns of multivariate sexual selection in a wild insect population. *American Naturalist* 175:401–414.

Puurtinen M, Ketola T, Kotiaho JS (2005) Genetic compatibility and sexual selection. *Trends in Ecology and Evolution* 20:157–158.

Puurtinen M, Ketola T, Kotiaho JS (2009) The good-genes and compatible-genes benefits of mate choice. *American Naturalist* 174:741–752.

Queller DC (1997) Why do females care more than males? *Proceedings of the Royal Society of London B* 264:1555–1557.

Qvarnstrom A, Bailey RI (2009) Speciation through evolution of sex-linked genes. *Heredity* 102:4–15.

Qvarnstrom A, Brommer JE, Gustafsson L (2006) Testing the genetics underlying the co-evolution of mate choice and ornament in the wild. *Nature* 441:84–86.

Råberg L, Stjernman M (2003) Natural selection on immune responsiveness in blue tits *Parus caeruleus*. *Evolution* 57:1670–1678.

Råberg L, Sim D, Read AF (2007) Disentangling genetic variation for resistance and tolerance to infectious diseases in animals. *Science* 318:812–814.

Råberg L, Graham AL, Read AF (2009) Decomposing health: tolerance and resistance to parasites in animals. *Philosophical Transactions of the Royal Society of London B* 364:37–49.

Radesäter T, Halldórsdóttir H (1993) Fluctuating asymmetry and forceps size in earwigs, *Forficula auricularia*. *Animal Behaviour* 45:626–628.

Radhakrishnan P, Pérez-Staples D, Weldon CW, Taylor PW (2009) Multiple mating and sperm depletion in male Queensland fruit flies: effects on female remating behaviour. *Animal Behaviour* 78:839–846.

Ram KR, Wolfner MF (2007) Seminal influences: *Drosophila* Acps and the molecular interplay between males and females during reproduction. *Integrative and Comparative Biolology* 47:427–445.

Randerson JP, Smith NG, Hurst LD (2000) The evolutionary dynamics of male-killers and their hosts. *Heredity* 84:152–160.

Rantala MJ, Vainikka A, Kortet R (2003) The role of juvenile hormone in immune function and pheromone production trade-offs: a test of the immunocompetence handicap hypothesis. *Proceedings of the Royal Society of London B* 270:2257–2261.

Rascón B, Mutti NS, Tolfsen C, Amdam GV (2011) Honey bee life history plasticity: development, behavior, and aging. In T Flatt, A Heyland (eds), *Mechanisms of Life History Evolution*. Oxford University Press, Oxford.

Rasmussen J (1994) The influence of horn and body size on the reproduvctive behavior of the horned rainbow scarab beetle *Phanaeus difformis* (Coleoptera: Scarabaeidae). *Journal of Insect Behavior* 7:67–82.

Ratnieks FLW, Boomsma JJ (1995) Facultative sex allocation by workers and the evolution of polyandry by queens in social Hymenoptera. *American Naturalist* 145:969–993.

Ratnieks FLW, Foster KR, Wenseleers T (2006) Conflict resolution in insect societies. *Annual Review of Entomology* 51:581–608.

Refsnider JM, Janzen FJ (2010) Putting eggs in one basket: ecological and evolutionary hypotheses for variation in oviposition-site choice. *Annual Review in Ecology, Evolution and Systematics* 41:39–57.

Reid ML, Stamps JA (1997) Female mate choice tactics in a resource-based mating system: field tests of alternative models. *American Naturalist* 150:98–121.

Reinhardt K (2001) Determinants of ejaculate size in a grasshopper (*Chorthippus parallelus*). *Behavioral Ecology and Sociobiology* 50:503–510.

Reinhardt K (2006) Sperm numbers vary between inter- and intra-population matings of the grasshopper *Chorthippus parallelus*. *Biology Letters* 2:239–241.

Reinhardt K, Arlt D (2003) Ejaculate size variation in the migratory locust, *Locusta migratoria*. *Behaviour* 140:319–332.

Reinhardt K, Siva-Jothy MT (2007) Biology of the bed bugs (Cimicidae). *Annual Review of Entomology* 52:351–374.

Reinhardt K, Naylor R, Siva-Jothy MT (2003) Reducing a cost of traumatic insemination: female bedbugs evolve a unique organ. *Proceedings of the Royal Society of London B* 270:2371–2375.

Reinhardt K, Naylor AA, Siva-Jothy MT (2009) Ejaculate components delay reproductive senescence while elevating female reproductive rate in an insect. *Proceedings of the National Acadamy of Sciences of the USA* 106:21743–21747.

Reinhold K, Kurtz J, Engqvist L (2002) Cryptic male choice: sperm allocation strategies when female quality varies. *Journal of Evolutionary Biology* 15:201–209.

Resh VH, Cardé R (eds), (2009) *Encylopedia of Insects*, 2nd edn. Academic Press, Amsterdam.

Reynolds JD (1996) Animal breeding systems. *Trends in Ecology and Evolution* 11:68–72.

Reynolds JD, Gross MR (1990) Costs and benefits of female mate choice: is there a lek paradox? *American Naturalist* 136:230–243.

Reynolds JD, Székely T (1997) The evolution of parental care in shorebirds: life histories, ecology, and sexual selection. *Behavioral Ecology* 8:126–134.

Rhainds M (2010) Female mating failures in insects. *Entomologia Experimentalis et Applicata* 136:211–226.

Rice WR (1984) Sex chromosomes and the evolution of sexual dimorphism. *Evolution* 38:735–742.

Rice WR (1987) The accumulation of sexually antagonistic genes as a selective agent promoting the evolution of reduced recombination between primitive sex chromosomes. *Evolution* 41:911–914.

Rice WR (1996) Sexually antagonistic male adaptation triggered by experimental arrest of female evolution. *Nature* 381:232–234.

Rideout EJ, Billeter JC, Goodwin SF (2007) The sex-determination genes *fruitless* and *doublesex* specify a neural substrate required for courtship song. *Current Biology* 17:1473–1478.

Riemann JG, Moen DO, Thorson BJ (1967) Female monogamy and its control in the housefly, *Musca domestica* L. *Journal of Insect Physiology* 13:407–418.

Ringo J, Sharon G, Segal D (2011) Bacteria-induced sexual isolation in *Drosophila*. *Fly* 5:310–315.

Rion S, Kawecki TJ (2007) Evolutionary biology of starvation resistance: what we have learned from *Drosophila*. *Journal of Evolutionary Biology* 20:1655–1664.

Risdill-Smith JT, Simmons LW (2011) *The Ecology and Evolution of Dung Beetles*. Wiley–Blackwell, Oxford.

Ritchie MG (1996) The shape of female mating preferences. *Proceedings of the National Academy of Sciences of the USA* 93:14628–14631.

Ritchie MG (2000) The inheritance of female preference functions in a mate recognition system. *Proceedings of the Royal Society of London B* 267:327–332.

Ritchie MG (2007) Sexual selection and speciation. *Annual Review of Ecology, Evolution and Systematics* 38:79–102.

Ritchie MG, Kyriacou CP (1994) Reproductive isolation and the period gene of *Drosophila*. *Molecular Ecology* 3:595–599.

Ritchie MG, Halsey EJ, Gleason JM (1999) *Drosophila* song as a species-specific mating signal, and the behavioural importance of Kyriacou & Hall cycles in *D. melanogaster* song. *Animal Behaviour* 58:649–657.

Ritchie MG, Saarikettu M, Hoikkala A (2005) Variation, but no covariance, in female preference functions and male song in a natural population of *Drosophila montana*. *Animal Behaviour* 70:849–854.

R'kha S, Moreteau B, Coyne JA, David JR (1997) Evolution of a lesser fitness trait: egg production in the specialist *Drosophila sechellia*. *Genetical Research* 69:17–23.

Robertson KA, Monteiro A (2005) Female *Bicyclus anynana* butterflies choose males on the basis of their dorsal UV-reflective eyespot pupils. *Proceedings of the Royal Society of London B* 272:1541–1546.

Robertson HM, Wanner KW (2006) The chemoreceptor superfamily in the honey bee, *Apis mellifera*: expansion of the odorant, but not gustatory, receptor family. *Genome Research* 16:1395–1403.

Robison GW, Jr (1972) Microtubular patterns in spermatozoa of coccid insects in relation to bending. *Journal of Cell Biology* 52:66–83.

Rockman MV (2012) The QTN program and the alleles that matter for evolution: all that's gold does not glitter. *Evolution* 66:1–17.

Rodríguez RL, Greenfield MD (2003) Genetic variance and phenotypic plasticity in a component of female mate choice in an ultrasonic moth. *Evolution* 57:1304–1313.

Rodríguez-Muñoz R, Bretman A, Slate J, et al. (2010) Natural and sexual selection in a wild insect population. *Science* 328:1269–1272.

Roelofs WL, Du J-W, Linn C, et al. (1986) The potential for genetic manipulation of the redbanded leafroller moth sex pheromone blend. In MD Huettel (ed.), *Evolutionary Genetics of Invertebrate Behavior – Progress and Prospects*, pp. 263–272. Plenum Press, New York.

Roelofs W, Glover T, Tang XH, et al. (1987) Sex pheromone production and perception in European corn borer moths is determined by both autosomal and sex-linked genes. *Proceedings of the National Academy of Sciences of the USA* 84:7585–7589.

Roff DA (1996) The evolution of threshold traits in animals. *Quarterly Review of Biology* 71:3–35.

Roff DA, Fairbairn DJ (2007) The evolution of trade-offs: where are we? *Journal of Evolutionary Biology* 20:433–447.

Rogers DW, Greig D (2009) Experimental evolution of a sexually selected display in yeast. *Proceedings of the Royal Society of London B* 276:543–549.

Rolff JO, Reynolds S (eds) (2010) *Insect Infection and Immunity: Evolution, Ecology and Mechanisms*. Oxford University Press, Oxford.

Romo-Beltran A, Macias-Ordonez R, Cordoba-Aguilar A (2009) Male dimorphism, territoriality and mating success in the tropical damselfly, *Paraphlebia zoe* Selys (Odonata: Megapodagrionidae). *Evolutionary Ecology* 23:699–709.

Ronacher B, Stange N (2013) Processing of acoustic signals in grasshoppers–a neuroethological approach towards female choice. *Journal of Physiology* 107:41–50.

Rönn J, Katvala M, Arnqvist G (2007) Coevolution between harmful male genitalia and female resistance in seed beetles. *Proceedings of the National Acadamy of Sciences of the USA* 104:10921–10925.

Rönn JL, Katvala M, Arnqvist G (2011) Correlated evolution between male and female primary reproductive characters in seed beetles. *Functional Ecology* 25:634–640.

Röseler PF (1973) Die Anzahl der Spermien im Receptaculum seminis von Hummelköniginnen (Hymenoptera, Apoidea, Bombinae). *Apidologie* 4:267–274.

Ross L, Pen I, Shuker DM (2010) Genomic conflict in scale insects: the causes and consequences of bizarre genetic systems. *Biological Reviews* 85:807–827.

Ross L, Shuker DM, Pen I (2011) The evolution and suppression of male suicide under paternal genome elimination. *Evolution* 65:554–563.

Ross L, Hardy NB, Okusu A, Normark BB (2013) Large population size predicts the distribution of asexuality in scale insects. *Evolution* 67:196–206.

Rosvall KA (2011) Intrasexual competition in females: evidence for sexual selection? *Behavioral Ecology* 22:1131–1140.

Rotenberry JT, Zuk M, Simmons LW, Hayes C (1996) Phonotactic parasitoids and cricket song structure: an evaluation of alternative hypotheses. *Evolutionary Ecology* 10:233–243.

Roth LM, Stay B (1962) A comparative study of oöcyte development in false ovoviviparous cockroaches. *Psyche* 69:165–208.

Roughgarden J, Oishi M, Akçay E (2006) Reproductive social behavior: cooperative games to replace sexual selection. *Science* 311:965–969.

Rowe L, Arnqvist G (2012) Sexual selection and the evolution of genital shape and complexity in water striders. *Evolution* 66:40–54.

Rowe L, Houle D (1996) The lek paradox and the capture of genetic variance by condition dependent traits. *Proceedings of the Royal Society of London B* 263:1415–1421.

Rowe L, Cameron E, Day T (2005) Escalation, retreat, and female indifference as alternative outcomes of sexually antagonistic coevolution. *American Naturalist* 165:S5–S18.

Royer M (1973) Rôle de la polysepermie dans la formation des lignées cellulaires haploïde et diploïde chez l'insecte Icerya purchasi. *Annales d'Embryologie et de Morphogenèse* 6:243–252.

Royle NJ, Smiseth PT, Kölliker M (eds) (2012) *The Evolution of Parental Care*. Oxford University Press, Oxford.

Rozen DE, Engelmoer DJP, Smiseth PT (2008) Antimicrobial strategies in burying beetles breeding on carrion. *Proceedings of the National Academy of Sciences of the USA* 105:17890–17895.

Rudin FS, Briffa M (2011) The logical polyp: assessments and decisions during contests in the beadlet anemone *Actinia equina*. *Behavioral Ecology* 22:1278–1285.

Rugman-Jones PF, Eady PE (2007) Conspecific sperm precedence in *Callosobruchus subinnotatus* (Coleoptera: Bruchidae): mechanisms and consequences. *Proceedings of the Royal Society of London* B 274:983–988.

Rundle HD, Chenoweth SF, Doughty P, Blows MW (2005) Divergent selection and the evolution of signal traits and mating preferences. *PLoS Biology* 3:e368.

Rundle HD, Odeen A, Mooers AØ (2007) An experimental test for indirect benefits in *Drosophila melanogaster*. *BMC Evolutionary Biology* 7:36.

Russell JE, Stouthamer R (2011) The genetics and evolution of obligate reproductive parasitism in *Trichogramma pretiosum* infected with parthenogenesis-inducing *Wolbachia*. *Heredity* 106:58–67.

Ruttner F (1956) The mating of the honey bee. *Bee World* 3:2–15.

Ruttner F (1976) *The Instrumental Insemination of the Queen Bee*. Apimondia, Bucharest.

Ryan MJ (1990) Sexual selection, sensory systems and sensory exploitation. *Oxford Surveys in Evolutionary Biology* 7:157–195.

Ryan MJ (1998) Sexual selection, receiver biases, and the evolution of sex differences. *Science* 281:1999–2003.

Ryan MJ, Keddy-Hector A (1992) Directional patterns of female mate choice and the role of sensory biases. *American Naturalist* 139:S4–S35.

Ryan MJ, Rand AS (1990) The sensory basis of sexual selection for complex calls in the Túngara frog, *Physalaemus pustulosus* (sexual selection for sensory exploitation). *Evolution* 44:305–314.

Saeki Y, Kruse KC, Switzer PV (2005) Physiological costs of mate guarding in the Japanese beetle (*Popillia japonica* Newman). *Ethology* 111:863–877.

Sahara K, Takemura Y (2003) Application of artificial insemination technique to eupyrene and/or apyrene sperm in *Bombyx mori*. *Journal of Experimental Zoology* 297A:196–200.

Sakaluk SK (1984) Male crickets feed females to ensure complete sperm transfer. *Science* 223:609–610.

Sakaluk SK (1985) Spermatophore size and its role in the reproductive behavior of the cricket, *Gryllodes supplicans* (Orthoptera, Gryllidae). *Canadian Journal of Zoology* 63:1652–1656.

Sakaluk SK (1997) Cryptic female choice predicated on wing dimorphism in decorated crickets. *Behavioral Ecology* 8:326–331.

Sakaluk SK (2000) Sensory exploitation as an evolutionary origin to nuptial food gifts in insects. *Proceedings of the Royal Society of London B* 267:339–343.

Sakaluk SK, Eggert AK (1996) Female control of sperm transfer and intraspecific variation in sperm precedence: antecedents to the evolution of a courtship food gift. *Evolution* 50:694–703.

Sakaluk SK, Avery RL, Weddle CB (2006) Cryptic sexual conflict in gift-giving insects: chasing the chase-away. *American Naturalist* 167:94–104.

Sakurai T, Nakagawa T, Mitsuno H, et al. (2004) Identification and functional characterization of a sex pheromone receptor in the silkmoth *Bombyx mori*. *Proceedings of the National Academy of Sciences of the USA* 101:16653–16658.

Salt G (1937) The egg-parasite of *Sialis lutaria*: a study of the influence of the host upon a dimorphic parasite. *Parasitology* 29:539–553.

Sanchez-Gracia A, Vieira FG, Rozas J (2009) Molecular evolution of the major chemosensory gene families in insects. *Heredity* 103:208–216.

Sandrock C, Vorburger C (2011) Single-locus recessive inheritance of asexual reproduction in a parasitoid wasp. *Current Biology* 21:433–437.

Sappington TW, Taylor OR (1990) Disruptive sexual selection on *Colias eurytheme* butterflies. *Proceedings of the National Academy of Sciences of the USA* 87:6132–6135.

Sarikaya DP, Belay AA, Ahuja A, et al. (2012) The roles of cell size and cell number in determining ovariole number in *Drosophila*. *Developmental Biology* 363:279–289.

Sato H, Imamori M (2008) Nesting behaviour of a subsocial African ball-roller *Kheper platynotus* (Coleoptera: Scarabaeidae). *Ecological Entomology* 12:415–425.

Schäfer MA, Berger D, Jochmann R, et al. (2013) The developmental plasticity and functional significance of an additional sperm storage compartment in female yellow dung flies. *Functional Ecology* 27:1392–1402.

Schärer L, Vizoso DB (2007) Phenotypic plasticity in sperm production rate: there's more to it than testis size. *Evolutionary Ecology* 21:295–306.

Schärer L, Da Lage J-L, Joly D (2008) Evolution of testicular architecture in the Drosophilidae: a role for sperm length. *BMC Evolutionary Biology* 8:143.

Schaus JM, Sakaluk SK (2001) Ejaculate expenditures of male crickets in response to varying risk and intensity of sperm competition: not all species play games. *Behavioral Ecology* 12:740–745.

Scheirs J, De Bruyn L, Verhagen R (2000) Optimization of adult performance determines host choice in a grass miner. *Proceedings of the Royal Society of London B* 267:2065–2069.

Schmid-Hempel R, Schmid-Hempel P (2000) Female mating frequencies in *Bombus* spp. from Central Europe. *Insectes Sociaux* 47:36–41.

Schmidt ED, Dorn A (2004) Structural polarity and dynamics of male germline stem cells in the milkweed bug (*Oncopeltus fasciatus*). *Cell and Tissue Research* 318:383–394.

Schöfl G, Taborsky M (2002) Prolonged tandem formation in firebugs (*Pyrrhocoris apterus*), serves mate-guarding. *Behavioral Ecology and Sociobiology* 52:426–433.

Schowalter TD (2006) *Insect Ecology: An Ecosystem Approach*. Elsevier, Burlington, MA.

Schrader F, Hughes-Schrader S (1931) Haploidy in metazoa. *Quarterly Review of Biology* 6:411–438.

Schurko AM, Neiman M, Logsdon Jr JM (2009) Signs of sex: what we know and how we know it. *Trends in Ecology and Evolution* 24:208–217.

Schutze MK, Yeates DK, Graham GC, Dodson G (2007) Phylogenetic relationships of antlered flies, *Phytalmia Gerstaecker* (Diptera: Tephritidae): the evolution of antler shape and mating behaviour. *Australian Journal of Entomology* 46:281–293.

Schwander T, Crespi BJ (2009a) Multiple direct transitions from sexual reproduction to apomictic parthenogenesis in *Timema* stick insects. *Evolution* 63:84–103.

Schwander T, Crespi BJ (2009b) Twigs on the tree of life? Neutral and selective models for integrating macroevolutionary patterns with microevolutionary processes in the analysis of asexuality. *Molecular Ecology* 18:28–42.

Schwander T, Vuilleumier S, Dubman J, Crespi BJ (2010) Positive feedback in the transition from sexual reproduction to parthenogenesis. *Proceedings of the Royal Society of London B* 277:1435–1442.

Schwander T, Henry L, Crespi BJ (2011) Molecular evidence for ancient asexuality in *Timema* stick insects. *Current Biology* 21:1129–1134.

Scott D (1994) Genetic variation for female mate discrimination in *Drosophila melanogaster*. *Evolution* 48:112–121.

Serrano-Meneses MA, Córdoba-Aguilar A, Méndez V, et al. (2007) Sexual size dimorphism in the American rubyspot: male body size predicts male competition and mating success. *Animal Behaviour* 73:987–997.

Servedio MR, Lande R (2006) Population genetic models of male and mutual mate choice. *Evolution* 60:674–685.

Sharma MD, Tregenza T, Hosken DJ (2010) Female mate preference in *Drosophila simulans*: evolution and costs. *Journal of Evolutionary Biology* 23:1672–1678.

Shaw KL (2000) Interspecific genetics of mate recognition: inheritance of female acoustic preference in Hawaiian crickets. *Evolution* 54:1303–1312.

Shaw KL, Herlihy DP (2000) Acoustic preference functions and song variability in the Hawaiian cricket *Laupala cerasina*. *Proceedings of the Royal Society of London B* 267:577–584.

Shaw KL, Lesnick SC (2009) Genomic linkage of male song and female acoustic preference QTL underlying a rapid species divergence. *Proceedings of the National Academy of Sciences of the USA* 106:9737–9742.

Shaw KL, Parsons YM, Lesnick SC (2007) QTL analysis of a rapidly evolving speciation phenotype in the Hawaiian cricket *Laupala*. *Molecular Ecology* 16:2879–2892.

Shaw RG, Geyer CJ (2010) Inferring fitness landscapes. *Evolution* 64:2510–2520.

Sheldon BC (2000) Differential allocation: tests, mechanisms and implications. *Trends in Ecology and Evolution* 15:397–402.

Sheldon BC, Verhulst S (1996) Ecological immunity: costly parasite defense and trade-offs in evolutionary ecology. *Trends in Ecology and Evolution* 11:317–321.

Shellman-Reeve JS (2001) Genetic relatedness and partner preference in a monogamous, wood-dwelling termite. *Animal Behaviour* 61:869–876.

Shelly TE, Greenfield MD (1985) Alternative mating strategies in a desert grasshopper—a transitional analysis. *Animal Behaviour* 33:1211–1222.

Shelly TE, Greenfield MD (1989) Satellites and transients—ecological constraints on alternative mating tactics in male grasshoppers. *Behaviour* 109:200–221.

Shelly TE, Wittier TS (1997) Lekking in insects. In JC Choe, BJ Crespi (eds), *The Evolution of Mating Systems in Insects and Arachnids*, pp. 273–293. Cambridge University Press, Cambridge.

Shirangi TR, Dufour HD, Williams TM, Carroll SB (2009) Rapid evolution of sex pheromone-producing enzyme expression in *Drosophila*. *PLoS Biology* 7:e1000168.

Shuker DM (2010) Sexual selection: endless forms or tangled bank? *Animal Behaviour* 79:e11–e17.

Shuker DM, Phillimore AJ, Burton-Chellew MN, et al. (2007) The quantitative genetic basis of polyandry in the parasitoid wasp, *Nasonia vitripennis*. *Heredity* 98:69–73.

Shuster SM (2009) Sexual selection and mating systems. *Proceedings of the National Academy of Sciences of the USA* 106:10009–10016.

Shuster SM, Wade MJ (2003) *Mating Systems and Strategies*. Princeton University Press, Princeton, NJ.

Simmons LW (1986a) Intermale competition and mating success in the field cricket, *Gryllus bimaculatus* (Degeer). *Animal Behaviour* 34:567–579.

Simmons LW (1986b) Female choice in the field cricket, *Gryllus bimaculatus* (De Geer). *Animal Behaviour* 34:1463–1470.

Simmons LW (1987) Sperm Competition as a mechanism of female choice in the field cricket, *Gryllus bimaculatus*. *Behavioral Ecology and Sociobiology* 21:197–202.

Simmons LW (1988) The calling song of the field cricket, *Gryllus bimaculatus* (De Geer): constraints on transmission and its role in intermale competition and female choice. *Animal Behaviour* 36:380–394.

Simmons LW (1990) Nuptial feeding in tettigoniids: male costs and the rates of fecundity increase. *Behavioral Ecology and Sociobiology* 27:43–47.

Simmons LW (1992) Quantification of role reversal in relative parental investment in a bush cricket. *Nature* 358:61–63.

Simmons LW (2001) *Sperm Competition and its Evolutionary Consequences in the Insects*. Princeton University Press, Princeton, NJ.

Simmons LW (2004) Genotypic variation in calling song and female preferences of the field cricket *Teleogryllus oceanicus*. *Animal Behaviour* 68:313–322.

Simmons LW (2005) The evolution of polyandry: sperm competition, sperm selection, and offspring viability. *Annual Review of Ecology, Evolution, and Systematics* 36:125–146.

Simmons LW, Beveridge M (2010) The strength of postcopulatory sexual selection within natural populations of field crickets. *Behavioral Ecology* 21:1179–1185.

Simmons LW, Beveridge M (2011) Seminal fluid affects sperm viability in a cricket. *PLoS One* 6:e17975.

Simmons LW, Emlen DJ (2006) Evolutionary trade-off between weapons and testes. *Proceedings of the National Acadamy of Sciences of the USA* 103:16349–16351.

Simmons LW, García-González F (2008) Evolutionary reduction in testes size and competitive fertilization success in response to the experimental removal of sexual selection in dung beetles. *Evolution* 62:2580–2591.

Simmons LW, García-González F (2011) Experimental coevolution of male and female genital morphology. *Nature Communications* 2:374.

Simmons LW, Gwynne DT (1991) The refractory period of female katydids (Orthoptera: Tettigoniidae): sexual conflict over the remating interval? *Behavioral Ecology* 2:276–282.

Simmons LW, Kotiaho JS (2007) Quantitative genetic correlation between trait and preference supports a sexually selected sperm process. *Proceedings of the National Academy of Sciences of the USA* 104:16604–16608.

Simmons LW, Parker GA (1989) Nuptial feeding in insects: mating effort versus paternal investment. *Ethology* 81:332–343.

Simmons LW, Ridsdill-Smith TJ (eds) (2011) *Ecology and Evolution of Dung Beetles*. Wiley–Blackwell, Oxford.

Simmons LW, Roberts B (2005) Bacterial immunity traded for sperm viability in male crickets. *Science* 309:2031.

Simmons LW, Siva-Jothy MJ (1998) Sperm competition in insects: mechanisms and the potential for selection. In TR Birkhead, AP Møller (eds), *Sperm Competition and Sexual Selection*, pp. 341–434. Academic Press, London.

Simmons LW, Tomkins JL (1996) Sexual selection and the allometry of earwig forceps. *Evolutionary Ecology* 10:97–104.

Simmons LW, Parker GA, Stockley P (1999) Sperm displacement in the yellow dung fly, *Scatophaga stercoraria*: an investigation of male and female processes. *American Naturalist* 153:302–314.

Simmons LW, Tomkins JL, Alcock J (2000) Can minor males of Dawson's burrowing bee, *Amegilla dawsoni* (Hymenoptera: Anthophorini) compensate for reduced access to virgin females through sperm competition? *Behavioral Ecology* 11:319–325.

Simmons LW, Zuk M, Rotenberry JT (2001) Geographic variation in female preference functions and male songs of the field cricket *Teleogryllus oceanicus*. *Evolution* 55:1386–1394.

Simmons LW, Beveridge M, Krauss S (2004) Genetic analysis of parentage within experimental populations of a male dimorphic beetle, *Onthophagus taurus*, using amplified fragment length polymorphism. *Behavioral Ecology and Sociobiology* 57:164–173.

Simmons LW, Beveridge M, Wedell N, Tregenza T (2006) Postcopulatory inbreeding avoidance by female crickets only revealed by molecular markers. *Molecular Ecology* 15:3817–3824.

Simmons LW, Denholm A, Jackson C, et al. (2007a) Male crickets adjust ejaculate quality with both risk and intensity of sperm competition. *Biology Letters* 3:520–522.

Simmons LW, Emlen DJ, Tomkins JL (2007b). Sperm competition games between sneaks and guards: a comparative analysis using dimorphic male beetles. *Evolution* 61:2684–2692.

Simmons LW, House CM, Hunt J, Garcia-Gonzalez F (2009) Evolutionary response to sexual selection in male genital morphology. *Current Biology* 19:1442–1446.

Simpson SJ, Sword GA, Lo N (2011) Polyphenism in insects. *Current Biology* 21:R738–R749.

Sinervo B, Basolo AL (1996) Testing adaptation using phenotypic manipulations. In MR Rose, G Lauder (eds), *Adaptation*, pp. 149–185. Academic Press, New York.

Sinervo B, Lively CM (1996) The rock-paper-scissors game and the evolution of alternative male strategies. *Nature* 380:240–243.

Sirot LK, Brockmann HJ (2001) Costs of sexual interactions to females in Rambur's forktail damselfly, *Ischnura ramburi* (Zygoptera: Coenagrionidae). *Animal Behaviour* 61:415–424.

Sirot LK, Buehner N, Fiumera AC, Wolfner MF (2009) Seminal fluid protein depletion and replenishment in the fruit fly, *Drosophila melanogaster*: an ELISA-based method for tracking individual ejaculates. *Behavioral Ecology and Sociobiology* 63:1505–1513.

Sirot LK, Wolfner MF, Wigby S (2011) Protein-specific manipulation of ejaculate composition in response to female mating status in *Drosophila melanogaster*. *Proceedings of the National Academy of Sciences of the USA* 108:9922–9926.

Siva-Jothy MJ (1999) Male wing pigmentation may affect reproductive success via female choice in a calopterygid damselfly (Zygoptera). *Behaviour* 136:1365–1377.

Siva-Jothy MJ (2000) A mechanistic link between parasite resistance and expression of a sexually selected trait in a damselfly. *Proceedings of the Royal Society of London B* 267:2523–2527.

Siva-Jothy MT, Stutt AD (2003) A matter of taste: direct detection of female mating status in the bedbug. *Proceedings of the Royal Society of London B* 270:649–652.

Siva-Jothy MT, Moret Y, Rolff J (2005) Insect immunity: an evolutionary ecology perspective. *Advances in Insect Physiology* 32:1–48.

Sivinski J (1984) Sperm in competition. In R Smith (ed.), *Sperm Competition and the Evolution of Animal Mating Systems*, pp. 86–149. Academic Press, New York.

Skroblin A, Blows MW (2006) Measuring natural and sexual selection on breeding values of male display traits in *Drosophila serrata*. *Journal of Evolutionary Biology* 19:35–41.

Slate J (2013) From Beavis to beak color: a simulation study to examine how much QTL mapping can reveal about the genetic architecture of quantitative traits. *Evolution* 67:1251–1262.

Slatyer RA, Mautz, BS, Backwell PRY, Jennions MD (2012) Estimating genetic benefits of polyandry from experimental studies: a meta-analysis. *Biological Reviews* 87:1–33.

Smadja C, Butlin RK (2009) On the scent of speciation: the chemosensory system and its role in premating isolation. *Heredity* 102:77–97.

Smadja CM, Butlin RK (2011) A framework for comparing processes of speciation in the presence of gene flow. *Molecular Ecology* 20:5123–5140.

Smadja C, Shi P, Butlin RK, Robertson HM (2009) Large gene family expansions and adaptive evolution for odorant and gustatory receptors in the pea aphid, *Acyrthosiphon pisum*. *Molecular Biology and Evolution* 26:2073–2086.

Smadja CM, Canback B, Vitalis R, et al. (2012) Large-scale candidate gene scan reveals the role of chemoreceptor genes in host plant specialization in the pea aphid. *Evolution* 66:2723–2738.

Smiseth PT, Moore AJ (2002) Does resource availability affect offspring begging and parental provisioning in a partially begging species? *Animal Behaviour* 63:577–585.

Smiseth PT, Moore AJ (2004a) Signalling of hunger when offspring forage by both begging and self-feeding. *Animal Behaviour* 67:1083–1088.

Smiseth PT, Moore AJ (2004b) Behavioral dynamics between caring males and females in a beetle with facultative biparental care. *Behavioral Ecology* 15:621–628.

Smiseth PT, Darwell CT, Moore AJ (2003) Partial begging: an empirical model for the early evolution of offspring signalling. *Proceedings of the Royal Society of London B* 270:1773–1777.

Smiseth PT, Dawson C, Varley E, Moore AJ (2005) How do caring parents respond to mate loss? Differential response by males and females. *Animal Behaviour* 69:551–559.

Smiseth PT, Royle NJ, Kölliker M (2012) What is parental care? In NJ Royle, PT Smiseth PT, Kölliker M (eds), *The Evolution of Parental Care*, pp. 1–17. Oxford University Press, Oxford.

Smith CC, Fretwell SD (1974) The optimal balance between size and number of offspring. *American Naturalist* 108:499–506.

Smith DT, Hosken DJ, Rostant WG, et al. (2011) DDT resistance, epistasis and male fitness in flies. *Journal of Evolutionary Biology* 24:1351–1362.

Smith G, Fang YX, Liu X, et al. (2013) Transcriptome-wide expression variation associated with environmental plasticity and mating success in cactophilic *Drosophila mojavensis*. *Evolution* 67:1950–1963.

Smith JM (1956) Fertility, mating behaviour and sexual selection in *Drosophila subobscura*. *Journal of Genetics* 54:261–279.

Smith MS, Milton I, Strand MR (2010) Phenotypically plastic traits regulate caste formation and soldier function in polyembryonic wasps. *Journal of Evolutionary Biology* 23:2677–2684.

Smith NGC (2000) The evolution of haplodiploidy under inbreeding. *Heredity* 84:186–192.

Smith PH, Browne LB, van Gerwen ACM (1989) Causes and correlates of loss and recovery of sexual receptivity in *Lucilia cuprina* females after their first mating. *Journal of Insect Behavior* 2:325–338.

Smith PH, Gillott C, Barton Browne L, van Gerwen ACM (1990) The mating-induced refractoriness of *Lucilia cuprina* females: manipulating the male contribution. *Physiological Entomology* 15:469–481.

Smith RL (1974) Life history of *Abedus herberti* in Central Arizona (Hemiptera: Belostomatidae). *Psyche* 81:272–283.

Smith RL (1976) Male brooding behavior of the water bug *Abedus herberti* (Hemiptera: Belostomatidae). *Annals of the Entomological Society of America* 69:740–747.

Smith RL (1979) Paternity assurance and altered roles in the mating behaviour of a giant water bug *Abedus herberti* (Heteroptera: Belostomatidae). *Animal Behaviour* 27:716–725.

Smith RL (1980) Evolution of exclusive postcopulatory paternal care in the insects. *Florida Entomologist* 63:65–78.

Smith RL (1984) *Sperm Competition and the Evolution of Animal Mating Systems*. Academic Press, London.

Smith RL (1997) Evolution of paternal care in the giant water bugs (Heteroptera: Belostomatidae). In JC Choe, BJ Crespi (eds), *The Evolution of Social Behavior in Insects and Arachnids*, pp. 116–149. Cambridge University Press, Cambridge.

Snook RR (2013) Sexual selection and its impact on mating systems. In J Losos (ed.), *The Princeton Guide to Evolution*, pp. 632–640. Princeton University Press, Princeton, NJ.

Snook RR, Cleland SY, Wolfner MF, Karr TL (2000) Offsetting effects of *Wolbachia* infection and heat shock on sperm production in *Drosophila simulans*: analyses of fecundity, fertility and accessory gland proteins. *Genetics* 155:167–178.

Snyder BF, Gowaty PA (2013) A reappraisal of Bateman's classic selection study of intrasexual selection. *Evolution* 61:2457–2468.

Solensky MJ, Oberhauser KS (2009) Male monarch butterflies, *Danaus plexippus*, adjust ejaculates in response to intensity of sperm competition. *Animal Behaviour* 77:465–472.

Som C, Reyer HU (2007) Hemiclonal reproduction slows down the speed of Muller's ratchet in the hybridogenetic frog *Rana esculenta*. *Journal of Evolutionary Biology* 20:650–660.

Sota T, Kubota K (1998) Genital lock-and-key as a selective agent against hybridization. *Evolution* 52:1507–1513.

Spencer M, Blaustein L, Cohen JE (2002) Oviposition habitat selection by mosquitoes (*Culiseta longiareolata*) and consequences for population size. *Ecology* 83:669–679.

Spradling A, Fuller MT, Braun RE, Yoshida S (2011) Germline stem cells. *Cold Spring Harbor Perspectives in Biology* 3:a002642

Staerkle M, Kölliker M (2008) Maternal food regurgitation to nymphs in earwigs (*Forficula auricularia*). *Ethology* 114:844–850.

Starr CK (1984) Sperm competition, kinship and sociality in the aculeate Hymenoptera. In RL Smith (ed.), *Sperm Competition and the Evolution of Animal Mating Systems*, pp. 427–464. Academic Press, London.

Stay B, Coop AC (1974) 'Milk' secretion for embryogenesis in a viviparous cockroach. *Tissue and Cell* 6:669–693.

Stearns SC (1992) *Evolution of Life Histories*. Oxford University Press, Oxford.

Stenberg P, Lundmark M, Knutelski S, Saura A (2003) Evolution of clonality and polyploidy in a weevil system. *Molecular Biology and Evolution* 20:1626–1632.

Stern DL, Orgogozo V (2008) The loci of evolution: how predictable is genetic evolution? *Evolution* 62:2155–2177.

Stewart AD, Morrow EH, Rice WR (2005) Assessing putative interlocus sexual conflict in *Drosophila melanogaster* using experimental evolution. *Proceedings of the Royal Society of London B* 272:2029–2035.

Stewart AD, Hannes AM, Mirzatuny A, Rice WR (2008) Sexual conflict is not counterbalanced by good genes in the laboratory *Drosophila melanogaster* model system. *Journal of Evolutionary Biology* 21:1808–1813.

Stinchcombe JR, Hoekstra HE (2007) Combining population genomics and quantitative genetics: finding the genes underlying ecologically important traits. *Heredity* 100:158–170.

Stireman JOI (2005) The evolution of generalization? Parasitoid flies and the perils of inferring host range evolution from phylogenies. *Journal of Evolutionary Biology* 18:325–336.

Stiver KA, Alonzo SH (2009) Parental and mating effort: is there necessarily a trade-off? *Ethology* 115:1101–1126.

Stjernman M, Råberg L, Nilsson J (2008) Maximum host survival at intermediate parasite infection intensities. *PLoS One* 3:e2463.

Stockley P, Seal NJ (2001) Plasticity in reproductive effort of male dung flies (*Scatophaga stercoraria*) as a response to larval density. *Functional Ecology* 15:96–102.

Stone G, French V (2003) Evolution: have wings come, gone, and come again? *Current Biology* 13:R436–R438.

Stouthamer R, Luck RF, Hamilton WD (1990) Antibiotics cause parthenogenetic *Trichogramma* (Hymenoptera/Trichogrammatidae) to revert to sex. *Proceedings of the National Academy of Sciences of the USA* 87:2424–2427.

Strand MR (2009) Polyembryony. In VH Resh, RT Cardé (eds), *Encyclopedia of Insects*, pp. 821–825. Academic Press, Burlington, VT.

Strassmann J (2001) The rarity of multiple mating by females in the social Hymenoptera. *Insectes Sociaux* 48:1–13.

Stutt AD, Siva-Jothy MT (2001) Traumatic insemination and sexual conflict in the bed bug *Cimex lectularius*. *Proceedings of the National Academy of Sciences of the USA* 98:5683–5687.

Sullivan-Beckers L, Cocroft RB (2010) The importance of female choice, male–male competition, and signal transmission as causes of selection on male mating signals. *Evolution* 64:3158–3171.

Suomalainen E, Saura A, Lokki J (1987) *Cytology and Evolution in Parthenogenesis*. CRC Press, Boca Raton.

Sutherland WJ (1985) Chance can produce a difference in variance in mating success and explain Bateman's data. *Animal Behaviour* 33:1349–1352.

Sutherland WJ (1987) Random and deterministic components of variance in mating success. In JW Bradbury, MB Andersson (eds), *Sexual Selection: Testing the Alternatives*, pp. 207–219. John Wiley & Sons, Chichester.

Suzuki S, Nagano M (2009) To compensate or not? Caring parents respond differentially to mate removal and mate handicapping in the burying beetle, *Nicrophorus quadripunctatus*. *Ethology* 115:1–6.

Suzuki S, Kitamura M, Matsubayashi K (2005) Matriphagy in the hump earwig, *Anechura harmandi* (Dermaptera: Forficulidae), increases the survival rates of the offspring. *Journal of Ethology* 23:211–213.

Suzuki Y, Nijhout HF (2008) Genetic basis of adaptive evolution of a polyphenism by genetic accommodation. *Journal of Evolutionary Biology* 21:57–66.

Svensson EI, Gosden TP (2007) Contemporary evolution of secondary sexual traits in the wild. *Functional Ecology* 21:422–433.

Svensson EI, Eroukhmanoff F, Friberg M (2006) Effects of natural and sexual selection on adaptive population divergence and premating isolation in a damselfly. *Evolution* 60:1242–1253.

Swanson WJ, Vacquier VD (2002) Reproductive protein evolution. *Annual Review of Ecology and Systematics* 33:161–179.

Swarup S, Huang W, Mackay TFC, Anholt RRH (2013) Analysis of natural variation reveals neurogenetic networks for *Drosophila* olfactory behavior. *Proceedings of the National Academy of Sciences of the USA* 110:1017–1022.

Szulkin M, Stopher KV, Pemberton JM, Reid JM (2013) Inbreeding avoidance, tolerance, or preference in animals? *Trends in Ecology and Evolution* 28:205–211.

Takada H (1988) Interclonal variation in the photoperiodic response for sexual morph production of Japanese *Aphis gossypii* Glover (Homoptera, Aphididae). *Journal of Applied Entomology* 106:188–197.

Takami Y (2007) Spermatophore displacement and male fertilization success in the ground beetle *Carabus insulicola*. *Behavioral Ecology* 18:628–634.

Tallamy DW (1984) Insect parental care. *BioScience* 34:20–24.

Tallamy DW (2000) Sexual selection and the evolution of exclusive paternal care in arthropods. *Animal Behaviour* 60:559–567.

Tallamy DW (2001) Evolution of exclusive paternal care in arthropods. *Annual Review of Entomology* 46:139–165.

Tallamy DW, Denno RF (1981) Maternal care in *Gargaphia solani* (Hemiptera: Tingidae). *Animal Behaviour* 29:771–778.

Tallamy DW, Wood TK (1986) Convergence patterns in subsocial insects. *Annual Review of Entomology* 31:369–390.

Tallamy DW, Powell BE, McClafferty JA (2002) Male traits under cryptic female choice in the spotted cucumber beetle (Coleoptera: Chrysomelidae). *Behavioral Ecology* 13:511–518.

Tallamy DW, Darlington MB, Pesek JD, Powell BE (2003) Copulatory courtship signals male genetic quality in cucumber beetles. *Proceedings of the Royal Society of London* B 270:77–82.

Tallamy DW, Walsh E, Peck DC (2004) Revisiting paternal care in the assassin bug, *Atopozelus pallens* (Heteroptera: Reduviidae). *Journal of Insect Behavior* 17:431–436.

Tallamy DW, Denno RF (1981) Maternal care in *Gargaphia solani* (Hemiptera: Tingidae). *Animal Behaviour* 29:771–778.

Tallamy DW, Wood TK (1986) Convergence patterns in subsocial insects. *Annual Review of Entomology* 31:369–390.

Tang-Martinez Z (2010) Bateman' s principles: original experiment and modern data for and against. In MD Breed, J Moore (eds), *Encyclopedia of Animal Behavior*, pp. 166–176. Academic Press, Oxford.

Tarpy DR (2003) Genetic diversity within honeybee colonies prevents severe infections and promotes colony growth. *Proceedings of the Royal Society of London B* 270:99–103.

Tatarnic NJ, Cassis G (2010) Sexual coevolution in the traumatically inseminating plant bug genus *Coridromius*. *Journal of Evolutionary Biology* 23:1321–1326.

Tatarnic NJ, Cassis G, Hochuli DT (2006) Traumatic insemination in the plant bug genus *Coridromius* Signoret (Heteroptera: Miridae). *Biology Letters* 2:58–61.

Tauber CA, Tauber MJ (1977) A genetic model for sympatric speciation through habitat diversification and seasonal isolation. *Nature* 268:702–705.

Taylor DR, Jaenike J (2002) Sperm competition and the dynamics of X chromosome drive: stability and extinction. *Genetics* 160:1721–1751.

Taylor BJ, Whitman DW (2010) A test of three hypotheses for ovariole number determination in the grasshopper *Romalea microptera*. *Physiological Entomology* 35:214–221.

Taylor PD, Williams GC (1982) The lek paradox is not resolved. *Theoretical Population Biology* 22:392–409.

Teal PEA, Gomez-Simuta Y, Proveaux A (2000) Mating experience and juvenile hormone enhance sexual signaling and mating in male Caribbean fruit flies. *Proceedings of the National Academy of Sciences of the USA* 97:3708–3712.

Tebb G, Thoday JM (1956) Reversal of mating preference by crossing strains of *Drosophila melanogaster*. *Nature* 177:707.

Thomas LK, Manica A (2005) Intrasexual competition and mate choice in assassin bugs with uniparental male and female care. *Animal Behaviour* 69:275–281.

Thomas ML, Simmons LW (2007) Male crickets adjust the viability of their sperm in response to female mating status. *American Naturalist* 170:190–195.

Thomas ML, Simmons LW (2008) Male-derived cuticular hydrocarbons signal sperm competition intensity and affect ejaculate expenditure in crickets. *Proceedings of the Royal Society of London B* 276:383–388.

Thomas ML, Simmons LW (2009) Sexual selection on cuticular hydrocarbons in the Australian field cricket, *Teleogryllus oceanicus*. *BMC Evolutionary Biology* 9:162.

Thomas ML, Gray B, Simmons LW (2011) Male crickets alter the relative expression of cuticular hydrocarbons when exposed to different acoustic environments. *Animal Behaviour* 82:49–53.

Thornhill R (1976) Sexual selection and nuptial feeding behavior in *Bittacus apicalis* (Insecta: Mecoptera). *American Naturalist* 110:529–548.

Thornhill R (1983) Cryptic female choice and its implications in the scorpionfly *Harpobittacus nigriceps*. *American Naturalist* 122:765–788.

Thornhill R (1984) Alternative female choice tactics in the scorpionfly *Hylobittacus apicalis* (Mecoptera) and their implications. *American Zoologist* 24:367–383.

Thornhill R, Alcock J (1983) *The Evolution of Insect Mating Systems*. Harvard University Press, Cambridge, MA.

Tibbetts EA, Sheehan MJ (2011) Facial patterns are a conventional signal of agonistic ability in *Polistes exclamans* paper wasps. *Ethology* 117:1138–1146.

Tibbetts EA, Shorter JR (2009) How do fighting ability and nest value influence usurpation contests in *Polistes* wasps? *Behavioral Ecology and Sociobiology* 63:1377–1385.

Tinghitella RM (2008) Rapid evolutionary change in a sexual signal: genetic control of the mutation 'flatwing' that renders male field crickets (*Teleogryllus oceanicus*), mute. *Heredity* 100:261–267.

Tinghitella RM, Wang JM, Zuk M (2009) Preexisting behavior renders a mutation adaptive: flexibility in male phonotaxis behavior and the loss of singing ability in the field cricket *Teleogryllus oceanicus*. *Behavioral Ecology* 20:722–728.

Tomioka K, Matsumoto A (2010) A comparative view of insect circadian clock systems. *Cellular and Molecular Life Sciences* 67:1397–1406.

Tomkins JL (1999) Environmental and genetic determinants of the male forceps length dimorphism in the European earwig *Forficula auricularia* L. *Behavioral Ecology and Sociobiology* 47:1–8.

Tomkins JL, Brown GS (2004) Population density drives the local evolution of a threshold dimorphism. *Nature* 431:1099–1103.

Tomkins JL, Hazel WN (2007) The status of the conditional evolutionarily stable strategy. *Trends in Ecology and Evolution* 22:522–528.

Tomkins JL, Hazel WN (2011) Explaining phenotypic diversity: the conditional strategy and threshold trait expression. In LW Simmons, TJ Ridsdill-Smith (eds), *Dung Beetle Ecology and Evolution*, pp. 107–125. Wiley–Blackwell, Oxford.

Tomkins JL, Simmons LW (1996) Dimorphisms and fluctuating asymmetry in the forceps of male earwigs. *Journal of Evolutionary Biology* 9:753–770.

Tomkins JL, Simmons LW (1998) Female choice and manipulations of forceps size and symmetry in the earwig *Forficula auricularia* L. *Animal Behaviour* 56:347–356.

Tomkins JL, Simmons LW, Alcock J (2001) Brood-provisioning strategies in Dawson's burrowing bee, *Amegilla dawsoni* (Hymenoptera: Anthophorini). *Behavioral Ecology and Sociobiology* 50:81–89.

Tomkins JL, Radwan J, Kotiaho JS, Tregenza T (2004) Genic capture and resolving the lek paradox. *Trends in Ecology and Evolution* 19:323–328.

Tomkins JL, Kotiaho JS, Le Bas N (2005) Phenotypic plasticity in the developmental integration of morphological trade-offs and secondary sexual trait compensation. *Proceedings of the Royal Society of London B* 272:543–551.

Torchio PF, Tepedino VJ (1980) Sex ratio, body size and seasonality in a solitary bee, *Osmia lignaria propinqua* Cresson (Hymenoptera: Megachelidae). *Evolution* 34:993–1003.

Torres-Vila LM, Jennions MD (2005) Male mating history and female fecundity in the Lepidoptera: do male virgins make better partners? *Behavioral Ecology and Sociobiology* 57:318–326.

Trautwein MD, Wiegmann BM, Beutel R, Kjer KM, Yeates DK (2012) Advances in insect phylogeny at the dawn of the postgenomic era. *Annual Review of Entomology* 57:449–468.

Travis J, Reznick D (1998) Experimental approaches to the study of evolution. In WJ Resitarits, J Bernardo (eds), *Experimental Ecology: Issues and Perspectives*, pp. 437–460. Oxford University Press, New York.

Travisano M, Shaw RG (2013) Lost in the map. *Evolution* 67:305–314.

Tregenza T, Wedell N (1998) Benefits of multiple mates in the cricket, *Gryllus bimaculatus*. *Evolution* 52:1726–1730.

Tregenza T, Wedell N (2000) Genetic compatibility, mate choice and patterns of parentage: invited review. *Molecular Ecology* 9:1013–1027.

Tregenza T, Wedell N (2002) Polyandrous females avoid costs of inbreeding. *Nature* 415:71–73.

Tregenza T, Wedell N, Chapman T (2006) Sexual conflict: a new paradigm? *Philosophical Transactions of the Royal Society of London B* 361:229–234.

Triplehorn CA, Johnson NF (2005) *Borror and DeLong's Introduction to the Study of Insects*. Thomson Brooks/Cole, Belmont, CA.

Trivers R (1972) Parental investment and sexual selection. In B Campbell (ed.), *Sexual Selection and the Descent of Man 1871–1971*, pp. 139–179. Aldine Press, Chicago.

Trivers RL (1974) Parent–offspring conflict. *American Zoologist* 14:249–264.

Trivers RL, Hare H (1976) Hapoldiploidy and the evolution of the social insects. *Science* 191:249–263.

Trivers RL, Willard DE (1973) Natural-selection of parental ability to vary sex-ratio of offspring. *Science* 179:90–92.

Trumbo ST (1996) Parental care in invertebrates. *Advances in the Study of Behavior* 25:3–51.

Trumbo ST (2007) Defending young biparentally: female risk-taking with and without a male in the burying beetle, *Nicrophorus pustulatus*. *Behavioral Ecology and Sociobiology* 61:1717–1723.

Trumbo ST (2012) Patterns of parental care in invertebrates. In NJ Royle, PT Smiseth, M Kölliker (eds), *The Evolution of Parental Care*, pp. 81–100. Oxford University Press, Oxford.

Tsubaki Y (2003) The genetic polymorphism linked to mate-securing strategies in the male damselfly *Mnais costalis* Selys (Odonata: Calopterygidae). *Population Ecology* 45:263–266.

Tsubaki Y, Hooper RE (2004) Effects of eugregarine parasites on adult longevity in the polymorphic damselfly *Mnais costalis* Selys. *Ecological Entomology* 29:361–366.

Tsubaki Y, Ono T (1986) Competition for territorial sites and alternative mating tactics in the dragonfly, *Nannophya pygmaea* Rambur (Odonata, Libellulidae). *Behaviour* 97:234–252.

Tsubaki Y, Ono T (1987) Effects of age and body size on the male territorial system of the dragonfly, *Nannophya pygmaea* Rambur (Odonata, Libellulidae). *Animal Behaviour* 35:518–525.

Tsubaki Y, Hooper RE, Siva-Jothy MT (1997) Differences in adult and reproductive lifespan in the two male forms of *Mnais pruinosa costalis* Selys (Odonata: Calopterygidae). *Researches on Population Ecology* 39:149–155.

Tuni C, Albo MJ, Bilde T (2013a) Polyandrous females acquire indirect benefits in a nuptial feeding species. *Journal of Evolutionary Biology* 26:1307–1316.

Tuni C, Beveridge M, Simmons LW (2013b) Female crickets assess relatedness during mate guarding and bias storage of sperm toward unrelated males. *Journal of Evolutionary Biology* 26:1261–1268.

Tuttle EM (2003) Alternative reproductive strategies in the white-throated sparrow: behavioral and genetic evidence. *Behavioral Ecology* 14:425–432.

Úbeda F, Normark BB (2006) Male killers and the origins of paternal genome elimination. *Theoretical Population Biology* 70:511–526.

Ueda T (1979) Plasticity of the reproductive behaviour of dragonfly, *Sympetrum parvulum* Bartneff, with reference to the social relationships of males and the density of territories. *Researches on Population Ecology* 21:135–152.

Vahed K (1998) The function of nuptial feeding in insects: a review of empirical studies. *Biological Reviews* 73:43–78.

Vahed K (2007a) All that glisters is not gold: sensory bias, sexual conflict and nuptial feeding in insects and spiders. *Ethology* 113:105–127.

Vahed K (2007b) Comparative evidence for a cost to males of manipulating females in bushcrickets. *Behavioral Ecology* 18:499–506.

Vala F, Egas M, Breeuwer JAJ, Sabelis MW (2004) *Wolbachia* affects oviposition and mating behaviour of its spider mite host. *Journal of Evolutionary Biology* 17:692–700.

Välimäki P, Kaitala A (2010) Properties of male ejaculates do not generate geographical variation in female mating tactics in a butterfly *Pieris napi*. *Animal Behaviour* 79:1173–1179.

Välimäki P, Kivelä SM, Jääskeläinen L, et al. (2008) Divergent timing of egg-laying may maintain life history polymorphism in potentially multivoltine insects in seasonal environments. *Journal of Evolutionary Biology* 21:1711–1723.

Vancassel M (1984) Plasticity and adaptive radiation of dermapteran parental behavior: results and perspectives. *Advances in the Study of Behavior* 14:51–80.

Vandel A (1928) La parthénogenèse géographique: contribution à l'étude biologique et cytologique de la parthénogenèse naturelle. *Bulletin Biologique de la France et de la Belgique* 62:164–281.

van Driesche RG, Hoddle M, Center T (2008) *Control of Pests and Weeds by Natural Enemies: an Introduction to Biological Control*. Wiley–Blackwell, Hoboken, NJ.

van Dyck H, Wiklund C (2002) Seasonal butterfly design: morphological plasticity among three developmental pathways relative to sex, flight and thermoregulation. *Journal of Evolutionary Biology* 15:216–225.

van Homrigh A, Higgie M, McGuigan K, Blows MW (2007) The depletion of genetic variance by sexual selection. *Current Biology* 17:528–532.

van Lieshout E, Elgar MA (2011) Longer exaggerated male genitalia confer defensive sperm-competitive benefits in an earwig. *Evolutionary Ecology* 25:351–362.

van Noordwijk AJ, de Jong G (1986) Acquisition and allocation of resources: their influence on variation in life history tactics. *American Naturalist* 128:137–142.

van Staaden MJ, Romer H (1997) Sexual signalling in bladder grasshoppers: tactical design for maximizing calling range. *Journal of Experimental Biology* 200:2597–2608.

Vargo EL, Labadie PE, Matsuura K (2012) Asexual queen succession in the subterranean termite *Reticulitermes virginicus*. *Proceedings of the Royal Society of London B* 279:813–819.

Vehrencamp SL, Bradbury JW (1984) Mating systems. In JR Krebs, NB Davies (eds), *Behavioural Ecology: An Evolutionary Approach*, pp. 251–278. Blackwell, Oxford.

Viney ME, Riley EM, Buchanan KL (2005) Optimal immune responses: immunocompetence revisited. *Trends in Ecology and Evolution* 20:665–669.

von Helversen D, von Helversen O (1991) Pre-mating sperm removal in the bush-cricket *Metaplastes ornatus* Ramme 1931 (Orthoptera, Tettigonoidea, Phaneropteridae). *Behavioral Ecology and Sociobiology* 28:391–396.

Waage JK (1979a) Adaptive significance of postcopulatory guarding of mates and non-mates by male *Calopteryx maculata* (Odonata). *Behavioral Ecology and Sociobiology* 6:147–154.

Waage JK (1979b) Dual function of the damselfly penis: sperm removal and transfer. *Science* 203:916–918.

Wade MJ (1984) *The Evolution of Insect Mating Systems* by Thornhill, R. and Alcock, J. *Evolution* 38:706–708.

Wade MJ, Kalisz S (1989) The additive partitioning of selection gradients. *Evolution* 43:1567–1569.

Wade MJ, Shuster SM (2002) The evolution of parental care in the context of sexual selection: a critical reassessment of parental investment theory. *American Naturalist* 160:285–292.

Wagner WE (1998) Measuring female mating preferences. *Animal Behaviour* 55:1029–1042.

Wagner WE, Kelley RJ, Tucker KR, Harper CJ (2001) Females receive a life-span benefit from male ejaculates in a field cricket. *Evolution* 55:994–1001.

Wagstaff BJ, Begun DJ (2007) Adaptive evolution of recently duplicated accessory gland protein genes in desert *Drosophila*. *Genetics* 177:1023–1030.

Walters JR, Harrison RG (2010) Combined EST and proteomic analysis identifies rapidly evolving seminal fluid proteins in *Heliconius* butterflies. *Molecular Biology and Evolution* 27:2000–2013.

Wang J, Wurm Y, Nipitwattanaphon M, et al. (2013) A Y-like social chromosome causes alternative colony organization in fire ants. *Nature* 493:664–668.

Wang Q, Yang LH, Hedderley DC (2008) Function of prolonged copulation in *Nysius huttoni* White (Heteroptera: Lygaeidae) under male-biased sex ratio and high population density. *Journal of Insect Behavior* 21:89–99.

Ward PI (2000) Cryptic female choice in the yellow dung fly *Scathophaga stercoraria* (L.). *Evolution* 54:1680–1686.

Ward PI (2007) Postcopulatory selection in the yellow dung fly *Scathophaga stercoraria* (L.), and the mate-now-choose-later mechanism of cryptic female choice. *Advances in the Study of Behavior* 37:343–369.

Ward PI, Wilson AJ, Reim C (2008) A cost of cryptic female choice in the yellow dung fly. *Genetica* 134:63–67.

Ward RJS, Cotter SC, Kilner RM (2009) Current brood size and residual reproductive value predict offspring desertion in the burying beetle *Nicrophorus vespilloides*. *Behavioral Ecology* 20:1274–1281.

Watson NL, Simmons LW (2011) Unravelling the effects of differential maternal allocation and male genetic quality on offspring viability in the dung beetle, *Onthophagus sagittarius*. *Evolutionary Ecology* 26:139–147.

Watson PJ, Arnqvist G, Stallmann RR (1998) Sexual conflict and the energetic costs of mating and mate choice in water striders. *American Naturalist* 151:46–58.

Weatherhead PJ, Robertson RJ (1979) Offspring quality and the polygyny threshold: 'the sexy son hypothesis'. *American Naturalist* 113:201–208.

Weber NA (1972) *Gardening Ants: the Attines*. American Philosophical Society, Philadelphia.

Weddle C, Steiger S, Hamaker CG, et al. (2013) Cuticular hydrocarbons as a basis for chemosensory self-referencing in crickets: a potentially universal mechanism facilitating polyandry in insects. *Ecology Letters* 16:346–353.

Wedell N (1998) Sperm protection and mate assessment in the bushcricket *Coptaspis* sp. 2. *Animal Behaviour* 56:357–363.

Wedell N (2013) The dynamic relationship between polyandry and selfish genetic elements. *Philosophical Transactions of the Royal Society of London B* 368:20120048.

Wedell N, Cook PA (1999) Strategic sperm allocation in the small white butterfly *Pieris rapae* (Lepidoptera: Pieridae). *Functional Ecology* 13:85–93.

Wedell N, Gage MJG, Parker GA (2002a) Sperm competition, male prudence and sperm-limited females. *Trends in Ecology and Evolution* 17:313–320.

Wedell N, Wiklund C, Cook PA (2002b) Monandry and polyandry as alternative lifestyles in a butterfly. *Behavioral Ecology* 13:450–455.

Wedell N, Wiklund C, Bergström J (2009) Coevolution of non-fertile sperm and female receptivity in a butterfly. *Biology Letters* 5:678–681.

Wenninger EJ, Averill AL (2006) Influence of body and genital morphology on relative male fertilization success in oriental beetle. *Behavioral Ecology* 17:656–663.

Went DF (1982) Egg activation and parthenogenetic reproduction in insects. *Biological Reviews* 57:319–344.

Werren JH (2011) Selfish genetic elements, genetic conflict, and evolutionary innovation. *Proceedings of the National Academy of Sciences of the USA* 108:10863–10870.

West S (2009) *Sex Allocation*. Princeton University Press, Princeton, NJ.

West-Eberhard MJ (1983) Sexual selection, social competition, and speciation. *Quarterly Review of Biology* 58:155–183.

West-Eberhard MJ (1989) Phenotypic plasticity and the origins of diversity. *Annual Review of Ecology and Systematics* 20:249–278.

West-Eberhard MJ (2003) *Developmental Plasticity and Evolution*. Oxford University Press, Oxford.

Westneat DF, Birkhead TR (1998) Alternative hypotheses linking the immune system and mate choice for good genes. *Proceedings of the Royal Society of London B* 265:1065–1073.

Westneat DF, Sherman PW, Morton ML (1990) The ecology and evolution of extra-pair copulations in birds. *Current Ornithology* 7:330–369.

Wheeler DA, Kyriacou CP, Greenacre ML, et al. (1991) Molecular transfer of a species-specific behavior from *Drosophila simulans* to *Drosophila melanogaster*. *Science* 251:1082–1085.

White-Cooper H, Bausek N (2010) Evolution and spermatogenesis. *Philosophical Transactions of the Royal Society of London B* 365:1465–1480.

Whitman DW, Ananthakrishnan TN (2009) *Phenotypic Plasticity of Insects: Mechanisms and Consequences*. Science Publishers, Enfield.

Wigby S, Chapman T (2005) Sex peptide causes mating costs in female *Drosophila melanogaster*. *Current Biology* 15:316–321.

Wigby S, Sirot LK, Linklater JR, et al. (2009) Seminal fluid protein allocation and male reproductive success. *Current Biology* 19:751–757.

Wiklund C, Fagerström T (1977) Why do males emerge before females? A hypothesis to explain the incidence of protandry in butterflies. *Oecologia* 31:153–158.

Wiklund C, Kaitala A, Lindfors V, Abenius J (1993) Polyandry and its effect on female reproduction in the green-veined white butterfly (*Pieris napi* L.). *Behavioral Ecology and Sociobiology* 33:25–33.

Wiley C, Ellison CK, Shaw KL (2012) Widespread genetic linkage of mating signals and preferences in the Hawaiian cricket *Laupala*. *Proceedings of the Royal Society of London B* 279:1203–1209.

Wiley RH, Poston J (1996) Indirect mate choice, competition for mates, and coevolution of the sexes. *Evolution* 50:1371–1381.

Wilhelm M, Chhetri, M, Rychtar J, Rueppell O (2011) A game theoretical analysis of the mating sign behavior in the honey bee. *Bulletin of Mathematical Biology* 73:626–638.

Wilkinson GS, Dodson GN (1997) Function and evolution of antlers and eye stalks in flies. In JC Choe, BJ Crespi (eds), *The Evolution of Mating Systems in Insects and Arachnids*, pp. 310–328. Cambridge University Press, Cambridge.

Wilkinson GS, Reillo PR (1994) Female choice response to artificial selection on an exaggerated male trait in a stalk-eyed fly. *Proceedings of the Royal Society of London B* 255:1–6.

Wilkinson GS, Taper M (1999) Evolution of genetic variation for condition-dependent traits in stalk-eyed flies. *Proceedings of the Royal Society of London B* 266:1685–1690.

Wilkinson GS, Presgraves DC, Crymes L (1998) Male eye span in stalk-eyed flies indicates genetic quality by meiotic drive suppression. *Nature* 391:276–279.

Wilkinson GS, Swallow JG, Christense SJ, Madden K (2003) Phylogeography of sex ratio and multiple mating in stalk-eyed flies from southeast Asia. *Genetica* 117:37–46.

Williams GC (1966) *Adaptation and Natural Selection*. Princeton University Press, Princeton, NJ.

Williams GC (1975) *Sex and Evolution*. Princeton University Press, Princeton, NJ.

Wilson EO (1971) *The Insect Societies*. Belknap Press, Cambridge, MA.

Wilson EO (1975) *Sociobiology. The New Synthesis*. Belknap Press, Cambridge, MA.
Wilson N, Tubman SC, Eady PE, Robertson GW (1997) Female genotype affects male success in sperm competition. *Proceedings of the Royal Society of London B* 264:1491–1495.
Winston ML (1991) *The Biology of the Honey Bee*. Harvard University Press. Cambridge, MA.
Wolf JB, Brodie III ED, Cheverud JM, et al. (1998) Evolutionary consequences of indirect genetic effects. *Trends in Ecology and Evolution* 13:64–69.
Wolfner MF (1997) Tokens of love: functions and regulation of *Drosophila* male accessory gland products. *Insect Biochemistry and Molecular Biology* 27:179–192.
Wolfner MF (2002) The gifts that keep on giving: physiological functions and evolutionary dynamics of male seminal proteins in *Drosophila*. *Heredity* 88:85–93.
Wolfner MF (2009) Battle and ballet: molecular interactions between the sexes in *Drosophila*. *Journal of Heredity* 100:399–410.
Wolschin F, Mutti NS, Amdam GV (2011) Insulin receptor substrate influences female caste development in honeybees. *Biology Letters* 7:112–115.
Wong A, Turchin M, Wolfner MF, Aquadro CF (2012) Temporally variable selection on proteolysis-related reproductive tract proteins in *Drosophila*. *Molecular Biology and Evolution* 29:229–238.
Wong JWY, Meunier J, Kölliker M (2013) The evolution of parental care in insects: the roles of ecology, life history and the social environment. *Ecological Entomology* 38:123–137.
Woyke J (1962) Natural and artificial insemination of queen honeybees. *Bee World* 43:21–25.
Wyatt GR (1997) Juvenile hormone in insect reproduction—a paradox? *European Journal of Entomology* 94:323–335.
Wyman MJ, Agrawal AF, Rowe L (2010) Condition-dependence of the sexually dimorphic transcriptome in *Drosophila melanogaster*. *Evolution* 64:1836–1848.
Xu J, Wang Q (2010a) Thiotepa, a reliable marker for sperm precedence measurement in a polyandrous moth. *Journal of Insect Physiology* 56:102–106.
Xu J, Wang Q (2010b) Mechanisms of last male precedence in a moth: sperm displacement at ejaculation and storage sites. *Behavioral Ecology* 21:714–721.
Yamamura N, Tsuji N (1989) Postcopulatory guarding strategy in a finite mating period. *Theoretical Population Biology* 35:36–50.
Yamane T, Miyatake T (2005) Intra-specific variation in strategic ejaculation according to level of polyandry in *Callosobruchus chinensis*. *Journal of Insect Physiology* 51:1240–1243.
Yamane T, Miyatake T (2012) Evolutionary correlation between male substances and female remating frequency in a seed beetle. *Behavioral Ecology* 23:715–722.
Yamane T, Okada K, Nakayama S, Miyatake T (2010) Dispersal and ejaculatory strategies associated with exaggeration of weapon in an armed beetle. *Proceedings of the Royal Society of London B* 277:1705–1710.
Yamauchi K, Oguchi S, Nakamura Y, et al. (2001) Mating behavior of dimorphic reproductives of the ponerine ant, *Hypoponera nubatama*. *Insectes Sociaux* 48:83–87.
Yapici N, Kim YJ, Ribeiro C, Dickson BJ (2008) A receptor that mediates the post-mating switch in *Drosophila* reproductive behaviour. *Nature* 451:33–37.
Yasui Y (1998) The 'genetic benefits' of female multiple mating reconsidered. *Trends in Ecology and Evolution* 13:246–250.
Yee WKW, Sutton KL, Dowling DK (2013) In vivo male fertility is affected by naturally occurring mitochondrial haplotypes. *Current Biology* 23:R55–R56.
Yen JH, Barr AR (1971) New hypothesis of the cause of cytoplasmic incompatibility in *Culex pipiens* L. *Nature* 232:657–658.
Yew JY, Dreisewerd K, Luftmann H, et al. (2009) A new male sex pheromone and novel cuticular cues for chemical communication in *Drosophila*. *Current Biology* 19:1245–1254.
Zahavi A (1975) Mate selection: a selection for a handicap. *Journal of Theoretical Biology* 53:205–214.
Zchori-Fein E, Roush RT, Rosen D (1998) Distribution of parthenogenesis-inducing symbionts in ovaries and eggs of *Aphytis* (Hymentoptera: Aphelinidae). *Current Microbiology* 36:1–8.
Zeh DW, Smith RL (1985) Paternal investment by terrestrial arthropods. *American Zoologist* 25:785–805.
Zeh DW, Zeh JA (1992) Dispersal-generated sexual selection in a beetle-riding pseudoscorpion. *Behavioral Ecology and Sociobiology* 30:135–142.

Zeh DW, Zeh JA, Smith RL (1989) Ovipositors, amnions and eggshell architecture in the diversification of terrestrial arthropods. *Quarterly Review of Biology* 64:147–168.

Zeh DW, Zeh JA, Tavakilian G (1992) Sexual selection and sexual dimorphism in the harlequin beetle *Acrocinus longimanus*. *Biotropica* 24:86–96.

Zeh JA, Zeh DW (1994) Last male sperm precedence breaks down when females mate with 3 males. *Proceedings of the Royal Society of London B* 257:287–292.

Zeh JA, Zeh DW (1996) The evolution of polyandry I: Intra-genomic conflict and genetic incompatibility. *Proceedings of the Royal Society of London B* 263:1711–1717.

Zeh JA, Zeh DW (1997) The evolution of polyandry II. Post-copulatory defences against genetic incompatibility. *Proceedings of the Royal Society of London B* 264:69–75.

Zeh JA, Zeh DW (2001a) Reproductive mode and the genetic benefits of polyandry. *Animal Behaviour* 61:1051–1063.

Zeh JA, Zeh DW (2001b) Spontaneous abortion depresses female sexual receptivity in a viviparous arthropod. *Animal Behaviour* 62:427–433.

Zeh JA, Zeh DW (2003) Toward a new sexual selection paradigm: polyandry, conflict and incompatibility. *Ethology* 109:929–950.

Zeng ZB (1988) Long-term correlated response, interpopulation covariation, and interspecific allometry. *Evolution* 42:363–374.

Zera AJ, Harshman LG (2001) The physiology of life history trade-offs in animals. *Annual Review of Ecological Systems* 32:95–126.

Zhang ZQ (2011) Animal biodiversity: an outline of higher-level classification and survey of taxonomic richness. *Zootaxa* 3148:1–237.

Zhu DH, Tanaka S (2002) Prolonged precopulatory mounting increases the length of copulation and sperm precedence in *Locusta migratoria* (Orthoptera: Acrididae). *Annals of the Entomological Society of America* 95:370–373.

Zimmer M, Diestelhorst O, Lunau K (2003) Courtship in long-legged flies (Diptera: Dolichopodidae): function and evolution of signals. *Behavioral Ecology* 14:526–530.

Zink AG (2003) Quantifying the costs and benefits of parental care in female treehoppers. *Behavioral Ecology* 14:687–693.

Zonneveld C, Metz JAJ (1991) Models on butterfly protandry: virgin females are at risk to die. *Theoretical Population Biology* 40:308–321.

Zuk M (1987) The effects of gregarine parasites, body size, and time of day on spermatophore production and sexual selection in field crickets. *Behavioral Ecology and Sociobiology* 21:65–72.

Zuk M (1990) Reproductive strategies and disease susceptibility: an evolutionary viewpoint. *Parasitology Today* 6:231–233.

Zuk M (1994) Immunology and the evolution of behavior. In L Real (ed.), *Behavioral Mechanisms in Ecology*, pp. 354–368. University of Chicago Press, Chicago.

Zuk M, Kolluru GR (1998) Exploitation of sexual signals by predators and parasitoids. *Quarterly Reviews of Biology* 73:415–438.

Zuk M, Simmons LW (1997) Reproductive strategies of the crickets (Orthoptera: Gryllidae). In JC Choe, BJ Crespi (eds), *The Evolution of Mating Systems in Insects and Arachnids*, pp. 89–109. Cambridge University Press, Cambridge.

Zuk M, Stoehr AM (2002) Immune defense and host life history. *American Naturalist* 160:S9–S22.

Zuk M, Simmons LW, Cupp L (1993) Calling characteristics of parasitized and unparasitized populations of the field cricket *Teleogryllus oceanicus*. *Behavioral Ecology and Sociobiology* 33:339–343.

Zuk M, Rotenberry JT, Simmons LW (1998) Calling songs of field crickets (*Teleogryllus oceanicus*), with and without phonotactic parasitoid infection. *Evolution* 52:166–171.

Zuk M, Rotenberry JT, Simmons LW (2001) Geographical variation in calling song of the field cricket *Teleogryllus oceanicus*: the importance of spatial scale. *Journal of Evolutionary Biology* 14:731–741.

Zuk M, Rotenberry JT, Tinghitella RM (2006) Silent night: adaptive disappearance of a sexual signal in a parasitized population of field crickets. *Biology Letters* 2:521–524.

Index

A

Adedus herberti 222, **227**, 228, 232, 233, 235
Acanthocephala declivis **93**
accessory gland proteins (ACPs), *see* seminal fluid
Acheta domesticus 135, **138**, 145, 248
Achroia grisella 136, 147
Acrea encedon 256
Acrocinus longimanus **93**, 95, **96**
Acromyrmex echinatior 273
Acyrthosiphon pisum 76
Adalia bipunctata 149, 151, 187
Alsophila pometaria **4**
alternative mating phenotypes (tactics) 28, 106
  and development 125
  and evolution of novel traits 128
  and sperm competition 197
  crickets 107, **108**
  determinant of 124
  environmental threshold 113, **114**
  of female 127
  parental effects 126
  phylogenetic distribution **118**, **119**, 120
  theoretical models 110
Amegilla dawsoni 113, 126
Ampulex compressa 223, 230
Anechura harmandi 222, 230
anisogamy 23, 159, 231, 233
Anomala albopilosa sakishimana 190, 191
Antipularia urichi 222, 226
apomixis 1, 2
  and heterozygosity 3
Asobara tabida 258
Atrophaneura alcinous **194**, 223, 225
Atta 267
  columbica 273
automixis 1, 2

B

B chromosome 15
Bacillus rossius 3, **4**
Bactrocera cucurbitae 73
Bateman gradient 44, **45**, 48, 159
  and mate searching 53
  and mating rates 55
  and OSR 49
Bombus
  hypnorum 270
  terrestris 143, **266**
Bombyx mori 64, 75, 203
Bullacris membracioides **118**, 126

C

Calligrapha 12
Callosobruchus
  chinensis 164, 200, 210
  maculatus 165, 170, 195, 210, **217**
  phaseoli **217**
  rhodesianus **217**
  subinnotatus 185, **217**
Calopteryx
  maculata (*maculatum*) 122, 188
  splendens xanthostoma 136, 247
Cardiocondyla 267
Cimex lectularius 185, 194, 200, **249**
circadian clock **74**
Coccinella septempuncta 223, 225
Coelopa
  frigida 112, 161, 148, 164
  nebularum 112
Coenagrion puella 188
competition
  access to mates 22, 95, 100
  and alternative mating tactics 106
  and weaponry 92
  arms race 100
  contest theory 27, **29**
  hormonal regulation of 89
  postcopulatory, *see* sperm competition
  scramble 27, 100
condition dependence 36, 60
  alternative mating tactics 123
  and gene networks 73
  and mate choice 131, 135
conditional strategy 110, 111, 113, **116**
  see also alternative mating phenotypes
contest competition 28
  and alternative mating strategies 123
  and mating systems 45
  arms race 100
  resource defence 92, **94**, 95, **96**
  resource holding power (RHP) 28
  theory 28, **29**
Copidosoma
  bakeri 91
  floridanum 90
Copris lunaris 230
copula duration
  and ejaculate transfer 200
  and mate guarding **190**, 191
  and nuptial gifts 204
  and sperm displacement **182**, 183
Cordylochernes scorpioides 97
Coridromius 195
costs, of mating 56; *see also* polyandry
courtship 36, 63, **67**
  copulatory 212, **213**
  effect on gene expression 68
  song 107, 141
cryptic female choice (CFC) 204
  and genital morphology **211**, 216, **217**
  and polyandry 169
  and selfish genetic elements 252
  and seminal fluid proteins 215
  and sperm competition 207, 215
  and sperm morphology 211, 216
  copulatory courtship 212, **213**
  definition 206
  detecting CFC 210
  forms of selection via CFC 208
  male relatedness 212
  selection on male song 145
  sperm storage 214
cryptic male choice 218
Culiseta longiareolata 223, 226
cuticular hydrocarbons (CHCs) 63, 74
  and mate choice 107, 133, 141, 145
  cue to sperm competition level **199**
  divergence in 153
  multivariate selection on 154
Cyclommatus elaphus **93**

Cyrtodiopsis dalmanni **93**, 102, 135, 164, 255; *see also* stalk-eyed fly
cytoplasmic incompatibility 253

D

Diabrotica undecimpunctata 133, 209
Dicranocephalus bourgoini **93**
dimorphism, male, *see* alternative mating phenotypes
Diploptera punctata 229
Drosophila 12, 14, 44
  bifurca 201
  bipectinata 195
  melanogaster
    circadian clock **74**, 81
    courtship **67**
    cryptic female choice 210, 213, 215
    cuticular hydrocarbons 63
    DDT resistance 257, **258**
    gametogenesis 81
    gene expression, post mating **69**, 70, 174
    mate choice 140, 151
    ovary number 86
    polyandry 164, 174
    seminal fluid 23, 44, 67, 174, 215
    sexual conflict 166, **167**, 170
    sperm competition 79, 183, 201, **202**
    Wolbachia and mate choice 251
  mojavensis 70
  montana 133, **134**
  neotestacea 251
  paulistorum 251, 255
  pseudoobscura 73, 168, 173, 203, 252, **253**, 254, 256, **257**
  quantitative trait loci (QTL) 62
  serrata 153
  seychellia 86
  simulans 86, 251, 253
  subobscura 164
  testacea 243
dung beetles
  alternative mating phenotypes **108**, 109
  biparental care 229, 232
  mate competition 101, **102**
  status dependent selection **114**, 115
  see also *Onthophagus*
Dytiscidae 127

E

ecological immunology 246; *see also* immunity
egg
  activation 12
  brooding 228
  provisioning **224**, 225
  trophic 230
ejaculate expenditure, *see* strategic ejaculation
Elaphrothrips tuberculatus 125
endosymbionts
  and haplodiploidy 16
  and parthenogenisis 13
  and sex determination 18
  see also *Wolbachia*
endurance rivalry 27
Euoniticellus intermedius 248
Euryades corethrus **194**
evolutionarily stable strategy (ESS), *see* game theory
experimental evolution
  polyandry 172, 253
  seminal fluid productivity 184, 192, 197
  testes size 197

F

female
  choice, *see* mate choice
  competition 24
  preference, *see* mate choice
Ferrisia virgata **15**
Fisher process 31, 33, **34**, 61, 129; *see also* mate choice
Forficula auricularia **118**, 226, **227**, 233, **235**

G

Gargaphia solani 229
game theory
  alternative mating strategies 109
  contest theory 28
  parental care 231
  sperm competition 196, **197**, **199**
  threshold traits **114**, 115
gametogenesis 80
  ovariole number 85
  stem cells 80
  testes tubule number 87
genetic compatibility
  and polyandry 169, 178
  inbreeding avoidance 37
  mutual mate choice 38, 133
genetics (genomics) 59
  and alternative mating strategies 112, 115
  and mate choice 146
  and polyandry 163, 171, 174
  gene expression 66
  major genes 63, 117
  quantitative genetic variation
    in mating frequency 163
    maintenance of 39, 107
  quantitative trait loci (QTL) 61
    and mate choice 151
    and nutritional plasticity 86
genic capture, *see* lek paradox
genital morphology 89, **139**, **165**, 195
  and cryptic female choice **211**, 216, **217**
genital tract
  female **81**
  male **82**
genotype × environment interactions (GEIs) 37, 60
  and lek paradox 136
  and mate choice 134
  and polyandry 177
Gerris
  gracilicornis **165**, **166**
  lacustris **139**, 146, **165**
  lateralis **211**
  odontogaster **139**
Glossina 223, **227**, 229
Gnatocerus cornutus 196, 243
Goniozus legneri 24
good genes 36
  and genetics 60
  and polyandry 170
  see also mate choice
Graphium sarpedon **194**
Gryllodes sigillatus 137, **138**, 145, 171, **172**
Gryllus
  bimaculatus 171, 202, 212
  campestris 63, **108**, 162, 248
  integer **138**, 244
  pennsylvanicus 244
  veletis **138**, 244

H

haplodiploidy 14
  and fertilization 17, **18**
  and inbreeding avoidance 37
  origins of **6**, 16
Harpobittacus
  nigriceps 204
  similis **205**
Heliconius 62, 190, 191, 192
Heliothis subflexa 66
Hemideina crassidens 126
Hetaerina americana **247**
hermaphroditism 17
heritability, *see* genetics
honey bees
  male reproductive tract **272**
  mating swarms 267
  plasticity in reproduction 89

hormones 69
  and alternative mating phenotypes 117
  and caste determination 90
  reproductive physiology and trade-offs 88
  targets of selection 89
Hylobittacus apicalis 204
Hypolimnas bolina 256
Hypothenemus 14

I

Icerya purchasi 17
immunity
  and mate choice 134, 136
  and parasite-mediated sexual selection 246
  and sperm production 89
  and *Wolbachia* 251
  immunocompetence 39
  mating response 68
  to selfish genetic elements 250
  trade-off with mating effort 71
insulin signalling 74
  and caste determination 90
  and female attractiveness 89
  and gametogenesis 84, 86
  and secondary sexual traits 73
Ips acuminatus **4**

J

juvenile hormone, *see* hormones

K

Kawanaphila nartee 168, 222, 225
kin selection 261

L

Labidomera clivicollis 244
Lasioglossum 125
Lasiorhyncus barbiornis **93**
Laupala 62
  cerasina 152, 156
  kohalensis 152
  paranigra 152, **155**
Leistotrophus versicolor 115
lek paradox 36, 39, 60, 130, 135
  and genotype × environment interactions 136
  and multivariate traits 154
  and parasites 135, 243
Leptothorax gredleri 271
Libellula luctuosa 146
Ligurotettix 122
Linepithema humile 271
Locusta migratoria 190, 200
Lucanidae 98, **99**
Lucilia cuprina 193
Lycaena hippothoe 122
Lytta vesicatoria 223, 225

M

Manduca sexta 117
marginal value theorem (MVT) 27
  and copula duration **182**
mate
  availability
    mate searching 53
    mating pool 45, **50**
  guarding 55, 139, 183, 188
    and alternative mating tactics 109
    and operational sex ratio 189, **190**, **191**
    and parental care
    and selection 145
    biochemical 191
    in social insects 270
    precopulatory 190
    see also sperm competition
  searching 46, 51
    and mating rate 55
    and nuptial gifts 53
    and sperm competition 53
mate choice 23, 26, 31, 129
  and divergence in male sexual traits 152
  and genotype × environment interactions 136
  and mating rates 55
  and selection on male traits 141, **144**
  and selfish genetic elements 255
  and sexual conflict 32
  and speciation 70, 155
  competitive 27
  costs of 35
  Fisher process 31, 33, **34**, 61, 129
  genetic covariance with male traits 33, 62, 146
  good genes 36, 129, **134**
  maintenance of 33
  male mate choice 56, 62, 79, 146
    cryptic 218
  mate compatibility 133
  mechanisms of 140
  mutual 38
  null model 40
  origin of 32
  postcopulatory, *see* cryptic female choice
  quantitative trait loci for 151
  sensory bias 32, 137
  theory 131
mate competition, *see* competition
mate preference, *see* mate choice
mate rejection 55, **140**
mating plug 193, **194**, 270, **271**, 272
mating rate
  null model 54
  sexual differences in 55
mating response
  gene expression 68, **69**
mating system 20, 42
  and parental care 233
  plasticity in 175
  sex roles 23
  terminology 42
  theory 44
Megarhyssa atrata 223, 226
meiotic drive 250, 253, 255
Melittobia 127
Merosargus cingulatus 185, 200, 212
Mimosestes amicus 223, 225
Mnais costalis 111, 116, **118**, 124
multiple mating, *see* polyandry
Musca domestica 210, 258

N

Nannophya pygmaea 122
Nasonia vitripennis 66, 163
Nauphoeta cinerea 79, 147
Nicrophorus vespilloides 176, 221, **227**, 229, 230, 232, 238, **239**, **240**
Notonecta triguttata 165
nuptial gifts
  and cryptic male choice 218
  and polyandry 168
  as paternal care 225
  ejaculates **132**, 225
  prey items 204, **205**
  sensory traps 137
  spermatophores 131, **132**

O

Oncopeltus fasciatus 80, **81**, **82**, **83**, 84
Oncychiurus **4**
Onthophagus **108**, 109, **197**
  nigriventris 88, **93**, **102**, 196
  rangifer **93**, **102**
  taurus **114**, 115, **118**, 126, 197, 202, 223, 229, 232
operational sex ratio (OSR) 45, 49, **52**
  and Bateman gradient 49
  and mate guarding 189
  and sex ratio 50
Orgyia antiqua **10**
Ormia ochracea 107, 128
Ormiini 244

ornaments 20, 31, 33; *see also* secondary sexual traits
Ostrinia
furnacalis 65
nubilalis 64, **65**, 148
ovaries
oogenesis 80, **81**
ovariole number 86
oviposition 225
ovoviviparity 228
Oxysternon conspicillatum **102**

P

Panorpa 185, 218
Paraphlebia zoe 122
parasites 39, 242
and cost of sex 7
and host immunity 246
and lek paradox 243
and mate choice 135
and selfish genetic elements 250
parasitoids 107, 244
Wolbachia and mate choice 250
Parastrachia japonensis 230
Paratrichocladius rufiventris 244
parental care 221
and mating systems 233
egg attendance 226, **228**, **237**
egg brooding 228, 236
egg provisioning 224
forms of 222, 230
game theory 231
nest building 226
offspring guarding 229, **235**
offspring feeding 229, 235, 239, **240**
viviparity 228
parental investment 44
and mating system theory 44
and OSR 50
see also parental care
parthenogenesis 2
and endosymbionts 13
and population size 9
ecological correlates of 9
facultative 4
obligate 3
origins of 13
phylogenetic distribution of **6**
paternal genome elimination (PGE) 14
pathogens, *see* parasites
Periplaneta americana 230
Phanaeus igneus **102**
phenotypic plasticity
alternative mating phenotypes 106, 113, 122
and gene expression 69
in egg size **224**
in germ line stem cell production 83
in mate guarding 189
in mating systems 176
in reproduction 79, 85
limits to 87
ovariole number 86
social insects 89
see also strategic ejaculation
pheromones
and insulin signalling 74
and mate choice 141
and nuptial gifts 132
as anti-aphrodisiacs 192
evolutionary divergence in 192
major genes 64
odorant receptors 75
social insects 267
Philotrypesis 125
Phlebonotus pallens 229
Phyllomorpha laciniata 185, 200, **227**, 228
phylogeny of insects **118**
Phymata americana 143, 145
Phytalmia antilocapra **93**, 98, **99**
Pieris napi 176, **177**, 192, 200
Pisaura mirabilis 170
Plodia interpunctella 197
Polistes dominula **15**, 125, 226, **227**, 267, **270**
polyandry 159
and inbreeding avoidance 171, 173, **174**
and life history **178**
and mating system plasticity 175, **177**
and selfish genetic elements 173, 252
and sexual conflict 164
benefits 168
convenience 164
economics of 170
evolutionary causes 162
fertility assurance 168
genetics 171, **172**
genomics 174
mating costs 164
sexy/good sperm 169
social insects **264**
postcopulatory sexual selection, *see* cryptic female choice; sperm competition
protandry 31, 54
Publilia concava 221, 222, 226, **228**, 231
Pyrrhochoris apterus 189, **190**

Q

quantitative trait loci (QTL), *see* genetics

R

resource defence 92, **94**, **96**
resource holding power (RHP) 28; *see also* game theory
Reticulitermes 5
Rhinocoris 222, 232, 236
Ribautodelphax pungens **4**

S

Saissetia coffeae **10**
satellite behaviour 107, 126
Scatophaga (*Scathophaga*) *stercoraria* 161, 165, 181, **182**, 200, 214
secondary sexual traits
and alternative mating strategies 123
evolution of 141
hormonal regulation of 88
in females 219
in social insects 270
selection on 142
selfish genetic elements 173, 250
and fertility 251
and mate choice 255
and polyandry 252, **257**
and sex ratio 251, 256
seminal fluid proteins
and cryptic female choice 215
and mate guarding 192
and sperm viability 199
and strategic ejaculation 193
Drosophila 67, 174, 184, 192
female mating response 174, 187
in social insects 272
paternal effects 171
sperm incapacitation 273
Sepsis cynipsea 190, 200
sex, evolution of
cost of sex 6
population size 9
role of parasites 7
sex allocation 126
sex peptide 67, 174, 184, 192
sex ratio 40, 50
distortion of 251, 256, **257**
in polyembryonic wasps 91
sex ratio distorters 173
sex roles 23, 46, 159, 233, 256
sexual conflict
and alternative mating phenotypes 127
and cryptic female choice 215
and mate choice 32, 139
and polyandry 164
and selfish genetic elements 256, **258**
and seminal fluid proteins 67, 174

and sexual selection 25
and speciation 71
and sperm competition 192
chase-away models 37, 218
mating rates 55
over parental care 232
over sex ratio 91
sexual dimorphism
in gene expression 63
in courtship behaviour 141
in parental care 231
sexual selection 20
and alternative mating strategies 123
and hormones 89
and immunity 246, **247**
and kin selection 263
and natural selection 24
and parasites 39, 107, 135, 242
and sexual conflict 25
and speciation 61
and weaponry 92
Fisher process 31, 33, **34**
forms of 26
good genes 36, 129, **134**
in social insects 261, 266
opportunity for 48
postcopulatory 22, 27; *see also* cryptic female choice *and* sperm competition
strength of 48, 159
sexually antagonistic selection, *see* sexual conflict
social insects
and parthenogenesis 5
and sexual selection 261
ejaculates 272
mating biology 263
plasticity in castes 89
polyandry **264**, 265
sexually selected traits **268**, **270**
sperm competition in 187, 192
Solenopsis invicta 112, 168, 253
song
and gene expression 68
female choice of 133, **134**, 136, 141
pleiotropy with female preference 62, 152
selection on **144**
species divergence of 156
use in alternative mating tactics 107, 126
speciation
and sexual conflict 71
and sexual selection 61, 151, 155
circadian clock 73
sperm
and selfish genetic elements 251
apyrene 203
cooperation among 203
immunity trade-off 89
length 201
quality (viability) 197, **199**
spermatogenesis 82, **272**
storage by female 161, 195
ultrastructure 19
sperm competition 28, 181
and alternative mating tactics **197**
and cryptic female choice 207, 215
and gene expression 68
and mate searching 53
and parental care 231, 236
and selfish genetic elements 251, 253
and sperm viability 198
and testes size 196
behavioural mate guarding 183, 188, **189**, **190**
biochemical mate guarding 191
in social insects 265
intensity 198, **199**
paternity of second male (P_2) 184, **187**
risk 196, **197**
sex ratio drive **257**
sperm displacement **183**, 201, **202**
sperm length 201
sperm number **200**
see also mating plugs; strategic ejaculation
spermatophores
and cryptic female choice 145, 212, 214
and parental care 225
and sperm transfer 107, 138, 144
as nuptial gifts 131, **132**, 168
spermatophylax 137, **138**, 168
stalk-eyed flies
contests 30
exaggerated weapons **93**
female choice 135
hormonal control of eye-span 89
male competition 102
meiotic drive 253, 255
Stator limbatus 223, **224**
status dependent selection 113
strategic ejaculation 56, 79, 198
and cryptic female choice 212
and cryptic male choice 218
and polyandry 168
of seminal fluid proteins 193
of sperm **200**
Sympetrum parvulum 189
Synagris
cornuta **93**, **118**
fulva **93**, 98

T

Teleogryllus
commodus **144**, 145, 154
oceanicus 128, 152, 171, 187, **199**, **245**, 248
Teleopsis
pallifacies **135**
quinqueguttata 104
whitei 102
termites 5, 265
territoriality 92, **94**
testes
size and sperm competition 196, **197**
spermatogenesis 81, **83**
trade-off with secondary sexual traits 88
tubule number 87
Tetranychus urticae 255
Theodosia viridiaurata **93**
Timema 12
Trachymyrmex zeteki 273
trade-off, life history
immunity and ornaments 136, **247**, 248
immunity and sperm production 89, 248
reproductive and non-reproductive behaviour 89
testes and weapons 88, 196
traumatic insemination 194, 248, **249**
Tremex columba 226
Tribolium castaneum 146, 150, 173, **174**, 210, 212, **213**
Trichogramma 13, 200
Trypoxylon politum 223, 226
Trypoxylus dichotomus **72**, **93**

U

Uscana semifumipennis 225
Utetheisa ornatrix 131, **132**

V

viviparity 178, 223, 228

W

war of attrition 29
Warramaba virgo 3
Wasmannia auropunctata 15, 18
weapons 20, 28, 92, **93**
and alternative mating strategies 123
and insulin signalling **72**, 89
costs of 92, 123
Wolbachia 8, 13, 250, 253, 255